Density-Based Reactivity Theory

Shubin Liu

WILEY VCH

Author

Prof. Shubin Liu
The University of North Carolina
Chapel Hill, NC, USA

Cover Design: Wiley
Cover Image: Courtesy of Shubin Liu and American Chemical Society

Typesetting: Lumina Datamatics, Madurai, India.

Library of Congress Card No.: applied for

British Library Cataloguing-in-Publication Data
A catalogue record for this book is available from the British Library.

Bibliographic information published by the Deutsche Nationalbibliothek The Deutsche Nationalbibliothek lists this publication in the Deutsche Nationalbibliografie; detailed bibliographic data are available on the Internet at <http://dnb.d-nb.de>.

Print ISBN: 9783527355440
ePDF ISBN: 9783527853229
ePub ISBN: 9783527853212
oBook ISBN: 9783527853236

Printed and bound by CPI Group (UK) Ltd, Croydon, CR0 4YY

C9783527355440_010726

This book is dedicated to the exceptional teachers and mentors who paved the way for my intellectual and professional journey, fundamentally changing my life for the better.

With deepest gratitude to:

Guangxiao Yin (殷光晓), *Physics Teacher, Ningxiang No. 4 High School, Hunan, China*

Delong Liu (刘德龙), *Chemistry Teacher, Ningxiang No. 4 High School, Hunan, China*

Run-E Li (李润娥), *Organic Chemistry Professor, Hunan Normal University, China*

Robert G. Parr, *PhD Adviser, University of North Carolina at Chapel Hill, USA*

Contents

Foreword

Chemical reactivity lies at the heart of chemistry. From elementary transformations to complex catalytic networks, chemists seek not only to compute structures and energies but also to understand *why* chemical systems behave as they do. For much of modern chemistry, this understanding has been shaped by orbital-based thinking, which has provided a powerful and enduring conceptual framework. Yet, as our theoretical tools and computational reach have expanded, it has become increasingly clear that chemical insight can also emerge directly from the electron density itself—a fundamental descriptor of electronic structure.

I have had the privilege of knowing Dr. Shubin Liu since his time as a graduate student at the University of North Carolina at Chapel Hill with Professor Robert G. Parr, and later as a postdoctoral researcher at Duke University with me, and of following the intellectual development that ultimately led to *Density-Based Reactivity Theory*. Even at an early stage, it was evident that his interests went beyond applying density functional theory as a computational method. He was motivated instead by the question whether the electron density, treated as a primary object rather than a byproduct, could support a chemically intuitive theory of reactivity. This book represents the comprehensive realization of that vision.

In this monograph, Dr. Liu presents a density-based framework for understanding chemical behavior, integrating conceptual density functional theory, density-associated quantities, information-theoretic approaches, and orbital-free perspectives. These tools are developed with mathematical rigor, but always with an eye toward chemical meaning. Concepts such as hardness, electrophilicity, steric effects, noncovalent interactions, regioselectivity, cooperativity, and aromaticity are examined and clarified through the language of electron density.

What is particularly impressive is the consistency with which chemical intuition is preserved—and in many cases sharpened—throughout the book. The discussions demonstrate that density-based descriptions do not replace familiar chemical concepts, but rather provide them with a more fundamental and broadly applicable foundation. The treatment of both ground and excited states, weak and strong interactions highlights the generality of the approach.

This work will be valuable to a wide spectrum of chemists. Graduate students will find in it a coherent introduction to modern reactivity theory. Theoretical and computational chemists will appreciate the systematic development of density-based descriptors and their interrelations. Experimental chemists, too, will find that the analyses offered here illuminate trends and selectivities encountered in practice, offering a bridge between observable behavior and electronic structure.

Density-Based Reactivity Theory reflects many years of sustained effort. It stands as a significant contribution to theoretical chemistry and as a persuasive demonstration that electron density can serve not only as a computational quantity, but as the foundation of a robust and predictive chemical theory. I am confident that this book will influence how chemists think about reactivity for years to come.

Duke University
April 2026

Weitao Yang

About the Author

Dr. Shubin Liu is Senior Computational Scientist in the Research Computing Center and Adjunct Professor in the Department of Chemistry, both at the University of North Carolina at Chapel Hill (UNC), USA. He received his Ph.D. in 1996 under the supervision of the late Robert G. Parr of UNC and completed postdoctoral training with Weitao Yang of Duke University. Since 2000, he has been an independent researcher in the current position, focusing on developing a density-based reactivity theory within the framework of density functional theory. Dr. Liu has co/authored more than 300 peer-reviewed publications and has been recognized by several scientific awards, including the Wiley–IJQC Young Investigator Award. In addition to authoring this monograph, he has served as the editor of two books, *Conceptual Density Functional Theory* (2022) and *Exploring Chemical Concepts through Theory and Computation* (2024), each published by Wiley-VCH. Beginning in January 2026, he is Editor-in-Chief of the *International Journal of Quantum Chemistry*. Dr. Liu initiated and co/organized the international symposium series "Chemical Concepts from Theory and Computation" (CCTC), with meetings held in Changsha, China (2018); virtually during the COVID pandemic (2021); Lyon, France (2023); Hawaii, USA (2025); and an upcoming meeting in Colorado, USA (2027).

Preface

Chemistry is a science built on concepts. Long before the Schrödinger equation was formulated, chemists spoke of chemical bonds, acids and bases, functional groups, aromatic compounds, steric effects, and reaction mechanisms. These concepts, rooted in empirical observation and refined through abstraction, have guided generations of chemists in predicting reactivity, interpreting experiments, and designing molecules. As theoretical and computational chemistry advanced through wavefunction theory and density functional theory, these concepts were neither discarded nor reduced; they were deepened, sharpened, and enriched with theoretical insights.

This book grew from a simple but persistent question that has accompanied me throughout my career: *How can chemical reactivity be systematically understood from electron density?* Its justification rests on the two basic theorems of Hohenberg and Kohn in density functional theory, yet turning these foundational theorems into a systematic understanding has remained elusive. My scientific lineage—shaped especially by Robert G. Parr and Weitao Yang—taught me how to view chemistry through density, information, and conceptual structure. Over the past two decades, the field has expanded dramatically, yielding a constellation of new density-based descriptors, information-theoretic measures, and real-space frameworks for analyzing bonding, acidity, electrophilicity, steric effects, aromaticity, homochirality, cooperativity, frustration, and excited-state behavior. Nevertheless, these advances have remained scattered across the literature. What was missing was a unified and coherent conceptual framework—a way to see density-based reactivity theory as a system rather than merely a collection of tools.

This book attempts to establish that system.

This book is divided into three parts: Frameworks, Applications, and Perspectives. In Part I, I outline four density-based frameworks, including conceptual density functional theory, density-associated quantities, information-theoretic approach, and orbital-free density functional theory, and discuss their interrelationships and recent advances. In Part II, across 10 chapters, I apply these frameworks to classical chemical concepts, such as bonding, reactivity, aromaticity, acidity/basicity, electrophilicity/nucleophilicity, regioselectivity, stereoselectivity, homochirality, and catalysis, and reinterpret them through the unifying lens of the density-based framework. These chapters reflect not only technical progress but also a philosophical stance: electron density offers an ontologically grounded and epistemologically transparent foundation for chemical understanding.

Yet density-based reactivity theory, as applied to traditional chemical concepts, represents only part of the story. Chemistry today stands at the threshold of a profound transformation. Machine learning (ML) and quantum computing (QC) are reshaping how we simulate molecules, how we explore chemical space, and how we derive knowledge from data. These developments both revive and reframe Dirac's dilemma: the fundamental laws are known, but their exact application is intractable for realistic systems. ML and QC do not solve this dilemma in the traditional sense—they transcend it. ML provides many-to-many mappings between structure and properties, uncovering latent patterns that classical theories cannot reach. QC promises quantum-native solutions to quantum problems, aligning the substrate of computation with the substrate of nature itself. Together, they challenge us to rethink chemical ontology, chemical epistemology, and the very meaning of chemical understanding.

For this reason, the final two chapters, serving as Part III of the book, look ahead. They articulate how density-based reactivity theory connects to ML and QC, how feature-based and qubit-based representations extend traditional orbital and density paradigms, and how hierarchical modeling may become the organizing framework for theoretical chemistry in the decades to come. If multiscale modeling dominated the 20th century, hierarchical modeling—blending density, orbitals, features, qubits, and emergent patterns—may well define the 21st.

This book is aimed at graduate students, researchers in theoretical chemistry, computational chemists, and practitioners in related fields such as chemical biology, materials science, and artificial intelligence (AI). It is also intended for those who wish to go beyond prediction to pursue understanding—a word that has guided every page of this work. Prediction computes; understanding explains. Prediction tells us *what* will happen; understanding tells us *why*. Density-based reactivity theory seeks to bring these two together seamlessly.

Writing a book of this scope would not have been possible without the encouragement and support of many individuals, including my colleagues, collaborators, students, and family. My scientific teachers—especially the four to whom this book is dedicated—planted the seeds of conceptual thinking; my collaborators sharpened the ideas; my students inspired new questions; and my family provided patience and grounding. AI tools were used in creating an image, and large language models assisted in polishing some English prose. I am equally grateful to the editorial and production teams at Wiley-VCH, whose professionalism, patience, and commitment have shepherded this project from conception to completion. To all of them, I extend my sincere thanks.

As chemistry enters the next phase of the quantum era, surrounded by unprecedented amounts of data, empowered by new computational paradigms including ML and QC, and energized by conceptual innovation, I hope this book offers a coherent framework for thinking about reactivity, concepts, and understanding in a unified way. May it serve as both a guide and an invitation to engage with the layered structures of chemical concepts, to explore the nested hierarchies of chemical behavior, and to deepen our collective pursuit of chemical understanding.

Chapel Hill, NC, USA
December 2025

Shubin Liu

Part I

Frameworks

1

Introduction

TABLE OF CONTENTS

1.1 Theoretical and Computational Chemistry

Theoretical and computational chemistry, a vibrant nexus of chemical disciplines, seamlessly melds chemical principles with the rigor of physics, the intricacy of biology, and the power of computer science. Through sophisticated mathematical models, fundamental physical laws, and cutting-edge computational techniques, this subdiscipline of chemistry delves into the electronic-structure realm of matter, simulating complex chemical processes, predicting physicochemical properties, and unlocking profound insights into molecules, biological systems, and materials.

The evolution of modern theoretical and computational chemistry traces back to the early twentieth century with the advent of quantum physics. Trailblazers like Heitler and London [1] pioneered the application of quantum mechanical principles to chemical systems, beginning with the hydrogen molecule H_2, by establishing valence bond theory (VBT) [2–4]. The 1940s marked a turning point as electronic computers, notably ENIAC, enabled rudimentary quantum chemical calculations and transformed theoretical chemistry from a domain of applied mathematics to computational exploration.

In the 1950s and 1960s, the field advanced significantly with the development of ab initio methods within the framework of molecular orbital theory (MOT) [5–7], such as Hartree–Fock (HR) and post-Hartree–Fock approaches, alongside semi-empirical techniques, facilitating more precise modeling of molecular electronic structures. The 1970s and 1990s introduced density functional theory (DFT) [8–10], parallel computing, and multiscale modeling techniques [11, 12], enabling the study of increasingly complex molecular systems.

By the 2000s, high-performance computing had further expanded the field's capabilities, providing profound insights into chemical reactions, material properties, and atomistic interactions. The early 2010s ushered in the integration of machine learning (ML) techniques, which has been revolutionizing predictive and analytical approaches in computational chemistry [13–15]. Most recently, developments since the early 2020s have centered on quantum computers, artificial intelligence (AI), and deep learning, beginning a new era of unparalleled computational power and precision in molecular modeling [16–19].

Theoretical and computational chemistry encompasses a diverse range of physical methodologies, integrating quantum mechanics (QM), classical mechanics or molecular mechanics (MM), and statistical mechanics to simulate complex systems through advanced multiscale modeling approaches. The recent three Nobel Prizes in Chemistry awarded to the field—John Pople and Walter Kohn in 1998; Martin Karplus, Michael Levitt, and Arieh Warshel in 2013; and David Baker, Demis Hassabis, and John Jumper in 2024—underscore the pivotal role of theoretical and computational chemistry as a cornerstone of interdisciplinary research. This indispensable tool empowers academia and industry alike to unravel complex chemical behaviors, design innovative materials, and explore biological functions with precision and depth.

We cannot just compute; we should also understand [20, 21]. The true value of theoretical and computational chemistry lies not only in generating data but in interpreting it through the lens of chemical concepts. Computational results serve as a bridge between abstract models and real-world phenomena, offering insights that deepen our understanding of molecular behavior, reaction mechanisms, and material properties. However, without a strong grasp of underlying chemical principles, the interpretation of numerical results risks becoming superficial or misleading. Chemical concepts provide the framework for translating numerical output into meaningful knowledge that can guide experiments, predict outcomes, and inspire innovation. Better understanding of chemical concepts through theory and computation fosters the ability to critically evaluate computational predictions, identify limitations, and refine methodologies, ensuring that our models remain robust, reliable, and truly impactful. In this context, Charles Coulson's seminal statement that "we want insight, not numbers" [22, 23] can be nuanced to "we want insight, not *just* numbers," acknowledging that numerical data and deep understanding can work hand in hand.

In this chapter, we will briefly introduce multiscale modeling, followed by three frameworks to solve the Schrödinger equation, VBT, MOT, and DFT, together with the insights from the first two theories, which are both orbital-based, in enhancing our understanding. We will spend a lot more space on the insight from DFT in the ensuing chapters, as it constitutes the central theme of this book.

1.2 Multiscale Modeling

Multiscale modeling [11, 12] is a computational approach that combines systems at multiple spatial and temporal scales to simulate structure, dynamics, reactivity, and functions for complex systems with unprecedented accuracy and detail (Figure 1.1). By bridging the gap between macroscopic observations and microscopic interactions, multiscale modeling enables us to gain insights into systems that cannot be fully understood using a single-scale approach. Multiscale modeling has been well developed as an integral method in theoretical and computational chemistry. The Nobel Prize of Chemistry in 2013 with the citation "*for the development of multiscale models for complex chemical systems*" confirmed just that.

Three key features of multiscale modeling distinguish it from others. (i) It incorporates different simulation models for different spatial scales, from the atomic level (angstrom scale using QM), the mesoscopic level (nanometer scale using classical mechanics), to the macroscopic scale (meter scale using finite element methods). (ii) It couples different models at different scales through mathematical relationships or numerical techniques to pass information from smaller to larger scales; (iii) It is often a hierarchical and bottom-up approach, where one starts at the smallest scale (e.g., atomic or quantum) and builds up to larger scales by averaging or coarse-graining.

At the microscopic scale, which spans from Angstroms and nanometers, simulation methodologies are dominantly QM by solving the Schrödinger equation to describe the electronic structure of molecules and condensed phase materials. MM and molecular dynamics to simulate atomistic motions using Newton's law via classical force fields, and Monte Carlo approaches, which employ stochastic statistics to investigate thermodynamic and kinetic properties, can also be employed. At the mesoscopic level (from nanometers to micrometers), coarse-grained models simplify atomistic details by grouping atoms and functional groups into "beads"

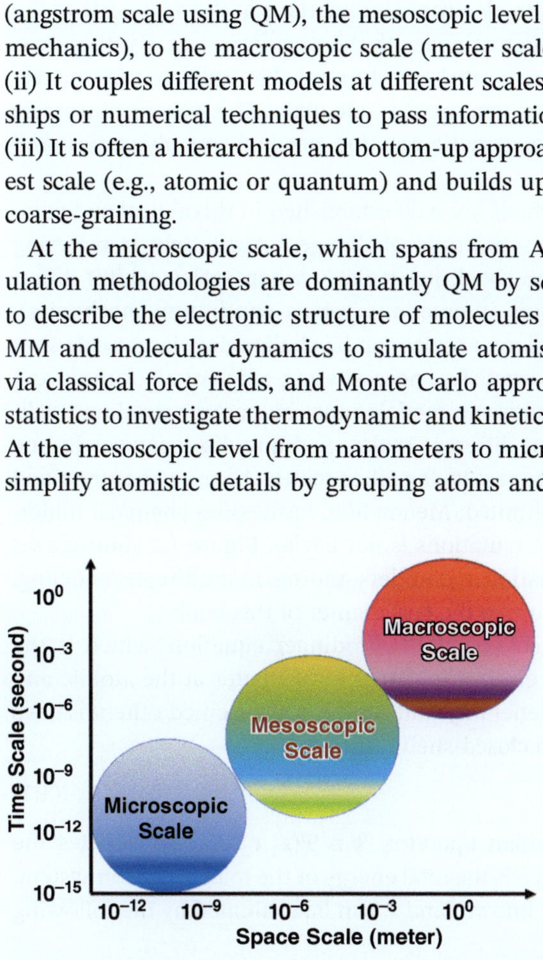

Figure 1.1 Schematic representation of multiscale modeling with respect to time and space scales. *Source*: Reproduced from Ref. [20] with permission of American Chemical Society.

(e.g., via Martini force field or dissipative particle dynamics) to study emergent structural and dynamical behaviors for colloids, polymers, or cellular systems. At the macroscopic level (millimeters to meters and beyond), finite element analysis, fluid dynamics, and continuum mechanics models are applied to study stress, strain, heat transfer, and other macroscopic properties and behaviors for materials.

Depending on the coupling scheme among different scales, which can be sequential, concurrent, and hybrid, we might have different implementations in various disciplines. In computational chemistry, combining quantum mechanical calculations for the region of interest with MM for the rest, that is, QM/MM, is the most popular implementation [24, 25]. QM is applied to describe the part of the system where electronic-level details are crucial, whereas MM is employed to simulate the rest of the system as the environment, for which QM treatment is computationally too costly. One excellent example of such systems is enzymes, because their function spans multiple scales, from electronic interactions at the active site to the dynamics of protein folding and interactions with their environment. For these systems, reactions at the active site are treated quantum mechanically, but for the rest of the enzyme, classical mechanics treatment is adequate. This way of multiscale treatment allows us to simulate enzymatic reaction mechanisms, dynamics, and their roles in biological systems.

While multiscale modeling methods are well established in theoretical and computational chemistry, it still faces multiple challenges. Accurately transferring information between scales is nontrivial, particularly when scales exhibit different behaviors (e.g., discrete atomic interactions versus continuum mechanics). Developing coupling models at different scales can be challenging; large-scale simulations can be computationally expensive, necessitating efficient algorithms and high-performance computing; multiscale models must be parametrized and validated against experimental data at different scales, and any discrepancies in one scale can propagate and impact the results for other scales; Access to accurate and reliable data at all scales is often limited. Meanwhile, harnessing chemical understanding from different scales of simulations is not trivial. Figure 1.2 summarizes the two challenging tasks, computation and understanding, in multiscale modeling. We will come back to this big picture in the last chapter of this book.

In what follows, we focus on solving the Schrödinger equation, which is the foundation of QM governing the electronic structure of matter at the atomic and subatomic level. Without losing generality and unless it is specified otherwise, we consider its time-independent and closed-shell form,

$$H\Psi = E\Psi \tag{1.1}$$

where H represents the Hamiltonian operator, $\Psi \equiv \Psi(\mathbf{r}_1,\mathbf{r}_2,\ldots,\ \mathbf{r}_N)$ denotes the N-electron total wavefunction, and E is the total energy of the many-electron system. For a given system, H is explicitly known, and E can be evaluated by the following expectation value,

$$E = \langle\Psi|H|\Psi\rangle/\langle\Psi|\Psi\rangle \tag{1.2}$$

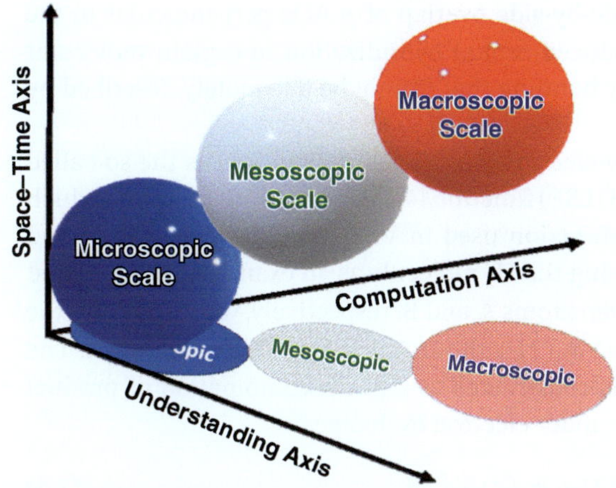

Figure 1.2 Schematic representation of computation and understanding as two paramount and challenging tasks in multiscale modeling. *Source*: Reproduced from Ref. [20] with permission of American Chemical Society.

so in principle there is only one unknown in Eq. (1.1), for example, Ψ. Unfortunately, for a system with N electrons, Ψ is too complicated a function, with $4N$ ($3N$ spatial + N spin) variables. Except for the one-electron case, such as a hydrogen atom or a hydrogen molecular cation H_2^+, only approximate solutions of Ψ are possible. Among the three theoretical frameworks in the literature to approximately solve the Schrödinger equation, that is, VBT, MOT, and DFT, the first two are orbital-based, and the last is density-based. In orbital-based theories, each electron is assumed to be described by an individual wavefunction called an orbital, and the many-electron wavefunction Ψ constitutes a proper combination of the atomic orbital (AO) and/or molecular orbital (MO). In the next section, we quickly overview these two orbital-based theories together with the chemical insights derived from them.

1.3 Orbital-based Theories

1.3.1 Valence Bond Theory

Introduced in the 1920s by Walter Heitler and Fritz London [1] and later refined by Linus Pauling [26], who introduced hybridization and resonance, VBT offers the first framework to approximate the many-electron wavefunction of the Schrödinger equation by considering how atoms combine to form molecules. Through the interactions of AOs, it explains bond formation in terms of the overlap of wavefunctions and the pairing of electrons. VBT states that a chemical bond forms when AOs of two atoms overlap, allowing electrons to be shared between them. The greater the overlap, the stronger the bond. The overlapping orbitals must have compatible symmetry, similar energy levels, and opposite spins. There are two major types of orbital overlapping: sigma (σ) bonds via head-on overlap of AOs along the internuclear axis and pi (π) bonds through side-by-side overlap of p AOs perpendicular to the internuclear axis. VBT also introduced orbital hybridization to explain molecular geometry and resonance to justify bonding that cannot be adequately described by a single Lewis structure.

The building block of the many-electron wavefunction Ψ in VBT is the so-called Heitler–London–Pauling–Slater (HLSP) function [4], also called VB function, which refers to the many-electron wavefunction used in VBT to describe covalent bond formation between two atoms. Using the H_2 molecule as an example, let $\phi_A(1)$ and $\phi_B(2)$ represent the AOs of hydrogen atoms A and B, respectively, where $\phi_A(1)$ is the 1s AO of atom A for electron 1, and $\phi_B(2)$ is the 1s AO of atom B for electron 2. The spatial part of the wavefunction is constructed as a linear combination of product states, ensuring proper symmetry under electron exchange,

$$\Psi_{\text{spatial}} = \frac{1}{\sqrt{2 \pm S}} [\phi_A(1)\phi_B(2) \pm \phi_A(2)\phi_B(1)], \tag{1.3}$$

where the "+" sign corresponds to the symmetric spatial wavefunction (bonding orbital), the "−" sign corresponds to the antisymmetric spatial wavefunction (anti-bonding orbital), and S is the overlap integral between two nonorthogonal AOs,

$$S = \int \phi_A(\mathbf{r})\phi_B(\mathbf{r})d\mathbf{r}. \tag{1.4}$$

Since electrons are fermions, the many-electron wavefunction (spatial × spin) must be antisymmetric under spin exchange between spin-up (α) and spin-down (β) electrons. Therefore, we have the spin part of the total wavefunction,

$$\Psi_{\text{spin}} = \frac{1}{\sqrt{2}} [(\alpha(1)\beta(2) \mp \beta(1)\alpha(2))]. \tag{1.5}$$

The HLSP function Ψ_{HLSP} is the product of spatial and spin parts,

$$\Psi_{\text{HLSP}} = \Psi_{\text{spatial}} * \Psi_{\text{spin}}. \tag{1.6}$$

In general, in VBT, the many-electron wavefunction Ψ of a molecule can be expressed in terms of a linear combination of HLSP functions, as

$$\Psi = \sum\nolimits_K C_K \Psi^K_{\text{HLSP}}, \tag{1.7}$$

where Ψ^K_{HLSP} is a VB function (e.g., HLSP function) corresponding to a specific Lewis structure, and C_K is its corresponding coefficient. A general HLSP function can be written as

$$\Psi^K_{\text{HLSP}} = \hat{A}\,\Omega_K \Theta_K, \tag{1.8}$$

where $\hat{A}$ is the antisymmetrization operator and Ω_K is a direct product of occupied spatial AOs $\{\phi_{K_i}\}$,

$$\Omega_K = \phi_{K_1}(1)\phi_{K_2}(2)\cdots\phi_{K_N}(N), \tag{1.9}$$

and the spin contribution Θ_K is

$$\Theta_K = \prod_{(ij)} \frac{1}{\sqrt{2}} [\alpha(i)\beta(j) - \beta(i)\alpha(j)] \prod_k \alpha(k), \tag{1.10}$$

where (ij) runs over all n covalent bonds and k runs over all unpaired electrons, which implies that an HLSP function is expanded as a linear combination of 2^n Slater determinants.

With the above approximate total wave function Ψ, one can solve the generalized eigenvalue equation,

$$\mathbf{HC} = \mathbf{SC}\boldsymbol{\varepsilon}, \tag{1.11}$$

where $\mathbf{H}$ and $\mathbf{S}$ are Hamiltonian and overlap matrices constructed in the basis of HLSP functions, respectively, $\mathbf{C}$ and $\boldsymbol{\varepsilon}$ are eigenvectors and eigenvalues. This equation must be numerically solved self-consistently.

The key advantage of VBT is that each HLSP function exactly corresponds to a classical Lewis structure, and thus it provides unambiguous and intuitive insights into the nature of chemical bonding and reactivity mechanisms. Additionally, Pauling's hybridization theory successfully explains molecular geometry and resonance structures. On the other hand, VBT cannot easily explain delocalized electron systems, does not account well for excited states, and becomes mathematically complex for larger molecules. Recent developments in VBT have resolved some of these issues with methods like generalized valence bond theory [27], ab initio VBT [4], and block-localized wavefunction theory [28].

1.3.2 Molecular Orbital Theory

MOT is the second electronic structure theory to approximately solve the Schrödinger equation. Developed in the 1930s by Robert Mulliken [29] and Friedrich Hund [30, 31], and later by many others, MOT introduces the concept of MOs, which are formed from the linear combination of atomic orbitals (LCAOs). Unlike VBT, which focuses on using AOs to form localized covalent bonds, MOT emphasizes the delocalized nature of electrons in MOs. This delocalization enables MOT to describe a broader range of chemical phenomena, from conjugated systems and aromaticity to bonding in metals and coordination complexes.

To form an approximate many-electron wavefunction Ψ for Eq. (1.1), MOT starts with the Born–Oppenheimer approximation to separate the motion of electrons from that of atomic nuclei within a molecule because atomic nuclei are much heavier than electrons and then uses the LCAO approximation to generate MOs,

$$\psi_i = a \sum\nolimits_j c_{ij} \phi_j \tag{1.12}$$

where ψ_i are MOs, ϕ_j AOs, c_{ij} linear coefficients, and a the normalization constant. Since electrons are fermions, the many-electron wavefunction represented by MOs should satisfy the antisymmetric condition, which is done through the Slater determinant,

$$\Psi(\mathbf{x}_1,\mathbf{x}_2,\cdots,\mathbf{x}_N)=\frac{1}{\sqrt{N!}}\begin{vmatrix}\psi_1(\mathbf{x}_1) & \psi_2(\mathbf{x}_1) & \cdots & \psi_N(\mathbf{x}_1)\\ \psi_1(\mathbf{x}_2) & \psi_2(\mathbf{x}_2) & \cdots & \psi_N(\mathbf{x}_2)\\ \vdots & \vdots & \ddots & \vdots\\ \psi_1(\mathbf{x}_N) & \psi_2(\mathbf{x}_N) & \cdots & \psi_N(\mathbf{x}_N)\end{vmatrix}, \tag{1.13}$$

where $\mathbf{x}$ denotes the position and spin coordinate of a single electron for an N-electron system. If the many-electron wavefunction in Eq. (1.1) is expressed by the above single Slater determinant, this method is called the Hartree–Fock (HF) method. In this case, plugging Eq. (1.13) into Eq. (1.2) to variationally optimize all linear coefficients in Eq. (1.12), one obtains the so-called Hartree–Fock–Roothaan equation with the basis set $\{\phi_i\}$,

$$\mathbf{Fc}=\mathbf{Sc\varepsilon}, \tag{1.14}$$

where $\mathbf{F}$ is the Fock matrix, $\mathbf{S}$ is the overlap matrix of the basis functions, $\mathbf{c}$ is the coefficient matrix, and $\boldsymbol{\varepsilon}$ is the diagonal matrix of orbital energies. In the case of an orthonormalized basis set, the overlap matrix, $\mathbf{S}$, reduces to the identity matrix. Same as Eq. (1.11) in VBT, Eq. (1.14) should be solved numerically in the self-consistent-field (SCF) manner. The obtained MOs are occupied from the lowest orbital energy, each by two antiparallel electrons, forming the ground state of the system.

Since the HF method uses only one Slater determinant in approximating the many-electron wavefunction, it only includes the static correlation stemming from the exchange effect among spin-parallel electrons. The configuration interaction (CI) method goes beyond the HF method by including the dynamic correlation effect from spin-antiparallel electrons. It does so by expressing the many-electron wavefunction as a linear combination of multiple Slater determinants,

$$\Psi_{\mathrm{CI}}=c_0\Psi_0+\sum\nolimits_i c_i\Psi_i+\sum\nolimits_{i<j} c_{ij}\Psi_{ij}+\cdots \tag{1.15}$$

where Ψ_0 is the HF ground-state determinant; $\Psi_i, \Psi_{ij}, \cdots$, are excited determinants obtained by promoting one, two, or more electrons from occupied MOs to unoccupied (virtual) orbitals; and $c_0, c_i, c_{ij}, \cdots$, are the coefficients determined variationally. If only single excitation is considered, that is, the right-hand side of Eq. (1.15) is truncated at the second term, we have CIS (CI singles); if two electrons are promoted (i.e., Eq. (1.15) is truncated up to the third term), we have CISD (CI singles and doubles); if it includes all possible excitations in Eq. (1.15), the full CI (FCI) method is obtained. FCI represents the most accurate results that can be obtained by MOT. Because FCI is exceedingly costly in computational expenses, it is limited only to small systems. CI methods improve the accuracy of molecular energy calculations, capture dynamic electron correlation effects, and provide a systematic way to improve results.

Figure 1.3 depicts the relationship between HF and FCI methods. Other multi-determinant approaches are available in the literature, such as coupled cluster theory, complete active space SCF theory, and multi-reference CI. Also included in Figure 1.3 is the series of perturbation theories, such as MP2 (second-order Møller–Plesset perturbation theory) and CASPT2 (complete active space second-order

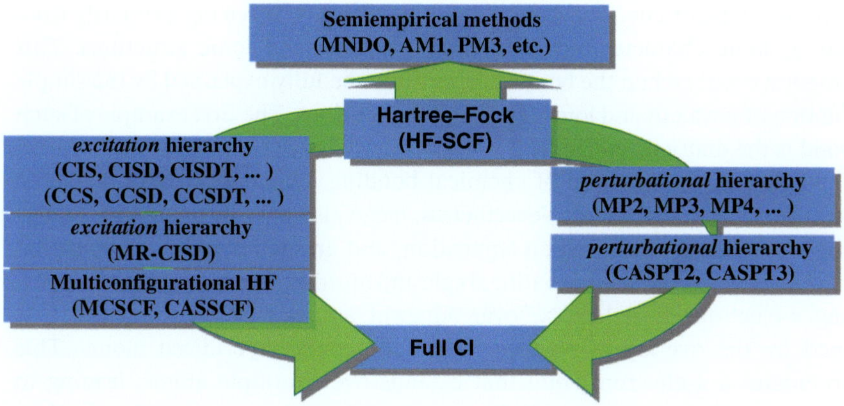

Figure 1.3 Available approaches in MOT.

perturbation theory). On top of the figure are semiempirical approaches that simplify the ab initio wavefunction theory by (i) employing one Slater determinant, (ii) focusing only on valence electrons, and (iii) incorporating experimental parameters to neglect or approximate certain two-electron integrals. These simplifications significantly reduce computational costs and thus make them suitable for treating large systems, such as biomolecules and polymers.

1.3.3 Concepts from Orbital-based Theories

The two frameworks of orbital-based theories, VBT and MOT, not only provide practical pathways to approximately solve the Schrödinger equation numerically for many-electron systems but also offer insights and understanding about various chemical concepts [32, 33].

In VBT, how AOs of individual atoms overlap to form chemical bonds is the focus [32]. A chemical bond forms when the AOs of two atoms overlap, resulting in a region of shared electron density between the two nuclei. Greater overlapping leads to stronger bonding and lower potential energy of the molecular system. When the two AOs are overlapped head-on, a sigma (σ) bond is formed. When they overlap side-by-side, a pi (π) bond is formed, which often accompanies a sigma bond in double and triple bonds.

The hybridization concept introduced by Linus Pauling explains how AOs can combine to form new hybrid AOs with different shapes, such as sp, sp^2, sp^3, and sp^3d^2, which can be used to predict bond angles and molecular geometry, for example, tetrahedral and trigonal planar.

When a single VB structure cannot fully represent the true electronic structure of a molecule, such as benzene, the concept of resonance is introduced by Pauling and Hund to describe the situation. In this case, multiple VB structures, called *resonance structures*, contribute to the overall bonding and properties of the molecule. The actual molecule is considered a *resonance hybrid*, which is a weighted average of these resonance structures. First proposed by Shaik et al. [34, 35], a *charge-shift*

(CS) bond is a type of chemical bond where the stability arises not primarily from covalent or ionic character but from resonance between ionic structures. This phenomenon occurs when the bond energy cannot be fully explained by the simple combination of covalent and ionic bonding contributions. The first example of such a CS bond is the diatomic molecule F_2.

VBT emphasizes localization of chemical bonding, where electrons are shared between specific pairs of atoms. Nevertheless, it can also provide qualitative understanding about conjugation, hyperconjugation, and aromaticity [32]. Conjugation occurs when a molecule has alternating single and multiple bonds (e.g., C=C—C=C), allowing π-electrons to delocalize over adjacent atoms. In VBT, conjugation is explained by the overlap of adjacent p-orbitals on sp^2-hybridized atoms. This overlap creates a π-electron cloud that extends over multiple atoms, leading to electron delocalization and overall decreased energy. Hyperconjugation involves the delocalization of electrons from a σ-bond (usually C—H or C—C) into an adjacent empty or partially filled p-orbital or π-system, whereas aromaticity refers to the exceptional stability of cyclic, planar molecules with delocalized π-electrons, governed by Hückel's rule ($4n + 2$ π-electrons). With overlapping between σ- or π-bonds, VBT can provide qualitative descriptions about conjugation, hyperconjugation, and aromaticity, but its understanding is often less quantitative than MOT.

MOT provides a comprehensive understanding of chemical bonding, electronic structure, and molecular properties [33, 36–38]. Since an MO is formed by the LCAOs from different atoms in a molecule, electrons in a molecule occupying such MOs are usually delocalized over the entire molecule rather than being associated with a single atom or localized between a pair of atoms. Also, MOs, either head-on (σ overlap) or side-on (π overlap), can be *bonding* (in-phase overlap, lowering energy, e.g., σ or π) or *antibonding* (out-of-phase overlap, raising energy, e.g., σ^* or π^*) or *nonbonding* (no overlap, no change in energy). These different bonding natures of MOs can be represented by the MO diagram, as shown in Figure 1.4 for O_2 as an example. It elucidates the relative energy of MOs with respect to

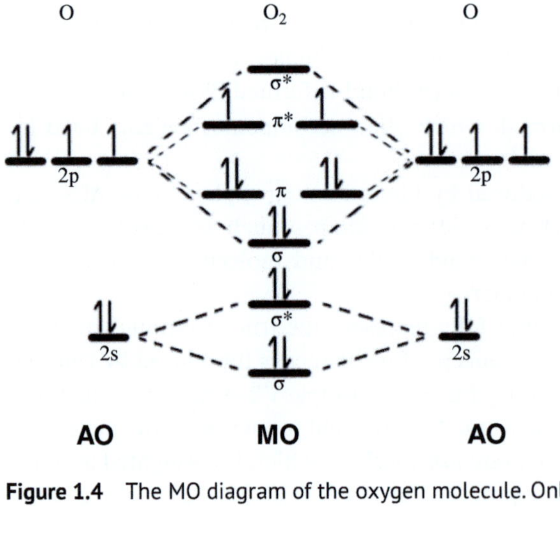

Figure 1.4 The MO diagram of the oxygen molecule. Only valence orbitals are shown.

those of AOs. Electrons fill MOs in the diagram according to the *Aufbau principle* (electrons occupy MOs in order of increasing energy), *Pauli exclusion principle* (each MO can only occupy two antiparallel electrons), and *Hund's rule* (when electrons occupy degenerate MOs, they will first fill each orbital singly with parallel spins before pairing up). Using the diagram, one can evaluate bond orders, determine paramagnetic and diamagnetic properties, and quantitatively describe conjugation, hyperconjugation, and aromaticity.

Even though localization and hybridization are concepts from VBT, it can be integrated with MOT to explain molecular geometry and bonding through, for example, the *natural bond orbital* analysis [39, 40]. Localized molecular orbitals (LMOs) are obtained through the unitary transformation of the conventional, delocalized MOs. LMOs are spatially concentrated in specific regions of a molecule and thus provide an intuitive picture of bonding that resembles the traditional Lewis structure, where bonds are localized between pairs of atoms and lone pairs are confined to specific atoms.

Molecules have symmetry, so do AOs and MOs. AOs that have the same symmetry with respect to a molecule's symmetry elements can mix and form MOs, whereas those with different symmetries do not interact, due to orthogonality. LCAO preserves the symmetry property of the molecule, leading to the *orbital symmetry conservation principle* in MOT [38]. As a result, MOs inheriting the symmetry of a molecule can be classified as σ, π, δ, or with symmetry labels such as a_1 and b_2. Based on this principle for chemical reactions, Woodward and Hoffmann proposed a set of rules, called the *Woodward–Hoffmann rules* [37, 38], to predict the outcome of pericyclic reactions.

The frontier molecular orbital (FMO) theory proposed by Kenichi Fukui is a conceptual framework in MOT to explain and predict chemical reactivity and selectivity [36, 41, 42]. It focuses on the interaction between the highest-occupied MO and the lowest-unoccupied molecular orbital of the reactants. These orbitals, referred to as the "frontier orbitals," are most important in governing the reactivity of molecules. Because of this work, Fukui shared the 1981 Nobel Prize in Chemistry with Roald Hoffmann for his contribution to establishing the Woodward–Hoffmann rules.

1.4 Density Functional Theory

DFT [9] has been the workhorse in the past few decades as the most widely used computational method to simulate the electronic structure of molecules and condensed matter alike [43]. It has become a cornerstone of materials science, chemistry, and condensed matter physics due to its balance of remarkable accuracy and computational efficiency. At its core, unlike VBT and MOT, DFT avoids directly solving the Schrödinger equation, Eq. (1.1), with an approximate many-electron wavefunction Ψ. Instead, it rewrites Eq. (1.2) as

$$E = E[\rho], \tag{1.16}$$

where $\rho \equiv \rho(\mathbf{r})$ stands for the electron density of a many-electron system in the ground state, which is a function of only three spatial coordinates and is normalized to N,

$$\int \rho(\mathbf{r})d\mathbf{r} = N, \tag{1.17}$$

where N is the total number of electrons in the system. This electron density $\rho(\mathbf{r})$ can be obtained experimentally from X-ray crystallography. It can also be obtained from wavefunction theory from the N-body wavefunction, $\Psi(\mathbf{r}_1,\mathbf{r}_2,\ldots,\mathbf{r}_N)$,

$$\rho(\mathbf{r}) = N\iint\cdots\int|\Psi(\mathbf{r},\mathbf{r}_2,\ldots,\mathbf{r}_N)|^2\,d\mathbf{r}_2\,d\mathbf{r}_3\cdots d\mathbf{r}_N, \tag{1.18}$$

or MOs $\{\psi_i(\mathbf{r})\}$,

$$\rho(\mathbf{r}) = \sum_{i=1}^{\mathrm{occ}}|\psi_i(\mathbf{r})|^2, \tag{1.19}$$

where the sum runs over all occupied MOs in the single-particle approximation, such as in Hartree–Fock theory.

The idea of describing an electronic system by its density goes back to the Thomas–Fermi model developed separately by Thomas [44] and Fermi [45]. In these early works, the electron density was used as a central variable. However, the notation was not entirely standardized then. Many authors used symbols like $n(\mathbf{r})$ or $\rho(\mathbf{r})$ interchangeably to denote the electron density. To the best of the author's knowledge, it was Dirac's 1930 paper "Note on Exchange Phenomena in the Thomas Atom" [46] that derived an expression for the exchange energy of an electron gas and formulated it in terms of the local electron density, denoted by $\rho(\mathbf{r})$. This paper can be credited as the first influential instance where the symbol $\rho(\mathbf{r})$ was used explicitly to represent the electron density.

Equation (1.16) suggests that the total energy of the system is a functional of the electron density, as opposed to Eq. (1.2), where the total energy is a functional of the many-electron wavefunction,

$$E \equiv E[\Psi(\mathbf{r}_1,\mathbf{r}_2,\ldots,\mathbf{r}_N)]. \tag{1.20}$$

This reformulation of Eq. (1.16) is based on the two Hohenberg–Kohn theorems [8]. The first one states that there is a one-to-one mapping between the electron density $\rho(\mathbf{r})$ and the external potential $v_{\mathrm{ext}}(\mathbf{r})$ in the ground state. This suggests that all ground state properties of the system are functionals of the electron density, including the total energy,

$$E[\rho] = F[\rho] + \int \rho(\mathbf{r})v_{\mathrm{ext}}(\mathbf{r})d\mathbf{r}, \tag{1.21}$$

where $F[\rho]$ is the universal density functional. The second theorem asserts that the correct ground-state electron density should minimize the total energy functional, thus providing a variational principle for Eq. (1.21) and yielding the following Euler equation,

$$\mu = \frac{\delta E[\rho]}{\delta \rho(\mathbf{r})} = \frac{\delta F[\rho]}{\delta \rho(\mathbf{r})} + v_{\text{ext}}(\mathbf{r}), \tag{1.22}$$

where μ is the chemical potential and $\delta/\delta\rho(\mathbf{r})$ is the functional derivative with respect to the electron density [9].

According to these theorems, in principle, any energetic component of the total electronic energy can be rigorously expressed by a density functional, but in practice it is impossible to do so. In DFT, a conventional partition of the total electronic energy is the following,

$$E[\rho] = T_S[\rho] + V_{ne}[\rho] + J[\rho] + E_{xc}[\rho], \tag{1.23}$$

where T_S is the noninteracting total kinetic energy, V_{ne} the electron–nucleus attraction, J the electron–electron Coulombic repulsion, and E_{xc} the exchange–correlation energy, which captures the quantum mechanical effects of electron exchange and correlation. Note that when calculating the total energy, one should add the electrostatic repulsion among nuclei, V_{nn}. Among the four components in Eq. (1.23), only two terms are known to have an explicit expression in terms of the electron density. The first is $V_{ne}[\rho]$, which is the second term at the right-hand side of Eq. (1.21), and the other is the classical electron–electron Coulombic repulsion $J[\rho]$,

$$J[\rho] = \frac{1}{2}\iint \frac{\rho(\mathbf{r})\rho(\mathbf{r}')}{|\mathbf{r}-\mathbf{r}'|} d\mathbf{r}d\mathbf{r}'. \tag{1.24}$$

For $T_S[\rho]$ and $E_{xc}[\rho]$, their exact expression in terms of the electron density is unknown.

To overcome this difficulty, Kohn and Sham [47] introduced a fictitious system of noninteracting electrons moving in the so-called Kohn–Sham orbitals $\{\psi_i(\mathbf{r})\}$ with an effective potential $v_{eff}(\mathbf{r})$, so the same ground-state electron density $\rho(\mathbf{r})$ as the interacting system is reproduced

$$\rho(\mathbf{r}) = \sum_{i=1}^{N} |\psi_i(\mathbf{r})|^2, \tag{1.25}$$

the noninteracting total kinetic energy is expressed as

$$T_S[\rho] = -\frac{1}{2}\sum_{i=1}^{N} \int \psi_i(\mathbf{r})\nabla^2\psi_i(\mathbf{r})d\mathbf{r}, \tag{1.26}$$

and then the Kohn–Sham equation is obtained as,

$$\left(-\frac{1}{2}\nabla^2 + v_{eff}(\mathbf{r})\right)\psi_i(\mathbf{r}) = \epsilon_i\psi_i(\mathbf{r}), \tag{1.27}$$

where the effective Kohn–Sham potential is composed of several terms,

$$v_{eff}(\mathbf{r}) = v_{ext}(\mathbf{r}) + v_J(\mathbf{r}) + v_{xc}(\mathbf{r}), \tag{1.28}$$

with the classical Coulombic potential

$$v_J(\mathbf{r}) = \int \frac{\rho(\mathbf{r}')}{|\mathbf{r} - \mathbf{r}'|} d\mathbf{r}', \tag{1.29}$$

and the exchange–correlation potential

$$v_{xc}(\mathbf{r}) = \frac{\delta E_{xc}[\rho]}{\delta \rho(\mathbf{r})}. \tag{1.30}$$

In the Kohn–Sham scheme, the total kinetic energy is obtained by Eq. (1.26), so there is only one unknown left in Eq. (1.23) to approximate, which is the exchange–correlation energy density functional $E_{xc}[\rho]$.

The development and history of DFT are closely tied to the search for better exchange–correlation functionals because $E_{xc}[\rho]$ is the key approximation in the Kohn–Sham equation that determines its accuracy and applicability. The local density approximation (LDA) was the first form proposed, and it is the simplest exchange–correlation functional. It assumes that the exchange–correlation energy at any point in space depends only on the local electron density at that point,

$$E_{xc}^{\text{LDA}} = \int \rho(\mathbf{r}) e_{xc}(\rho(\mathbf{r})) d\mathbf{r}, \tag{1.31}$$

where $e_{xc}(\rho)$ is the exchange–correlation energy per particle. LDA works surprisingly well for systems with slowly varying electron densities (e.g., simple metals) but fails for systems with strong inhomogeneities (e.g., atoms, molecules, surfaces). To address the limitations of LDA, the generalized gradient approximation (GGA) [48] was developed, including not only the local electron density but also its gradient,

$$E_{xc}^{\text{GGA}} = \int \rho(\mathbf{r}) e_{xc}(\rho(\mathbf{r}), \nabla \rho(\mathbf{r})) d\mathbf{r}. \tag{1.32}$$

GGA functionals, such as B88 (Becke 88) [49], Lee–Yang–Parr (LYP) [50], and Perdew–Burke–Ernzerhof [51], significantly improved the accuracy of DFT for molecular systems, bond energies, and structural properties. Another popular category is the hybrid functional, such as B3LYP (Becke 3-parameter Lee-Yang-Parr), mix a portion of exact exchange from Hartree–Fock theory with GGA exchange and correlation,

$$E_{xc}^{\text{Hybrid}} = aE_x^{\text{HF}} + (1 - a)E_x^{\text{GGA}} + E_c^{\text{GGA}}, \tag{1.33}$$

where E_x^{HF} is the exact exchange energy from Hartree–Fock theory.

Other kinds of approximate exchange–correlation functionals in DFT are (i) meta-GGA functionals, such as Tao–Perdew–Staroverov–Scuseria [52] and Strongly Constrained and Appropriately Normed [53], which go beyond GGA by including higher-order derivatives of the electron density (e.g., the kinetic energy density), and (ii) range-separated hybrid and double hybrid functionals [54], which further refine the treatment of exchange and correlation energies by incorporating long-range corrections and perturbative correlation, respectively.

Unlike ab initio wavefunction theory methods in Figure 1.3, where a clear pathway is available to systematically improve the accuracy, DFT does not have such a route.

Figure 1.5 The Jacob's ladder in DFT (Generated with AI using DALL·E–OpenAI).

Still, the Jacob's ladder metaphor in DFT, shown in Figure 1.5, represents the hierarchy of increasingly accurate density functionals. This "ladder" metaphor, introduced by John P. Perdew, compares functional development to climbing a ladder toward the "heaven" of chemical accuracy. Each rung of the ladder corresponds to a level of approximation, with higher rungs providing greater accuracy but also increasing computational cost. It serves as a framework to understand the tradeoff between accuracy and computational cost in DFT, guiding researchers in choosing the appropriate functional for their systems.

DFT has become immensely popular over the last few decades due to its unique combination of accuracy, versatility, and computational efficiency. While DFT is powerful and widely used, its limitations arise primarily from approximations in the exchange–correlation functional and its inability to handle specific phenomena like strong correlation, excited states, or van der Waals interactions. Manifested by, for example, the self-interaction error [55], it overestimates electron delocalization, leading to incorrect prediction of properties like band gaps, dissociation energies, and ionization potentials. On the other hand, it underestimates electron localization, causing a poor description of charge transfer, oxidation states, spin states, or electronic configurations of localized systems such as transition metal complexes. These errors arise because DFT functionals do not fully capture the complex exchange and correlation effects. Recent developments, such as double hybrid functionals (e.g., B2PLYP), combine a portion of exact exchange with a perturbative treatment of correlation, providing a more balanced description of delocalized and localized states. Additionally, ML techniques are being used to develop new functionals that better capture the complexities of electron delocalization and localization [56–58].

1.5 Scope of This Book

We want insight, not just numbers [20]. Any electronic structure theory should provide a new understanding of traditional chemical concepts. As highlighted in Section 1.3, VBT did that with the insight into chemical bonding, and MOT provided the FMO theory, orbital symmetry conservation principle, and Woodward–Hoffmann rules. They are both orbital-based. In DFT, electron density is the basic variable. To the best of the author's knowledge, there is no systematic work to present all conceptual frameworks to harness chemical understanding using density-based ideas [59–61]. Conceptual DFT (CDFT) [9, 62–65] is the first effort in DFT to analyze chemical reactivity in terms of global and local descriptors such as electronegativity, hardness, and electrophilicity. Other efforts such as quantum theory of atoms in molecules (QTAIM) [66], electron localization function (ELF) [67], and noncovalent interaction (NCI) [68, 69] analyses are also available in the literature. However, their scope is limited, and controversies are abundant. The purpose of this book is to fill the knowledge gap by providing a systematic overview of how to employ density-based ideas to harvest chemical insights, including covalent bonding, weak interactions, conformational stability, acidity/basicity, electrophilicity, nucleophilicity, aromaticity, stereoselectivity, and catalysis, with density-based descriptors.

In what follows, in Part I of this book, we will present the four density-based frameworks, including CDFT (Chapter 2), density-associated quantities (Chapter 3), information-theoretic approach (Chapter 4), and orbital-free DFT (Chapter 5), and the interrelationships among these frameworks as well as a few recent developments (Chapter 6).

In Part II, applications of these four frameworks to real chemical problems are discussed, including covalent and noncovalent interactions (Chapter 7), cooperativity and frustration (Chapter 8), principle of chirality hierarchy and homochirality (Chapter 9), electrophilicity and nucleophilicity (Chapter 10), steric effect and stereoselectivity (Chapter 11), acidity and basicity (Chapter 12), aromaticity and antiaromaticity (Chapter 13), excited states (Chapter 14), catalysis (Chapter 15), and miscellaneous applications (Chapter 16).

In Part III, based on the ideas from this book, in the final two chapters (Chapters 17 and 18), we provide perspectives on how chemical understanding is, in general, obtained with chemical concepts (Chapter 17) and how to expand the territory of chemical reactivity theory by harvesting chemical understanding with ML and quantum computing (Chapter 18).

References

1 Heitler W, London F. Wechselwirkung neutraler Atome und homopolare Bindung nach der Quantenmechanik. Z Phys. 1927 Jun;44(6–7):455–72.

2 Gerratt J, Cooper DL, Karadakov PB, Raimondi M. Modern valence bond theory. Chem Soc Rev. 1997;26(2):87.

3 Shaik S, Hiberty PC. Valence bond theory, its history, fundamentals, and applications: a primer. In: Lipkowitz KB, Larter R, Cundari TR, editors. Reviews in Computational Chemistry. Wiley; 2004. pp. 1–100.

4 Zhou C, Wu X, Ying F, Wu W. Ab initio valence bond theory for strongly correlated systems. J Chem Theory Comput. 2025 Jul 8;21(13):6243–57.

5 Roothaan CCJ. New developments in molecular orbital theory. Rev Mod Phys. 1951 Apr 1;23(2):69–89.

6 Hehre WJ. Ab initio molecular orbital theory. Acc Chem Res. 1976 Nov 1;9(11): 399–406.

7 Dewar MJS. A molecular orbital theory of organic chemistry. I. General principles. J Am Chem Soc. 1952;74(13):3341–5.

8 Hohenberg P, Kohn W. Inhomogeneous electron gas. Phys Rev. 1964 Nov 9; 136(3B):B864–71.

9 Parr RG, Weitao Y. Density-Functional Theory of Atoms and Molecules. Oxford University Press; 1995.

10 Cohen AJ, Mori-Sánchez P, Yang W. Challenges for density functional theory. Chem Rev. 2012 Jan 11;112(1):289–320.

11 Tozzini V. Multiscale modeling of proteins. Acc Chem Res. 2010 Feb 16;43(2): 220–30.

12 Horstemeyer MF. Multiscale modeling: a review. In: Leszczynski J, Shukla MK, editors. Practical aspects of computational chemistry. Dordrecht: Springer Netherlands; 2010. pp. 87–135.

13 Mjolsness E, DeCoste D. Machine learning for science: state of the art and future prospects. Science. 2001 Sep 14;293(5537):2051–5.

14 Rupp M, Tkatchenko A, Müller K-R, von Lilienfeld OA. Fast and accurate modeling of molecular atomization energies with machine learning. Phys Rev Lett. 2012 Feb 3;108(5):058301.

15 LeCun Y, Bengio Y, Hinton G. Deep learning. Nature. 2015 May 28;521 (7553):436–44.

16 Huang B, von Lilienfeld OA. Ab initio machine learning in chemical compound space. Chem Rev. 2021 Aug 25;121(16):10001–36.

17 Keith JA, Vassilev-Galindo V, Cheng B, Chmiela S, Gastegger M, Müller K-R, et al. Combining machine learning and computational chemistry for predictive insights into chemical systems. Chem Rev. 2021 Aug 25;121(16):9816–72.

18 Kulichenko M, Nebgen B, Lubbers N, Smith JS, Barros K, Allen AEA, et al. Data generation for machine learning interatomic potentials and beyond. Chem Rev. 2024 Nov 21;124:13681–714.

19 Meuwly M. Machine learning for chemical reactions. Chem Rev. 2021 Aug 25;121(16):10218–39.

20 Liu S. Harvesting chemical understanding with machine learning and quantum computers. ACS Phys Chem Au. 2024 Mar 27;4(2):135–42.

21 Liu S. Exploring Chemical Concepts Through Theory and Computation. Wiley; 2024.

22 Agranat I. Aromaticity – a theoretical notion. Struct Chem. 2024 Jun;35(3):715–20.

23 Neese F, Atanasov M, Bistoni G, Maganas D, Ye S. Chemistry and quantum mechanics in 2019: give us insight and numbers. J Am Chem Soc. 2019 Feb 20;141(7):2814–24.

24 Vennelakanti V, Nazemi A, Mehmood R, Steeves AH, Kulik HJ. Harder, better, faster, stronger: large-scale QM and QM/MM for predictive modeling in enzymes and proteins. Curr Opin Struct Biol. 2022 Feb;72:9–17.

25 Tzeliou CE, Mermigki MA, Tzeli D. Review on the QM/MM methodologies and their application to metalloproteins. Molecules. 2022 Apr 20;27(9):2660.

26 Pauling L. The nature of the chemical bond—1992. J Chem Educ. 1992 Jul;69(7):519.

27 Dunning TH, Xu LT, Takeshita TY, Lindquist BA. Insights into the electronic structure of molecules from generalized valence bond theory. J Phys Chem A. 2016 Mar 24;120(11):1763–78.

28 Mo Y, Song L, Lin Y. Block-localized wavefunction (BLW) method at the density functional theory (DFT) level. J Phys Chem A. 2007 Aug 30;111(34):8291–301.

29 Mulliken RS. Spectroscopy, molecular orbitals, and chemical bonding. Science. 1967 Jul 7;157(3784):13–24.

30 Hund F. Zur deutung der molekelspektren. IV. Z Phys. 1928 Nov;51(11–12):759–95.

31 Murrell JN. The origins and later developments of molecular orbital theory. Int J Quantum Chem. 2012 Sep 5;112(17):2875–9.

32 Zhou C, Ying F, Wu W. Chemical concepts from *ab initio* valence bond theory. In: Liu S, editor. Exploring Chemical Concepts Through Theory and Computation. Wiley; 2024. pp. 23–42.

33 Long Gu F, Yu J, Yang W. Chemical concepts from molecular orbital theory. In: Liu S, editor. Exploring Chemical Concepts Through Theory and Computation. Wiley; 2024. pp. 1–22.

34 Shaik S, Danovich D, Wu W, Hiberty PC. Charge-shift bonding and its manifestations in chemistry. Nat Chem. 2009 Sep;1(6):443–9.

35 Shaik S, Danovich D, Galbraith JM, Braïda B, Wu W, Hiberty PC. Charge-shift bonding: a new and unique form of bonding. Angew Chem. 2020 Jan 13;132(3):996–1013.

36 Fukui K. Role of frontier orbitals in chemical reactions. Science. 1982 Nov 19;218(4574):747–54.

37 Geerlings P, Ayers PW, Toro-Labbé A, Chattaraj PK, De Proft F. The Woodward-Hoffmann rules reinterpreted by conceptual density functional theory. Acc Chem Res. 2012 May 15;45(5):683–95.

38 Woodward RB, Hoffmann R. The conservation of orbital symmetry. Angew Chem Int Ed. 1969 Nov;8(11):781–853.

39 Reed AE, Weinhold F. Natural localized molecular orbitals. J Chem Phys. 1985 Aug 15;83(4):1736–40.

40 Glendening ED, Landis CR, Weinhold F. Natural bond orbital methods. WIREs Comput Mol Sci. 2012 Jan;2(1):1–42.

41 Yang W, Parr RG, Pucci R. Electron density, Kohn–Sham frontier orbitals, and Fukui functions. J Chem Phys. 1984;81(6):2862.

42 Huang Y, Rong C, Zhang R, Liu S. Evaluating frontier orbital energy and HOMO/LUMO gap with descriptors from density functional reactivity theory. J Mol Model. 2017 Jan;23(1):3.

43 Teale AM, Helgaker T, Savin A, Adamo C, Aradi B, Arbuznikov AV, et al. DFT exchange: sharing perspectives on the workhorse of quantum chemistry and materials science. Phys Chem Chem Phys. 2022 Dec 7;24(47):28700–81.

44 Thomas LH. The calculation of atomic fields. Math Proc Camb Phil Soc. 1927;23:542–8.

45 Fermi E. Eine statistische methode zur bestimmung einiger eigenschaften des atoms und ihre anwendung auf die theorie des periodischen systems der elemente. Z Phys. 1928 Jan;48(1–2):73–9.

46 Dirac PAM. Note on exchange phenomena in the thomas atom. Math Proc Camb Phil Soc. 1930 Jul;26(03):376.

47 Kohn W, Sham LJ. Self-consistent equations including exchange and correlation effects. Phys Rev. 1965 Nov 15; 140(4A):A1133–8.

48 Burke K, Perdew JP, Ernzerhof M. Why the generalized gradient approximation works and how to go beyond it. Int J Quantum Chem. 1997 Jan 1;61(2):287–93.

49 Becke AD. A multicenter numerical integration scheme for polyatomic molecules. J Chem Phys. 1988;88(4):2547.

50 Lee C, Yang W, Parr RG. Development of the Colle-Salvetti correlation-energy formula into a functional of the electron density. Phys Rev B Condens Matter. 1988 Jan 15;37(2):785–9.

51 Perdew JP, Burke K, Ernzerhof M. Generalized gradient approximation made simple. Phys Rev Lett. 1996 Oct 28;77(18):3865–8.

52 Tao J, Perdew JP, Staroverov VN, Scuseria GE. Climbing the density functional ladder: nonempirical meta-generalized gradient approximation designed for molecules and solids. Phys Rev Lett. 2003 Oct 3;91(14):146401.

53 Sun J, Ruzsinszky A, Perdew JP. Strongly constrained and appropriately normed semilocal density functional. Phys Rev Lett. 2015 Jul 17;115(3):036402.

54 Brémond É, Pérez-Jiménez ÁJ, Sancho-García JC, Adamo C. Range-separated hybrid and double-hybrid density functionals: a quest for the determination of the range-separation parameter. J Chem Phys. 2020 Jun 28;152(24):244124.

55 Cohen AJ, Mori-Sánchez P, Yang W. Insights into current limitations of density functional theory. Science. 2008 Aug 8;321(5890):792–4.

56 Voss J. Machine learning for accuracy in density functional approximations. J Comput Chem. 2024 Apr 26;45:1829–45.

57 Remme R, Kaczun T, Ebert T, Gehrig CA, Geng D, Gerhartz G, et al. Stable and accurate orbital-free density functional theory powered by machine learning. J Am Chem Soc. 2025 Aug 1; 147(32):28851–9.

58 Kumar S, Jing X, Pask JE, Medford AJ, Suryanarayana P. Kohn-Sham accuracy from orbital-free density functional theory via Δ-machine learning. J Chem Phys. 2023 Dec 28;159(24).

59 Rong C, Zhao D, He X, Liu S. Development and applications of the density-based theory of chemical reactivity. J Phys Chem Lett. 2022 Dec 8;13(48):11191–200.

60 He X, Li M, Rong C, Zhao D, Liu W, Ayers PW, et al. Some recent advances in density-based reactivity theory. J Phys Chem A. 2024 Feb 22;128(7):1183–96.

61 An X, Zhang W, He X, Li M, Rong C, Liu S. Density-based reactivity theory applied to excited states. AAPPS Bull. 2024 Feb 2;34(1):8.

62 Liu S. Conceptual density functional theory and some recent developments. Acta Phys-Chim Sin. 2009;25(3):590–600.

63 Liu S, editor. Conceptual Density Functional Theory: Towards a New Chemical Reactivity Theory. 1st ed. Wiley-VCH GmbH: Wiley; 2022.

64 Geerlings P, De Proft F, Langenaeker W. Conceptual density functional theory. Chem Rev. 2003 May;103(5):1793–873.

65 Geerlings P, Chamorro E, Chattaraj PK, De Proft F, Gázquez JL, Liu S, et al. Conceptual density functional theory: status, prospects, issues. Theor Chem Acc. 2020 Feb;139(2):36.

66 Bader RFW. Atoms in Molecules. Oxford University Press;1989.

67 Savin A, Nesper R, Wengert S, Fässler TF. ELF: the electron localization function. Angew Chem Int Ed. 1997 Sep 17;36(17):1808–32.

68 Johnson ER, Keinan S, Mori-Sánchez P, Contreras-García J, Cohen AJ, Yang W. Revealing noncovalent interactions. J Am Chem Soc. 2010 May 12;132(18):6498–506.

69 Contreras-García J, Johnson ER, Keinan S, Chaudret R, Piquemal J-P, Beratan DN, et al. NCIPLOT: a program for plotting non-covalent interaction regions. J Chem Theory Comput. 2011 Mar 8;7(3):625–32.

2

Conceptual Density Functional Theory

TABLE OF CONTENTS

2.1 Hypotheses and Brief History

2.1.1 Hypotheses of CDFT

Pioneered by the late Robert G. Parr and coworkers in the late 1970s, conceptual density functional theory (CDFT) [1–5] is the first systematic framework that applies the fundamental theorems of DFT to study chemical reactivity and physicochemical properties.

The basic hypotheses of CDFT are that (i) chemical reactivity of a system can be quantified by its response to changes in the total number of electrons N and the external potential $v(\mathbf{r})$ during a chemical process or transformation through a Taylor expansion series, and (ii) an electronic system can be transcribed in the spirit of classical thermodynamics, where the state functional of one ensemble can be converted to another one by the Legendre transformation with a set of conjugate state variables.

More specifically, CDFT assumes that a system can be characterized by four state variables, N, $v(\mathbf{r})$, electron density $\rho(\mathbf{r})$, and chemical potential μ, and its reactivity can be determined through a Taylor expansion of the state function(al) in terms of these variables using partial derivatives and functional derivatives. These derivatives are assumed to exist, be continuous, and exhibit well-behaved properties. The central task of CDFT over the past few decades has been to interpret these derivatives by providing physicochemical insights, leading to the development of various reactivity descriptors, such as electronegativity, hardness, softness, and Fukui function. These descriptors can be applied to qualitatively and quantitatively characterize the chemical behavior of molecules and condensed-phase matters alike.

2.1.2 Brief History

Birth of CDFT (1978). The key paper by Parr and coworkers in 1978 [6] marked the birth of CDFT. They identified the Lagrange multiplier (μ) in the DFT formalism, Eq. (1.22), as electronegativity, a fundamental concept in chemistry. This connection opened the door to interpreting the DFT variable directly in terms of chemical reactivity.

Four Ensembles of CDFT (1982). In the seminal article by Nalewajski and Parr [7], they laid the groundwork for transcribing DFT in analogy to classical thermodynamics by introducing four ensembles: canonical, grand canonical, isomorphic, and grand isomorphic. These four ensembles provided the conceptual framework to systematically introduce CDFT descriptors. Bridging the concepts of thermodynamics with DFT, this paper also offered opportunities to define such concepts as local temperature and Ghosh–Berkowitz–Parr entropy in later developments [8].

Hardness and Softness (1983). Pearson and Parr [9] introduced the concepts of chemical hardness (η) and softness (S), which describe the resistance of a system to changes in its electron distribution. These concepts became central to understanding reactivity and stability in chemical systems since they are closely related to the hard–soft acid–base (HSAB) principle [10, 11], first proposed by Pearson for inorganic systems.

Fukui Functions (1984). Fukui functions, introduced by Yang and Parr [12, 13], describe the site-specific response of electron density to changes in the number of electrons. They are formulated as the derivative of the electron density with respect to the number of electrons, with the external potential fixed. These functions became essential for predicting the regioselectivity of electrophilic and nucleophilic attacks in molecules.

Reactivity Descriptors (1990s–2010s). Throughout the 1990s to 2010s, numerous reactivity descriptors, such as local softness [14, 15], electrophilicity index [16, 17], dual descriptor [18–21], and philicity indices, were developed. These descriptors provided deeper insights into the reactivity and selectivity of chemical systems. CDFT matured significantly with the development of additional reactivity descriptors and their application to various chemical problems. A few review articles on CDFT were published during this period [2, 22].

Modern Developments (2010s–Present). CDFT continues to evolve, with ongoing research focused on refining reactivity descriptors, extending the theory to excited states and time-dependent phenomena, and applying it to complex systems such as biomolecules, surfaces, and materials. Advances in computational methods have also made it possible to apply CDFT to larger and more complex systems. The incorporation of machine learning and other advanced computational techniques continues to enhance its predictive power. A systematic overview of the status of CDFT by major players of the field was published [4]. The first book on CDFT with two volumes was released in 2022 [5].

CDFT has been referred to by various names, including chemical DFT and density functional reactivity theory. A literature survey indicates that the term "CDFT" was first casually mentioned in 1993 by Parr and Zhou [23]. Its formal introduction appeared in the 1995 *Annual Review of Physical Chemistry* article by Parr and Yang [24]. In this work, they concluded that "Conceptual (qualitative) DFT holds much untapped promise ... there appear to be sound grounds for an optimistic outlook concerning the ultimate value of conceptual DFT." Over the years, there have been complaints that the name "CDFT" is misleading, as it suggests two incorrect notions: (i) CDFT is purely conceptual and incapable of producing numerical results, and (ii) DFT is purely computational and lacks conceptual depth. Addressing this issue, the author once discussed the matter with Bob Parr, the founding father of CDFT. He firmly believed that "CDFT" was an appropriate name, particularly from a historical perspective. Moreover, he offered one compelling and surprising reason for his stance: "Walter Kohn likes it!"

2.2 Basic Formulation

2.2.1 Legendre Transform

A Legendre transform converts a function or functional of one set of variables to another function or functional of a conjugate set of variables. Let us consider a function with two independent variables, $F(x,y)$, whose differential is

$$dF(x,y) = \left(\frac{\partial F}{\partial x}\right)_y dx + \left(\frac{\partial F}{\partial y}\right)_x dy. \tag{2.1}$$

Defining $u \equiv (\partial F/\partial x)_y$ and $w \equiv (\partial F/\partial y)_x$, we have

$$dF(x,y) = u\ dx + w\ dy. \tag{2.2}$$

We call u and x a *conjugate* pair of variables and likewise w and y. A Legendre-transformed function G is defined as follows:

$$G \equiv F - \left(\frac{\partial F}{\partial y}\right)_x y = F - wy, \tag{2.3}$$

and has the differential

$$dG = dF - d(wy) = udx - ydw. \tag{2.4}$$

So, G can be regarded as the function of two new independent variables x and w, $G(x,w)$. The variable change of y in $F(x,y)$ to its conjugate variable w in $G(x,w)$ is made possible by the Legendre transform.

The Maxwell relation originates from the symmetrical requirement on the second-order derivatives of a twice-differentiable function $F(x,y)$; that is, $\frac{\partial}{\partial x}\left(\frac{\partial F}{\partial y}\right) = \frac{\partial}{\partial y}\left(\frac{\partial F}{\partial x}\right)$. These Maxwell relations are useful in connecting one set of derivatives with another set of derivatives.

In thermodynamics, using Legendre transforms, one can convert between the internal energy U, enthalpy H, Helmholtz free energy F, and Gibbs free energy G, using the conjugate pressure—volume pair (P,V), temperature–entropy pair (T,S), and the particle-number–chemical-potential pair (N,μ). According to the thermodynamic identity

$$dU = TdS - PdV + \mu\, dN \tag{2.5}$$

where T, P, and μ are therefore the variables conjugate to S, V, and N, respectively, and, from Eq. (2.5), we also have

$$\left(\frac{\partial U}{\partial S}\right)_{V,N} = T, \left(\frac{\partial U}{\partial V}\right)_{S,N} = -P \text{ and} \left(\frac{\partial U}{\partial N}\right)_{S,V} = \mu \tag{2.6}$$

We can use the Legendre transform to convert $U(N,S,V)$ to a new thermodynamic potential, enthalpy $H(N,S,P)$

$$H(N,S,P) \equiv U(N,S,V) - \left(\frac{\partial U}{\partial V}\right)_S V = U(N,S,V) + PV \tag{2.7}$$

so its differential is as follows after Eq. (2.5) is applied,

$$dH(N,S,P) = TdS + VdP + \mu\, dN \tag{2.8}$$

with

$$\left(\frac{\partial H}{\partial S}\right)_{P,N} = T, \left(\frac{\partial H}{\partial V}\right)_{S,N} = V \text{ and} \left(\frac{\partial H}{\partial N}\right)_{S,P} = \mu \tag{2.9}$$

Formulas for the Gibbs free energy $G(N,P,T)$ and the Helmholtz free energy $F(N,V,T)$ can be similarly obtained. In statistical mechanics, these thermodynamic quantities, U, H, F, and G, are the state functions in their respective ensembles, such as the canonical ensemble (N,V,T) and grand canonical ensemble (μ,V,T).

2.2.2 Four Ensembles of CDFT

In CDFT, the total energy E of a system can be thought of as a functional of two variables, N and $\upsilon(\mathbf{r})$,

$$E = E[N,\upsilon]. \tag{2.10}$$

The change in E due to the infinitesimal change of these two variables can be written as

$$dE = \left(\frac{\partial E}{\partial N}\right)_{\upsilon} dN + \int\left(\frac{\delta E}{\delta \upsilon(\mathbf{r})}\right)_{N} \delta\upsilon(\mathbf{r})d\mathbf{r}. \tag{2.11}$$

Assuming that E is differentiable with respect to N and $\upsilon(\mathbf{r})$ and that N can be treated as a continuous variable, one finds

$$\left(\frac{\partial E}{\partial N}\right)_{\upsilon} = \mu \text{ and} \left(\frac{\delta E}{\delta \upsilon(\mathbf{r})}\right)_{N} = \rho(\mathbf{r}). \tag{2.12}$$

From Eq. (2.11), according to Eq. (2.2), we can see that N and μ, and $\rho(\mathbf{r})$ and $\upsilon(\mathbf{r})$ are two pairs of conjugate variables. In analogy with classical thermodynamics, we can define four ensembles [7, 25]:

$$\text{Canonical ensemble: } E = E[N,\ \upsilon]; \tag{2.13}$$

$$\text{Grand canonical ensemble: } \Omega[\mu,\upsilon] \equiv E - \left(\frac{\partial E}{\partial N}\right)_{\upsilon} N = E - \mu N; \tag{2.14}$$

$$\begin{aligned}\text{Isomorphic ensemble: } F[N,\varrho] &\equiv E - \int\left(\frac{\delta E}{\delta \upsilon(\mathbf{r})}\right)_{N} \upsilon(\mathbf{r})d\mathbf{r} \\ &= E - \int \varrho(\mathbf{r})\upsilon(\mathbf{r})d\mathbf{r};\end{aligned} \tag{2.15}$$

$$\begin{aligned}\text{Grand isomorphic ensemble: } R[\mu,\varrho] &\equiv E - \left(\frac{\partial E}{\partial N}\right)_{\upsilon} \\ &\quad N - \int\left(\frac{\delta E}{\delta \upsilon(\mathbf{r})}\right)_{N} \upsilon(\mathbf{r})d\mathbf{r} \\ &= E - \mu N - \int \varrho(\mathbf{r})\upsilon(\mathbf{r})d\mathbf{r},\end{aligned} \tag{2.16}$$

where E, Ω, F, and R are state functionals of the four ensembles, with their natural variables defined through their respective Legendre transform.

2.2.3 Basic Relations in Canonical Ensemble

In the canonical ensemble, the total energy E is represented as a functional of the global variable N and the local variable $v(\mathbf{r})$, $E[N, v]$. Assuming E to be differentiable and using Eq. (2.11), we have the following differential expression for E:

$$dE = \mu dN + \int \rho(\mathbf{r})\delta v(\mathbf{r})d\mathbf{r}. \tag{2.17}$$

The Maxwell relation of Eq. (2.11) is the following:

$$\frac{\partial}{\partial N}\left(\frac{\delta E}{\delta v(\mathbf{r})}\right) = \frac{\delta}{\delta v(\mathbf{r})}\left(\frac{\partial E}{\partial N}\right). \tag{2.18}$$

With Eq. (2.12), the Maxwell relation associated with Eq. (2.17) is the following:

$$\left(\frac{\partial \rho(\mathbf{r})}{\partial N}\right)_v = \left(\frac{\delta \mu}{\delta v(\mathbf{r})}\right)_N = f(\mathbf{r}) \tag{2.19}$$

which are the definitions of the Fukui function.

Also, in the canonical ensemble, chemical potential μ and electron density $\rho(\mathbf{r})$ are functionals of N and $v(\mathbf{r})$, $\mu = \mu[\mathrm{N}, v(\mathbf{r})]$, and $\rho(\mathbf{r}) = \rho[\mathrm{N}, v(\mathbf{r})]$, so their differential forms are as follows:

$$d\mu = \eta dN + \int f(\mathbf{r})\delta v(\mathbf{r})d\mathbf{r}, \tag{2.20}$$

$$d\rho(\mathbf{r}) = f(\mathbf{r})dN + \int \varpi(\mathbf{r},\mathbf{r}')\delta v(\mathbf{r}')d\mathbf{r}', \tag{2.21}$$

where η is the hardness,

$$\eta = \left(\frac{\partial^2 E}{\partial N^2}\right)_v = \left(\frac{\partial \mu}{\partial N}\right)_v, \tag{2.22}$$

and response function $\varpi(\mathbf{r},\mathbf{r}')$ is defined as

$$\varpi(\mathbf{r},\mathbf{r}') = \left(\frac{\delta \rho(\mathbf{r})}{\delta v(\mathbf{r}')}\right)_N. \tag{2.23}$$

The Maxwell relations for Eqs. (2.20) and (2.21) are, respectively,

$$\left(\frac{\partial f(\mathbf{r})}{\partial N}\right)_v = \left(\frac{\delta \eta}{\delta v(\mathbf{r})}\right)_N \tag{2.24}$$

and

$$\left(\frac{\partial \varpi(\mathbf{r},\mathbf{r}')}{\partial N}\right)_v = \left(\frac{\delta f(\mathbf{r})}{\delta v(\mathbf{r})}\right)_N. \tag{2.25}$$

Now, let us consider the change of the total energy ΔE in the canonical ensemble when an attacking agent induces changes of the number of electrons and external potential of the system, ΔN and $\Delta v(\mathbf{r})$. Using the Taylor expansion up to the second order, we have

$$\begin{aligned}
\Delta E &\equiv E[N+\Delta N, v(\mathbf{r})+v(\mathbf{r})] - E[N, v(\mathbf{r})] \\
&= \left(\frac{\partial E}{\partial N}\right)_v \Delta N + \int \left(\frac{\delta E}{\delta v(\mathbf{r})}\right)_N \Delta v(\mathbf{r}) d\mathbf{r} \\
&+ \frac{1}{2!}\left\{ \left(\frac{\partial^2 E}{\partial N^2}\right)_v \Delta N^2 + 2\int \left(\frac{\partial}{\partial N}\left(\frac{\delta E}{\delta v(\mathbf{r})}\right)_N\right)_v \Delta N \Delta v(\mathbf{r}) d\mathbf{r} \right. \\
&+ \left. \iint \left(\frac{\delta^2 E}{\delta v(\mathbf{r})\delta v(\mathbf{r}')}\right)_N \Delta v(\mathbf{r})\Delta v(\mathbf{r}') d\mathbf{r} d\mathbf{r}' \right\},
\end{aligned} \tag{2.26}$$

where $(\partial E/\partial N)_v$ and $[\delta E/\delta v(\mathbf{r})]_N$ are the first-order partial derivatives of the total energy with respect to N and $v(\mathbf{r})$ with $v(\mathbf{r})$ and N fixed, respectively, and $(\partial^2 E/\partial N^2)_v$ and $[\delta^2 E/\delta v^2(\mathbf{r})]_N$ are the corresponding second-order terms, whereas $[\partial/\partial N(\delta E/\delta v(\mathbf{r}))_N]_v$ is the second-order cross term. With the quantities introduced earlier, we have

$$\begin{aligned}
\Delta E = \mu \Delta N + \int \rho(\mathbf{r})\Delta v(\mathbf{r}) d\mathbf{r} + \frac{1}{2}\{\eta \Delta N^2 + 2\Delta N \\
\int f(\mathbf{r})\Delta v(\mathbf{r}) d\mathbf{r} + \iint \varpi(\mathbf{r},\mathbf{r}')\Delta v(\mathbf{r})\Delta v(\mathbf{r}') d\mathbf{r} d\mathbf{r}'\}
\end{aligned} \tag{2.27}$$

2.2.4 Basic Relations in Grand Canonical Ensemble

In the grand canonical ensemble, the natural variables are the chemical potential μ and the external potential $v(\mathbf{r})$. The grand potential Ω defined in Eq. (2.14) takes the following differential form:

$$d\Omega = -N d\mu + \int \rho(\mathbf{r})\delta v(\mathbf{r}) d\mathbf{r}. \tag{2.28}$$

This equation was obtained by differentiating Eq. (2.14) and then employing Eq. (2.17), so

$$\left(\frac{\partial \Omega}{\partial \mu}\right)_v = -N \text{ and } \left(\frac{\delta \Omega}{\delta v(\mathbf{r})}\right)_\mu = \rho(\mathbf{r}). \tag{2.29}$$

The Maxwell relation associated with Eq. (2.28)

$$\left(\frac{\partial \rho(\mathbf{r})}{\partial \mu}\right)_v = -\left(\frac{\delta N}{\delta v(\mathbf{r})}\right)_\mu = s(\mathbf{r}) \tag{2.30}$$

leads to the definition of local softness $s(\mathbf{r})$. The other two variables, N and $\rho(\mathbf{r})$, can also be expressed as functionals of the chemical potential μ and external potential $v(\mathbf{r})$ in the grand canonical ensemble, $N = N[\mu, v(\mathbf{r})]$ and $\rho(\mathbf{r}) = \rho[\mu, v(\mathbf{r})]$, leading to their differential form as

$$dN = -Sd\mu + \int s(\mathbf{r})\delta v(\mathbf{r})d\mathbf{r}, \tag{2.31}$$

and

$$d\rho(\mathbf{r}) = s(\mathbf{r})d\mu - \int s(\mathbf{r},\mathbf{r}')\delta v(\mathbf{r}')d\mathbf{r}', \tag{2.32}$$

respectively, where the softness S is

$$S = \left(\frac{\partial N}{\partial \mu}\right)_v, \tag{2.33}$$

and the softness kernel $s(\mathbf{r},\mathbf{r}')$ is

$$s(\mathbf{r},\mathbf{r}') = -\left(\frac{\delta \rho(\mathbf{r})}{\delta v(\mathbf{r}')}\right)_\mu. \tag{2.34}$$

The Maxwell relations associated with Eqs. (2.31) and (2.32) are, respectively

$$-\left(\frac{\partial s(\mathbf{r})}{\partial \mu}\right)_v = -\left(\frac{\delta S}{\delta v(\mathbf{r})}\right)_\mu, \tag{2.35}$$

and

$$-\left(\frac{\partial s(\mathbf{r},\mathbf{r}')}{\partial \mu}\right)_v = -\left(\frac{\delta s(\mathbf{r})}{\delta v(\mathbf{r}')}\right)_\mu. \tag{2.36}$$

The Taylor series up to the second order in the grand canonical ensemble is the following:

$$\begin{aligned}\Delta\Omega &\equiv \Omega[\mu + \Delta\mu, v(\mathbf{r}) + v(\mathbf{r})] - \Omega[\mu, v(\mathbf{r})]\\ &= \left(\frac{\partial \Omega}{\partial \mu}\right)_v \Delta\mu + \int \left(\frac{\delta \Omega}{\delta v(\mathbf{r})}\right)_\mu \Delta v(\mathbf{r})d\mathbf{r}\\ &\quad + \frac{1}{2!}\left\{\left(\frac{\partial^2 \Omega}{\partial \mu^2}\right)_v \Delta\mu^2 + 2\int \left(\frac{\partial}{\partial \mu}\left(\frac{\delta \Omega}{\delta v(\mathbf{r})}\right)_\mu\right)_v \Delta\mu\Delta v(\mathbf{r})d\mathbf{r}\right.\\ &\quad \left. + \iint \left(\frac{\delta^2 \Omega}{\delta v(\mathbf{r})\delta v(\mathbf{r}')}\right)_\mu \Delta v(\mathbf{r})\Delta v(\mathbf{r}')d\mathbf{r}d\mathbf{r}'\right\}\end{aligned} \tag{2.37}$$

With the quantities introduced earlier, we have

$$\begin{aligned}\Delta\Omega &= -N\Delta\mu + \int \rho(\mathbf{r})\Delta v(\mathbf{r})d\mathbf{r}\\ &\quad - \frac{1}{2}\left\{S\Delta\mu^2 - 2\Delta\mu\int s(\mathbf{r})\Delta v(\mathbf{r})d\mathbf{r}\right.\\ &\quad \left. + \iint s(\mathbf{r},\mathbf{r}')\Delta v(\mathbf{r})\Delta v(\mathbf{r}')d\mathbf{r}d\mathbf{r}'\right\}\end{aligned} \tag{2.38}$$

2.2.5 Basic Relations in Isomorphic and Grand Isomorphic Ensembles

In the isomorphic ensemble, two natural variables N and $\rho(\mathbf{r})$ are dependent through the normalization condition in Eq. (1.17). A modified version for the isomorphic ensemble was provided by Baekelandt et al. [26] in terms of the shape function $\sigma(\mathbf{r})$

$$\sigma(\mathbf{r}) = \frac{\rho(\mathbf{r})}{N}, \tag{2.39}$$

so the state function becomes

$$\text{F}[N, \sigma] \equiv E - N\int \sigma(\mathbf{r})v(\mathbf{r})d\mathbf{r}, \tag{2.40}$$

leading to the introduction of a few new quantities and relationships. However, they are often difficult to evaluate, so we do not introduce them here. Should there be any interest in this ensemble, readers should refer to the original articles [26] and the CDFT book [5] for details.

In the grand isomorphic ensemble, whose state function is R and its natural variables are chemical potential μ and electron density $\rho(\mathbf{r})$. Differentiating Eq. (2.16) with the use of Eq. (2.11) yields

$$dR = -Nd\mu - N\int v(\mathbf{r})\delta\rho(\mathbf{r})d\mathbf{r}, \tag{2.41}$$

so we have

$$\left(\frac{\partial R}{\partial \mu}\right)_{\rho} = -N, \left(\frac{\delta R}{\delta \rho(\mathbf{r})}\right)_{\mu} = -v(\mathbf{r}). \tag{2.42}$$

The Maxwell relations associated with Eq. (2.41) is

$$\left(\frac{\partial v(\mathbf{r})}{\partial \mu}\right)_{\rho} = \left(\frac{\delta N}{\delta \rho(\mathbf{r})}\right)_{\mu} = 1. \tag{2.43}$$

To express the other independent variables, N and $v(\mathbf{r})$, in the grand isomorphic ensemble, one has the following differential forms:

$$\text{d}N = \left(\frac{\partial N}{\partial \mu}\right)_{\rho} d\mu + \int \left(\frac{\delta N}{\delta \rho(\mathbf{r})}\right)_{\mu} \delta\rho(\mathbf{r})\, d\mathbf{r} = \int \delta\rho(\mathbf{r})\, d\mathbf{r}, \tag{2.44}$$

and

$$\text{d}v(\mathbf{r}) = \left(\frac{\partial v(\mathbf{r})}{\partial \mu}\right)_{\rho} d\mu + \int \left(\frac{\delta v(\mathbf{r})}{\delta \rho(\mathbf{r}')}\right)_{\mu} \delta\rho(\mathbf{r}')\, d\mathbf{r}'. \tag{2.45}$$

With Eq. (2.43), Eq. (2.45) becomes

$$\text{d}v(\mathbf{r}) = d\mu + \int \eta(\mathbf{r},\mathbf{r}')\delta\rho(\mathbf{r}')\, d\mathbf{r}', \tag{2.46}$$

where the hardness kernel $\eta(\mathbf{r},\mathbf{r}\ )$ is defined as

$$\eta(\mathbf{r},\mathbf{r}') = \left(\frac{\delta v(\mathbf{r})}{\delta \rho(\mathbf{r}')}\right)_{\mu} = \frac{\delta^2 F[\rho]}{\delta \rho(\mathbf{r})\delta \rho(\mathbf{r}')}. \tag{2.47}$$

The Taylor expansion in the grand isomorphic ensemble up to the second order becomes

$$\begin{aligned}
\Delta R &\equiv R[\mu + \Delta\mu,\rho(\mathbf{r}) + \rho(\mathbf{r})] - R[\mu,\rho(\mathbf{r})] \\
&= \left(\frac{\partial R}{\partial \mu}\right)_{\rho} \Delta\mu + \int \left(\frac{\delta R}{\delta \sigma(\mathbf{r})}\right)_{\mu} \Delta\rho(\mathbf{r})d\mathbf{r} \\
&+ \frac{1}{2!}\left\{\left(\frac{\partial^2 R}{\partial \mu^2}\right)_{\rho} \Delta\mu^2 + 2\int \left(\frac{\partial}{\partial \mu}\left(\frac{\delta R}{\delta \rho(\mathbf{r})}\right)_{\mu}\right)_{\rho} \Delta\mu\Delta\rho(\mathbf{r})d\mathbf{r}\right. \\
&+ \left.\iint \left(\frac{\delta^2 R}{\delta \rho(\mathbf{r})\delta \rho(\mathbf{r}')}\right)_{\mu} \Delta\rho(\mathbf{r})\Delta\rho(\mathbf{r}')d\mathbf{r}d\mathbf{r}'\right\},
\end{aligned} \tag{2.48}$$

so we have

$$\Delta R = -\int v(\mathbf{r})\Delta\rho(\mathbf{r})d\mathbf{r} + \frac{1}{2}\iint \eta(\mathbf{r},\mathbf{r}')\Delta\rho(\mathbf{r})\Delta\rho(\mathbf{r}')d\mathbf{r}d\mathbf{r}'. \tag{2.49}$$

2.2.6 Identities in CDFT

In this section, we summarize a few well-known relationships among CDFT quantities introduced earlier. The first one is the relationship between hardness η defined in Eq. (2.22) and softness S defined in Eq. (2.33),

$$S = 1/\eta. \tag{2.50}$$

The second one is the normalization condition of the Fukui function $f(\mathbf{r})$,

$$\int f(\mathbf{r})d\mathbf{r} = 1. \tag{2.51}$$

Using their definitions, the proof of Eqs. (2.50) and (2.51) should be straightforward. The third well-known relationship [14, 27] is the identities between softness S, local softness $s(\mathbf{r})$, and softness kernel $s(\mathbf{r},\mathbf{r}')$,

$$S = \int s(\mathbf{r})d\mathbf{r}, \tag{2.52}$$

and

$$s(\mathbf{r}) = \int s(\mathbf{r},\mathbf{r}')d\mathbf{r}', \tag{2.53}$$

as well as the relationship among Fukui function, softness, and local softness

$$s(\mathbf{r}) = S\,f(\mathbf{r}). \tag{2.54}$$

Interestingly, no clear relationship can be established for hardness, local hardness, and the hardness kernel [27, 28]. This is because local hardness cannot be defined without ambiguity. Although several approaches have been explored in the literature, the issue remains unresolved.

Finally, the following identity is called the Berkowitz–Parr relation [27],

$$\varpi(\mathbf{r},\mathbf{r}') = \frac{s(\mathbf{r})s(\mathbf{r}')}{S} - s(\mathbf{r},\mathbf{r}'), \tag{2.55}$$

which links the response function $\varpi(\mathbf{r},\mathbf{r}\,)$ defined in Eq. (2.23) with softness, local softness, and softness kernel. The modified version of the isomorphic ensemble removed the interdependence of $\rho(\mathbf{r})$ and N by using the shape function and introduced a few new relationships [27]. More details are included in the CDFT book [5].

2.3 Basic Concepts and Principles

2.3.1 Electronegativity and Hardness

It is no exaggeration to say that the history of CDFT is fundamentally a quest to understand and interpret partial and functional derivatives introduced earlier. This process started in 1978 by Parr and co-workers who published a seminal paper [6] in the *Journal of Chemical Physics* discussing the equivalence among chemical potential μ as the Lagrange multiplier from DFT, Eq. (1.22), the partial derivative $\mu = \left(\frac{\partial E}{\partial N}\right)_v$ in Eq. (2.12), and electronegativity χ, which was first introduced by Pauling "as the power of an atom in a molecule to attract electrons to itself,"

$$\mu = -\chi = \left(\frac{\partial E}{\partial N}\right)_v = \frac{\delta E_v}{\delta \rho(\mathbf{r})}. \tag{2.56}$$

This connection links the concept of electronegativity, a qualitative chemical property, to a rigorous quantitative framework.

With a finite difference approximation and a quadratic model of the total energy versus the number of electrons with the external potential fixed, $E_v(N)$, μ and χ can be expressed using the ionization potential (I) and electron affinity (A),

$$\mu = -\chi = -\frac{I + A}{2}, \tag{2.57}$$

which ties chemical potential and electronegativity to experimentally measurable quantities. With Koopman's frozen orbital approximation, they can be formulated by the highest-occupied molecular orbital (HOMO) and lowest-unoccupied molecular orbital (LUMO) energies,

$$\mu = -\chi \approx \frac{\varepsilon_{\mathrm{HOMO}} + \varepsilon_{\mathrm{LUMO}}}{2} \tag{2.58}$$

This equation makes both concepts, chemical potential μ and electronegativity χ, directly accessible from typical computational chemistry calculations.

The second chemical concept quantified by CDFT is hardness. Hardness has a long history in inorganic chemistry, exemplified by the HSAB principle [11], which measures the resistance of a chemical species toward charge transfer. Parr and Pearson [9] quantified hardness as the second derivative of the total energy with respect to the number of electrons, Eq. (2.22). Closely related to hardness is softness, defined as the reciprocal of the hardness, Eqs. (2.33) and (2.50), which has been shown to be related to polarizability. The corresponding operational definition for hardness is the finite difference formula

$$\eta = I - A, \tag{2.59}$$

and through Koopman's approximation, it becomes

$$\eta = \varepsilon_{\mathrm{LUMO}} - \varepsilon_{\mathrm{HOMO}}, \tag{2.60}$$

indicating that hardness is just the HOMO–LUMO gap. Chattaraj, Lee, and Parr discussed the validity of HSAB from the CDFT perspective [10].

2.3.2 Fukui Function and FMO Theory

CDFT's next milestone is the formulation of the Fukui function by Yang and Parr in 1984 [12, 13], Eq. (2.19), with the normalization condition in Eq. (2.51). It was named as such because it aligns with Fukui's FMO theory [29], where HOMO and LUMO play paramount roles in determining reactivity. For isolated molecules at zero temperature, however, the Fukui function is ill-defined because of the derivative discontinuity [30]. To resolve this difficulty, directional derivatives are employed by approaching from the left-handed and right-hand sides, yielding two forms: $f^{+}(\mathbf{r})$ and $f^{-}(\mathbf{r})$. The $f^{+}(\mathbf{r})$ function measures the response of the electron density change following the addition of an electron. The larger the $f^{+}(\mathbf{r})$ value, the more electrons it can accept during a nucleophilic (electron-gain) attack. Therefore, a molecule is susceptible to nucleophilic attack at sites where $f^{+}(\mathbf{r})$ is large. Similarly, a molecule is susceptible to electrophilic attack at sites where $f^{-}(\mathbf{r})$ is large. In the finite difference approximation,

$$f^{+}(\boldsymbol{r}) = \rho_{N+1}(\boldsymbol{r}) - \rho_{N}(\boldsymbol{r}), \tag{2.61}$$

$$f^{-}(\boldsymbol{r}) = \rho_{N}(\boldsymbol{r}) - \rho_{N-1}(\boldsymbol{r}), \tag{2.62}$$

where $\rho_{N+1}(\boldsymbol{r})$, $\rho_{N}(\boldsymbol{r})$, and $\rho_{N-1}(\boldsymbol{r})$ denote the electron density of a system with $N+1$, N, and $N-1$ electrons, respectively. With the frozen orbital approximation, we have

$$f^{+}(\boldsymbol{r}) \approx |\psi_{\mathrm{LUMO}}(\boldsymbol{r})|^{2} = \rho_{\mathrm{LUMO}}(\boldsymbol{r}) \tag{2.63}$$

$$f^{-}(\boldsymbol{r}) \approx |\psi_{\mathrm{HOMO}}(\boldsymbol{r})|^{2} = \rho_{\mathrm{HOMO}}(\boldsymbol{r}) \tag{2.64}$$

with ρ_{LUMO} and ρ_{HOMO} standing for HOMO and LUMO densities, respectively. These final two equations emphasize that, under the frozen orbital approximation, the FMO results are effectively reproduced within the CDFT framework. A decent

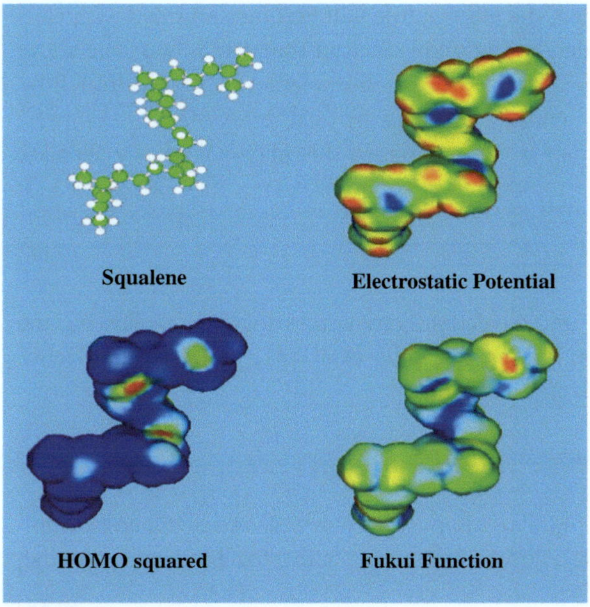

Figure 2.1 Comparison of nucleophilic reactivity prediction using electrostatic potential, HOMO, and Fukui function for squalene. Courtesy of Lee Bartolotti, East Carolina University, NC, USA.

example to illustrate the applicability and reliability of the Fukui function in predicting chemical reactivity is squalene, as shown in Figure 2.1. A linear hydrocarbon in which one C=C double bond is separated by two C—C single bonds. It is not a conjugate system per se, yet its C=C double bonds can serve as electron-donating (i.e., nucleophilic) sites with different propensities. We applied three quantities to predict the possible reaction site, electrostatic potential, HOMO, and Fukui function (with Eq. (2.62)). Among the three approaches to predict the nucleophilic propensity of these C=C double bonds, the electrostatic potential predicts nearly uniform values for all double bonds, and FMO (via HOMO) forecasts that inner C=C double bonds are more reactive. It is only the Fukui function that judges that the reaction sites should be the two C=C bonds at the terminals. Experimental evidence confirmed that only the Fukui function made the correct prediction.

2.3.3 Electrophilicity and Nucleophilicity

CDFT's next noticeable advance occurred in 1999 when Parr, Szentpály, and I [16] quantified electrophilicity as the stabilization energy when atoms or molecules in their ground states acquire the maximal amount of electron transfer from the environment. The question we asked is the following: How much could the total energy be lowered if the maximal flow of electrons is assumed? Starting with Eq. (2.27) and assuming a fixed external potential, we have

$$\Delta E = \mu \Delta N + \tfrac{1}{2}\, \eta\, (\Delta N)^2 . \tag{2.65}$$

If enough electrons are provided, the electrophile will become saturated with electrons when the energy stops decreasing upon electron transfer. Solving the above equation with $d(\Delta E)/d(\Delta N) = 0$ gives the maximum amount of electron flow, $\Delta N_{\max} = -\mu/\eta$ and thus the maximum decrease in total energy, $\Delta E_{\min} = -\mu^2/2\eta$, which is always negative because $\eta > 0$. We named this maximal energy decrease the *electrophilicity index*

$$\omega \equiv \frac{\mu^2}{2\eta}. \tag{2.66}$$

The pair of electrophilicity and nucleophilicity comes together. Following the electrophilicity work described earlier, Jaramillo et al. [31] proposed an electrophilic partner–dependent nucleophilicity scale,

$$\omega^- = \frac{1}{2}\left(\frac{\mu_A - \mu_B}{\eta_A + \eta_B}\right)^2 \eta_A, \tag{2.67}$$

assuming that A is the nucleophile and B is the electrophile. They did so by minimizing Eq. (2.38) from the grand canonical ensemble with a fixed external potential,

$$\Delta\Omega = -N\Delta\mu - ½\, S(\Delta\mu)^2. \tag{2.68}$$

Other schemes to quantify nucleophilicity within the CDFT framework are available in the literature [32–35]. These formulations of electrophilicity and nucleophilicity have demonstrated strong correlations with experimental data for a few systems. However, a fundamental conceptual issue remains. Formulas in Eqs. (2.65) and (2.68) are thermodynamic in nature, so these indices should be interpreted as *thermodynamic predispositions* rather than kinetic rate descriptors, but, in the experimental scale, electrophilicity and nucleophilicity are quantified by the rate constant, which is kinetic in nature. In this regard, the previous formulations are more appropriate for Lewis's acidity and basicity. We will come back to this issue in the following sections.

2.3.4 Dual Descriptor and Woodward–Hoffmann Rules

All Taylor series in Section 2.2 are up to the second order. Nevertheless, studies show that amazing results can be obtained when third-order derivatives are taken into consideration. One example is the dual descriptor [18]. It can "be at one and the same time an indicator for both the nucleophilic and electrophilic regions of a molecule," and it recovers the Woodward–Hoffman rules [19, 36] from the CDFT perspective.

Morell, Grand, and Toro-Labbe proposed in 2005 [18] the concept of dual descriptor using the third-order derivatives:

$$f^{(2)}(\mathbf{r}) = \left(\frac{\partial^2}{\partial^2 N}\left(\frac{\delta E}{\delta v(\mathbf{r})}\right)_N\right)_v = \left(\frac{\delta}{\delta v(\mathbf{r})}\left(\frac{\partial^2 E}{\partial^2 N}\right)_v\right)_N = \left(\frac{\partial f(\mathbf{r})}{\partial N}\right)_v = \left(\frac{\delta \eta}{\delta v(\mathbf{r})}\right)_N \tag{2.69}$$

Since the Fukui function is normalized to 1, the normalization condition for $f^{(2)}(\mathbf{r})$ is as follows:

$$\int f^{(2)}(\mathbf{r})d\mathbf{r} = 0. \tag{2.70}$$

Under the finite difference approximation and using Eqs. (2.61) and (2.62), one has

$$f^{(2)}(\mathbf{r}) = f^{+}(\mathbf{r}) - f^{-}(\mathbf{r}). \tag{2.71}$$

Within the frozen-orbital approximation, it becomes

$$f^{(2)}(\mathbf{r}) \approx \rho_{\mathrm{LUMO}}(\mathbf{r}) - \rho_{\mathrm{HOMO}}(\mathbf{r}). \tag{2.72}$$

These results show that $f^{(2)}(\mathbf{r})$ will be positive in electrophilic regions as $\rho_{\mathrm{LUMO}}(\mathbf{r})$ is large and negative in nucleophilic regions as $\rho_{\mathrm{HOMO}}(\mathbf{r})$ dominates the regions. Compared to the Fukui function, one benefit of the dual descriptor is that it can illustrate electrophilic and nucleophilic propensities simultaneously. We applied it to an enzymatic system and observed a proton-coupled electron transfer mechanism over a long distance [37, 38]. When applied to the butadiene + ethene [4+2] cycloaddition reaction [39], the dual descriptor correctly predicts the alignment of the two molecules. The favorable interactions originate from the different nature of molecular regions, nucleophilic $\left(f^{(2)}(\mathbf{r}) < 0\right)$ and electrophilic $\left(f^{(2)}(\mathbf{r}) > 0\right)$. This result clearly shows that orbital and symmetry are not necessary to interpret the validity of the Woodward–Hoffmann rules.

2.3.5 CDFT Principles

With the link between μ and χ established, it naturally follows the principle of electronegativity equalization, originally introduced by Sanderson [40–43], who postulated that, upon molecule formation, the electronegativities of the constituent atoms become equal, yielding the molecular electronegativity as the geometric mean of the atomic electronegativity. With the relationship in Eq. (2.56), a solid proof of the electronegativity equalization principle (EEP) was initially given by Donnelly and Parr [6, 44] and later by Politzer and Weinstein [45]. This principle has since been extensively applied to calculate atomic charges and other properties in the literature [6, 46].

Electronegativity alone cannot properly describe the reactivity of chemical systems. The corresponding values for hardness need to be considered as well to describe the reactivity of a system. Pearson proposed the concept of hardness. The usefulness of this concept was verified while trying to describe the characteristics of acid–base reactions through his HSAB principle [11]. The HSAB principle is stated as follows: "Hard acids prefer to coordinate with hard bases and soft acids prefer to coordinate with soft bases for both their kinetic and thermodynamic properties."

With quantifications of hardness and softness, Eqs. (2.22) and (2.33), from CDFT, proofs of HSAB were proposed originally by Chattaraj, Lee, and Parr [10] and later elaborated in more detail by Gázquez [47]. Ayers, Parr, and Pearson [48, 49] reexamined the issues related to the meaning and conditions of validity of the HSAB principle, elucidating that when electron-transfer effects are important and other effects are negligible, the HSAB principle is driven by the surpassing stability of the soft acid–soft base product. When electrostatic effects dominate the reactivity, the HSAB principle is driven by the surpassing stability of the hard acid–hard base product.

Closely related to the HSAB principle is the maximum hardness principle (MHP) [50–52]. Since hardness is the HOMO–LUMO gap, and it's well known that the bigger the gap, the more stable the molecule, it is conceivable that "there seems to be a rule of nature that molecules arrange themselves to be as hard as possible" as first postulated by Pearson [51]. In 1991, a formal proof of MHP was proposed by Parr and Chattaraj [50, 53] based on a combination of the fluctuation dissipation theorem from statistical mechanics and density functional theory under the condition that both the external and chemical potentials are both held fixed, a severe restriction putting heavy constraints on the verifiability and applicability of the principle.

Recent studies have seemingly shown that "the natural direction of a chemical reaction is toward a state of minimum electrophilicity," the so-called minimum electrophilicity principle (MEP) [54–56]. Although demonstrations on specific systems have been confirmative, conclusive proof is still lacking. These principles (EEP, HSAB, MHP, and MEP) are more of empirical rules than fundamental principles, and their validity is bound to specific conditions. They lack robust theoretical derivations from first principles, thus making their universality questionable [57].

2.4 Extensions

2.4.1 Fractional Electron Numbers

In DFT, the ground-state energy $E(N)$ is a piecewise linear function of the number of electrons, where for integer N, $E(N)$ corresponds to the ground-state energy of the system with exactly N electrons, which is called a pure state. For fractional N, $E(N)$ interpolates linearly between the energies of two adjacent integer-electron states with N_0 and $N_0 + 1$ electrons, $E(N_0)$ and $E(N_0 + 1)$,

$$E(N) = (1 - \alpha)E(N_0) + \alpha E(N_0 + 1), \tag{2.73}$$

where $\alpha = N - N_0$. For fractional N, the system is treated as an ensemble of states, with their probabilities determined by chemical potential μ, so it is no longer a pure state. The grand canonical ensemble is characterized by the chemical potential μ, which controls the average number of electrons in the system, and the external potential $v(\mathbf{r})$, with the grand canonical potential Ω defined in Eq. (2.14). In this regard, the grand canonical ensemble provides the natural and consistent framework for handling fractional electron numbers in CDFT.

In 1982, Perdew, Parr, Levy, and Balduz [30] examined the fractional charge system in the grand canonical ensemble. In this seminal article, they showed that the energy $E(N)$ function as a function of N is a series of straight-line segments with slope discontinuities at integer N. That is, $\mu = (\partial E/\partial N)_v$ is discontinuous at N. These discontinuities are linked to the ionization potential I and electron affinity A or HOMO and LUMO energies in the frozen orbital approximation.

$$\mu = \begin{cases} -I \approx \varepsilon_{\text{HOMO}} & (N_0 - 1 < N < N_0) \\ -A \approx \varepsilon_{\text{LUMO}} & (N_0 < N < N_0 + 1) \end{cases} \tag{2.74}$$

Their average value gives rise to the formal quantification of chemical potential and electronegativity in Eqs. (2.57) and (2.58). Because of the discontinuous nature of energy derivatives, hardness has the same behavior.

The fractional charge system was further examined in connection with degenerate ground states by Yang, Zhang, and Ayers in their influential 2000 paper [58]. They proved that energy is a convex function of the electron number, and chemical potential is equalized across subsystems in equilibrium. The study revealed that fractional charges cannot exist in isolated systems; however, fractional charges can appear in subsystems that exchange electrons with each other, such as molecular fragments interacting in a molecule or subsystems exchanging electrons with reservoirs. The relationship between electronegativity equalization and fractional charges is clarified, reinforcing the idea that electronegativity drives charge transfer until equilibrium is reached. This framework has recently been extended to treat excited states [59].

Additionally, fractional electron numbers are closely related to the intrinsic problem of the delocalization error in DFT [60–62]. This error arises because approximate exchange–correlation functionals fail to properly describe the piecewise linear behavior of the total energy with respect to the number of electrons, leading to incorrect predictions for systems with fractional charges. As a result, the computed energies tend to be overly delocalized, distorting the electronic structure and reducing the accuracy of calculations for properties such as ionization potentials, electron affinities, and reaction barriers. Addressing delocalization error is critical for improving the predictive power of density functional theory, particularly for systems involving charge transfer or mixed-valence states.

2.4.2 Finite-temperature CDFT

Finite-temperature CDFT was first introduced by Gázquez and coworkers [63–65]. It extended the zero-temperature CDFT in the grand canonical ensemble for ground-state systems to incorporate temperature effects, enabling the analysis of chemical reactivity descriptors (e.g., chemical potential, Fukui function, and hardness) under thermal conditions. Their approach was built upon the finite-temperature DFT framework established by Mermin in 1965 [66] but specifically focused on bridging temperature-dependent statistical mechanics with chemical reactivity concepts. Subsequent developments by Nagy and Parr in 1996 [67] further refined the ideas. Since CDFT originally focuses on ground-state and zero-temperature systems, this extension accounts for thermal excitations and statistical ensembles, making it applicable to real-world conditions where temperature plays a role (e.g., catalysis, biological systems, or high-temperature reactions).

At finite temperatures, electrons populate excited states according to the Fermi–Dirac distribution. This thermal smearing affects reactivity indices. This method integrates statistical mechanics via Mermin's 1965 finite-temperature DFT formalism and applies it to analyze CDFT reactivity descriptors. The natural variables of the grand canonical ensemble at finite temperature are the chemical potential of the reservoir μ, external potential $v_{ext}(\mathbf{r})$, and temperature T, so

$$\Omega(\mu, v_{\text{ext}}, T) = \sum_N \sum_i e^{-\beta(E_{N,i} - \mu N)}, \tag{2.75}$$

where $E_{N,i}$ is the energy of the i-th N-electron excited state ($i = 0$ corresponds to the ground state), $\beta = 1/k_B\ T$, and k_B is Boltzmann's constant. With this, we can define other variables in the ensemble using their average,

$$\langle N \rangle = \sum_N \sum_i w_{N,i} N, \tag{2.76}$$

$$\langle E \rangle = \sum_N \sum_i w_{N,i} E_{N,i}, \tag{2.77}$$

$$\langle \rho(\mathbf{r}) \rangle = \sum_N \sum_i w_{N,i}\, \varrho_{N,i}(\mathbf{r}), \tag{2.78}$$

with the weighting factor $w_{N,i}$ expressed in terms of the grand partition function as

$$w_{N,i} = \frac{1}{\Omega} e^{-\beta(E_{N,i} - \mu N)}, \tag{2.79}$$

so, the chemical potential for the electronic system at finite temperature is

$$\mu = \left(\frac{\partial \langle E \rangle}{\partial \langle N \rangle} \right)_{v_{\text{ext}}, T} = \frac{\langle EN \rangle - \langle E \rangle \langle N \rangle}{\langle N^2 - \langle N \rangle \langle N \rangle \rangle}, \tag{2.80}$$

and Fukui function

$$f(\mathbf{r}) = \left(\frac{\partial \langle \rho(\mathbf{r}) \rangle}{\partial \langle N \rangle} \right)_{v_{\text{ext}}, T} = \frac{\langle \rho(\mathbf{r}) N \rangle - \langle \rho(\mathbf{r}) \rangle \langle N \rangle}{\langle N^2 - \langle N \rangle \langle N \rangle \rangle}. \tag{2.81}$$

They defined and evaluated other derivatives in finite-temperature CDFT in a similar manner.

2.4.3 Spin-polarized CDFT

Spin-polarized CDFT [20, 68–70] is an extension of CDFT to explicitly consider spin degrees of freedom. In spin-polarized systems, the electronic density is separated into spin-up (ρ^α) and spin-down (ρ^β) components. This enables a description of spin-dependent properties and reactions in open-shell systems such as radicals, spintronics, transition-metal complexes, and magnetic materials.

For an open-shell system with N^α and N^β electrons whose density is $\rho^\alpha(\mathbf{r})$ and $\rho^\beta(\mathbf{r})$, and their responding external potential $v_{\text{ext}}^\alpha(\mathbf{r})$ and $v_{\text{ext}}^\beta(\mathbf{r})$, respectively, the total energy of the system reads

$$E[\rho^\alpha, \rho^\beta] = F[\rho^\alpha, \rho^\beta] + \int \rho^\alpha(\mathbf{r}) v_{\text{ext}}^\alpha(\mathbf{r}) d\mathbf{r} + \int \rho^\beta(\mathbf{r}) v_{\text{ext}}^\beta(\mathbf{r}) d\mathbf{r}. \tag{2.82}$$

The Euler–Lagrange equations for the two spins are

$$\mu_\alpha = \left\{\frac{\delta F\left[\rho^\alpha,\rho^\beta\right]}{\delta\rho^\alpha(\mathrm{r})}\right\}_{\rho^\beta} + v_{\mathrm{ext}}^\alpha(\mathbf{r}) \text{ and } \mu_\beta = \left\{\frac{\delta F\left[\rho^\alpha,\rho^\beta\right]}{\delta\rho^\beta(\mathrm{r})}\right\}_{\rho^\alpha} + v_{\mathrm{ext}}^\beta(\mathbf{r}), \tag{2.83}$$

with

$$\mu_\alpha = \left\{\frac{\delta E\left[\rho^\alpha,\rho^\beta\right]}{\delta N^\alpha}\right\}_{N^\beta,v_{\mathrm{ext}}^\alpha(\mathbf{r}),v_{\mathrm{ext}}^\beta(\mathbf{r})} \text{ and } \mu_\beta = \left\{\frac{\delta E\left[\rho^\alpha,\rho^\beta\right]}{\delta N^\beta}\right\}_{N^\alpha,v_{\mathrm{ext}}^\alpha(\mathbf{r}),v_{\mathrm{ext}}^\beta(\mathbf{r})}. \tag{2.84}$$

Other derivatives can be similarly defined in this manner. There is an alternative representation of the system using another pair of variables, N and N_S, where $N = N^\alpha + N^\beta$ and $N_S = N^\alpha - N^\beta$. Using this representation, one can define CDFT reactivity descriptors accordingly. For example, for hardness, a total of four definitions with the two new variables are possible:

$$\eta_{NN} = \left(\frac{\partial\mu_N}{\partial N}\right)_{N_S,v_{\mathrm{ext}}^N(\mathbf{r}),v_{\mathrm{ext}}^{N_S}(\mathbf{r})};\ \eta_{NN_S} = \left(\frac{\partial\mu_N}{\partial N_S}\right)_{N,v_{\mathrm{ext}}^N(\mathbf{r}),v_{\mathrm{ext}}^{N_S}(\mathbf{r})}, \tag{2.85}$$

and

$$\eta_{N_SN} = \left(\frac{\partial\mu_{N_S}}{\partial N}\right)_{N_S,v_{\mathrm{ext}}^N(\mathbf{r}),v_{\mathrm{ext}}^{N_S}(\mathbf{r})};\ \eta_{N_SN_S} = \left(\frac{\partial\mu_S}{\partial N_S}\right)_{N,v_{\mathrm{ext}}^N(\mathbf{r}),v_{\mathrm{ext}}^{N_S}(\mathbf{r})}. \tag{2.86}$$

Fukui function and other local spin-polarized CDFT descriptors can be defined in the same spirit. From the previous illustration, we can see that with these spin-resolved descriptors, spin-polarized CDFT enables the study of reactivity and selectivity in systems with unpaired electrons, providing insights into spin effects on chemical properties and processes. However, for a descriptor of second-order derivatives, such as Eqs. (2.85) and (2.86), a combination of four possibilities should be considered, but how to interpret these spin-resolved reactivity descriptors can be complex, system-dependent, and thus challenging.

2.4.4 Excited State CDFT

Same as DFT, CDFT is a ground-state theory. While tremendous progress has been achieved in the last few decades to extend DFT to treat excited states [71], the counterpart in CDFT is still not well accomplished. The reasons behind it are complicated [72, 73]. For instance, extending CDFT to excited states requires ensemble DFT or time-dependent DFT (TD-DFT). Reactivity indices derived from energy derivatives in excited states could become ambiguous or less physically intuitive because excited-state energies do not obey variational stability conditions in the same way as ground-state DFT. Also, many excited states involve multi-configurational character, challenging for single-reference DFT calculations and thus impacting the effectiveness of CDFT interpretations. Several people contributed to the effort to extend ground-state CDFT to excited states, such as Nagy [74], Chattaraj [75],

Morell [72], De Proft [76], and others. TD-DFT is a feasible approach to accessing the structure for the first single or triplet excited state, so global and local reactivity descriptors from the ground state CDFT can be directly borrowed to describe excited states; however, its validity is limited to the linear response regime.

Another approach is through so-called ensemble state theory [77, 78], where one alternatively considers an ensemble of M states with different weights w_i (where $w_0 \geq w_1 \geq \ldots \geq 0$) standing for the probability of the i-th state, with $\sum_i w_i = 1$. The associated ensemble total energy is minimized

$$E_{en} = \sum_{i=0}^{M} w_i E_i, \tag{2.87}$$

yielding the total electron density

$$\rho(\mathbf{r}) = \sum_{i=0}^{M} w_i \rho_i(\mathbf{r}). \tag{2.88}$$

With this scheme, one can define the ensemble chemical potential

$$\mu_{en} = \left(\frac{\partial E_{en}}{\partial N}\right)_{v_{\text{ext}}(\mathbf{r})}, \tag{2.89}$$

and excited state hardness

$$\eta_{en} = \left(\frac{\partial^2 E_{en}}{\partial N^2}\right)_{v_{\text{ext}}(\mathbf{r})}. \tag{2.90}$$

In the case of an ensemble constituted by the ground state and first excited state, the excitation energy is obtained by

$$E_1 - E_0 = \varepsilon_{N+1} - \varepsilon_N + \frac{\partial E_{xc}[\rho]}{\partial w}, \tag{2.91}$$

where w is the weight of the first excited state, ε_N denotes the orbital energy of the highest occupied in the ground state, ε_{N+1} stands for the energy of the first vacant orbital, and E_{xc} is the exchange–correlation energy for the ensemble state. These formulations might be employed for understanding and designing photoactive molecules, light-harvesting systems, and fluorescent probes through their excited-state properties.

2.4.5 Conceptual Density Matrix Functional Theory

Conceptual density matrix functional theory (CDMFT) [79, 80] is an extension of CDFT that incorporates the one-body reduced density matrix (1-RDM) $\gamma_1(\mathbf{r};\mathbf{r}')$, instead of the electron density $\rho(\mathbf{r})$, as a fundamental variable,

$$\gamma_1(\mathbf{r}_1;\mathbf{r}_1') = N \iint \cdots \int \Psi(\mathbf{r}_1,\mathbf{r}_2,\ldots,\mathbf{r}_N)\Psi^*(\mathbf{r}_1',\mathbf{r}_2,\ldots,\mathbf{r}_N)\mathrm{d}\mathbf{r}_2\cdots\mathrm{d}\mathbf{r}_N. \tag{2.92}$$

Employing the second quantization expression for the Hamiltonian,

$$\hat{H} = \sum_{pq} h_{pq} a_p^\dagger a_q + \sum_{pqrs} g_{pqrs} a_p^\dagger a_q^\dagger a_r a_s \tag{2.93}$$

we can rewrite the state function in the canonical ensemble in CDMFT as

$$E \equiv E[N, h_{pq}], \tag{2.94}$$

so the corresponding Taylor expansion, up to the second order, becomes

$$\begin{aligned} \delta E[n, h_{pq}] = & \left(\frac{\partial E}{\partial N}\right)_{h_{pq}} dN + \sum_{pq} \left(\frac{\delta E}{\delta h_{pq}}\right)_N \delta h_{pq} + \frac{1}{2}\left(\frac{\partial^2 E}{\partial N^2}\right)_{h_{pq}} d^2 N \\ & + \sum_{pq} \left[\frac{\partial}{\partial N}\left(\frac{\delta E}{\delta h_{pq}}\right)_N\right]_{h_{pq}} dN \delta h_{pq} \\ & + \frac{1}{2}\sum_{pqrs}\left(\frac{\partial^2 E}{\delta h_{pq}\,\delta h_{rs}}\right)_N \delta h_{pq}\,\delta h_{rs}. \end{aligned} \tag{2.95}$$

In analogy to CDFT, reactivity descriptors can be defined accordingly:

$$\text{Chemical potential: } \mu = \left(\frac{\partial E}{\partial N}\right)_{h_{pq}} \tag{2.96}$$

$$\text{Hardness: } \eta = \left(\frac{\partial^2 E}{\partial N^2}\right)_{h_{pq}} \tag{2.97}$$

$$\text{Fukui matrix: } \boldsymbol{F} = \left[\frac{\partial}{\partial N}\left(\frac{\delta E}{\delta h_{pq}}\right)_N\right]_{h_{pq}} \tag{2.98}$$

$$\text{Linear response matrix: } \boldsymbol{\varpi} = \left(\frac{\partial^2 E}{\delta h_{pq}\,\delta h_{rs}}\right)_N \tag{2.99}$$

In addition, other ensembles in CDMFT can be defined according to the Legendre transform in CDFT. For example, in the grand canonical ensemble, the grand potential Ω,

$$\Omega[\mu, h_{pq}] = E[N, h_{pq}] - N\mu. \tag{2.100}$$

In the grand canonical ensemble, we can obtain the analogue of the Betkowitz–Parr relation [27], Eq. (2.55), connecting softness kernel matrix s_{pqrs} with linear response matrix, Fukui matrix, and hardness,

$$s_{pqrs} = \frac{\boldsymbol{F}_{pq}\boldsymbol{F}_{rs}}{\eta} - \boldsymbol{\varpi}_{pqrs}. \tag{2.101}$$

2.4.6 Miscellaneous Extensions

Nuclear Descriptors. All reactivity descriptors considered so far are for electrons, but CDFT provides descriptors for nuclei. These nuclear reactivity descriptors emerge naturally when considering the energy functional $E[\rho(\mathbf{r}), \mathbf{R}]$, which depends explicitly on both the electron density $\rho(\mathbf{r})$ and nuclear coordinates $\mathbf{R}$. For instance, the *nuclear Fukui function* $\boldsymbol{\Phi}_k$ [15, 81, 82] quantifies the response of the electronic chemical potential $\mu = (\partial \mathrm{E}/\partial \mathrm{N})_{\mathbf{R}}$ to a displacement of the k-th nucleus:

$$\boldsymbol{\Phi}_k = \left(\frac{\partial \boldsymbol{F}_k}{\partial N}\right)_{v_{\text{ext}}(\mathbf{r})} = -\left(\frac{\delta \mu}{\delta \boldsymbol{R}_k}\right)_N, \tag{2.102}$$

where $\boldsymbol{F}_k$ is the Hellmann–Feynman force on the k-th nucleus,

$$\boldsymbol{F}_k = Z_k\left[\int \rho(\mathbf{r})\frac{\mathbf{r}_k}{r_k^3}d\mathbf{r} - \sum_{\alpha \neq k}\frac{\mathbf{R}_{\alpha k}}{\mathrm{R}_{\alpha k}^3}\right], \tag{2.103}$$

with $\mathbf{r}_k$ is denoted as the displacement vector between the position of nucleus k and the point $\mathbf{r}$ of the electron density, and $\mathbf{R}_{\alpha k}$ stands for the displacement vector between the nucleus k with charge Z_k and nucleus α with charge Z_α. Same as the electronic Fukui function, one can use the finite difference approximation to describe a nucleophilic or electrophilic attack,

$$\boldsymbol{\Phi}_k^+ = \boldsymbol{F}_k^+ - \boldsymbol{F}_k^0, \tag{2.104}$$

$$\boldsymbol{\Phi}_k^- = \boldsymbol{F}_k^0 - \boldsymbol{F}_k^-, \tag{2.105}$$

where $\boldsymbol{\Phi}_k^+$ and $\boldsymbol{\Phi}_k^-$ are the nucleophilic and electrophilic nuclear Fukui functions, respectively, with $\boldsymbol{F}_k^+$, $\boldsymbol{F}_k^0$, and $\boldsymbol{F}_k^-$ referring to the Hellmann–Feynman force acting on the same nucleus k of anionic, neutral, and cationic species, respectively. SP-CDFT quantities can be defined accordingly.

Reaction Force. The concept of the reaction force in CDFT was first proposed by Toro–Labbé and coworkers [83, 84] to analyze the energy change along an intrinsic reaction coordinate (IRC) ξ,

$$F(\xi) = -\frac{dE(\xi)}{d\xi}, \tag{2.106}$$

where $E(\xi)$ is the energy as a function of the IRC ξ and $F(\xi)$ represents the force acting on the system as it evolves along ξ. The reaction force can be divided into three distinct regions along the reaction path: (i) reaction region $(\xi < \xi^\ddagger)$ with $F(\xi) < 0$, indicating a driving force that pushes the system toward the transition state; (ii) transition state region $(\xi = \xi^\ddagger)$ with $F(\xi) = 0$, corresponding to the maximum in the energy profile; and (iii) product region $(\xi > \xi^\ddagger)$ with $F(\xi) > 0$, acting as a restoring force, stabilizing the product. A relevant concept is the reaction electronic flux $J(\xi)$,

$$J(\xi) = \frac{d\mu}{d\xi} \tag{2.107}$$

where $\mu = \left(\frac{\partial E}{\partial N}\right)_{v_{\text{ext}}(\mathbf{r}),\xi}$ is the chemical potential along IRC. A positive $J(\xi)$ indicates an influx of the electron density, thus meaning that electronic charge is accumulating in the system, and vice versa.

Time-Dependent CDFT. TD-CDFT [4] provides a dynamic perspective on chemical reactivity, extending conventional static CDFT reactivity descriptors into the time domain. This extension is particularly useful for understanding electronic rearrangements in ultrafast chemical reactions and external field-induced processes. The foundation lies in the Runge–Gross theorem [85], which establishes a one-to-one correspondence between a time-dependent external potential $v_{\text{ext}}(\mathbf{r}, t)$ and the time-dependent electron density $\rho(\mathbf{r}, t)$, analogous to the Hohenberg–Kohn theorem for static systems. There are a few possible pathways to extend CDFT descriptors into the time domain, including (i) directly solving the time-dependent Kohn–Sham equations numerically to track electron density $\rho(\mathbf{r}, t)$ over time via *real-time propagation*; (ii) assuming the time-dependent exchange–correlation functional depends *instantaneously* on the density $\rho(\mathbf{r}, t)$ with memory effects ignored via *adiabatic approximation*; (iii) incorporating memory effects to describe electron–nuclear coupling beyond the Born–Oppenheimer approximation through *non-adiabatic treatment*; and (iv) other approaches, for example, quasi-static method and action formalism.

Condensed Phases Extension. To apply CDFT to crystals and surfaces where heterogeneous catalysis takes place, Huang and Zhuang [86, 87] proposed the concepts of Fermi softness S_F and local Fermi softness $s_F(\mathbf{r})$,

$$s_F(\mathbf{r}) = \int g(\mathbf{r}, \epsilon) s(\epsilon) d\epsilon, \tag{2.108}$$

where $g(\mathbf{r}, \epsilon_F)$ is the local density of states (LDOS) near the Fermi level ϵ_F, representing the number of electronic states per unit energy at position $\mathbf{r}$, and $s(\epsilon)$ is the local softness, which can be approximated by the derivative of the Fermi–Dirac distribution $1/\left(1 + \exp\left(\frac{\epsilon - \mu}{kT}\right)\right)$, yielding

$$s(\epsilon) = -\frac{\exp\left(\frac{\epsilon - \mu}{kT}\right)}{\left[\exp\left(\frac{\epsilon - \mu}{kT}\right) + 1\right]^2} \tag{2.109}$$

so

$$s_F(\mathbf{r}) = -\int \frac{\exp\left(\frac{\epsilon - \mu}{kT}\right)}{\left[\exp\left(\frac{\epsilon - \mu}{kT}\right) + 1\right]^2} g(\mathbf{r}, \epsilon) d\epsilon, \tag{2.110}$$

and

$$S_F = \int s_F(\mathbf{r}) d\mathbf{r}. \tag{2.111}$$

With this formulation, they demonstrated that step edges and defects on transition metal surfaces (e.g., Pt(111)) exhibit high local Fermi softness, correlating with enhanced catalytic activity for reactions like CO oxidation, and they also explained why certain sites preferentially adsorb reactants (e.g., O_2 or CO) due to elevated LDOS and Fermi softness. In another work, Li et al. [88, 89] extended the Fukui function $f(\mathbf{r})$ to study the properties of crystals by integrating $f(\mathbf{r})$ over crystal faces (e.g., (100), (010)), this face-condensed CDFT quantity can predict which crystallographic planes are more reactive, linking electronic structure to anisotropic crystal growth behavior.

Bridging Valence-Bond Theory. Shaik et al. [90–92] have recently forged a novel theoretical framework that unites CDFT with VBT, creating a synergistic approach to chemical reactivity and bonding. By integrating CDFT's quantitative reactivity descriptors, such as electronegativity (χ), chemical hardness (η), and Fukui functions (f^+ and f^-), with VBT's mechanistic insights into resonance structures and electron-pair dynamics, their work offers a dual perspective on molecular interactions. For example, they map VB resonance contributions onto spatially resolved Fukui functions by expressing $f(\mathbf{r}) = \Sigma_i w_i f_i(\mathbf{r})$, where w_i and $f_i(\mathbf{r})$ represent the weights and reactivity of individual VB configurations. This integration clarifies how electron delocalization (captured by VB resonance) reduces chemical hardness, aligning soft systems with delocalized electrons and hard systems with localized bonding. Their analysis of reactions like the Diels–Alder process demonstrates how VB's diabatic states (e.g., reactant/product-like configurations) correlate with CDFT's charge-redistribution profiles, predicting regioselectivity and transition-state energetics. By bridging these theories, a unified lens to explain both where reactions occur (via CDFT's descriptors) and why they proceed (via VB's bonding motifs), offering deeper mechanistic insights into catalysis, bond activation, and enzymatic processes.

2.5 Concluding Remarks

CDFT extends the foundational structure of DFT into a rigorous and chemically meaningful framework for describing reactivity. By formulating the total energy in four ensembles and employing Legendre transforms, CDFT provides a systematic route to define global and local reactivity descriptors as energy derivatives with respect to the electron number and external potential. Central concepts such as electronegativity, chemical hardness/softness, Fukui functions, electrophilicity, nucleophilicity, and the dual descriptor arise naturally from this formulation. Their associated principles—electronegativity equalization, HSAB, maximum hardness, and minimum electrophilicity—offer intuitive interpretations of stability, selectivity, and chemical behavior across diverse systems.

A major strength of CDFT is its ability to connect measurable quantities, frontier molecular orbitals, and real-space electron distributions through a unified formalism. Its extensions to fractional electron numbers, finite temperatures, spin polarization, excited states, and density-matrix representations further broaden its scope, allowing the theory to address multireference systems, time-dependent processes, and condensed phases. Together, these developments have established CDFT as a versatile and mature framework capable of capturing how molecules respond to perturbations and how electrons reorganize during chemical transformations.

Within the broader context of density-based reactivity theory, CDFT provides the first and most classical lens through which reactivity is viewed—one rooted in thermodynamic analogies, response theory, and functional differentiation. The next chapters introduce complementary frameworks that emphasize real-space electron density (Chapter 3), information-theoretic quantities (Chapter 4), and orbital-free kinetic-energy functionals (Chapter 5). Each offers a distinct yet synergistic perspective on chemical behavior. In this way, CDFT not only stands as a complete framework in its own right but also forms an essential cornerstone for the density-based approaches that follow.

References

1 Parr RG, Yang W. Density-Functional Theory of Atoms and Molecules. Oxford University Press; 1989.

2 Geerlings P, De Proft F, Langenaeker W. Conceptual density functional theory. Chem Rev. 2003 May;103(5):1793–873.

3 Liu S. Conceptual density functional theory and some recent developments. Acta Phys-Chim Sin. 2009;25(03):590–600.

4 Geerlings P, Chamorro E, Chattaraj PK, De Proft F, Gázquez JL, Liu S, et al. Conceptual density functional theory: status, prospects, issues. Theor Chem Acc. 2020 Feb;139(2):36.

5 Liu S, editor. Conceptual Density Functional Theory: Towards a New Chemical Reactivity Theory. 1st ed. Wiley-VCH GmbH: Wiley; 2022.

6 Parr RG, Donnelly RA, Levy M, Palke WE. Electronegativity: the density functional viewpoint. J Chem Phys. 1978 Apr 15;68(8):3801–7.

7 Nalewajski RF, Parr RG. Legendre transforms and Maxwell relations in density functional theory. J Chem Phys. 1982 Jul 1;77(1):399–407.

8 Ghosh SK, Berkowitz M, Parr RG. Transcription of ground-state density-functional theory into a local thermodynamics. Proc Natl Acad Sci USA. 1984 Dec;81(24):8028–31.

9 Parr RG, Pearson RG. Absolute hardness: companion parameter to absolute electronegativity. J Am Chem Soc. 1983 Dec;105(26):7512–6.

10 Chattaraj PK, Lee H, Parr RG. HSAB principle. J Am Chem Soc. 1991 Feb;113(5):1855–6.

11 Ho T-L. Hard soft acids bases (HSAB) principle and organic chemistry. Chem Rev. 1975 Feb 1;75(1):1–20.

12 Yang W, Parr RG. Hardness, softness, and the Fukui function in the electronic theory of metals and catalysis. Proc Natl Acad Sci USA. 1985 Oct;82(20):6723–6.

13 Yang W, Parr RG, Pucci R. Electron density, Kohn–Sham frontier orbitals, and Fukui functions. J Chem Phys. 1984;81(6):2862.

14 De Proft F, Liu S, Parr RG. Chemical potential, hardness, hardness and softness kernel and local hardness in the isomorphic ensemble of density functional theory. J Chem Phys. 1997 Aug 22;107(8):3000–6.

15 De Proft F, Liu S, Geerlings P. Calculation of the nuclear Fukui function and new relations for nuclear softness and hardness kernels. J Chem Phys 1998 Oct 1;108(18):7549–54.

16 Parr RG, Szentpály Lv, Liu S. Electrophilicity index. J Am Chem Soc. 1999 Mar 1;121(9):1922–4.

17 Chattaraj PK, Sarkar U, Roy DR. Electrophilicity index. Chem Rev. 2006 Jun;106(6):2065–91.

18 Morell C, Grand A, Toro-Labbé A. New dual descriptor for chemical reactivity. J Phys Chem A. 2005 Jan 13;109(1):205–12.

19 Geerlings P, Ayers PW, Toro-Labbé A, Chattaraj PK, De Proft F. The Woodward-Hoffmann rules reinterpreted by conceptual density functional theory. Acc Chem Res. 2012 May 15;45(5):683–95.

20 Alcoba DR, Lain L, Torre A, Oña OB, Chamorro E. Fukui and dual-descriptor matrices within the framework of spin-polarized density functional theory. Phys Chem Chem Phys. 2013 Jun 28;15(24):9594–604.

21 Tognetti V, Morell C, Ayers PW, Joubert L, Chermette H. A proposal for an extended dual descriptor: a possible solution when frontier molecular orbital theory fails. Phys Chem Phys. 2013 Sep 14;15(34):14465–75.

22 Geerlings P, Fias S, Boisdenghien Z, De Proft F. Conceptual DFT: chemistry from the linear response function. Chem Soc Rev. 2014 Jul 21;43(14):4989–5008.

23 Parr RG, Zhou Z. Absolute hardness: unifying concept for identifying shells and subshells in nuclei, atoms, molecules, and metallic clusters. Acc Chem Res. 1993 May 1;26(5):256–8.

24 Parr RG, Yang W. Density-functional theory of the electronic structure of molecules. Annu Rev Phys Chem. 1995;46:701–28.

25 Liu S, Parr RG. Second-order density-functional description of molecules and chemical changes. J Chem Phys. 1997 Apr 1;106(13):5578–86.

26 Baekelandt BG, Cedillo A, Parr RG. Reactivity indices and fluctuation formulas in density functional theory: isomorphic ensembles and a new measure of local hardness. J Chem Phys. 1995 Nov 15;103(19):8548–56.

27 Berkowitz M, Parr RG. Molecular hardness and softness, local hardness and softness, hardness and softness kernels, and relations among these quantities. J Chem Phys. 1988 Feb 15;88(4):2554–7.

28 Ayers PW, Liu S, Li T. Stability conditions for density functional reactivity theory: an interpretation of the total local hardness. Phys Chem Chem Phys. 2011 Mar 14;13(10):4427–33.

29 Fukui K. Role of frontier orbitals in chemical reactions. Science. 1982 Nov 19;218(4574):747–54.

30 Perdew JP, Parr RG, Levy M, Balduz JL. Density-functional theory for fractional particle number: derivative discontinuities of the energy. Phys Rev Lett. 1982 Dec 6;49(23):1691–4.

31 Jaramillo P, Pérez P, Contreras R, Tiznado W, Fuentealba P. Definition of a nucleophilicity scale. J Phys Chem A. 2006 Jul 6;110(26):8181–7.

32 Zhang J, Yang J-D, Cheng J-P. A nucleophilicity scale for the reactivity of diazaphospholenium hydrides: structural insights and synthetic applications. Angew Chem Int Ed. 2019 Apr 23;58(18):5983–7.

33 Aroule O, Morell C, Chermette H, Hoffmann G. A conceptual density functional theory-based maximum nucleophilicity principle. J Phys Chem A. 2025 Sep 3;129(37):8714–20

34 Ríos-Gutiérrez M, Saz Sousa A, Domingo LR. Electrophilicity and nucleophilicity scales at different DFT computational levels. J Phys Org Chem. 2023 Jul;36(7):e4503.

35 Hoffmann G, Chermette H, Morell C. Revisiting nucleophilicity: an index for chemical reactivity from a CDFT approach. J Mol Model. 2024 Jun 28;30(7):232.

36 Woodward RB, Hoffmann R. The conservation of orbital symmetry. Angew Chem Int Ed. 1969 Nov;8(11):781–853.

37 Liu S, Ess DH, Schauer CK. Density functional reactivity theory characterizes charge separation propensity in proton-coupled electron transfer reactions. J Phys Chem A. 2011 May 12;115(18):4738–42.

38 Kumar N, Liu S, Kozlowski PM. Charge separation propensity of the coenzyme B_{12}–tyrosine complex in adenosylcobalamin-dependent methylmalonyl–CoA mutase enzyme. J Phys Chem Lett. 2012 Apr 19;3(8):1035–8.

39 Ayers PW, Morell C, De Proft F, Geerlings P. Understanding the Woodward-Hoffmann rules by using changes in electron density. Chem Eur J. 2007;13(29):8240–7.

40 Sanderson RT. An interpretation of bond lengths and a classification of bonds. Science. 1951 Dec 21;114(2973):670–2.

41 Ray NK, Samuels L, Parr RG. Studies of electronegativity equalization. J Chem Phys. 1979 Apr 15;70(8):3680–4.

42 Mortier WJ, Van Genechten K, Gasteiger J. Electronegativity equalization: application and parametrization. J Am Chem Soc. 1985 Feb;107(4):829–35.

43 Mortier WJ, Leuven KU. Electronegativity equalization and its applications. In: Sen KD, Jørgensen CK, editors. Structure and Bonding. Berlin, Heidelberg: Springer Berlin Heidelberg; 1987. pp. 125–43.

44 Donnelly RA, Parr RG. Elementary properties of an energy functional of the first-order reduced density matrix. J Chem Phys. 1978 Nov 15;69(10):4431–9.

45 Politzer P, Weinstein H. Some relations between electronic distribution and electronegativity. J Chem Phys. 1979 Dec 1;71(11):4218–20.

46 Nalewajski RF. Normal (decoupled) representation of electronegativity equalization equations in a molecule. Int J Quantum Chem. 1991 Aug;40(2):265–85.

47 Gazquez JL, Mendez F. The hard and soft acids and bases principle: an atoms in molecules viewpoint. J Phys Chem. 1994 Apr;98(17):4591–3.

48 Ayers PW, Cárdenas C. Communication: a case where the hard/soft acid/base principle holds regardless of acid/base strength. J Chem Phys. 2013 May 14;138(18):181106.

49 Ayers PW, Parr RG, Pearson RG. Elucidating the hard/soft acid/base principle: a perspective based on half-reactions. J Chem Phys. 2006 May 21;124(19):194107.

50 Chattaraj PK, Ayers PW. The maximum hardness principle implies the hard/soft acid/base rule. J Chem Phys. 2005 Aug 22;123(8):086101.

51 Pearson RG. The principle of maximum hardness. Acc Chem Res. 1993 May;26(5):250–5.

52 Pearson RG. Principle of maximum physical hardness. J Phys Chem. 1994 Feb;98(7):1989–92.

53 Parr RG, Chattaraj PK. Principle of maximum hardness. J Am Chem Soc. 1991 Feb;113(5):1854–5.

54 Noorizadeh S. Is there a minimum electrophilicity principle in chemical reactions? Chin J Chem. 2007 Oct;25(10):1439–44.

55 Pan S, Solà M, Chattaraj PK. On the validity of the maximum hardness principle and the minimum electrophilicity principle during chemical reactions. J Phys Chem A. 2013 Feb 28;117(8):1843–52.

56 Miranda-Quintana RA. Note: The minimum electrophilicity and the hard/soft acid/base principles. J Chem Phys. 2017 Jan 28;146(4):046101.

57 Chakraborty D, Chattaraj PK. Conceptual density functional theory based electronic structure principles. Chem Sci. 2021 Mar 31;12(18):6264–79.

58 Yang W, Zhang Y, Ayers PW. Degenerate ground states and a fractional number of electrons in density and reduced density matrix functional theory. Phys Rev Lett. 2000 May 29;84(22):5172–5.

59 Yang W, Fan Y. Fractional charges, linear Conditions and chemical potentials for excited states in ΔSCF theory. arXiv. 2024; arXiv:2408.08443.

60 Cohen AJ, Mori-Sánchez P, Yang W. Challenges for density functional theory. Chem Rev. 2012 Jan 11;112(1):289–320.

61 Bryenton KR, Adeleke AA, Dale SG, Johnson ER. Delocalization error: the greatest outstanding challenge in density-functional theory. WIREs Comput Mol Sci. 2022 Jul; 13(2):e1631.

62 Cohen AJ, Mori-Sánchez P, Yang W. Insights into current limitations of density functional theory. Science. 2008 Aug 8;321(5890):792–4.

63 Gázquez JL, Franco-Pérez M. Finite temperature conceptual density functional theory. In: Liu S, editor. Conceptual Density Functional Theory: Towards a New Chemical Reactivity Theory. Wiley; 2022. pp. 137–60.

64 Franco-Pérez M, Gázquez JL, Ayers PW, Vela A. Revisiting the definition of the electronic chemical potential, chemical hardness, and softness at finite temperatures. J Chem Phys. 2015 Oct 21;143(15):154103.

65 Franco-Pérez M, Ayers PW, Gázquez JL, Vela A. Local and linear chemical reactivity response functions at finite temperature in density functional theory. J Chem Phys. 2015 Dec 28;143(24):244117.

66 Mermin ND. Thermal properties of the inhomogeneous electron gas. Phys Rev. 1965 Mar 1;137(5A):A1441–3.

67 Nagy Á, Parr RG, Liu S. Local temperature in an electronic system. Phys Rev A. 1996 May;53(5):3117–21.

68 Galvan M, Vela A, Gazquez JL. Chemical reactivity in spin-polarized density functional theory. J Phys Chem. 1988 Nov;92(22):6470–4.

69 Richer M, Heidar-Zadeh F, Ríos-Gutiérrez M, Yang XD, Ayers PW. Spin-polarized conceptual density functional theory from the convex hull. J Chem Theory Comput. 2024 May 31;20(11):4616–28.

70 Feng X-T, Yu J-G, Liu R-Z, Lei M, Fang W-H, De Proft F, et al. Why iron? A spin-polarized conceptual density functional theory study on metal-binding specificity of porphyrin. J Phys Chem A. 2010 Jun 3;114(21):6342–9.

71 Görling A. Density-functional theory for excited states. Phys Rev A. 1996 Nov;54(5):3912–5.

72 Morell C, Labet V, Grand A, Ayers PW, De Proft F, Geerlings P, et al. Characterization of the chemical behavior of the low excited states through a local chemical potential. J Chem Theory Comput. 2009 Sep 8;5(9):2274–83.

73 An X, Zhang W, He X, Li M, Rong C, Liu S. Density-based reactivity theory applied to excited states. AAPPS Bull. 2024 Feb 2;34(1):8.

74 Nagy Á. Relative information in excited-state orbital-free density functional theory. Int J Quantum Chem. 2020 Dec 5;120(23):e26405.

75 Chattaraj PK, Ghosh SK, Liu S, Parr RG. Exchange? correlation potential and excited-state density functional theory. Int J Quantum Chem. 1996 Oct 5;60(1):535–43.

76 Guégan F, Pigeon T, De Proft F, Tognetti V, Joubert L, Chermette H, et al. Understanding chemical selectivity through well selected excited states. J Phys Chem A. 2020 Jan 30;124(4):633–41.

77 Nagy A, Liu S, Bartolloti L. Generalized density functional theory for degenerate states. J Chem Phys. 2005 Apr 1;122(13):134107.

78 Liu S. Constant temperature from phase space view of ensembles of excited states. Acta Phys-Chim Sin. 2018;34(5):455.

79 Acke G, Van Hende D, De Vriendt X, Bultinck P. Extending conceptual density functional theory toward first-order reduced density matrices: an open subsystems viewpoint on the Fukui matrix. J Chem Theory Comput. 2023 Aug 22;19(16):5418–26.

80 Wang B, Geerlings P, Liu S, De Proft F. Extending the scope of conceptual density functional theory with second order analytical methodologies. J Chem Theory Comput. 2024 Feb 13;20(3):1169–84.

81 Baekelandt BG. The nuclear Fukui function and Berlin's binding function in density functional theory. J Chem Phys. 1996 Sep 15;105(11):4664–7.

82 Chamorro E, De Proft F, Geerlings P. Generalized nuclear Fukui functions in the framework of spin-polarized density-functional theory. J Chem Phys. 2005 Aug 22;123(8):084104.

83 Toro-Labbé A, Gutiérrez-Oliva S, Murray JS, Politzer P. The reaction force and the transition region of a reaction. J Mol Model. 2009 Jun;15(6):707–10.

84 Cárdenas C, Ayers PW, Chakraborty D, Gómez T, Echeverri A, Munoz F, et al. On the link between the reaction force constant and conceptual DFT. J Mol Model. 2024 Sep 14;30(10):332.

85 Runge E, Gross EKU. Density-functional theory for time-dependent systems. Phys Rev Lett. 1984 Mar 19;52(12):997–1000.

86 Huang B, Xiao L, Lu J, Zhuang L. Spatially resolved quantification of the surface reactivity of solid catalysts. Angew Chem Int Ed. 2016 May 17;55(21):6239–43.

87 Huang B, Zhuang L. Fermi softness: a local perspective on surface activity. In: Liu S, editor. Conceptual Density Functional Theory: Towards a New Chemical Reactivity Theory. Wiley; 2022. pp. 573–86.

88 Li T, Liu S, Feng S, Aubrey CE. Face-integrated Fukui function: understanding wettability anisotropy of molecular crystals from density functional theory. J Am Chem Soc. 2005 Feb 9;127(5):1364–5.

89 Li T, Ayers PW, Liu S, Swadley MJ, Aubrey-Medendorp C. Crystallization force—a density functional theory concept for revealing intermolecular interactions and molecular packing in organic crystals. Chem Eur J. 2009;15(2):361–71.

90 Stuyver T, De Proft F, Geerlings P, Shaik S. How do local reactivity descriptors shape the potential energy surface associated with chemical reactions? The valence bond delocalization perspective. J Am Chem Soc. 2020 Jun 3;142(22):10102–13.

91 Stuyver T, Shaik S. Promotion energy analysis predicts reaction modes: nucleophilic and electrophilic aromatic substitution reactions. J Am Chem Soc. 2021 Mar 24;143(11):4367–78.

92 Stuyver T, Shaik S. Bridging conceptual density functional and valence bond theories. In: Liu S, editor. Conceptual Density Functional Theory: Towards a New Chemical Reactivity Theory. Wiley; 2022. pp. 391–415.

3

Density-associated Quantities

TABLE OF CONTENTS

In this chapter, we present a density-based framework that directly employs density-associated quantities (DAQs), including the electron density $\rho(\mathbf{r})$, its gradient $\nabla\rho(\mathbf{r})$, the Laplacian $\nabla^2\rho(\mathbf{r})$, and their various combinations thereof, to probe both strong bonding and weak interactions among atoms in molecules or crystals. Unlike density functionals, these quantities are purely three-dimensional, local functions stemming from the electron distribution. This methodology was pioneered in the 1960s by Bader and coworkers [1, 2], who applied topological analysis of $\rho(\mathbf{r})$ to distinguish between covalent and ionic bonds and introduced the *zero-flux* condition to partition atoms in molecules. More recent studies, including the introduction of noncovalent interaction (NCI) [3, 4] and ultra-strong interaction (USI) [5] indices, have leveraged combinations of these DAQ to not only identify weak van der Waals interactions but also characterize ultra-strong covalent bonds. These DAQ indices illustrate why we avoid framing them within density functional reactivity theory [6–13] because they are not true functionals but rather purely density-based local descriptors. In what follows, we will overview the underlying formulation, discuss local features, and highlight applications of this density-based framework to various systems, including excited states.

3.1 Electron Density

In density functional theory (DFT) [14], the electron density $\rho(\mathbf{r})$ serves as the fundamental variable, playing the crucial role in evaluating properties of quantum systems such as atoms, molecules, and solids. For a system with N electrons, the electron density is a spatial function that describes the probability of finding an electron at a given three-dimensional point $\mathbf{r}$ in space and must satisfy the normalization condition:

$$\int \rho(\mathbf{r})d\mathbf{r} = N. \tag{3.1}$$

According to the fundamental theorems of DFT, there is a unique one-to-one correspondence between the ground-state electron density $\rho(\mathbf{r})$ and the external potential $v_{\text{ext}}(\mathbf{r})$ generated by the nuclei of all atoms within a system. Consequently, all ground-state properties, including structure, stability, and reactivity, can be determined entirely by the electron density. Unlike traditional wavefunction-based methods, which require solving a complex many-body wavefunction that depends on $4N$ coordinates ($3N$ spatial and N spin), DFT simplifies the problem by expressing it in terms of $\rho(\mathbf{r})$, which depends only on three spatial coordinates, x, y, and z.

In this section, we explore two key aspects: (i) methods for obtaining electron density and (ii) fundamental properties of the electron density for atoms and molecules.

3.1.1 How to Obtain Electron Density

We have at least the following four ways to obtain electron density, from (i) the many-body wavefunction Ψ; (ii) one-body $\gamma_1(\mathbf{r};\mathbf{r}')$ and two-body $\gamma_2(\mathbf{r}_1,\mathbf{r}_2;\mathbf{r}_1',\mathbf{r}_2')$ reduced density matrices (RDMs); (iii) molecular orbitals $\{\psi_i(\mathbf{r})\}$; and (iv) X-ray crystallographic experiments.

In quantum mechanics, electron density is fundamentally related to the many-body wavefunction Ψ of the system through the following equation,

$$\rho(\mathbf{r}) = \langle\Psi|\hat{\rho}|\Psi\rangle, \tag{3.2}$$

where Ψ is the N-electron wavefunction and $\hat{\rho}$ stands for the particle-density operator. For a pure quantum state, $\hat{\rho} \equiv |\Psi\rangle\langle\Psi|$, whose trace is unity: $\text{Tr}(\hat{\rho}) = 1$. For a non-relativistic system of N electrons, the electron density can be written as

$$\rho(\mathbf{r}) = N\iint \cdots \int |\Psi(\mathbf{r},\mathbf{r}_2,\ldots,\mathbf{r}_N)|^2\, d\mathbf{r}_2\, d\mathbf{r}_3 \cdots d\mathbf{r}_N. \tag{3.3}$$

This expression means that if the full many-electron wavefunction Ψ is given, the electron density at point $\mathbf{r}$ can be obtained by integrating the square of the wavefunction (which gives the probability density) over the coordinates of all electrons except the one at $\mathbf{r}$. For simplicity, we bound the spin coordinate together with the spatial coordinate.

Alternatively, electron density can be obtained from RDMs. For an N-particle system, the m-body RDM $\gamma_m(\mathbf{r}_1,\mathbf{r}_2,\ldots,\mathbf{r}_m;\mathbf{r}'_1,\mathbf{r}'_2,\ldots,\mathbf{r}'_m)$ is defined by

$$\gamma_m(\mathbf{r}_1,\mathbf{r}_2,\ldots,\mathbf{r}_m;\mathbf{r}'_1,\mathbf{r}'_2,\ldots,\mathbf{r}'_m) = \frac{N!}{(N-m)!}\iint\cdots\int\Psi(\mathbf{r}_1,\mathbf{r}_2,\ldots,\mathbf{r}_m,\mathbf{r}_{m+1},\ldots,\mathbf{r}_N)$$
$$\Psi^*(\mathbf{r}'_1,\mathbf{r}'_2,\ldots,\mathbf{r}'_m,\mathbf{r}_{m+1},\ldots,\mathbf{r}_N)d\mathbf{r}_{m+1}\cdots d\mathbf{r}_N. \tag{3.4}$$

For the special case of a one-body reduced density matrix (1-RDM) $\gamma_1(\mathbf{r}_1;\ \mathbf{r}_1)$, in the pure state, it is defined as

$$\gamma_1(\mathbf{r};\mathbf{r}') = N\iint\cdots\int\Psi(\mathbf{r},\mathbf{r}_2,\ldots,\mathbf{r}_N)\Psi^*(\mathbf{r}',\mathbf{r}_2,\ldots,\mathbf{r}_N)d\mathbf{r}_2\cdots d\mathbf{r}_N. \tag{3.5}$$

And for two-body RDM $\gamma_2(\mathbf{r}_1,\mathbf{r}_2;\mathbf{r}'_1,\mathbf{r}'_2)$ (2-RDM), we have

$$\gamma_2(\mathbf{r},\mathbf{r}_2;\mathbf{r}',\mathbf{r}'_2) = N(N-1)\iint\cdots\int\Psi(\mathbf{r},\mathbf{r}_2,\mathbf{r}_3,\ldots,\mathbf{r}_N)\Psi^*(\mathbf{r}',\mathbf{r}'_2,\mathbf{r}_3,\ldots,\mathbf{r}_N)$$
$$d\mathbf{r}_3\cdots d\mathbf{r}_N. \tag{3.6}$$

Again, we put together the spin coordinate with the spatial coordinate in these formulations for simplicity. With the formulation of RDM, the electron density is just the diagonal of the 1-RDM

$$\rho(\mathbf{r}) = \gamma_1(\mathbf{r};\mathbf{r}')\big|_{\mathbf{r}'=\mathbf{r}} = \gamma_1(\mathbf{r};\mathbf{r}), \tag{3.7}$$

and for 2-RDM, one has

$$\rho(\mathbf{r}) = \frac{1}{N-1}\int\gamma_2(\mathbf{r},\mathbf{r}_2;\mathbf{r},\mathbf{r}_2)d\mathbf{r}_2. \tag{3.8}$$

In Kohn–Sham scheme [15], the interacting electron system is mapped onto a fictitious system of noninteracting electrons moving in an effective Kohn–Sham potential $v_{\mathrm{eff}}(\mathbf{r})$, yielding the set of Kohn–Sham orbitals $\{\psi_i(\mathbf{r})\}$, and the Kohn–Sham equation,

$$\left[-\frac{1}{2}\nabla^2 + v_{\mathrm{eff}}(\mathbf{r})\right]\psi_i(\mathbf{r}) = \epsilon_i\psi_i(\mathbf{r}), \tag{3.9}$$

where the effective Kohn–Sham potential is composed of several terms,

$$v_{\mathrm{eff}}(\mathbf{r}) = v_{\mathrm{ext}}(\mathbf{r}) + v_J(\mathbf{r}) + v_{xc}(\mathbf{r}), \tag{3.10}$$

with the classical Coulombic potential

$$v_J(\mathbf{r}) = \int\frac{\rho(\mathbf{r}')}{|\mathbf{r}-\mathbf{r}'|}d\mathbf{r}', \tag{3.11}$$

and the exchange–correlation potential is

$$v_{xc}(\mathbf{r}) = \frac{\delta E_{xc}[\rho]}{\delta\rho(\mathbf{r})}. \tag{3.12}$$

The electron density can be expressed as

$$\rho(\mathbf{r}) = \sum_{i=1}^{\text{occ}} |\psi_i(\mathbf{r})|^2, \tag{3.13}$$

where the sum runs over all occupied Kohn–Sham orbitals. This Kohn–Sham equation must be solved numerically with the same self-consistent-field manner as Hartree–Fock and other wavefunction theory-based methods.

Finally, electron density can also be obtained experimentally by using techniques like X-ray diffraction. In a crystal, the diffraction pattern provides information about the Fourier transform of the electron density. By applying an inverse Fourier transform and using appropriate phase information (which may be obtained by various methods), one can reconstruct the electron density distribution in the crystal (e.g., proteins, small organic and inorganic compounds). The X-ray diffraction experiment collects intensities $I(hkl)$ corresponding to various sets of Miller indices (h, k, l). These intensities are directly related to the square of the magnitude of the structure factors $|F(hkl)|^2$,

$$I(hkl) \propto |F(hkl)|^2. \tag{3.14}$$

The structure factor $F(hkl)$ is a complex number that contains both an amplitude and a phase $\phi(hkl)$,

$$F(hkl) = |F(hkl)|e^{i\phi(hkl)}, \tag{3.15}$$

The electron density $\rho(\mathbf{r})$ within the unit cell can be obtained from the structure factors by an inverse Fourier transform:

$$\rho(\mathbf{r}) = \frac{1}{V}\sum_{hkl} F(hkl)e^{-2\pi i(hx+ky+lz)}, \tag{3.16}$$

where V is the volume of the unit cell and (x, y, z) are fractional coordinates within the unit cell. Here, each term in the sum represents the contribution of a particular set of lattice planes to the overall electron density.

3.1.2 Properties of Electron Density

Let us first consider the electron density for atoms. Besides the property of the normalization condition in Eq. (3.1), electron density distributions in atoms exhibit a monotonically exponential decay pattern [16], elucidating how electron probability diminishes with the increasing distance from the nucleus. Figure 3.1a and b shows 1D and 2D distributions of the electron density for the argon atom, confirming its monotonic and exponential decaying characteristic.

However, if we replot Figure 3.1a with the logarithmic scale for the y-axis, as shown in Figure 3.1c, the piecewise behavior of the atomic electron density decay becomes apparent. This behavior was first discovered by Wang and Parr in 1977 [16]. This piecewise behavior can also be manifested by the radial distribution function, shown in Figure 3.1d as an example for Ar, where the atomic shell structure is clearly shown yet well-known [14].

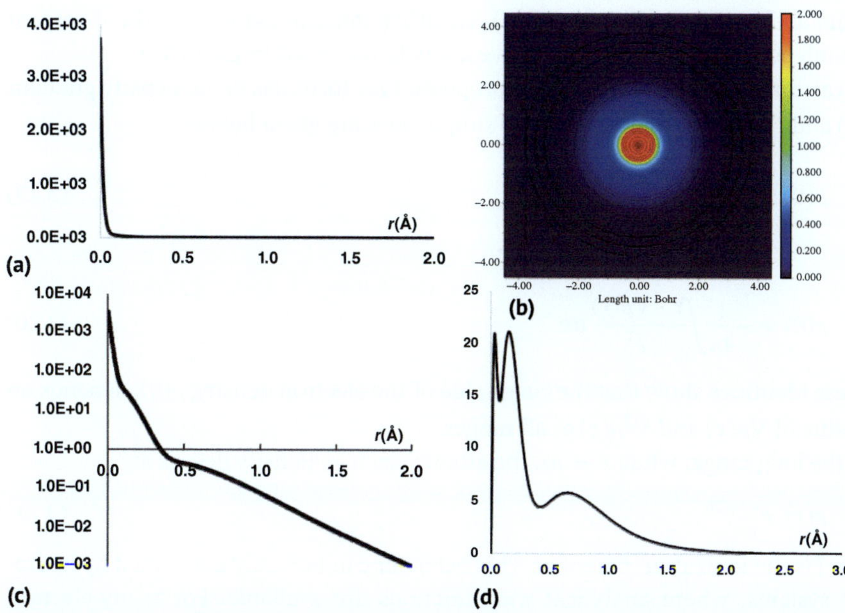

Figure 3.1 (a) Monotonically decaying nature of the atomic electron density in one dimension; (b) monotonically decaying nature of the same atomic electron density in two dimensions; (c) piecewise exponential decay of the electron density, and (d) atomic shell structure in the radial distribution function of the electron density, all for the argon atom. The position of the nucleus is located at the origin of the coordinate system.

This piecewise exponential decay behavior of atomic electron densities arises from the superposition of contributions from different atomic orbitals in different spatial regions, each region characterized by a distinct exponent. For example, for an Ar atom, we may have

$$\rho(\mathrm{r}) = \begin{cases} A_1 e^{-k_1 r} & \text{(core/cusp region)} \\ A_2 e^{-k_2 r} & \text{(valence region)} \\ A_3 e^{-k_3 r} & \text{(asymptotic region)} \end{cases} \tag{3.17}$$

where $k_1 > k_2 > k_3$ reflect three decay rates with respect to the radial distance away from the nucleus.

Near the nuclear cusp, Kato's cusp condition [17, 18] for the electron density is as follows:

$$\left.\frac{d\rho(r)}{dr}\right|_{r=0} = -2Z\rho(0), \tag{3.18}$$

where $\rho(r)$ is the spherically averaged atomic electron density, $\rho(0)$ denotes the electron density at the nucleus (at $r = 0$), and $\left.\frac{d\rho(r)}{dr}\right|_{r=0}$ stands for the radial derivative of the electron density at $r = 0$. Eq. (3.18) shows that the electron density does not form a sharp cusp (discontinuous derivative) at the nucleus. Instead, the derivative is finite and proportional to $-2Z\rho(0)$, ensuring smoothness despite the divergent Coulomb potential. With Eq. (3.18), we know that $k_1 = 2Z$ in Eq. (3.17).

To calculate $\rho(0)$, in 1995 [19], we proposed four formulas using density gradient $\nabla\rho(\mathbf{r})$ and Laplacian $\nabla^2\rho(\mathbf{r})$. The two simple ones are given below:

$$\rho(0) = -\frac{1}{4\pi}\int\frac{\nabla^2\rho(\mathbf{r})}{|\mathbf{r}|}d\mathbf{r}, \tag{3.19}$$

and

$$\rho(0) = -\frac{1}{4\pi}\int\frac{\mathbf{r}\cdot\nabla\rho(\mathbf{r})}{r^3}d\mathbf{r} \tag{3.20}$$

These identities show that the cusp value of the electron density, $\rho(0)$, depends on the value of $\nabla\rho(\mathbf{r})$ and $\nabla^2\rho(\mathbf{r})$ at all ranges.

At the long range, when $r \to \infty$, the atomic electron density decays as

$$\rho(r) \sim e^{-2\sqrt{2I}r}, \tag{3.21}$$

where I is the ionization potential. This behavior can be easily confirmed by hydrogenic systems, where analytical wavefunctions are available. For many-electron atoms, although the situation is more complex due to electron–electron interactions, the asymptotic behavior of electron density is still controlled by the highest occupied orbital and therefore decays exponentially with an exponent determined by the ionization energy. With Eq. (3.21), one obtains that $k_3 = 2\sqrt{2I}$ in Eq. (3.17).

For molecules, electron density is no longer a spherically averaged, monotonically exponential decay function. Instead, it reflects the intricate interplay of multiple nuclei and the resulting chemical bonds, leading to regions of electron accumulation (such as in bonding and lone pair areas) and depletion. This anisotropic distribution means that the density varies significantly in different directions, influenced by molecular geometry and the local environment. These variations also underscore how different atoms within the same molecule can exhibit unique electronic behaviors. For example, atoms with localized regions of high electron density may act as nucleophiles, while areas of electron deficiency can become electrophilic centers, thereby directly influencing reactivity. The asymptotic decay also varies spatially: it is slower along covalent bond axes or lone-pair directions (governed by the molecule's ionization potential) and faster in nonbonding regions.

Shown in Figure 3.2 are two representations of the electron density for the benzene molecule. The first (Figure 3.2a) is the contour map with a density gradient line exhibited, and the other is the difference map using promolecular density as the reference. A promolecular density refers to a theoretical electron density calculated by simply summing the spherically averaged electron densities of individual atoms in a molecule, positioned at their respective nuclear coordinates. From Figure 3.2a, we can confirm that the molecular electron density is no longer monotonically exponentially decaying. The difference map in Figure 3.2b shows which parts of the molecule electrons are accumulated (solid line) or depleted (dashed line) because of chemical bonding and interactions between atoms. This information is invaluable

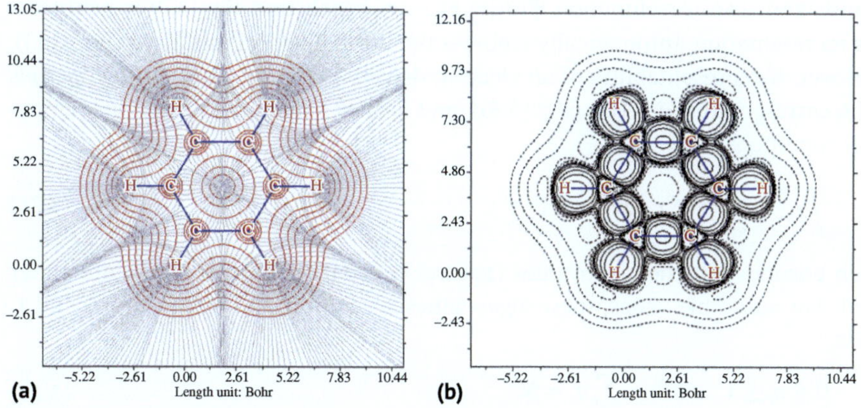

Figure 3.2 (a) Contour map of electron density for benzene with gradient map shown, and (b) the deformation map of the electron density for benzene using promolecular density as the reference.

for understanding molecular reactivity, as areas of high electron density might act as nucleophilic centers and regions of low electron density as electrophilic sites.

Another important property of electron density is related to coordinate scaling [20, 21]. Under a uniform scaling transformation of the spatial coordinates, $\mathbf{r} \rightarrow \lambda\mathbf{r}$, with $\lambda > 0$, the electron density should transform as follows:

$$\rho_\lambda(\mathbf{r}) = \lambda^3 \rho(\lambda\mathbf{r}). \tag{3.22}$$

This is because we need to ensure that the normalization condition of Eq. (3.1) remains intact:

$$\int \rho_\lambda(\mathbf{r})d\mathbf{r} = \lambda^3 \int \rho(\lambda\mathbf{r})d\mathbf{r} = \int \rho(\mathbf{r}')d\mathbf{r}' = N, \tag{3.23}$$

where the change of variable $\mathbf{r}' = \lambda\mathbf{r}$ has been used. This scaling behavior is not only crucial for maintaining physical consistency but also informs how energy components scale. For example, in the Kohn–Sham scheme, the noninteracting kinetic energy density functional $T_S[\rho]$ scales as [22]

$$T_S[\rho_\lambda] = \lambda^2 T_S[\rho], \tag{3.24}$$

suggesting that the kinetic energy density functional is homogeneous of degree two in coordinate scaling. For the exchange energy density functional $E_x[\rho]$, it becomes [23]

$$E_x[\rho_\lambda] = \lambda\, E_x[\rho], \tag{3.25}$$

so it is homogeneous of degree one in coordinate scaling. More discussions about scaling properties are provided [24–27].

N-representability [14, 28, 29] is the last property we want to discuss in this section. Not every nonnegative, integrable function can be a valid electron density; it must be "*N*-representable," meaning that it must come from an antisymmetric *N*-electron wavefunction. Specifically, an electron density $\rho(\mathbf{r})$ is *N*-representable if there exists an antisymmetric wavefunction $\Psi(\mathbf{r}_1,\mathbf{r}_2,\mathbf{r}_3,\ldots,\mathbf{r}_N)$ such that Eq. (3.3) is held true. This representation automatically enforces the normalization condition (Eq. (3.1)). Moreover, N-representability of an electron density is closely tied to the properties of the corresponding 1-RDM (Eq. (3.5)), and it must satisfy the following idempotency condition

$$\int \gamma_1(\mathbf{r},\mathbf{r}'')\gamma_1(\mathbf{r}'',\mathbf{r}')d\mathbf{r}'' = \gamma_1(\mathbf{r},\mathbf{r}'). \tag{3.26}$$

which ensures that the eigenvalues (natural occupation numbers) of γ are either 0 or 1. For ensemble states, these eigenvalues are required to lie between 0 and 1, that is,

$$0 \le n_i \le 1 \quad \text{with} \quad \sum_i n_i = N. \tag{3.27}$$

These conditions ensure that the electron density and its associated 1-RDM are physically meaningful and originate from a valid antisymmetric N-electron wavefunction.

3.2 Density Gradient and Laplacian

The gradient of the electron density, $\nabla\rho(\mathbf{r})$, is a vector field that indicates the direction and rate of the steepest decrease of exponential decay in $\rho(\mathbf{r})$. It is defined by taking the first partial derivatives with respect to the Cartesian coordinates,

$$\nabla\rho(\mathbf{r}) = \left(\frac{\partial\rho(\mathbf{r})}{\partial x}, \frac{\partial\rho(\mathbf{r})}{\partial y}, \frac{\partial\rho(\mathbf{r})}{\partial z}\right). \tag{3.28}$$

The gradient norm $|\nabla\rho(\mathbf{r})|$ of the density gradient is defined as the Euclidean norm,

$$|\nabla\rho(\mathbf{r})| = \sqrt{\left(\frac{\partial\rho(\mathbf{r})}{\partial x}\right)^2 + \left(\frac{\partial\rho(\mathbf{r})}{\partial y}\right)^2 + \left(\frac{\partial\rho(\mathbf{r})}{\partial z}\right)^2}. \tag{3.29}$$

Physically, $\nabla\rho(\mathbf{r})$ can be used to locate the critical points—points where the density is locally maximal, minimal, or a saddle point (i.e., where $\nabla\rho(\mathbf{r}) = \mathbf{0}$) [1]. In many chemical systems, the maximum of $\nabla\rho(\mathbf{r})$ coincides with nuclear positions.

On the other hand, the Laplacian of the electron density is the divergence of the gradient of $\rho(\mathbf{r})$ and provides a measure of the local curvature of the electron density,

$$\nabla^2\rho(\mathbf{r}) = \nabla \cdot \nabla\rho(\mathbf{r}) = \frac{\partial^2\rho(\mathbf{r})}{\partial x^2} + \frac{\partial^2\rho(\mathbf{r})}{\partial y^2} + \frac{\partial^2\rho(\mathbf{r})}{\partial z^2}. \tag{3.30}$$

It can distinguish between regions of electron density accumulation $(\nabla^2\rho < 0)$ and depletion $(\nabla^2\rho > 0)$, linked to bonding interactions. Figure 3.3 demonstrates the local behaviors of $|\nabla\rho(\mathbf{r})|$ and $\nabla^2\rho(\mathbf{r})$ for the Ar atom and the benzene molecule. A similar piecewise exponential decay behavior for the density gradient is seen in Figure 3.3a. For $\nabla^2\rho(\mathbf{r})$, however, no such piecewise behavior is observed

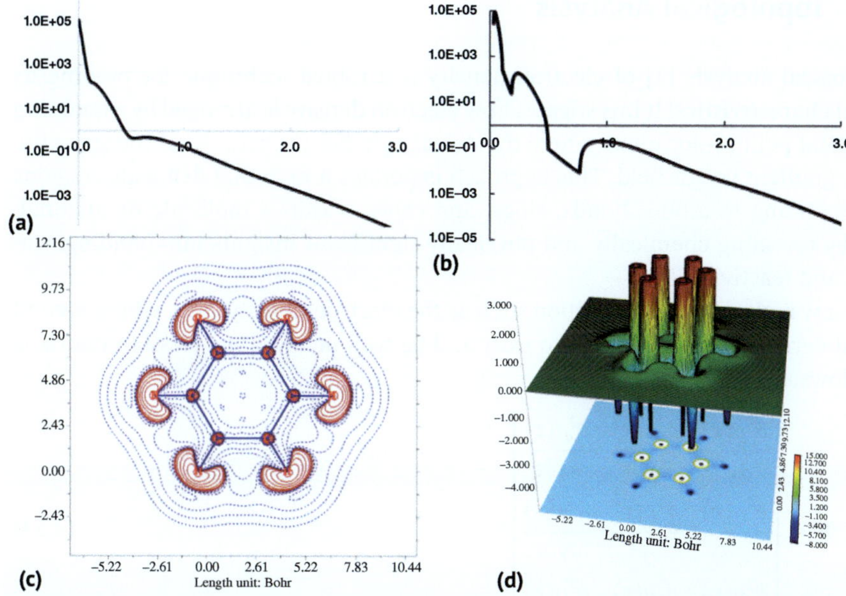

Figure 3.3 (a) Piecewise exponential decay of density gradient for Ar atom; (b) 1D curve of Laplacian $\nabla^2\rho(\mathbf{r})$ for the Ar atom; (c) the deformation map of the gradient norm $|\nabla\rho(\mathbf{r})|$ for benzene molecule; and (d) 3D and 2D representations of Laplacian $\nabla^2\rho(\mathbf{r})$ for the benzene molecule.

(in Figure 3.3b). Figure 3.3c illustrates the deformation map of the gradient norm $|\nabla\rho(\mathbf{r})|$ for the benzene molecule using the promolecular density as the reference. Figure 3.3d includes 3D and 2D representations of Laplacian $\nabla^2\rho(\mathbf{r})$ for the benzene molecule.

While $\rho(\mathbf{r})$ is expressed in electrons per cubic angstrom (e/Å^3), the unit of $\nabla\rho(\mathbf{r})$ and $\nabla^2\rho(\mathbf{r})$ has electrons per fourth power of length (e/Å^4) and electrons per fifth power of length (e/Å^5), respectively.

The quantities of $\nabla\rho(\mathbf{r})$ and $\nabla^2\rho(\mathbf{r})$ played important roles in the development of exchange–correlation functionals in DFT. In GGA [14], the exchange–correlation energy $E_{xc}[\rho]$ includes the contribution from the gradient of the electron density,

$$E_{xc}^{\mathrm{GGA}}[\rho] = \int e_{xc}(\rho(\mathbf{r}), \nabla\rho(\mathbf{r}))d\mathbf{r}. \tag{3.31}$$

In meta-GGA functionals [30, 31], the Laplacian (or related kinetic energy density) is incorporated to capture additional details about the local electronic structure,

$$E_{xc}^{\mathrm{meta\text{-}GGA}}[\rho] = \int e_{xc}\left(\rho(\mathbf{r}), \nabla\rho(\mathbf{r}), \nabla^2\rho(\mathbf{r})\right)d\mathbf{r}. \tag{3.32}$$

Including the Laplacian helps improve the description of regions where the density changes rapidly (for example, in bonding regions), thereby enhancing the accuracy of simulations.

3.3 Topological Analysis

Topological analysis [1] of electron density is a robust technique for probing its spatial characteristics. It investigates how electron density is arranged by examining its critical points—locations where the density gradient is zero—and the structure of the gradient vector field. This approach generates a map that delineates regions corresponding to atoms, bonds, rings, and cages within a molecule or material, thereby revealing chemically and physically significant insights into bonding, stability, and reactivity.

For any real-space local function such as the electron density, $\rho(\mathbf{r})$, whose second partial derivatives are assumed to exist and be well-behaved, its Hessian matrix is as follows,

$$\mathbf{H} = \begin{bmatrix} \frac{\partial^2\rho(\mathbf{r})}{\partial x^2} & \frac{\partial^2\rho(\mathbf{r})}{\partial x\partial y} & \frac{\partial^2\rho(\mathbf{r})}{\partial x\partial z} \\ \frac{\partial^2\rho(\mathbf{r})}{\partial y\partial x} & \frac{\partial^2\rho(\mathbf{r})}{\partial y^2} & \frac{\partial^2\rho(\mathbf{r})}{\partial y\partial z} \\ \frac{\partial^2\rho(\mathbf{r})}{\partial z\partial x} & \frac{\partial^2\rho(\mathbf{r})}{\partial z\partial y} & \frac{\partial^2\rho(\mathbf{r})}{\partial z^2} \end{bmatrix}. \tag{3.33}$$

After diagonalizing the above matrix, the eigenvectors and three eigenvalues, λ_1, λ_2, and λ_3, are obtained with $\lambda_1 \le \lambda_2 \le \lambda_3$ characterizing the signs of the Hessian eigenvalues

$$\mathbf{H} = \begin{bmatrix} \frac{\partial^2\rho(\mathbf{r})}{\partial x'^2} & 0 & 0 \\ 0 & \frac{\partial^2\rho(\mathbf{r})}{\partial y'^2} & 0 \\ 0 & 0 & \frac{\partial^2\rho(\mathbf{r})}{\partial z'^2} \end{bmatrix} = \begin{bmatrix} \lambda_1 & 0 & 0 \\ 0 & \lambda_2 & 0 \\ 0 & 0 & \lambda_3 \end{bmatrix}. \tag{3.34}$$

The critical points (cps) are characterized by a pair of numbers, (m, n), where rank m is the number of nonzero eigenvalues of the above diagonalized Hessian matrix and signature n is the number of positive eigenvalues minus the number of negative eigenvalues.

For DAQs, m is often, though not always, 3, meaning that all eigenvalues in Eq. (3.34) are nonzero. For cps of rank 3, there are the following four categories,

- (3,−3), all three eigenvalues negative, often at the position of a nucleus, so it is called "*nuclear critical point*" (*ncp*);
- (3,−1), a saddle point with two negative and one positive eigenvalue, often lying between two atoms, so it is called "*bond critical point*" (*bcp*);
- (3,+1), a saddle point with one negative and two positive eigenvalues, often showing up at the center of a cyclic structure, so it is called "*ring critical point*" (*rcp*);

- (3,+3), all three eigenvalues positive, often located at the center of a polyhedral structure, so it is called "*cage critical point*" (*ccp*).

These cps can be used to locate nuclear positions, identify covalent bonding between atoms, and provide insights into delocalized or multicenter interactions. In addition, bcp values of $\rho(\mathbf{r}_{bcp})$ can distinguish a covalent bond from an ionic bond or somewhere in between. Subtle differences in cps reveal not only strong covalent bonds but also weaker interactions like hydrogen bonds, π–π stacking, and van der Waals interactions. The connectivity of cps shows how electrons are shared or localized between atoms, clarifying the structure–property relationships within the molecule.

Besides electron density, topological analysis can also be performed for other DAQs [32–35] such as density gradient and Laplacian. In that case, we will still have these four cps of rank 3. However, the number and location of these four cps can be significantly different.

As illustrative examples, shown in Figure 3.4 are topological analysis results of electron density (Figure 3.4a) and Laplacian (Figure 3.4b) for benzene. For the density case in Figure 3.4a, all ncp (red) are at the C and H nuclei, bcp (blue) on C—C and C—H bonds, and rcp (orange) at the center of the ring. For Laplacian (Figure 3.4b), however, there are many more cps, and the meaning of these cps could be different from what we specified above.

3.4 Quantum Theory of Atoms in Molecules

The quantum theory of atoms in molecules (QTAIM) [1] represents a theoretical framework that revolutionizes our understanding of chemical bonding through a quantum-mechanical lens with density-based quantities. At its core, QTAIM employs topological analysis of molecular electron density to illuminate chemical

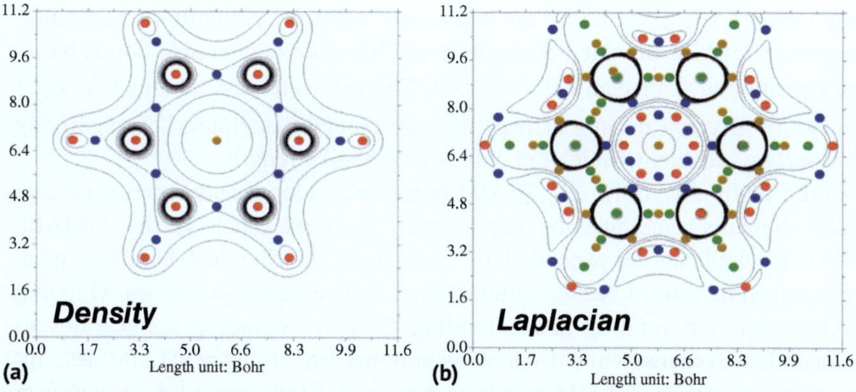

Figure 3.4 Topological analysis results for (a) electron density $\rho(\mathbf{r})$ and (b) Laplacian $\nabla^2\rho(\mathbf{r})$ of the benzene molecule. The color code is the following for the four cps: red is ncp, (3,−3); blue is bcp, (3,−1); orange is rcp, (3,+1); and green is ccp, (3,+3).

bonding. This innovative approach emerged from the pioneering work of the late Canadian theoretical chemist Richard F.W. Bader, whose vision and contributions were instrumental in its development.

The genesis of QTAIM lies in Bader's pursuit of a rigorous quantum mechanical definition for "atoms in molecules" and chemical bonding. From the late 1960s, Bader centered on the electron density, $\rho(\mathbf{r})$, and its topological properties. He introduced the concept of analyzing the density gradient, $\nabla\rho(\mathbf{r})$, defining *atoms in molecules* as *atomic basins* ($\boldsymbol{\Omega}_A$), which are real-space regions bounded by *zero-flux surfaces*. Mathematically, the boundary $\partial\boldsymbol{\Omega}_A$ of an atomic basin $\boldsymbol{\Omega}_A$ is determined by the following *zero-flux condition*:

$$\mathbf{n}(\mathbf{r}) \cdot \nabla\boldsymbol{\rho}(\mathbf{r}) = 0 \quad \text{for all } \mathbf{r} \in \partial\boldsymbol{\Omega}_{A,} \tag{3.35}$$

where $\mathbf{n}(\mathbf{r})$ is the unit normal vector. Chemical bonds were proposed to be indicated by *bond paths*, which are lines of maximum electron density connecting nuclei. This pioneering work signified a new paradigm for understanding chemical bonding, rooted purely in terms of the electron density $\rho(\mathbf{r})$.

The 1970s and 1980s marked the rigorous development and formalization of QTAIM. This period saw the establishment of the mathematical framework that enabled property partitioning. Atomic properties, P_A, are obtained by integrating three-dimensional property functions, $P(\mathbf{r})$, over atomic basins Ω_A:

$$P_A = \int_{\Omega_A} P(\mathbf{r})\, d\mathbf{r} \tag{3.36}$$

For example, the atoms-in-molecules (AIM) charge q_A in an atomic basin Ω_A is given by

$$q_A = Z_A - \int_{\Omega_A} \rho(\mathbf{r})\, d\mathbf{r}, \tag{3.37}$$

where Z_A is the nuclear charge. Unlike Mulliken/Löwdin charges, which are basis-set dependent, AIM charges are derived solely from the electron density, thereby yielding subsystems that are rigorously defined by quantum mechanics. Furthermore, QTAIM provided descriptors to characterize chemical bonds based on the density and Laplacian values at bcp. The values of $\rho\left(\mathbf{r}_{\text{bcp}}\right)$ and $\nabla^2\rho\left(\mathbf{r}_{\text{bcp}}\right)$ (Laplacian of electron density at bcp) became crucial descriptors for classifying bond types (e.g., covalent, ionic) and quantifying bond strength.

From the 1990s onward [2, 36], QTAIM witnessed wider adoption across diverse scientific domains, facilitated by user-friendly software (such as the AIMALL package). Its applications expanded to understanding chemical bonding, reaction mechanisms, intermolecular interactions, and material properties. Ongoing research focuses on refining and extending QTAIM, including applications to time-dependent systems, while discussions and interpretations of QTAIM descriptors continue to evolve. QTAIM stands as a powerful tool grounded in quantum mechanics, providing unique insights into chemical bonding. Despite debates (e.g., whether bond paths always signify "chemical bonds," especially in weak interactions), QTAIM's mathematical rigor—rooted in $\rho(\mathbf{r})$'s topology—remains pivotal for analyzing covalent, ionic, and noncovalent interactions.

The relationship between DFT and QTAIM merits careful examination because both frameworks build on electron density as their fundamental basis for exploring molecular bonding and reactivity. This connection once prompted the author of this book to consult Bob Parr, CDFT's founder. Instead of directly addressing my inquiry, he responded with a thought-provoking question: "Where does the electron density of QTAIM come from?" As previously discussed in Section 3.1.1, electron density can be derived through multiple approaches—from experimental methods to ab initio calculations, indicating that QTAIM need not be exclusively tied to DFT. This observation highlights QTAIM's potential autonomy as an independent theoretical framework. However, it also raises another question: whether electron density—whose derivatives include the Laplacian—alone determines all ground state properties of an electronic system, without requiring additional quantities. Fundamental theorems of DFT suggest an affirmative answer, implying that QTAIM should be viewed as a subset of DFT. This apparent contradiction in QTAIM's status—where QTAIM appears simultaneously autonomous and subordinate to DFT—is what we term *the QTAIM paradox*.

My view on this matter from a philosophical standpoint is multilayered. On the one hand, DFT provides the ontological foundation through the Hohenberg–Kohn theorems, as the ground-state electron density uniquely determines all properties of a system. On the other hand, QTAIM plays an epistemological role by analyzing the topology of this density or its gradient to extract chemical meanings such as atoms, bonds, rings, and cages. Methodologically, QTAIM is independent because it can be applied to densities obtained from DFT, wavefunction theory, or experiment, yet conceptually it remains subordinate since it adds no new ontological content beyond what the density already encodes. This tension, the above so-called *QTAIM paradox*—can be resolved by viewing DFT as the physics of density, concerned with laws and functionals, while QTAIM represents the chemistry of density, providing the interpretive grammar that renders density and its gradients intelligible in chemical terms.

Finally, it is important to recognize that QTAIM naturally falls under density-associated quantities (DAQ) because all of its descriptors are defined directly from $\rho(\mathbf{r})$ and its derivatives. It thus exemplifies how real-space features of the density can encode chemical structure, and the paradox simply reflects the broader interpretational challenges associated with DAQ rather than any conceptual inconsistency unique to QTAIM.

3.5 DAQ-based Indices

In earlier sections of this chapter, we explored how electron density $\rho(\mathbf{r})$, density gradient $\nabla\rho(\mathbf{r})$, and Laplacian $\nabla^2\rho(\mathbf{r})$ can be directly utilized to analyze bonding and interaction properties. However, these quantities do not need to be used in isolation. By combining them to form dimensionless indices through coordinate scaling, new descriptors can be generated, offering newer and deeper insights into physicochemical properties using density-based approaches. Given the available variables—$\mathbf{r}$ (vector), $\rho(\mathbf{r})$ (scalar), $\nabla\rho(\mathbf{r})$ (vector), and $\nabla^2\rho(\mathbf{r})$ (scalar)—there are multiple ways to construct meaningful, yet dimensionless combinations based on their coordinate scaling properties. Some notable examples include:

$$\frac{|\nabla\rho(\mathbf{r})|}{\rho^{4/3}(\mathbf{r})}, \frac{\nabla^2\rho(\mathbf{r})}{\rho^{5/3}(\mathbf{r})}, \frac{\mathbf{r}\cdot\nabla\rho(\mathrm{r})}{\rho(\mathbf{r})}, \frac{\nabla^2\rho(\mathbf{r})}{(\mathbf{r}\cdot\nabla\rho(\mathrm{r}))^{5/3}}, \frac{|\nabla\rho(\mathbf{r})|\nabla^2\rho(\mathbf{r})}{\rho^3(\mathbf{r})}, \frac{\mathbf{r}\cdot\nabla\rho(\mathrm{r})\nabla^2\rho(\mathbf{r})}{\rho^{8/3}(\mathbf{r})}, \dots . \tag{3.38}$$

These combinations are dimensionless because the numerator and denominator scale identically under the λ^3 scaling transformation in Eq. (3.22). Among these, the first index is known as the noncovalent interaction (NCI) index [3, 4], while the second is referred to as the USI index [5]. These two DAQ indices are specifically designed to identify weak and strong interactions, respectively. While other combinations are theoretically possible, their effectiveness in providing chemical insights remains an open question for further examinations.

3.5.1 Noncovalent Interaction (NCI) Index

The NCI index introduced in 2010 by Johnson, Yang, and coworkers [3, 4] was developed as a DAQ descriptor to visualize and analyze weak, inter- and intramolecular interactions such as hydrogen bonds and van der Waals interactions utilizing the reduced density gradient (RDG), defined as

$$s(\mathbf{r}) = \frac{|\nabla\rho(\mathbf{r})|}{2(3\pi)^{1/3}\rho^{4/3}(\mathbf{r})}. \tag{3.39}$$

Except the coefficient in the denominator, Eq. (3.39) is the first term in Eq. (3.38). Recall that the Hessian matrix $\mathbf{H}$ in Eq. (3.33) has three eigenvalues, $\lambda_1 \le \lambda_2 \le \lambda_3$. For noncovalent interactions, λ_2 is critical:

- If $\lambda_2 < 0$, the interaction is *attractive* (e.g., hydrogen bonds);
- If $\lambda_2 > 0$, the interaction is *repulsive* (e.g., steric clashes).

NCI combines the above trend as $\mathrm{sign}(\lambda_2)\rho$ and then plots against RDG to generate 3D isosurfaces that spatially map interaction regions. By combining promolecular density approximations (for computational efficiency) or actual computed (SCF) densities (for accuracy) with color-coded RDG plots, the index distinguishes between attractive (blue/green) and repulsive (red) interactions. Its simplicity, low computational cost, and intuitive graphical output have facilitated widespread adoption in chemistry, biosystems, and materials science, enabling insights into molecular recognition, crystal packing, and reaction mechanisms. Figure 3.5 shows NCI plots for a few systems with weak inter- and intramolecular interactions. The color-coded isosurfaces can clearly identify the nature of the weak interactions, with either SCF or promolecular densities.

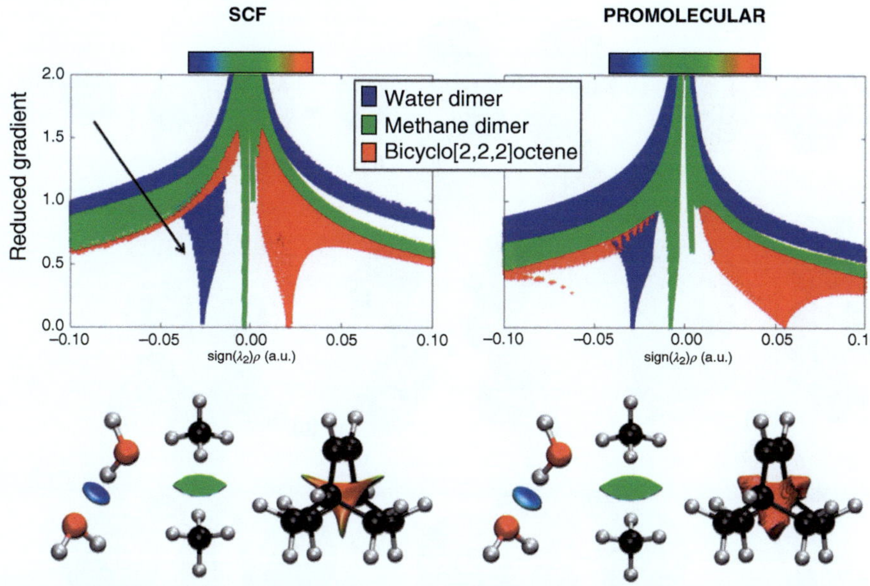

Figure 3.5 RDG plots versus sign$(\lambda_2)\rho$ for a few systems with inter- and intramolecular interactions. *Source*: Reproduced from Ref. [4] with permission of American Chemical Society.

One can also perform topological analysis for RDG, the same as what we did in Section 3.3 for electron density and Laplacian, with the help of the Hessian matrix of RDG. A critical point in the $\boldsymbol{s}(\mathbf{r})$ field is defined by

$$\nabla \boldsymbol{s}(\mathbf{r}_c) = \mathbf{0}. \tag{3.40}$$

Unlike the bond critical points (bcp) of electron density, which occur when $\nabla \boldsymbol{\rho}(\mathbf{r}) = \mathbf{0}$ and have a characteristic (3,−1) signature, a critical point in $\boldsymbol{s}(\mathbf{r})$ analogous to a bcp in $\rho(\mathbf{r})$ typically occurs in regions where the density is low and its variation is minimal—conditions often found in noncovalent interaction regions. Hence, while bcp from RDG topological analysis might signify "bonded" interactions, its cps often reveals nonbonded and weak interactions that do not necessarily agree with the results from electron density.

3.5.2 USI Index

In 2022, we introduced the USI index [5],

$$\text{USI} = \frac{\nabla^2 \rho(\mathbf{r})}{\rho^{5/3}(\mathbf{r})}, \tag{3.41}$$

which is the second term in Eq. (3.38), as an identifier of ultra-strong covalent bonds, such as quadruple, quintet, and even stronger bonds. Physically, Laplacian $\nabla^2\rho(\mathbf{r})$ measures the curvature of the electron density, so regions with large $\nabla^2\rho(\mathbf{r})$

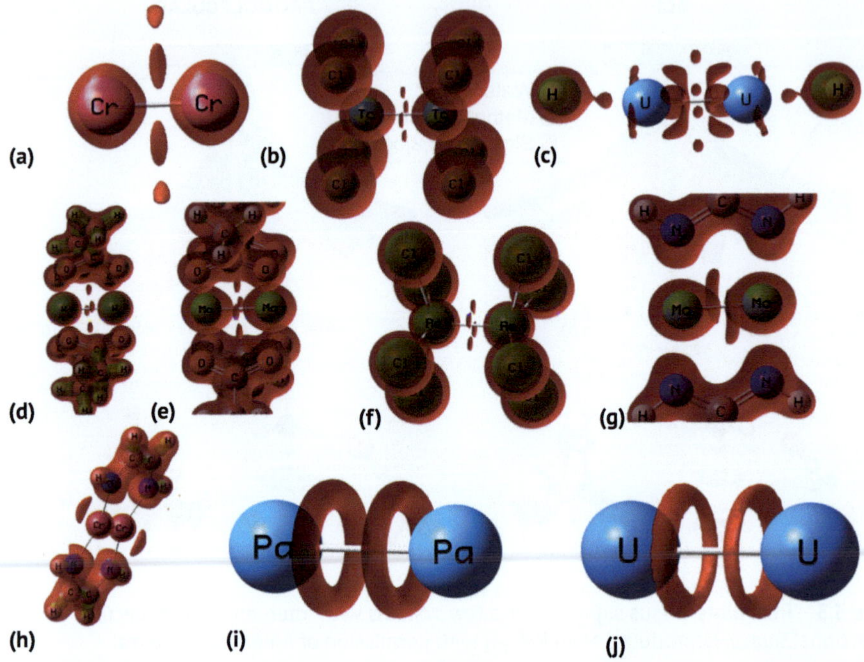

Figure 3.6 Signature isosurfaces of USI index for systems with quadruple (a–f), quintuple (g–h), and sextuple (i, j) covalent bonds: (a) Cr_2^{2+}; (b) $Tc_2Cl_8^{2-}$; (c) U_2H_2; (d) $W_2(CH_3CO_2)_4$; (e) $Mo_2(CH_3CO_2)_6$; (f) $Re_2Cl_8^{2-}$; (g) dimolybdenum complex with two diamido bridging ligands; (h) dichromium complex with diazadiene bridging ligands; (i) Pa_2; and (j) U_2^{2+}.
Source: Reproduced from Ref. [5] with permission of American Chemical Society.

values correspond to areas of strong density localization and large electron repulsion. In this regard, the USI index highlights "ultra-strong" interactions from multiple bonds (quadruple, quintuple, sextuple, etc.). As illustrative examples, shown in Figure 3.6 are USI isosurfaces for a few systems. For different compounds with the same bond multiplicity, the signature isosurface is always the same. For example, for quadruple bonds, the signature shape is four beads between the two bonding atoms. For quintuple bonds, we observed two corn beans, and for sextuple bonds, two parallel tori or donuts are discernible.

3.6 DAQs in Excited States

To examine the topological properties of DAQs in excited states, we need to obtain the electron density first. There are mainly three such approaches: (i) *ΔSCF* [37, 38] *method:* Directly compute an SCF solution for the excited state and then obtain $\rho_k(\mathbf{r})$ from the occupied orbitals,

$$\rho_k(\mathbf{r}) = \sum_{i \in \text{occ}} \left|\psi_i^k(\mathbf{r})\right|^2, \tag{3.42}$$

where $\{\psi_i^k(\mathbf{r})\}$ are the orbitals from the excited-state SCF solution. (ii) *TD-DFT (linear-response) approach* [39–41]: Use time-dependent DFT (TD-DFT) excitation amplitudes to construct a difference (or transition) density $\Delta\rho_k(\mathbf{r})$ added to the ground-state density $\rho_{GS}(\mathbf{r})$,

$$\rho_k(\mathbf{r}) \approx \rho_{GS}(\mathbf{r}) + \Delta\rho_k(\mathbf{r}), \tag{3.43}$$

where

$$\Delta\rho_k(\mathbf{r}) \approx \sum_{ia}\left(x_{ia}^k + y_{ia}^k\right)\psi_a(\mathbf{r})\psi_i^*(\mathbf{r}), \tag{3.44}$$

with x_{ia} and y_{ia} as the amplitudes of eigenvectors to characterize single excitations from occupied (index i) to virtual (index a) orbitals from the standard Casida formulation. Because TD-DFT densities are response densities rather than true eigenstate densities, they may not always satisfy the N-representability conditions discussed in Section 3.1.2. (iii) *Multi-configurational method* (e.g., CASSCF, MRCI, or CASPT2),

$$\rho_k(\mathbf{r}) = \langle \Psi_k|\hat{\rho}(\mathbf{r})|\Psi_k\rangle, \tag{3.45}$$

with

$$\Psi_k = \sum_I c_I^{(k)} D_I, \tag{3.46}$$

so

$$\rho_k(\mathbf{r}) = \sum_{I,J} c_I^{(k)*} c_J^{(k)} \langle \mathrm{D}_I|\hat{\rho}(\mathbf{r})|\mathrm{D}_J\rangle. \tag{3.47}$$

Recent studies employing topological analysis of the electron density and its Laplacian have been applied to a range of excited-state systems [42]. For example, changes from being predominantly covalent to more ionic upon excitation or vice versa are possible, important in understanding photochemical reaction mechanisms and charge-separation processes in dye–sensitized solar cells and organic light-emitting diodes (OLEDs) [43]. It is also essential for organic photovoltaic materials whose efficiency depends on subtle changes in bonding characteristics. In systems where nonadiabatic transitions occur, topological analysis of excited-state densities can be used to locate conical intersections for small organic molecules and even larger chromophores where photoisomerization takes place.

A challenge in the above methods is that excited-state densities obtained from the linear response method (such as TD-DFT) are not uniquely defined as true eigenstates of the Hamiltonian. Consequently, state-specific (e.g., ΔSCF or excited-state SCF methods) or multiconfigurational techniques should be used to yield a well-defined one-electron density for the excited state analysis. They allow for a more reliable topological analysis of the excited-state electron density and Laplacian.

3.7 DAQ in Momentum Space

While most traditional chemical topology (such as QTAIM) is based on the real-space electron density $\rho(\mathbf{r})$, the momentum-space analysis provides complementary information [44, 45]. In some cases, features that are "smeared out" in real space may appear as distinct topological characteristics in momentum space. The momentum-space electron density $\rho(\mathbf{p})$ is derived from the Fourier transform of the position-space counterpart $\rho(\mathbf{r})$,

$$\rho(\mathbf{p}) = \int \rho(\mathbf{r}) e^{-i\mathbf{p}\cdot\mathbf{r}/\hbar} \, d\mathbf{r}. \tag{3.48}$$

By mapping $\rho(\mathbf{p})$ in momentum space, one identifies its cps by solving

$$\nabla_p \rho(\mathbf{p}) = \mathbf{0}. \tag{3.49}$$

Diagonalizing the Hessian matrix at the cps allows classification (as maxima, minima, or saddle points) and thereby reveals directional features that are correlated with bonding propensities. Strictly speaking, $\rho(\mathbf{p})$ is obtained from the Fourier transform of the many-electron wavefunction, but for visualization and qualitative analysis, the Fourier transform of $\rho(\boldsymbol{r})$ is often used as an approximate one-body momentum density.

Moreover, experimental techniques such as angle-resolved photoemission spectroscopy (ARPES) [46] and its variants (e.g., photoemission orbital tomography) [47] directly probe momentum-space distributions and thereby offer a natural setting for such topological analysis. The Compton profile $J(p_z)$ is defined as

$$J(p_z) = \iint \rho(\mathbf{p}) dp_x \, dp_y, \tag{3.50}$$

where the three-dimensional electron momentum density $\rho(\mathbf{p})$ is projected along p_z momentum direction.

Although electron momentum density is vital for deciphering phenomena in chemical systems—from electron behavior in metals and semiconductors to the unique properties of warm dense matter—progress in this area remains constrained. At the same time, emerging techniques that merge position-space and momentum-space information, such as those employed in electron diffraction, are opening promising new pathways for the analysis of nanostructured and heterogeneous materials. Yet, significant obstacles persist in reaching the requisite resolution and in integrating disparate data sets from both spatial domains. Addressing these challenges will necessitate further refinement of high-precision experimental methods along with the creation of advanced computational approaches capable of accurately capturing and correlating both localized and extended electronic features. Ultimately, these developments are expected not only to deepen our fundamental understanding of electron dynamics but also to pave the way for the customized design of materials with innovative and enhanced functionalities.

3.8 Concluding Remarks

In this chapter, we have developed a systematic account of DAQs as local descriptors of electronic structure, bonding, and reactivity. Beginning with the electron density itself, we examined its methods of determination, mathematical properties, and scaling behavior, before extending to its derivatives—the gradient and Laplacian—that reveal local features of electron distribution. Topological analysis and the QTAIM demonstrated how these quantities can be used to partition real space into chemically meaningful regions and to define atoms, bonds, and higher-order structural motifs. We also highlighted how DAQ-based indices, such as the noncovalent interaction (NCI) index and the USI index, extend the descriptive power of density analysis to weak van der Waals interactions and ultra-multiple bonding, respectively. Finally, we discussed extensions of DAQ into excited states and momentum space, broadening their applicability beyond the ground-state paradigm and into spectroscopic and photochemical domains.

The philosophical discussion of the QTAIM–DFT relationship emphasized an important conceptual point: while DFT provides the ontological foundation by ensuring that the ground-state density determines all system properties, QTAIM and related topological approaches play the epistemological role of rendering this density chemically intelligible. This distinction underscores the complementary nature of physics-driven density theorems and chemistry-driven density interpretations.

DAQ form a powerful and versatile toolkit for understanding chemical bonding and interactions. Unlike density functionals, they are purely density-based and local, offering direct insights into how electrons accumulate, deplete, and organize themselves in molecules and materials. Their value lies not only in predicting energetics but also in translating the abstract language of quantum mechanics into the concrete concepts of chemistry. As DAQ analysis continues to expand—from real to momentum space, and from ground to excited states—it promises to deepen our ability to connect electronic structure with chemical reactivity, providing a conceptual bridge between theory and chemical understanding.

References

1 Bader RFW. Atoms in Molecules: A Quantum Theory. Oxford University Press; 1990.
2 Cortes-Guzman F, Bader R. Complementarity of QTAIM and MO theory in the study of bonding in donor-acceptor complexes. Coord Chem Rev. 2005 Mar;249 (5–6):633–62.
3 Johnson ER, Keinan S, Mori-Sánchez P, Contreras-García J, Cohen AJ, Yang W. Revealing noncovalent interactions. J Am Chem Soc. 2010 May 12;132(18):6498–506.

4 Contreras-García J, Johnson ER, Keinan S, Chaudret R, Piquemal J-P, Beratan DN, et al. NCIPLOT: a program for plotting non-covalent interaction regions. J Chem Theory Comput. 2011 Mar 8;7(3):625–32.
5 Zhong S, He X, Liu S, Wang B, Lu T, Rong C, et al. Toward density-based and simultaneous description of chemical bonding and noncovalent interactions with Pauli energy. J Phys Chem A. 2022 Apr 21;126(15):2437–44.
6 Yu D, Stuyver T, Rong C, Alonso M, Lu T, De Proft F, et al. Global and local aromaticity of acenes from the information-theoretic approach in density functional reactivity theory. Phys Chem Chem Phys. 2019 Aug 21;21(33):18195–210.
7 Liu S, Schauer CK. Origin of molecular conformational stability: perspectives from molecular orbital interactions and density functional reactivity theory. J Chem Phys. 2015 Feb 7;142(5):054107.
8 Liu S. Identity for Kullback-Leibler divergence in density functional reactivity theory. J Chem Phys. 2019 Oct 14;151(14):141103.
9 Cao X, Rong C, Zhong A, Lu T, Liu S. Molecular acidity: an accurate description with information-theoretic approach in density functional reactivity theory. J Comput Chem. 2018 Jan 15;39(2):117–29.
10 Wu W, Wu Z, Rong C, Lu T, Huang Y, Liu S. Computational study of chemical reactivity using information-theoretic quantities from density functional reactivity theory for electrophilic aromatic substitution reactions. J Phys Chem A. 2015 Jul 23;119(29):8216–24.
11 Liu S, Ess DH, Schauer CK. Density functional reactivity theory characterizes charge separation propensity in proton-coupled electron transfer reactions. J Phys Chem A. 2011 May 12;115(18):4738–42.
12 Liu S, Li T, Ayers PW. Potentialphilicity and potentialphobicity: reactivity indicators for external potential changes from density functional reactivity theory. J Chem Phys. 2009 Sep 21;131(11):114106.
13 Ayers PW, Liu S, Li T. Stability conditions for density functional reactivity theory: an interpretation of the total local hardness. Phys Chem Chem Phys. 2011 Mar 14;13(10):4427–33.
14 Parr RG, Yang W. Density-Functional Theory of Atoms and Molecules. Oxford University Press; 1989.
15 Kohn W, Sham LJ. Self-consistent equations including exchange and correlation effects. Phys Rev. 1965 Nov 15;140(4A):A1133–8.
16 Wang W-P, Parr RG. Statistical atomic models with piecewise exponentially decaying electron densities. Phys Rev A. 1977 Sep 1;16(3):891–902.
17 Kato T. On the eigenfunctions of many-particle systems in quantum mechanics. Commun Pure Appl Math. 1957;10(2):151–77.
18 Liu S. Conditions for accurate description of the electron density of atoms and molecules. Int J Quantum Chem. 2006 Jan;106(8):1762–8.
19 Liu S, Parr RG, Nagy A. Cusp relations for local strongly decaying properties in electronic systems. Phys Rev A. 1995 Oct;52(4):2645–51.
20 Levy M. Density-functional exchange correlation through coordinate scaling in adiabatic connection and correlation hole. Phys Rev A. 1991 May 1;43(9):4637–46.

21 Liu S, Parr RG. Expansions of the correlation-energy density functional Ec. Phys Rev A. 1996 Apr;53(4):2211–9.

22 Nagy Á. Hierarchy of equations for the energy functional of the density-functional theory. Phys Rev A. 1993 Apr;47(4):2715–9.

23 Levy M, Perdew JP. Hellmann-Feynman, virial, and scaling requisites for the exact universal density functionals. Shape of the correlation potential and diamagnetic susceptibility for atoms. Phys Rev A. 1985 Oct;32(4):2010–21.

24 Nagy Á, Liu S, Parr RG. Density-functional formulas for atomic electronic energy components in terms of moments of the electron density. Phys Rev A. 1999 May 1;59(5):3349–54.

25 Liu S, Nagy Á, Parr RG. Expansion of the density-functional energy components E_c and T_c in terms of moments of the electron density. Phys Rev A. 1999 Feb 1;59(2):1131–4.

26 Liu S, Parr RG. Expansions of density functionals in terms of homogeneous functionals: justification and nonlocal representation of the kinetic energy, exchange energy, and classical Coulomb repulsion energy for atoms. Phys Rev A. 1997 Mar 1;55(3):1792–8.

27 Liu S. Local-density approximation, hierarchy of equations, functional expansion, and adiabatic connection in current-density-functional theory. Phys Rev A. 1996 Aug;54(2):1328–36.

28 Ayers PW, Liu S, Li T. Chargephilicity and chargephobicity: two new reactivity indicators for external potential changes from density functional reactivity theory. Chem Phys Lett. 2009 Oct;480(4–6):318–21.

29 Ayers PW, Liu S. Necessary and sufficient conditions for the-representability of density functionals. Phys Rev A. 2007 Feb 26;75(2):022514.

30 Sun J, Ruzsinszky A, Perdew JP. Strongly constrained and appropriately normed semilocal density functional. Phys Rev Lett. 2015 Jul 17;115(3):036402.

31 Zahariev F, Leang SS, Gordon MS. Functional derivatives of meta-generalized gradient approximation (meta-GGA) type exchange-correlation density functionals. J Chem Phys. 2013 Jun 28;138(24):244108.

32 Suresh CH, Anila S. Molecular electrostatic potential topology analysis of noncovalent interactions. Acc Chem Res. 2023 Jul 4;56(13):1884–95.

33 He X, Lu T, Rong C, Liu S, Ayers PW, Liu W. Topological analysis of information-theoretic quantities in density functional theory. J Chem Phys. 2023 Aug 7;159(5):054112.

34 He X, Li M, Rong C, Zhao D, Liu W, Ayers PW, et al. Some recent advances in density-based reactivity theory. J Phys Chem A. 2024 Feb 22;128(7):1183–96.

35 He X, Liu W, Liu S. Using energetic information quantities from density functional theory to simultaneously identify both covalent and noncovalent interactions. ChemPhysChem. 2025 Apr 1;26(7):e202401068.

36 Jenkins S, Rong C, Kirk SR, Yin D, Liu S. Spanning set of silica cluster isomer topologies from QTAIM. J Phys Chem A. 2011 Nov 17;115(45):12503–11.

37 Yang W, Ayers PW. Foundation for the ΔSCF approach in density functional theory. arXiv. 2024;arXiv:2403.04604.

38 Paetow L, Neugebauer J. Excited state dipole moments from ΔSCF: a benchmark. Phys Chem Chem Phys. 2025 Aug 7;27(31):16354–70.

39 Teale AM, Helgaker T, Savin A, Adamo C, Aradi B, Arbuznikov AV, et al. DFT exchange: sharing perspectives on the workhorse of quantum chemistry and materials science. Phys Chem Chem Phys. 2022 Dec 7;24(47):28700–81.

40 Laurent AD, Jacquemin D. TD-DFT benchmarks: a review. Int J Quantum Chem. 2013 Sep 5;113(17):2019–39.

41 Casida ME, Huix-Rotllant M. Progress in time-dependent density-functional theory. Annu Rev Phys Chem. 2012 Jan 13;63:287–323.

42 Popelier P. On the full topology of the Laplacian of the electron density. Coord Chem Rev. 2000 Feb;197(1):169–89.

43 Zhang L, Cole JM. Dye aggregation in dye-sensitized solar cells. J Mater Chem A. 2017;5(37):19541–59.

44 Balanarayan P, Gadre SR. Atoms-in-molecules in momentum space: a Hirshfeld partitioning of electron momentum densities. J Chem Phys. 2006 May 28;124(20):204113.

45 Esquivel RO, Molina-Espíritu M, López-Rosa S. 3D information-theoretic analysis of the simplest hydrogen abstraction reaction. J Phys Chem A. 2023 Aug 3;127(30):6159–74.

46 Zhang H, Pincelli T, Jozwiak C, Kondo T, Ernstorfer R, Sato T, et al. Angle-resolved photoemission spectroscopy. Nat Rev Methods Primers. 2022 Dec;2(1):54.

47 Kern CS, Windischbacher A, Puschnig P. Photoemission orbital tomography for excitons in organic molecules. Phys Rev B. 2023 Aug 22;108(8):085132.

4

Information-theoretic Approach

TABLE OF CONTENTS

4.1 Overview

The central tenet of DFT [1], as established by the Hohenberg–Kohn theorems [2], is that the ground-state properties of a many-electron system are uniquely determined by its electron density rather than by its many-body wavefunction. Since electron density is, in effect, a probability distribution over space, it is natural to ask whether tools from information theory—which quantify uncertainty, information content, and localization—could be used to analyze and predict electronic structure and chemical reactivity. This line of inquiry eventually led to the establishment of the information-theoretic approach (ITA) [3, 4], which is the application of information theory to DFT. In essence, ITA treats the electron density $\rho(\mathbf{r})$ as if it were a probability density function and then employs various information-measures to extract insights. It goes beyond traditional DFT analysis, which primarily focuses on energy, forces, and related properties. ITA leverages the mathematical machinery of information theory to extract qualitative and quantitative, yet chemically intuitive, information about electronic structure, bonding, reactivity, and other phenomena.

ITA quantities possess several key features that make them valuable tools. First, ITA quantities are explicit yet often simple functionals of the electron density, emphasizing the "information content" of the electron density. This property contrasts with energy density functionals such as the exchange–correlation and total noninteracting kinetic energy density functionals, whose exact formulas are still unknown. This feature makes ITA quantities computationally efficient and compatible with standard DFT procedures. They can also uncover information encoded within the electron density distribution that might be missed by solely focusing on energy-based quantities. For instance, ITA quantities can be more sensitive to subtle changes in electron distribution that might not be strongly reflected in the total energy or its components.

Second, each ITA quantity has a distinct physical meaning, serving as a unique lens through which the characteristics of the electron density distribution can be viewed. For example, while Shannon entropy [5] provides a global measure of the overall delocalization and uniformity of the density distribution, Fisher information [6–8] offers a local measure that is highly sensitive to sharp variations and inhomogeneities. This complementary interplay not only deepens our understanding of electronic structure but also bridges the gap between abstract information-theoretic descriptors and observable chemical properties.

In addition, ITA quantities are gauge-invariant and size-extensive, and some of them are homogeneous in density and coordinate scaling [9–11]. These properties make ITA quantities robust and reliable descriptors of the electronic structure. Their invariance and scaling properties ensure that comparisons across different systems are meaningful, allowing them to serve as powerful tools for interpreting chemical reactivity, bonding, and stability in a generally applicable framework.

The roots of ITA can be traced back to the early days of DFT development [12]. The first series of studies was conducted by Sears and Parr in the early 1980s [13]. Since then, it has been extended to describe atomic populations, chemical bonding, features of atomic/molecular electron densities, chemical reactivity, electron correlation, and other chemical phenomena. Major players include Ayers [14–16], Borgoo [17, 18], Esquivel [19–21], Heidar-Zadeh [22], Liu [3, 4, 23, 24], Nagy [7, 25–27], Nalewajski [28–30], Parr [15, 28], Rong [3, 23, 31, 32], Sen [18, 33], and others. In recent years, we have seen renewed interest in ITA [3, 4, 34, 35]. Systematic studies have been conducted to accomplish its theoretical framework, identifying their physicochemical meanings, and applying to quantify regioselectivity, nucleophilicity, electrophilicity, and other reactivity propensities [24, 36].

4.2 ITA Quantities

4.2.1 Absolute ITA Quantities

The first ITA quantity that was extensively investigated and applied in DFT is Shannon entropy. Using the electron density $\rho(\mathbf{r})$ as the probability function, it reads,

$$S_S = \int s_S(\mathbf{r})d\mathbf{r} = -\int \rho(\mathbf{r})\ln\rho(\mathbf{r})d\mathbf{r}, \tag{4.1}$$

where $s_S(\mathbf{r})$ is the local Shannon entropy density. Shannon entropy quantifies the uncertainty or delocalization of the electron density distribution. It becomes maximal when electrons are uniformly distributed. Historically, Shannon entropy was applied to correlate with electron correlation, charge distribution, and other properties. From the functional perspective, Shannon entropy is a simple density functional in local density approximation (LDA), because the integrand in Eq. (4.1) depends only on the electron density.

The second ITA quantity widely examined in the literature is Fisher information.

$$I_F = \int i_F(\mathbf{r})d\mathbf{r} = \int \frac{|\nabla\rho(\mathbf{r})|^2}{\rho(\mathbf{r})}d\mathbf{r}, \tag{4.2}$$

where $i_F(\mathbf{r})$ is the Fisher information density, and $\nabla\rho(\mathbf{r})$ is the density gradient. This quantity quantifies the "sharpness," localization, and heterogeneity of the electron density by measuring how rapidly $\rho(\mathbf{r})$ changes in space. For a uniform electron gas, Fisher information vanishes. For this reason, it is no surprise that Fisher information can be employed to quantify the steric effect [37–39], a widely used concept in chemistry. From the functional perspective, Fisher information is an explicit density functional in the general gradient approximation (GGA) framework, because the integrand in Eq. (4.2) is a function of both the electron density and its gradient. Moreover, Eq. (4.2) is homogeneous of degree 2 in coordinate scaling and homogeneous of degree 1 in density scaling [10], suggesting that it has the same dimensionality as the kinetic energy and is size extensive. In addition, Fisher information is closely related to the Weizsäcker kinetic energy $T_W[\rho]$ [40], which differs from I_F by a factor of 1/8,

$$T_W[\rho] = \frac{1}{8}\int \frac{|\nabla\rho(\mathbf{r})|^2}{\rho(\mathbf{r})}d\mathbf{r} = \frac{1}{8}I_F. \tag{4.3}$$

Earlier, with the fact that the electron density decays rapidly at infinity, we proved that by integration by parts, an equivalent form of Fisher information, called *alternative Fisher information* I'_F, can be obtained [12].

$$I'_F = \int i'_F(\mathbf{r})d\mathbf{r} = -\int \nabla^2\rho(\mathbf{r})\ln\rho(\mathbf{r})d\mathbf{r}, \tag{4.4}$$

where $i'_F(\mathbf{r})$ is the alternative Fisher information density and $\nabla^2\rho(\mathbf{r})$ is the Laplacian. The quantity of Eqs. (4.2) and (4.4) should yield the same value, but their local behaviors are drastically different. Shown in Figure 4.1 as illustrative examples are two ITA quantities, Shannon entropy and Fisher information, for the benzene molecule. For Shannon entropy, depleted regions are apparent near nuclear cusps. For Fisher information, we observe spikes near the nuclei, but those spikes are much larger for carbon atoms and significantly smaller for hydrogen atoms, reflecting the heterogeneity of the electron density distribution between carbon and hydrogen atoms.

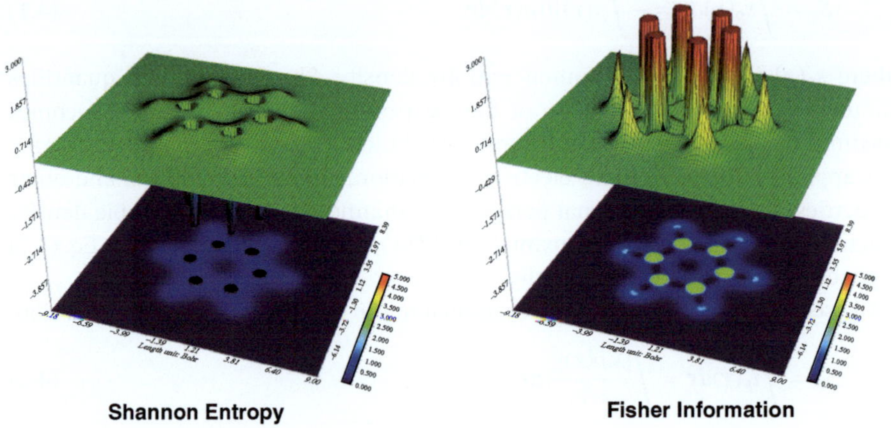

Figure 4.1 Shannon entropy and Fisher information for benzene.

The next absolute ITA quantity is the Ghosh–Berkowitz–Parr entropy S_{GBP} [41, 42],

$$S_{\text{GBP}} = \frac{3}{2}k\int \rho(\mathbf{r})\left[c + \ln\frac{t(\mathbf{r};\rho)}{t_{TF}(\mathbf{r};\rho)}\right]d\mathbf{r}, \tag{4.5}$$

where k is the Boltzmann constant, c is another constant, $t(\mathbf{r};\rho)$ is the total kinetic energy density, and $t_{TF}(\mathbf{r};\rho) = c_K\rho^{5/3}(\mathbf{r})$ is the Thomas–Fermi kinetic energy density. The GBP entropy was introduced by Parr and coworkers [41] to transcribe DFT into a local version of thermodynamics using a phase-space distribution function $f(\mathbf{r}, \mathbf{p})$,

$$f(\mathbf{r}, \mathbf{p}) = \left[\frac{\beta(\mathbf{r})}{2\pi}\right]^{3/2}\rho(\mathbf{r})\exp\left[-\frac{1}{2}p^2\beta(\mathbf{r})\right], \tag{4.6}$$

with

$$\beta(\mathbf{r}) = \frac{1}{kT(\mathbf{r})}, \tag{4.7}$$

and the local temperature $T(\mathbf{r})$ [43–45] defined as

$$T(\mathbf{r}) = \frac{2t(\mathbf{r};p)}{3k\rho(\mathbf{r})}. \tag{4.8}$$

From the functional perspective, GBP entropy is under the meta-GGA framework, because the integrand in Eq. (4.5) is a function of the total kinetic energy density $t(\mathbf{r};\rho)$. Shown in Figure 4.2 is the local distribution of GBP temperature, Eq. (4.8), and GBP entropy, Eq. (4.5), for the benzene molecule.

Other absolute ITA quantities include Rényi entropy [46] of order n,

$$R_n = \frac{1}{1-n}\ln\left[\int \rho^n(\mathbf{r})d\mathbf{r}\right], \tag{4.9}$$

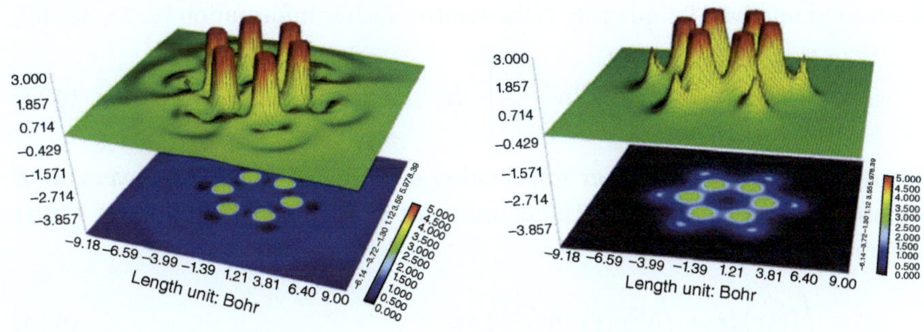

(a) GBP local temperature **(b) GBP local entropy**

Figure 4.2 Ghosh–Berkowitz–Parr temperature and Ghosh–Berkowitz–Parr entropy for benzene.

with the condition that $n \geq 0$ and $n \neq 1$. When n is closely approaching the value of 1, Rényi entropy is reduced to Shannon entropy. Tsallis entropy [46] of order n is defined as

$$T_n = \frac{1}{n-1}\ln\left[1 - \int \rho^n(\mathbf{r})d\mathbf{r}\right]. \tag{4.10}$$

It is a generalization of the standard Boltzmann–Gibbs entropy. The Onicescu information energy of order n [46] is formulated as

$$E_n = \frac{1}{n-1}\int \rho^n(\mathbf{r})d\mathbf{r}. \tag{4.11}$$

In all these cases, we have $n \geq 0$ and $n \neq 1$. These last three quantities are dependent on the electron density only, so they are LDA density functionals.

4.2.2 Relative ITA Quantities

The second category of ITA quantities is the relative one. In this category, a reference electron density $\rho_0(\mathbf{r})$, which satisfies the same normalization condition as $\rho(\mathbf{r})$, is used to define ITA quantities. The reference density can be the promolecular density, that is, the combination of constituting atoms of a molecule in the isolated state, or the density of the same molecule in a different conformation, or the reactant density when analyzing a transition state. The first relative ITA quantity is the information gain I_G, also called relative Shannon entropy, information divergence, or Kullback–Leibler divergence [15, 28],

$$I_G = \int i_G(\mathbf{r})d\mathbf{r} = \int \rho(\mathbf{r})\ln\frac{\rho(\mathbf{r})}{\rho_0(\mathbf{r})}d\mathbf{r}, \tag{4.12}$$

where $i_G(\mathbf{r})$ is the information gain density. It is the relative version of Shannon entropy in Eq. (4.1). Information gain is a quantitative measure of how one probability distribution diverges from the reference distribution, so it can find tremendous applications in determining reactivity differences for atoms in molecules.

The second relative ITA quantity is the relative Fisher information [8, 25, 47, 48],

$$ {}_{F}^{r}I \equiv \int {}_{F}^{r}i(\mathbf{r})d\mathbf{r} = \int \rho(\mathbf{r}) \left[\nabla \ln \frac{\rho(\mathbf{r})}{\rho_0(\mathbf{r})} \right]^2 d\mathbf{r}, \tag{4.13} $$

where ${}_{F}^{r}i(\boldsymbol{r})$ is the relative Fisher information density. It is the relative version of Fisher information in Eq. (4.2). The counterpart of the relative alternative Fisher information, Eq. (4.4), is as follows,

$$ {}_{F}^{r}I' \equiv \int {}_{F}^{r}i'(\mathbf{r})d\mathbf{r} = \int \nabla^2 \rho(\mathbf{r}) \ln \frac{\rho(\mathbf{r})}{\rho_0(\mathbf{r})} d\mathbf{r}, \tag{4.14} $$

with ${}_{F}^{r}i'(\mathbf{r})$ as the relative alternative Fisher information density. These two forms of relative Fisher information were recently proposed in the literature [47], so only a limited number of studies were dedicated to their applications.

The other relative ITA quantity is the relative Rényi entropy of order n [49],

$$ R_n^r = \frac{1}{n-1} \ln \left[\int \frac{\rho^n(\mathbf{r})}{\rho_0^{n-1}(\mathbf{r})} d\mathbf{r} \right], \tag{4.15} $$

which is the relative ITA version of Rényi entropy in Eq. (4.9). One can similarly define the relative counterpart for Tsallis entropy and Onicescu information energy of order n, respectively, in Eqs. (4.10) and (4.11).

4.3 Four ITA Representations

4.3.1 Shape-function Representation

Using the electron density $\rho(\mathbf{r})$ as the probability function in ITA, we obtained the first representation of ITA, as shown in Eqs. (4.1)–(4.15). There are three other representations of ITA. One is through the shape function $\sigma(\mathbf{r})$, which is related to the electron density $\rho(\mathbf{r})$ and the total number of electrons N through the following relationship,

$$ \rho(\mathbf{r}) = N\,\sigma(\mathbf{r}) \tag{4.16} $$

with the normalization condition to unit

$$ \int \sigma(\mathbf{r})d\mathbf{r} = 1. \tag{4.17} $$

Replacing $\rho(\mathbf{r})$ with $\sigma(\mathbf{r})$ in Eqs. (4.1)–(4.15), one obtains the set of ITA quantities in terms of the shape function. In this representation [31], we denote each ITA quantity by appending a superscript σ. For example, for Shannon entropy and two forms of Fisher information, one has

$$S_S^\sigma = -\int \sigma(\mathbf{r}) \ln \sigma(\mathbf{r}) d\mathbf{r}, \tag{4.18}$$

$$I_F^\sigma = \int \frac{|\nabla \sigma(\mathbf{r})|^2}{\sigma(\mathbf{r})} d\mathbf{r}, \tag{4.19}$$

and

$$I'^\sigma_F = -\int \nabla^2 \sigma(\mathbf{r}) \ln \sigma(\mathbf{r}) d\mathbf{r}, \tag{4.20}$$

respectively. For other absolute ITA quantities using the shape function representation, we can define them accordingly. We can also employ a reference shape function, $\sigma_0(\mathbf{r})$, to define the relative version of ITA quantities in this representation. For instance, the relative version of Eqs. (4.18)–(4.20) is the following:

$$I_G^\sigma = \int \sigma(\mathbf{r}) \ln \frac{\sigma(\mathbf{r})}{\sigma_0(\mathbf{r})} d\mathbf{r}, \tag{4.21}$$

$${}_F^rI^\sigma = \int \sigma(\mathbf{r}) \left[\nabla \ln \frac{\sigma(\mathbf{r})}{\sigma_0(\mathbf{r})} \right]^2 d\mathbf{r}, \tag{4.22}$$

and

$${}_F^rI'^\sigma = \int \nabla^2 \sigma(\mathbf{r}) \ln \frac{\sigma(\mathbf{r})}{\sigma_0(\mathbf{r})} d\mathbf{r}. \tag{4.23}$$

Since Eq. (4.16) connects the electron density with the shape function, the relationship of ITA quantities in these two representations can be determined explicitly [31]. For example,

$$S_S^\sigma = \frac{S_S}{N} + \ln N, \tag{4.24}$$

$$I_F^\sigma = \frac{I_F}{N}, \tag{4.25}$$

and

$$I_G^\sigma = \frac{I_G}{N}. \tag{4.26}$$

4.3.2 Atoms-in-molecules Representation

The third representation of ITA formulation is through the partition of atoms in molecules, where the total electron population N of the system is the summation of electron density in each atomic contribution, N_A,

$$N = \sum_A N_A = \sum_A \int_{\Omega_A} \rho_A(\mathbf{r}) d\mathbf{r}. \tag{4.27}$$

Three schemes to partition atoms in molecules are available in the literature: Becke's fuzzy atom approach [50], Bader's zero-flux atoms-in-molecules (AIM) approach [51], and Hirshfeld's stockholder approach [52]. Within this representation, for instance, Shannon entropy and information gain can be, respectively, rewritten as

$$S_S = -\sum_A \int_{\Omega_A} \rho_A(\mathbf{r}) \ln \rho_A(\mathbf{r}) d\mathbf{r}, \tag{4.28}$$

$$I_G = \sum_A \int_{\Omega_A} \rho_A(\mathbf{r}) \ln \frac{\rho_A(\mathbf{r})}{\rho_A^0(\mathbf{r})} d\mathbf{r}. \tag{4.29}$$

For other absolute and relative ITA quantities, we can formulate them accordingly within this AIM representation.

4.3.3 Finite-difference Representation

The fourth representation introduced recently by us [53] is the *finite difference representation*. In this representation, we consider the change of ITA quantities with respect to the change of the total number of electrons, following what we have examined in conceptual density functional theory (CDFT). Using Shannon entropy density $s_s(\mathbf{r})$ as an illustrative example, we take the partial derivative with respect to the total number of electrons N with the external field $v(\mathbf{r})$ held fixed,

$$\left(\frac{\partial s_s(\mathbf{r})}{\partial N}\right)_v = -\left(\frac{\partial \rho(\mathbf{r})}{\partial N}\right)_v \ln \rho(\mathbf{r}) - \left(\frac{\partial \rho(\mathbf{r})}{\partial N}\right)_v = -f(\mathbf{r})\ [\ln \rho(\mathbf{r}) + 1], \tag{4.30}$$

where the Fukui function is defined as

$$f(\mathbf{r}) = \left(\frac{\partial \rho(\mathbf{r})}{\partial N}\right)_v. \tag{4.31}$$

Integrating both sides of Eq. (4.30) results in

$$\left(\frac{\partial S_S}{\partial N}\right)_v = \int \left(\frac{\partial s_s(\mathbf{r})}{\partial N}\right)_v d\mathbf{r} = -\int f(\mathbf{r})\ \left[\ln \rho(\mathbf{r}) + 1\right] d\mathbf{r}. \tag{4.32}$$

Equation (4.32) shows that the change of Shannon entropy with respect to the number of electrons should be closely related to the Fukui function. Since the Fukui function has two distinct forms under the *finite difference approximation*, the nucleophilic attack $f^+(\mathbf{r})$ and electrophilic attack $f^-(\mathbf{r})$, which can be condensed to the atomic level, f_A^+ and f_A^-, in the same spirit, for Shannon entropy under the finite difference approximation, we have

$$s_S^+(\mathbf{r}) = s_S^{N+1}(\mathbf{r}) - s_S^N(\mathbf{r}), \tag{4.33}$$

$$s_S^-(\mathbf{r}) = s_S^N(\mathbf{r}) - s_S^{N-1}(\mathbf{r}), \tag{4.34}$$

where $s_S^{N+1}(\mathbf{r})$, $s_S^N(\mathbf{r})$, and $s_S^{N-1}(\mathbf{r})$ are Shannon entropy density $s_S(\mathbf{r})$ of the molecular system with $N + 1$, N, and $N - 1$ electrons, respectively. The corresponding condensed-to-atom form for atoms A in a molecule is

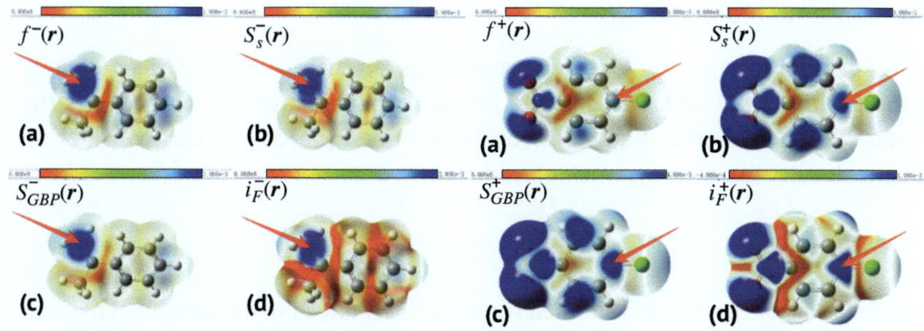

Figure 4.3 Comparison of (left) electrophilic (minus) and (right) nucleophilic (plus) attack propensities with (a) Fukui function, (b) Shannon entropy, (c) GBP entropy, and (d) Fisher information under the finite difference representation for (left) 2-propenylbenzene and (right) *p*-chloro-nitrobenzene. *Source*: Reproduced from Ref. [53] with permission of Springer Nature.

$$S_{S,A}^{+} = S_{S,N+1}^{A} - S_{S,N}^{A}, \tag{4.35}$$

$$S_{S,A}^{-} = S_{S,N}^{A} - S_{S,N-1}^{A}\,. \tag{4.36}$$

where $S_{S,N+1}^{A}$, $S_{S,N}^{A}$, and $S_{S,N-1}^{A}$ are Shannon entropy of Atom A of the molecule with $N+1$, N, and $N-1$ electrons, respectively. For other ITA quantities, their formulas in this representation can be defined accordingly.

Shown in Figure 4.3 is the comparison of ITA quantities with Fukui function for two molecules, 2-propenylbenzene (left) and *p*-chloro-nitrobenzene (right), to predict their electrophilic and nucleophilic regioselectivity. Qualitative agreement among these quantities was observed.

Notice that using the external potential $v(\mathbf{r})$ as the variable with the number of electrons N fixed, we can additionally formulate the change of ITA quantities with respect to the change of $v(\mathbf{r})$. For instance, for Shannon entropy,

$$\left(\frac{\partial S_S}{\partial v(\mathbf{r})}\right)_N = \int \left(\frac{\delta s_S(\mathbf{r}')}{\delta v(\mathbf{r})}\right)_N d\mathbf{r}' = -\int \varpi(\mathbf{r},\mathbf{r}')\,[\ln \rho(\mathbf{r}') + 1]d\mathbf{r}'. \tag{4.37}$$

where $\varpi(\mathbf{r},\mathbf{r}')$ is the response function in CDFT,

$$\varpi(\mathbf{r},\mathbf{r}') = \left(\frac{\delta \rho(\mathbf{r}')}{\delta v(\mathbf{r})}\right)_N = \frac{s(\mathbf{r})s(\mathbf{r}')}{S} - s(\mathbf{r},\mathbf{r}'). \tag{4.38}$$

The last equality is the Berkowitz–Parr relation [54] in CDFT using softness S, local softness $s(\mathbf{r})$, and softness kernel $s(\mathbf{r},\mathbf{r}')$. For other ITA quantities, we can derive the formulations in the same spirit. These relationships in the finite difference representation link the ITA framework with that of CDFT. More work is needed in the future to examine the behavior of this quantity and others.

4.3.4 Other Representations

Finally, in addition to the four representations discussed earlier, other density functions can also be introduced as the probability distribution in ITA. One is the energetic density function that we recently examined [55–57]. Let us consider an energetic density function $e(\mathbf{r})$, which yields

$$E = \int e(\mathbf{r})\ d\mathbf{r}. \tag{4.39}$$

where E stands for the energy obtained from the energetic density function. Dividing both sides of the above equation by E, we have

$$\int \frac{e(\mathbf{r})}{\mathbf{E}} d\mathbf{r} \equiv \int \varepsilon(\mathbf{r}) d\mathbf{r} = 1, \tag{4.40}$$

where $\varepsilon(\mathbf{r}) \equiv e(\mathbf{r})/E$ is the scaled energetic density. This equation shows that $\varepsilon(\mathbf{r})$ is normalized to a unit, so it can be regarded as a probability distribution function.

Recently, we also considered extending the ITA methodology to two-electron densities [16, 35]. This advancement broadens the scope of the ITA framework, facilitating the transition from individual ITA quantities to joint, conditional, and mutual ITA quantities. The pair-density is defined as

$$\Gamma(\mathbf{r}_1, \mathbf{r}_2) = N(N-1) \iint \cdots \int |\Psi(\mathbf{r}_1, \mathbf{r}_2, \cdots, \mathbf{r}_N)|^2 d\mathbf{r}_3\, d\mathbf{r}_4 \cdots d\mathbf{r}_N, \tag{4.41}$$

where Ψ is the system wavefunction and $\mathbf{r}$ is the space-spin coordinate. Like what we did in Eq. (4.40), we can introduce the unit-normalized pair density $\Gamma^\sigma(\mathbf{r}_1,\mathbf{r}_2)$,

$$\Gamma^\sigma(\mathbf{r}_1, \mathbf{r}_2) = \Gamma(\mathbf{r}_1, \mathbf{r}_2)/N(N-1). \tag{4.42}$$

with

$$\iint \Gamma^\sigma(\mathbf{r}_1, \mathbf{r}_2) d\mathbf{r}_1\, d\mathbf{r}_2 = 1. \tag{4.43}$$

In Chapter 6, we will explore these two additional options and discuss their properties and possible applications.

4.4 Three ITA Principles

The first ITA principle is the "principle of extreme physical information," as proposed by Nagy [8]. This principle states that the "physical information" K of a system, defined as the difference between the Fisher information I_F and a set of constraints J, must be extremized:

$$K = I_F - J = \text{extreme} \tag{4.44}$$

In this context, I_F is the Fisher information and J represents the specific constraints imposed on the system. According to Nagy, there are three constraints:

- The system's total wavefunction must be antisymmetric. (This requirement is shown by Nagy and others to produce the Pauli potential.)

- The electron density of the noninteracting reference system must equal that of the interacting system (mirroring how the Kohn–Sham potential is determined).
- The total electron density is normalized to N, which introduces the Lagrange multiplier μ, the chemical potential.

By minimizing the Fisher information subject to these constraints, one recovers the same Euler equation for orbital-free DFT (OF-DFT) [58–61], which is the framework to be introduced in the next chapter. Essentially, the principle of extreme physical information implies that, instead of relying on electronic energy, one can use information alone to determine all system properties. This idea is further supported by information functional theory (IFT) [62], introduced in Section 4.6.

The second principle in ITA—the *minimum information gain principle* by Nalewajski and Parr [28]—addresses how the formation of a molecule from its constituent atoms affects information content. They minimized the relative Shannon entropy in an AIM representation, Eq. (4.29), under the constraint that the atomic densities remain normalized to the total electron number N,

$$\int \rho(\mathbf{r})\, d\mathbf{r} = \sum_A \int \rho_A(\mathbf{r}) d\mathbf{r} = \sum_A \int \rho_A^0\, (\mathbf{r}) d\mathbf{r} = N. \tag{4.45}$$

This leads to the following equation:

$$\delta \left\{ I_G - \lambda \left[\sum_A \int \rho_A(\mathbf{r}) d\mathbf{r} - N \right] \right\} = 0, \tag{4.46}$$

where λ is the Lagrange multiplier. Functional differentiation of Eq. (4.46) with respect to all ρ_A yields

$$\rho_A(\mathbf{r}) = \frac{\rho_A^0(\mathbf{r})}{\sum_A \rho_A^0(\mathbf{r})} \rho(\mathbf{r}). \tag{4.47}$$

This result is the well-known "stockholder partition" proposed by Hirshfeld [52] and implies that, under the Hirshfeld partition, the formation of a molecule results in minimal information change—the atoms retain as much of their free-atom character (including electrophilic and nucleophilic properties) as possible. Because atomic charges are not directly observable, they can be defined in various ways; this work is notable for deriving atomic charges from a fundamental physicochemical principle.

The third principle, the *information conservation principle* [63, 64], extends the idea further. Since the *minimum information gain principle* implies that atoms in a molecule should closely resemble their reference states, the difference between the molecular atomic density ρ_A and its reference ρ_A^0 can be treated as a small perturbation. If we define

$$x = \frac{\rho_A - \rho_A^0}{\rho_A}, \tag{4.48}$$

then the overall information gain is expressed as

$$I_G = \sum_A \int \rho_A \ln\left(\frac{1}{1-x}\right) d\mathbf{r}. \tag{4.49}$$

Assuming that x is small and approximating $\ln\frac{1}{1-x} \approx x$, we find

$$I_G \approx \sum_A \int (\rho_A - \rho_A^0) d\mathbf{r} = -\sum_A q_A \tag{4.50}$$

where q_A is the Hirshfeld charge on atom (or group) A. This result shows that under the first-order approximation, the information gain simply gives rise to the Hirshfeld charge distribution. Meanwhile, since ρ_A and ρ_A^0 satisfy the same normalization condition, Eq. (4.45), the total information gain in Eq. (4.50) must vanish,

$$I_G \approx -\sum_A q_A \equiv 0 \tag{4.51}$$

suggesting that under the first-order approximation, *the information before and after a system is formed should be conserved*. We call this result the *information conservation principle*. This principle, stemming from the above first-order approximation, is a special case of the *minimum information gain principle*. Any net information gain would therefore come from higher-order terms in the Taylor expansion of $\ln\frac{1}{1-x}$, such as $x^2/2$, $x^3/3$.

To verify the validity of Eq. (4.50), the comparison of local behaviors of information gain with the deformation density for nitrobenzene is shown in Figure 4.4. No

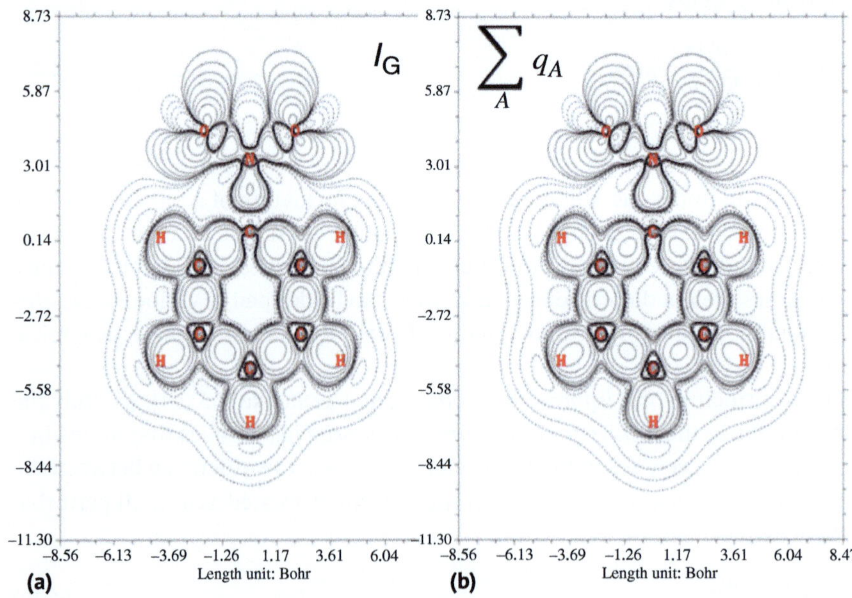

Figure 4.4 Comparison of local behaviors between (a) information gain and (b) deformation density for the nitrobenzene molecule. *Source*: Reproduced from Ref. [63] with permission of American Chemical Society.

discernible difference between these two plots is observed, suggesting that the first-order approximation for information gain should be adequate and accurate.

4.5 Two Identities Among ITA Quantities

In conventional information theory, quantities such as Shannon entropy and Fisher information measure completely different propensities of the probability distribution. For instance, Shannon entropy is a measure of homogeneity, whereas Fisher information gauges the heterogeneity of the distribution function. In ITA, however, the two basic theorems of DFT dictate that the ground-state electron density contains all the necessary information to determine all properties of the system. This means that ITA quantities contain redundant information, so ITA quantities should be closely related.

Two identities among ITA quantities have been theoretically proved and numerically verified in the literature [12, 65]. The first one is between the density of Shannon entropy and the density of two forms of Fisher information [12],

$$s_S(\mathbf{r}) = -\rho(\mathbf{r}) + \frac{1}{4\pi}\int \frac{i_F(\mathbf{r}')}{|\mathbf{r}-\mathbf{r}'|}d\mathbf{r}' - \frac{1}{4\pi}\int \frac{i'_F(\mathbf{r}')}{|\mathbf{r}-\mathbf{r}'|}d\mathbf{r}', \tag{4.52}$$

whose integration on both sides yields

$$S_S = -N + \frac{1}{4\pi}\iint \frac{i_F(\mathbf{r}')}{|\mathbf{r}-\mathbf{r}'|}d\mathbf{r}d\mathbf{r}' - \frac{1}{4\pi}\iint \frac{i'_F(\mathbf{r}')}{|\mathbf{r}-\mathbf{r}'|}d\mathbf{r}d\mathbf{r}'. \tag{4.53}$$

The other is between the density of information gain and the density of two relative forms of Fisher information [65],

$$i_G(\mathbf{r}) = -\frac{1}{4\pi}\int \frac{{}^{r}_{F}i(\mathbf{r}')}{|\mathbf{r}-\mathbf{r}'|}d\mathbf{r}' - \frac{1}{4\pi}\int \frac{{}^{r}_{F}i'(\mathbf{r}')}{|\mathbf{r}-\mathbf{r}'|}d\mathbf{r} - \frac{1}{4\pi}\int \frac{g_2(\mathbf{r}')}{|\mathbf{r}-\mathbf{r}'|}d\mathbf{r}', \tag{4.54}$$

where

$$g_2(\mathbf{r}') = \rho(\mathbf{r}')\left[\frac{\nabla^2\rho(\mathbf{r}')}{\rho(\mathbf{r}')} - \frac{\nabla^2\rho_0(\mathbf{r}')}{\rho_0(\mathbf{r}')}\right]. \tag{4.55}$$

The term $g_2(\mathbf{r})$ represents the difference in local curvature between the molecular density and the reference density. Integrating both sides of Eq. (4.54) gives

$$I_G = -\frac{1}{4\pi}\iint \frac{{}^{r}_{F}i(\mathbf{r}')}{|\mathbf{r}-\mathbf{r}'|}d\mathbf{r}d\mathbf{r}' - \frac{1}{4\pi}\iint \frac{{}^{r}_{F}i'(\mathbf{r}')}{|\mathbf{r}-\mathbf{r}'|}d\mathbf{r}d\mathbf{r}' - \frac{1}{4\pi}\iint \frac{g_2(\mathbf{r}')}{|\mathbf{r}-\mathbf{r}'|}d\mathbf{r}d\mathbf{r}'. \tag{4.56}$$

These identities suggest that ITA quantities are interrelated. When one ITA quantity is found to strongly correlate with a physicochemical property, others can serve the same purpose as well.

4.6 Information Functional Theory

In Section 4.4, Nagy's *principle of extreme physical information* [8] demonstrates that by starting with Fisher information and superimposing a few constraints, one can derive the Euler equation in OF-DFT—a rigorous and viable pathway for determining all ground-state physicochemical properties. In two other separate studies, Nagy [66, 67] showed that the Shannon entropy density alone encapsulates the necessary information to fully describe a Coulombic system. Although her theoretical proof is robust, it poses significant practical challenges. To overcome these, we put forward the IFT [62], which harnesses all ITA quantities to effectively evaluate a range of electronic properties, including total energy, its individual components, frontier orbital energies, and the HOMO/LUMO gap [68]. In the conventional DFT formulation, the total energy $E[\rho]$ [1]

$$E_v[\rho] = F[\rho] + \int \rho(\mathbf{r}) v_{\text{ext}}(\mathbf{r}) d\mathbf{r}, \tag{4.57}$$

where $v_{\text{ext}}(\mathbf{r})$ is the external potential and $F[\rho]$ is the universal density functional. In IFT, Eq. (4.57) can be reformulated as

$$E_v[\rho] = F[\{\text{ITA}\}] + \int \rho(\mathbf{r}) v_{\text{ext}}(\mathbf{r}) d\mathbf{r}. \tag{4.58}$$

with $\{\text{ITA}\}$ denoting the set of ITA quantities. In other words, the universal density functional can be evaluated by ITA quantities, which themselves are simple yet explicit density functionals.

As an illustrative example, Figure 4.5 shows that using all 50 data points from five polymeric systems, remarkably strong correlations between the frontier orbital properties (HOMO and HOMO/LUMO gap) and ITA quantities can be obtained, confirming the validity of IFT. Since ITA quantities are explicit density functionals, this idea is closely related to the earlier work by Parr and the author of this book to expand unknown density functionals using homogeneous functionals in coordinating or density scaling.

4.7 Applications for ITA Quantities

While early studies applied ITA quantities such as Shannon entropy to correlate with electron correlation energy, the seminal work by Nalewajski and Parr [15, 28] to minimize information gain subject to the constraint of density normalization to obtain the Hirshfeld partition of atoms in molecules was the first to apply ITA quantities to appreciate physicochemical properties for molecules. In 2014, we proposed using the same ITA quantity, information gain, and its first-order approximation, Hirshfeld charge, to quantify both electrophilicity and nucleophilicity [63, 69–72]. We will also discuss this in detail in Chapter 10. At the same time,

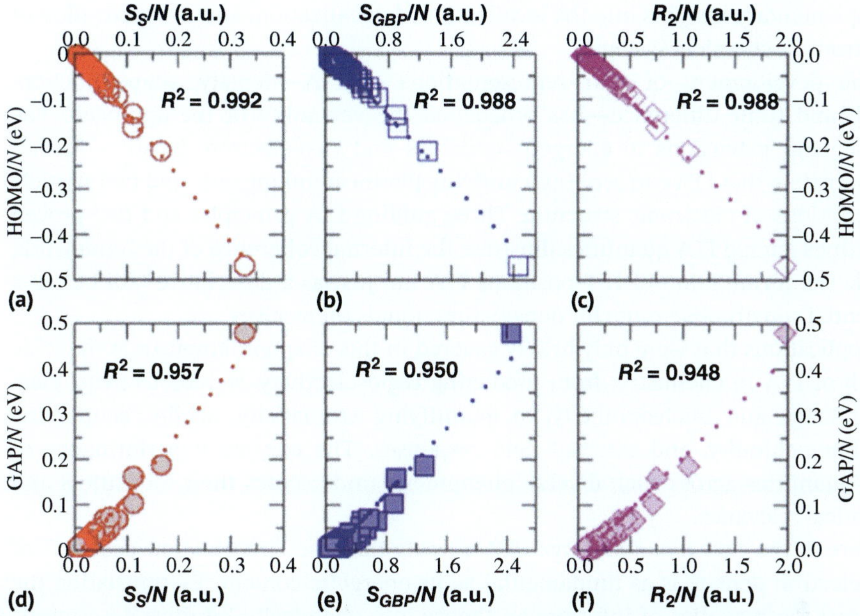

Figure 4.5 Strong linear correlations obtained at the atoms-in-molecules level between frontier orbital properties and ITA quantities with all data points from five polymeric systems put together. *Source*: Reproduced from Ref. [68] with permission of Springer Nature.

Fisher information, which is eight times the Weizsäcker kinetic energy, was previously proposed to quantify steric effect and stereoselectivity [19, 38, 39, 73–78]. We will examine this application in detail in Chapter 11. Besides these, ITA quantities have been applied to examine aromaticity and antiaromaticity [79–85], Brønsted–Lowry and Lewis acidity and basicity [83, 86, 87], conformational stability [88], anomeric effect [74, 89, 90], α-effect [91], cooperativity [32, 92–94], homochirality [95–97], impacts of external electric fields [83, 86, 98], excited states [99], and S_N2 reactions [100].

4.8 Conclusions and Outlook

In this chapter, we have delved into the ITA as a density-based framework for chemical reactivity. By treating electron density as a probability distribution, ITA imports concepts from information theory into DFT and thereby provides a new language to interpret electronic structure and molecular properties. Absolute ITA quantities, such as Shannon entropy, Fisher information, and their generalizations, quantify global and local features of electron density distributions, while relative ITA quantities, such as information gain, capture differences between molecular states or between a molecule and its reference. Together, these measures offer complementary insights into the localization, delocalization, and reorganization of electrons in chemical systems.

The development of four representations of ITA—density, shape-function, AIM, and finite-difference—has broadened the versatility of the approach. The subsequent extensions to energetic densities and two-electron densities further demonstrate that ITA can serve as a unifying platform linking one- and two-particle descriptions of electronic structure. Three guiding ITA principles and two proven identities among ITA quantities illustrate the internal coherence of the framework, while the formulation of IFT positions ITA not just as a descriptive tool but as a potential constructive route for density functionals themselves.

Applications that were only briefly covered in this chapter demonstrate the wide reach of ITA in chemistry: from predicting regioselectivity, stereoselectivity, electrophilicity, and nucleophilicity, to quantifying aromaticity, acidity, cooperativity, homochirality, and external field responses. The consistent performance of ITA quantities across such diverse phenomena underscores their robustness and chemical relevance.

Here is the *take-home message*: ITA reveals that the "information content" of the electron density is as fundamental as its energetic content. By translating the abstract mathematics of information theory into chemically intuitive descriptors, ITA complements traditional DFT analysis and enriches our ability to connect electronic structure with chemical reactivity.

Looking ahead, ITA research faces several exciting opportunities. On the theoretical side, extending ITA to two-particle densities, time-dependent regimes, and excited states promises to capture correlation and dynamics in new ways. On the computational side, the explicit and simple functional forms of ITA quantities make them attractive candidates for data-driven models, suggesting natural synergies with machine learning and quantum computing approaches. On the conceptual side, ITA can serve as a bridge between physics-based density theorems and chemistry-based reactivity principles, reinforcing the pursuit of chemical understanding as the central goal of density-based reactivity theory.

References

1 Parr RG, Yang W. Density-Functional Theory of Atoms and Molecules. Oxford University Press; 1989.

2 Hohenberg P, Kohn W. Inhomogeneous electron gas. Phys Rev. 1964 Nov 9; 136(3B):B864–71.

3 Rong C, Wang B, Zhao D, Liu S. Information-theoretic approach in density functional theory and its recent applications to chemical problems. WIREs Comput Mol Sci. 2020 Jul;10(4):e1461.

4 Liu S. Information-theoretic approach in density functional reactivity theory. Acta Phys-Chim Sin. 2016;32(1):98–118.

5 Shannon CE. A mathematical theory of communication. Bell Syst Techn J. 1948 Jul;27(3):379–423.

6 Fisher RA. Theory of statistical estimation. Math Proc Camb Phil Soc. 1925 Jul;22(05):700.

7 Nagy Á. Fisher information in density functional theory. J Chem Phys. 2003 Nov 8;119(18):9401–5.

8 Nagy Á. Fisher information and density functional theory. Int J Quantum Chem. 2022 Apr 15;122(8).

9 Levy M. Density-functional exchange correlation through coordinate scaling in adiabatic connection and correlation hole. Phys Rev A. 1991 May 1;43(9):4637–46.

10 Liu S, Parr RG. Expansions of the correlation-energy density functional Ec. Phys Rev A. 1996 Apr;53(4):2211–9.

11 Liu S. Expansions of the pair distribution function and the second-order density matrix in terms of homogeneous functionals. Phys Rev A. 1996 Dec;54(6):4863–7.

12 Liu S. On the relationship between densities of Shannon entropy and Fisher information for atoms and molecules. J Chem Phys. 2007 May 21;126(19):191107.

13 Sears SB, Parr RG, Dinur U. On the quantum-mechanical kinetic energy as a measure of the information in a distribution. Isr J Chem. 1980 Jan;19 (1–4):165–73.

14 Heidar-Zadeh F, Ayers PW, Verstraelen T, Vinogradov I, Vöhringer-Martinez E, Bultinck P. Information-theoretic approaches to atoms-in-molecules: Hirshfeld family of partitioning schemes. J Phys Chem A. 2018 May 3;122(17):4219–45.

15 Parr RG, Ayers PW, Nalewajski RF. What is an atom in a molecule? J Phys Chem A. 2005 May 5;109(17):3957–9.

16 Zhao Y, Zhao D, Rong C, Liu S, Ayers PW. Extending the information-theoretic approach from the (one) electron density to the pair density. J Chem Phys. 2025 Jun 28;162(24):244108.

17 Borgoo A, Jaque P, Toro-Labbé A, Van Alsenoy C, Geerlings P. Analyzing Kullback–Leibler information profiles: an indication of their chemical relevance. Phys Chem Chem Phys. 2009 Jan 21;11(3):476–82.

18 Borgoo A, Godefroid M, Sen KD, De Proft F, Geerlings P. Quantum similarity of atoms: a numerical Hartree–Fock and information theory approach. Chem Phys Lett. 2004 Dec;399(4–6):363–7.

19 Esquivel RO, Liu S, Angulo JC, Dehesa JS, Antolín J, Molina-Espíritu M. Fisher information and steric effect: study of the internal rotation barrier of ethane. J Phys Chem A. 2011 May 5;115(17):4406–15.

20 Esquivel RO, Molina-Espíritu M, López-Rosa S. 3D information-theoretic analysis of the simplest hydrogen abstraction reaction. J Phys Chem A. 2023 Aug 3;127(30):6159–74.

21 Esquivel RO, Carrera E. The separability problem in molecular quantum systems: information-theoretic framework for atoms in molecules. ChemPhysChem. 2024 Apr 22; e202400030.

22 Heidar Zadeh F, Fuentealba P, Cárdenas C, Ayers PW. An information-theoretic resolution of the ambiguity in the local hardness. Phys Chem Chem Phys. 2014 Apr 7;16(13):6019–26.

23 Rong C, Yu D, Liu S. Information-theoretic approach. In: Liu S, editor. Conceptual Density Functional Theory: Towards a New Chemical Reactivity Theory. Wiley; 2022. pp. 281–99.

24 He X, Li M, Rong C, Zhao D, Liu W, Ayers PW, et al. Some recent advances in density-based reactivity theory. J Phys Chem A. 2024 Feb 22;128(7):1183–96.

25 Nagy Á. Relative information in excited-state orbital-free density functional theory. Int J Quantum Chem. 2020 Dec 5;120(23):e26405.

26 Nagy Á, Romera E, Liu SB. Local coordinate, wave vector, Fisher and Shannon information in momentum representation. Phys Lett A. 2013 Jan;377 (3–4):286–90.

27 Nagy Á, Liu S. Local wave-vector, Shannon and Fisher information. Phys Lett A. 2008 Mar;372(10):1654–6.

28 Nalewajski RF, Parr RG. Information theory, atoms in molecules, and molecular similarity. Proc Natl Acad Sci USA. 2000 Aug 1;97(16):8879–82.

29 Nalewajski RF. Resultant information description of electronic states and chemical processes †. J Phys Chem A. 2019 Nov 14;123(45):9737–52.

30 Nalewajski RF. Information theoretic approach to fluctuations and electron flows between molecular fragments. J Phys Chem A. 2003 May 1;107(19):3792–802.

31 Rong C, Lu T, Ayers PW, Chattaraj PK, Liu S. Scaling properties of information-theoretic quantities in density functional reactivity theory. Phys Chem Chem Phys. 2015 Feb 21;17(7):4977–88.

32 Rong C, Zhao D, Yu D, Liu S. Quantification and origin of cooperativity: insights from density functional reactivity theory. Phys Chem Chem Phys. 2018 Jul 4;20(26):17990–8.

33 Sen KD, Panos CP, Chatzisavvas KCh, Moustakidis ChC. Net Fisher information measure versus ionization potential and dipole polarizability in atoms. Phys Lett A. 2007 Apr;364(3–4):286–90.

34 Teale AM, Helgaker T, Savin A, Adamo C, Aradi B, Arbuznikov AV, et al. DFT exchange: sharing perspectives on the workhorse of quantum chemistry and materials science. Phys Chem Chem Phys. 2022 Dec 7;24(47):28700–81.

35 Zhao Y, Zhao D, Rong C, Liu S, Ayers PW. Information theory meets quantum chemistry: a review and perspective. Entropy (Basel). 2025 Jun 16;27(6):644.

36 Rong C, Zhao D, He X, Liu S. Development and applications of the density-based theory of chemical reactivity. J Phys Chem Lett. 2022 Dec 8;13(48):11191–200.

37 Liu S. Steric effect: a quantitative description from density functional theory. J Chem Phys. 2007 Jun 28;126(24):244103.

38 Tsirelson VG, Stash AI, Liu S. Quantifying steric effect with experimental electron density. J Chem Phys. 2010 Sep 21;133(11):114110.

39 Liu S, Liu L, Yu D, Rong C, Lu T. Steric charge. Phys Chem Chem Phys. 2018 Jan 17;20(3):1408–20.

40 Weizsäcker CF. Zur Theorie der Kernmassen Z Phys. 1935;96:431–58; Bethe HA, Bacher RF. Nuclear physics A. Stationary states of nuclei. Rev Mod Phys. 1935;8(2):82.

41 Ghosh SK, Berkowitz M, Parr RG. Transcription of ground-state density-functional theory into a local thermodynamics. Proc Natl Acad Sci USA. 1984 Dec;81(24):8028–31.

42 Rong C, Lu T, Chattaraj PK, Liu S. On the relationship among Ghosh-Berkowitz-Parr entropy, Shannon entropy and Fisher information. Indian J Chem Sect A-Inorg Bio-Inorg Phys Theor Anal Chem. 2014;53(8–9):970–7.

43 Nagy Á, Parr RG, Liu S. Local temperature in an electronic system. Phys Rev A. 1996 May;53(5):3117–21.

44 Ayers PW, Parr RG, Nagy A. Local kinetic energy and local temperature in the density-functional theory of electronic structure. Int J Quantum Chem. 2002;90(1):309–26.

45 Guo C, He X, Rong C, Lu T, Liu S, Chattaraj PK. Local temperature as a chemical reactivity descriptor. J Phys Chem Lett. 2021 Jun 17;12(23):5623–30.

46 Liu S-B, Rong C-Y, Wu Z-M, Lu T. Rényi entropy, Tsallis entropy and Onicescu information energy in density functional reactivity theory. Acta Phys-Chim Sin. 2015;31(11):2057–63.

47 Wang B, Zhao D, Lu T, Liu S, Rong C. Quantifications and applications of relative Fisher information in density functional theory. J Phys Chem A. 2021 May 6;125(17):3802–11.

48 Zhao Y, Zhao D, Liu S, Rong C, Ayers PW. Why are information-theoretic descriptors powerful predictors of atomic and molecular polarizabilities. J Mol Model. 2024 Oct 3;30(11):361.

49 Nagy Á, Romera E. Relative Rényi entropy for atoms. Int J Quantum Chem. 2009 Jan;109(11):2490–4.

50 Becke AD. A multicenter numerical integration scheme for polyatomic molecules. J Chem Phys. 1988;88(4):2547.

51 Bader RFW. Atoms in Molecules: A Quantum Theory. Oxford University Press; 1990.

52 Hirshfeld FL. Bonded-atom fragments for describing molecular charge densities. Theor Chim Acta. 1977;44(2):129–38.

53 Guo C, Li M, Rong C, Liu S. Finite difference representation of information-theoretic approach in density functional theory. Theor Chem Acc. 2023 Sep;142(9):83.

54 Berkowitz M, Parr RG. Molecular hardness and softness, local hardness and softness, hardness and softness kernels, and relations among these quantities. J Chem Phys. 1988 Feb 15;88(4):2554–7.

55 Liu S. Quantifying energetic information in density functional theory. J Chem Phys. 2022 Sep 14;157(10):101103.

56 He X, Lu T, Rong C, Liu W, Ayers PW, Liu S. Energetic information from information-theoretic approach in density functional theory as quantitative measures of physicochemical properties. J Chem Theory Comput. 2024 Jul 12;20(14):6049–61.

57 He X, Liu W, Liu S. Using energetic information quantities from density functional theory to simultaneously identify both covalent and noncovalent interactions. ChemPhysChem. 2025 Apr 1;26(7):e202401068.
58 Majumdar S, Shi Z, Vignale G. Orbital-free density functional theory for periodic solids: construction of the Pauli potential. J Chem Theory Comput. 2025 May 30;21(12):6007–22.
59 Ryley MS, Withnall M, Irons TJP, Helgaker T, Teale AM. Robust all-electron optimization in orbital-free density-functional theory using the trust-region image method. J Phys Chem A. 2021 Jan 14;125(1):459–75.
60 Mi W, Luo K, Trickey SB, Pavanello M. Orbital-free density functional theory: an attractive electronic structure method for large-scale first-principles simulations. Chem Rev. 2023 Nov 8;123(21):12039–104.
61 Xu Q, Ma C, Mi W, Wang Y, Ma Y. Recent advancements and challenges in orbital-free density functional theory. WIREs Comput Mol Sci. 2024 May;14(3):e1724.
62 Zhou X-Y, Rong C, Lu T, Zhou P, Liu S. Information functional theory: electronic properties as functionals of information for atoms and molecules. J Phys Chem A. 2016 May 26;120(20):3634–42.
63 Liu S, Rong C, Lu T. Information conservation principle determines electrophilicity, nucleophilicity, and regioselectivity. J Phys Chem A. 2014 May 22;118(20):3698–704.
64 Fu J, Li M, Rong C, Zhao D, Liu S. Information-theoretic quantities as effective descriptors of electrophilicity and nucleophilicity in density functional theory. J Mol Model. 2024 Oct;30:341.
65 Liu S. Identity for Kullback–Leibler divergence in density functional reactivity theory. J Chem Phys. 2019 Oct 14;151(14):141103.
66 Nagy Á. Information theoretical and thermodynamic view of the excited-state density functional theory of Coulomb systems. J Chem Phys. 2020 Oct 21;153(15):154103.
67 Nagy Á. Shannon entropy density as a descriptor of Coulomb systems. Chem Phys Lett. 2013 Jan;556:355–8.
68 Huang Y, Rong C, Zhang R, Liu S. Evaluating frontier orbital energy and HOMO/LUMO gap with descriptors from density functional reactivity theory. J Mol Model. 2017 Jan;23(1):3.
69 Zhou X, Rong C, Lu T, Liu S. Hirshfeld charge as a quantitative measure of electrophilicity and nucleophilicity: nitrogen-containing systems. Acta Phys-Chim Sin. 2014;30(11):2055–62.
70 Wang B, Rong C, Chattaraj PK, Liu S. A comparative study to predict regioselectivity, electrophilicity and nucleophilicity with Fukui function and Hirshfeld charge. Theor Chem Acc. 2019 Dec;138(12):124.
71 Liu S. Where does the electron go? The nature of ortho/para and meta group directing in electrophilic aromatic substitution. J Chem Phys. 2014 Nov 21;141(19):194109.
72 Liu S. Quantifying reactivity for electrophilic aromatic substitution reactions with Hirshfeld charge. J Phys Chem A. 2015 Mar 26;119(12):3107–11.

73 Zhang J, He X, Wang B, Rong C, Zhao D, Liu S. Density-based quantification of steric effect: validation by Taft steric parameters from acid-catalyzed hydrolysis of esters. Phys Chem Chem Phys. 2024;26:23447–56.

74 Huang Y, Zhong A-G, Yang Q, Liu S. Origin of anomeric effect: a density functional steric analysis. J Chem Phys. 2011 Feb 28;134(8):084103.

75 Liu S, Hu H, Pedersen LG. Steric, quantum, and electrostatic effects on S(N)2 reaction barriers in gas phase. J Phys Chem A. 2010 May 13;114(18):5913–8.

76 Ess DH, Liu S, De Proft F. Density functional steric analysis of linear and branched alkanes. J Phys Chem A. 2010 Dec 16;114(49):12952–7.

77 Torrent-Sucarrat M, Liu S, De Proft F. Steric effect: partitioning in atomic and functional group contributions. J Phys Chem A. 2009 Apr 16;113(15):3698–702.

78 Wang B, Liu S, Lei M, De Proft F. Steric effect and intrinsic electrophilicity and nucleophilicity from a conceptual density functional theory and information-theoretic approach as quantitative probes of chemical reactions. Chem Eur J. 2024 Jul 3; e202401295.

79 Yu D, Rong C, Lu T, Geerlings P, De Proft F, Alonso M, et al. Switching between Hückel and Möbius aromaticity: a density functional theory and information-theoretic approach study. Phys Chem Chem Phys. 2020 Feb 28;22(8):4715–30.

80 Yu D, Stuyver T, Rong C, Alonso M, Lu T, De Proft F, et al. Global and local aromaticity of acenes from the information-theoretic approach in density functional reactivity theory. Phys Chem Chem Phys. 2019 Aug 21;21(33):18195–210.

81 Yu D, Rong C, Lu T, Chattaraj PK, De Proft F, Liu S. Aromaticity and antiaromaticity of substituted fulvene derivatives: perspectives from the information-theoretic approach in density functional reactivity theory. Phys Chem Chem Phys. 2017 Jul 19;19(28):18635–45.

82 Yu D, Rong C, Lu T, De Proft F, Liu S. Baird's rule in substituted fulvene derivatives: an information-theoretic study on triplet-state aromaticity and antiaromaticity. ACS Omega. 2018 Dec 31;3(12):18370–9.

83 Li M, Wan X, He X, Rong C, Liu S. Impacts of external fields on aromaticity and acidity of benzoic acid: a density functional theory, conceptual density functional theory and information-theoretic approach study. Phys Chem Chem Phys. 2023 Jan 18;25(3):2595–605.

84 He X, Yu D, Wu J, Wang B, Rong C, Chattaraj PK, et al. Towards understanding metal aromaticity in different spin states: a density functional theory and information-theoretic approach analysis. Chem Phys Lett. 2020 Dec;761:138065.

85 Yu DH, Rong CY, Lu T, De Proft F, Liu SB. Aromaticity study of benzene-fused fulvene derivatives using the information-theoretic approach in density functional reactivity theory. Acta Phys-Chim Sin. 2018;34(6):639–49.

86 Li M, Wan X, Rong C, Zhao D, Liu S. Directionality and additivity effects of molecular acidity and aromaticity for substituted benzoic acids under external electric fields. Phys Chem Chem Phys. 2023 Oct 18;25(40):27805–16.

87 Cao X, Rong C, Zhong A, Lu T, Liu S. Molecular acidity: an accurate description with information-theoretic approach in density functional reactivity theory. J Comput Chem. 2018 Jan 15;39(2):117–29.

88 Liu S, Schauer CK. Origin of molecular conformational stability: perspectives from molecular orbital interactions and density functional reactivity theory. J Chem Phys. 2015 Feb 7;142(5):054107.

89 Cao X, Liu S, Rong C, Lu T, Liu S. Is there a generalized anomeric effect? Analyses from energy components and information-theoretic quantities from density functional reactivity theory. Chem Phys Lett. 2017 Nov;687:131–7.

90 Zhou X, Yu D, Rong C, Lu T, Liu S. Anomeric effect revisited: perspective from information-theoretic approach in density functional reactivity theory. Chem Phys Lett. 2017 Sep;684:97–102.

91 Poddar A, Zhao D, Ayers PW, Liu S, Chattaraj PK. What dictates the α-effect in gas-phase SN2 reactions? A density functional theory study. J Phys Chem A. 2025 Feb 20;129(7):1847–55.

92 Rong C, Zhao D, Zhou T, Liu S, Yu D, Liu S. Homogeneous molecular systems are positively cooperative, but charged molecular systems are negatively cooperative. J Phys Chem Lett. 2019 Apr 18;10(8):1716–21.

93 Zhou T, Liu S, Yu D, Zhao D, Rong C, Liu S. On the negative cooperativity of argon clusters containing one lithium cation or fluorine anion. Chem Phys Lett. 2019 Feb;716:192–8.

94 Liu S, Rong C. Quantifying frustrations for molecular complexes with noncovalent interactions. J Phys Chem A. 2021 Jun 10;125(22):4910–7.

95 Liu S. Homochirality originates from the handedness of helices. J Phys Chem Lett. 2020 Oct 15;11(20):8690–6.

96 Li M, He X, Chen J, Wang B, Liu S, Rong C. Density functional theory and information-theoretic approach study on the origin of homochirality in helical structures. J Phys Chem A. 2021 Feb 11;125(5):1269–78.

97 Liu S. Principle of chirality hierarchy in three-blade propeller systems. J Phys Chem Lett. 2021 Sep 16;12(36):8720–5.

98 Chen J, Liu S, Li M, Rong C, Liu S. A density functional theory and information-theoretic approach study of chiral molecules in external electric fields. Chem Phys Lett. 2020 Oct;757:137858.

99 An X, Zhang W, He X, Li M, Rong C, Liu S. Density-based reactivity theory applied to excited states. AAPPS Bull. 2024 Feb 2;34(1):8.

100 Wu Z, Rong C, Lu T, Ayers PW, Liu S. Density functional reactivity theory study of SN2 reactions from the information-theoretic perspective. Phys Chem Chem Phys. 2015 Oct 28;17(40):27052–61.

5

Orbital-free Density Functional Theory

TABLE OF CONTENTS

5.1 Overview

Orbital-free density functional theory (OF-DFT) [1–6] is a computational method in DFT [7] that expresses the total energy *purely* as a functional of the electron density. Unlike Kohn–Sham DFT [8], where the total kinetic energy is expressed in terms of Kohn–Sham orbitals, OF-DFT approximates the kinetic energy using a kinetic energy density functional (KEDF) [9–11]. This approach eliminates the computationally expensive orbital diagonalization step, significantly enhancing efficiency, especially for large-scale systems containing millions of atoms. While the conventional Kohn–Sham scheme suffers from cubic scaling, limiting its applicability to extensive systems, OF-DFT's energy minimization over density typically scales linearly with the system size. This makes it particularly advantageous for studying bulk materials and large-scale solid-state systems in materials science and condensed matter physics.

However, achieving accurate simulation results heavily relies on developing reliable KEDF approximations, which still remains a huge challenge in OF-DFT [12–16]. The first and yet simplest of these is the Thomas–Fermi (TF) functional $T_{TF}[\rho]$ [17, 18],

$$T_{TF}[\rho] = \int \tau_{TF}(\mathbf{r})d\mathbf{r} = c_{TF} \int \rho^{5/3}(\mathbf{r})d\mathbf{r}, \tag{5.1}$$

with $c_{TF} = \frac{3}{10}(3\pi^2)^{2/3}$. This formula is exact for the uniform electron gas, that is, the density gradient $\nabla\rho(\mathbf{r})$ vanishes. For nonuniform electronic systems such as atoms, molecules, and condensed phase materials, gradient corrections such as the von Weizsäcker term, $T_W[\rho]$ [19],

$$T_W[\rho] = \frac{1}{8}\int \frac{|\nabla\rho(\mathbf{r})|^2}{\rho(\mathbf{r})} d\mathbf{r}, \tag{5.2}$$

are commonly incorporated. More advanced *semilocal* and *nonlocal* functionals account for density-gradient and nonlocal effects, offering better approximations to the KEDF. For example, the DePristo–Kress 87 formula [20] is a semilocal form,

$$T_{DK87}[\rho] = c\int \rho^{5/3}(\mathbf{r})F_{DK87}[s(\mathbf{r})]d\mathbf{r}, \tag{5.3}$$

with

$$F_{DK87}\left[s(\mathbf{r})\right] = 1 + 0.00132327s^2(\mathbf{r})\frac{1 + 0.861504s(\mathbf{r})}{1 + 0.44286s^2(\mathbf{r})}, \tag{5.4}$$

and the dimensionless reduced density gradient $s(\mathbf{r})$ defined by

$$s(\mathbf{r}) = \frac{|\nabla\rho(\mathbf{r})|}{\rho^{4/3}(\mathbf{r})}. \tag{5.5}$$

The Wang–Govind–Carter functional [21] and other more recent formulas are nonlocal forms, which take the following form in approximating the total kinetic energy functional $T_S[\rho]$,

$$T_S[\rho] = T_{TF}[\rho] + T_W[\rho] + T_{NL}[\rho] \tag{5.6}$$

with the last nonlocal term being a double integral,

$$T_{NL}[\rho] = \frac{1}{2}\iint \rho^{5/6}(\mathbf{r})w(|\mathbf{r}-\mathbf{r}'|;[\rho])\rho^{5/6}(\mathbf{r}')d\mathbf{r}d\mathbf{r}', \tag{5.7}$$

where $w(|\mathbf{r}-\mathbf{r}'|;[\rho])$ is the dimensionless density-dependent nonlocal kernel. Different kernel forms of $w(|\mathbf{r}-\mathbf{r}'|;[\rho])$ are available in the literature [15, 22, 23], and more forms can be introduced. One condition for it is that it must be dimensionless in coordinate scaling. Also, two exponents of the density function in Eq. (5.7) can be modified, just that their sum must be equal to 5/3, the dimensionality of the kinetic energy.

OF-DFT excels in systems where electron density is relatively smooth and nearly homogeneous, such as simple metals or materials with nearly free-electron behavior [1, 2, 5]. In these cases, approximate KEDFs effectively capture the dominant physical properties. However, its limitations become more evident in systems with localized electrons, including molecules, strongly covalent semiconductors, and transition-metal oxides [4, 24]. OF-DFT struggles with molecular systems because current KEDFs often fail to account for the subtle electron density variations associated with bonding and antibonding regions [11]. As a result, it is generally less accurate than Kohn–Sham DFT for small systems, where orbital-based methods provide more reliable results.

OF-DFT originated in the 1920s with the TF model, which pioneered the description of electronic systems using electron density only, bypassing the need for orbitals to simplify electronic structure calculations. Its accuracy was enhanced later by the von Weizsäcker correction after gradient effects were incorporated, though results were still far from chemical accuracy. Significant progress in the 1980s and 1990s focused on refining KEDFs through gradient corrections and nonlocal terms, leading to advanced formulations. Despite these improvements, developing highly accurate KEDFs remains a central challenge in OF-DFT, restricting its accuracy for complex systems. Recent advancements, including machine learning and high-performance computing [6, 25, 26], have enabled the introduction of more accurate and transferable KEDFs, expanding OF-DFT's applicability to complex materials and large-scale simulations. While challenges persist, OF-DFT remains a vibrant research field, striving to balance computational efficiency with ab initio-level accuracy—a trend underscored by recent reviews in the literature [1, 2].

Our perspective goes beyond using OF-DFT solely as a computational shortcut for large-scale simulations. We view it as a powerful conceptual framework that can reveal deeper insights into chemical reactivity [27, 28]. By applying quantities derived from OF-DFT to different chemical processes, we leverage its strengths to craft innovative density-based descriptors that directly relate electron density-derived quantities to physicochemical properties. This strategy not only fosters a more intuitive grasp of chemical reactivity in DFT language but also lays the groundwork for predictive density-based models that bridge the gap between chemical concepts and practical applications through in silico simulations.

OF-DFT thus occupies a unique position in DFT as an electronic structure theory. On the one hand, it offers a practical computational tool. By replacing orbital-dependent expressions with kinetic energy density functionals, OF-DFT achieves near-linear scaling and enables simulations of materials and biological systems at scales far beyond the reach of conventional Kohn–Sham DFT. On the other hand, OF-DFT also serves as a conceptual framework, because the explicit density-based quantities it employs—such as the Weizsäcker kinetic energy, Pauli energy [29], and their associated potentials, forces, and charges—provide new ways to dissect chemical effects. In this sense, OF-DFT bridges efficiency and insight: its descriptors such as electron localization function (ELF) [30], strong covalent interactions (SCI) [31, 32], and bonding and nonbonding interaction (BNI) [33–35] indices emerge directly from the density and allow one to analyze covalent bonding, noncovalent interactions, steric effects, and reactivity trends on equal footing.

This dual perspective motivates the structure of this chapter. We first lay out the theoretical foundation of OF-DFT and its energy partitioning schemes, then introduce the density-based descriptors derived from Pauli and steric contributions. Finally, we demonstrate how these descriptors clarify the relative roles of electrostatic, steric, and quantum effects in representative chemical processes such as bond rotation barriers and stereoselectivity. By treating OF-DFT simultaneously as a computational simplification and as a source of conceptual descriptors, we aim to show that its true value extends well beyond efficiency: it establishes a viable density-based framework to chemical understanding.

5.2 Theoretical Framework

The theoretical framework of OF-DFT must be understood on two complementary levels. On the one hand, it provides the formal machinery that replaces orbital-dependent expressions with density-only functionals, enabling efficient large-scale simulations without the bottleneck of orbital diagonalization. On the other hand, the same framework exposes uniquely density-based quantities—such as the Weizsäcker kinetic energy, Pauli energy, and their associated potentials and forces—that can be interpreted as conceptual probes of chemical effects. Thus, every equation developed below serves a dual purpose: it advances orbital-free computation while simultaneously supplying descriptors that clarify how steric, electrostatic, and quantum contributions shape chemical reactivity.

5.2.1 Euler–Lagrange Equation of OF-DFT

The variational principle of OF-DFT [36, 37] is the following: the total energy density functional $E[\rho]$ is minimized with respect to the electron density $\rho(\mathbf{r})$ under the constraint of particle conservation,

$$\frac{\delta}{\delta\rho(\mathbf{r})}\left\{E[\rho]-\mu\left[\int\rho(\mathbf{r})d\mathbf{r}-N\right]\right\}=0, \tag{5.8}$$

with N as the total number of electrons in the system and μ the Lagrange multiplier being the chemical potential. This minimization leads to the Euler–Lagrange equation of OF-DFT,

$$\mu=\frac{\delta E[\rho]}{\delta\rho(\mathbf{r})}. \tag{5.9}$$

In the noninteracting reference system of DFT [7],

$$E[\rho]=F[\rho]+\int\rho(\mathbf{r})v_{\text{ext}}(\mathbf{r})d\mathbf{r}=T_S[\rho]+J[\rho]+E_{xc}[\rho]+E_{ne}[\rho], \tag{5.10}$$

where the universal density functional $F[\rho]$ consists of the total noninteracting kinetic energy functional $T_S[\rho]$, the classical electron–electron Coulomb repulsion $J[\rho]$ (also called the Hartree energy $E_H[\rho]$), and exchange–correlation energy density functional $E_{xc}[\rho]$, which includes contributions from the exchange and correlation effects of quantum nature,

$$F[\rho] = T_S[\rho] + J[\rho] + E_{xc}[\rho], \tag{5.11}$$

$$J[\rho] \equiv E_H[\rho] = \frac{1}{2}\iint \frac{\rho(\mathbf{r})\rho(\mathbf{r}')}{|\mathbf{r} - \mathbf{r}'|} d\mathbf{r}d\mathbf{r}', \tag{5.12}$$

and

$$E_{ne}[\rho] = \int \rho(\mathbf{r})\upsilon_{\text{ext}}(\mathbf{r})d\mathbf{r}, \tag{5.13}$$

where $E_{ne}[\rho]$ represents the nuclear–electron attraction from the external potential $\upsilon_{\text{ext}}(\mathbf{r})$ generated by the atomic nuclei in a molecular system,

$$\upsilon_{\text{ext}}(\mathbf{r}) = -\sum_{\alpha} \frac{Z_\alpha}{|\mathbf{r} - \mathbf{R}_\alpha|}, \tag{5.14}$$

with Z_α as the nuclear charge of atom α and $\mathbf{R}_\alpha$ the position of nucleus α. With these, the Euler–Lagrange equation in Eq. (5.9) becomes

$$\mu = \frac{\delta T_S[\rho]}{\delta\rho(\mathbf{r})} + \upsilon_{KS}(\mathbf{r}), \tag{5.15}$$

where $\upsilon_{KS}(\mathbf{r})$ is the Kohn–Sham potential,

$$\upsilon_{KS}(\mathbf{r}) = \upsilon_{\text{ext}}(\mathbf{r}) + \upsilon_J(\mathbf{r}) + \upsilon_{xc}(\mathbf{r}), \tag{5.16}$$

with

$$\upsilon_J(\mathbf{r}) = \int \frac{\rho(\mathbf{r}')}{|\mathbf{r} - \mathbf{r}'|} d\mathbf{r}', \tag{5.17}$$

and

$$\upsilon_{xc}(\mathbf{r}) = \frac{\delta E_{xc}[\rho]}{\delta\rho(\mathbf{r})}. \tag{5.18}$$

Furthermore, the noninteracting kinetic energy can be split into two components,

$$T_S[\rho] = T_W[\rho] + E_P[\rho], \tag{5.19}$$

with $T_W[\rho]$ as the Weizsäcker kinetic energy defined in Eq. (5.2) and $E_P[\rho]$ being the Pauli energy [29, 37] arising from the contribution of the *Pauli exclusion principle* to the kinetic energy. With this decomposition, the Euler–Lagrange equation can be rewritten as

$$\mu = \frac{\delta T_W}{\delta\rho(\mathbf{r})} + \upsilon_P(\mathbf{r}) + \upsilon_{KS}(\mathbf{r}), \tag{5.20}$$

where $v_P(\mathbf{r})$ is the Pauli potential, defined as the functional of the derivative of E_P with respect to ρ,

$$v_P(\mathbf{r}) = \frac{\delta E_P[\rho]}{\delta \rho(\mathbf{r})}. \tag{5.21}$$

One can readily show that [7]

$$\frac{\delta T_W[\rho]}{\delta \rho(\mathbf{r})} = \frac{1}{8}\left(\frac{\nabla \rho(\mathbf{r})}{\rho(\mathbf{r})}\right)^2 - \frac{1}{4}\frac{\nabla^2 \rho(\mathbf{r})}{\rho(\mathbf{r})} = \rho^{-1/2}(\mathbf{r})\left(-\frac{1}{2}\nabla^2\right)\rho^{-1/2}(\mathbf{r}), \tag{5.22}$$

so Eq. (5.20) becomes

$$\left(-\frac{1}{2}\nabla^2 + v_P(\mathbf{r}) + v_{KS}(\mathbf{r})\right)\rho^{1/2}(\mathbf{r}) = \mu \rho^{1/2}(\mathbf{r}). \tag{5.23}$$

This equation resembles the conventional Kohn–Sham equation in DFT but avoids the explicit use of orbitals. It forms the foundation for many OF-DFT implementations.

5.2.2 Total Energy Partition Schemes

In DFT [7], we have the following conventional scheme to partition the total energy E,

$$E = T_S + V_{ne} + J + E_{xc} + V_{nn}, \tag{5.24}$$

where T_S, V_{ne}, J, E_{xc}, and V_{nn} stand for the noninteracting kinetic energy, nuclear–electron attraction, classical electron–electron Coulomb repulsion, exchange–correlation energy, and nuclear–nuclear repulsion, respectively. The nuclear–nuclear repulsion term V_{nn} has the form,

$$V_{nn} = \sum_{\alpha}\sum_{\beta \neq \alpha}\frac{Z_\alpha Z_\beta}{|\mathbf{R}_\alpha - \mathbf{R}_\beta|}, \tag{5.25}$$

where Z_α/Z_β stands for the nuclear charge of atom α/β and $\mathbf{R}_\alpha/\mathbf{R}_\beta$ the position of nucleus α/β, respectively. Since V_{ne}, J, and V_{nn} are of the electrostatic nature, they can be put together, yielding the electrostatic energy E_e,

$$E_e = V_{ne} + J + V_{nn}. \tag{5.26}$$

With this, Eq. (5.24) becomes

$$E = T_S + E_e + E_{xc} \tag{5.27}$$

This is the first scheme of total energy partition in DFT [38].

In 2007, we proposed another total energy partition scheme in DFT [38–41], assuming that the total energy comes from the following three independent physicochemical effects: steric E_s, electrostatic E_e, and fermionic quantum E_q due to the exchange–correlation interaction from the Pauli exclusion principle,

$$E[\rho] \equiv E_s[\rho] + E_e[\rho] + E_q[\rho]. \tag{5.28}$$

For the quantum part, we know that the contribution from the *Pauli exclusion principle* includes two components, one for the potential energy called the exchange–correlation energy, $E_{xc}[\rho]$, and the other for the kinetic energy, called the Pauli energy, $E_P[\rho]$, defined via Eq. (5.19),

$$E_P[\rho] = T_S[\rho] - T_W[\rho], \tag{5.29}$$

so

$$E_q[\rho] = E_{xc}[\rho] + E_P[\rho] = E_{xc}[\rho] + T_S[\rho] - T_W[\rho]. \tag{5.30}$$

With the help of Eqs. (5.27), (5.29), and (5.30), from Eq. (5.28), we arrive at [38]

$$E_s[\rho] \equiv T_W[\rho] = \frac{1}{8}\int \frac{|\nabla\rho(\mathbf{r})|^2}{\rho(\mathbf{r})} d\mathbf{r}. \tag{5.31}$$

This equation suggests that (i) if the assumption in Eq. (5.28) is valid, the result in Eq. (5.31) must also hold, (ii) the steric effect can be quantified in DFT by Weizsäcker kinetic energy, $T_W[\rho]$, and (iii) although the nature and origin of the steric effect have been widely debated—largely because it lacks a unique definition and direct physical observable—our previous finding clearly reveals its kinetic character, consistent with Weisskopf's idea of "kinetic pressure" [42].

Fisher information differs from the Weizsäcker kinetic energy only by a factor of 1/8, which allows it to characterize the steric effect within the information-theoretic approach framework [27, 28, 43–45]. Essentially, Fisher information quantifies the nonuniformity of the electron density, reaching its maximum at the nuclear cusp. High electron density indicates that nearby space is already occupied by atoms within a molecule. In this regard, Fisher information serves as a precise measure of the space occupied by atoms, reflecting their steric effect.

The previous quantification of the steric effect also has the following elegant properties. First, the steric energy $E_s[\rho]$ as a universal density functional is always nonnegative with a lower bound to the total kinetic energy $T_S[\rho]$. Second, $E_s[\rho]$ is homogeneous of degree one in density scaling [46–49], satisfying

$$E_s[\gamma\rho] = \gamma E_s[\rho] \tag{5.32}$$

with $0 \le \gamma \le 1$. This relationship suggests that $E_s[\rho]$ is extensive: The larger the system, the larger the steric energy. Third, for the limit case of a homogeneous electron gas with $\nabla\rho(\mathbf{r}) = 0$, $E_s[\rho] = 0$; and for one-electron cases such as the H atom and H_2^+ molecular cation, steric and total energies are equal: $E = -E_s$. Lastly, when Bader's zero-flux criterion [50] is employed to partition atoms in molecules, they are connected to regions of zero steric energy density, demonstrating that atoms within a molecule experience balanced steric repulsion.

From the physics perspective, if electrons in an atom or molecule behaved as bosons, they would collapse into the lowest orbital in the ground state [38]. If the electron density of this hypothetical bosonic state matches that of the fermionic state, $\rho(\mathbf{r})$, the total wave function of the system would be given by $\sqrt{[\rho(\mathbf{r})/N]}$. Neglecting electrostatic and quantum contributions, the total energy of this conjectured system is just $E_s[\rho]$. The spatial extent associated with $E_s[\rho]$ represents an intrinsic property of the system. In this sense, $E_s[\rho]$ serves as a measure of the system's intrinsic dimensions, independent of quantum and electrostatic effects, which is the basic assumption of Eq. (5.28).

While our partition schemes above focus on total energy, an alternative decomposition approach exists in the literature, dedicated specifically to analyzing the interaction energy between two molecules or fragments. Known as energy decomposition analysis (EDA) [51–55], this method breaks down interaction energy ΔE_{int} into distinct physical components, including electrostatics E_{elec} (the classical Coulombic interaction between the charge distributions of the two fragments), exchange-repulsion E_{exch} (the Pauli repulsion that arises due to the overlap of occupied orbitals, preventing electrons from occupying the same space), polarization E_{pol} (the stabilization due to mutual induction of charge densities, where each fragment's electron cloud distorts in response to the other's electric field), and dispersion E_{disp} (the attractive van der Waals interactions arising from correlated electron fluctuations between the fragments), that is,

$$\Delta E_{\text{int}} = E_{\text{elec}} + E_{\text{exch}} + E_{\text{pol}} + E_{\text{disp}} . \tag{5.33}$$

By isolating these contributions, EDA offers insights into the underlying forces governing molecular interactions, distinguishing whether they are dominated by electrostatic attraction, polarization effects, or dispersion forces. A variety of EDA implementations have been developed, each designed to suit different computational frameworks and theoretical levels, making them valuable tools for appreciating noncovalent interactions.

Steric effects and quantum interactions are well recognized as crucial factors in chemical processes and transformations, yet their relative contributions in different processes are less understood [38, 39, 41, 56–62]. For example, one controversial topic is their relative contributions to bond rotation barriers in ethane, which has two distinct conformers, staggered and eclipsed. Using the above two total energy partition schemes, we can quantify the contribution of these effects in the bond rotation barrier height ΔE [60]:

$$\Delta E = \Delta T_S + \Delta E_e + \Delta E_{xc} \tag{5.34}$$

$$\Delta E = \Delta E_s + \Delta E_e + \Delta E_q . \tag{5.35}$$

For six molecules with one single rotatable bond (Figure 5.1), we found that while steric repulsion and quantum exchange–correlation effects (which encapsulate the hyperconjugation effect) both contribute positively to the barrier, it is the electrostatic interaction that overwhelmingly governs its magnitude. For all six systems studied—spanning C—C, C—N, C—O, N—N, N—O, and O—O bonds—the bond rotation barrier height (ΔE) exhibits the strongest correlation with the electrostatic energy component (ΔE_e, $R^2 = 0.930$), indicating its governing influence (Figure 5.2a). Two-variable models (Figure 5.2b and c) further confirm this

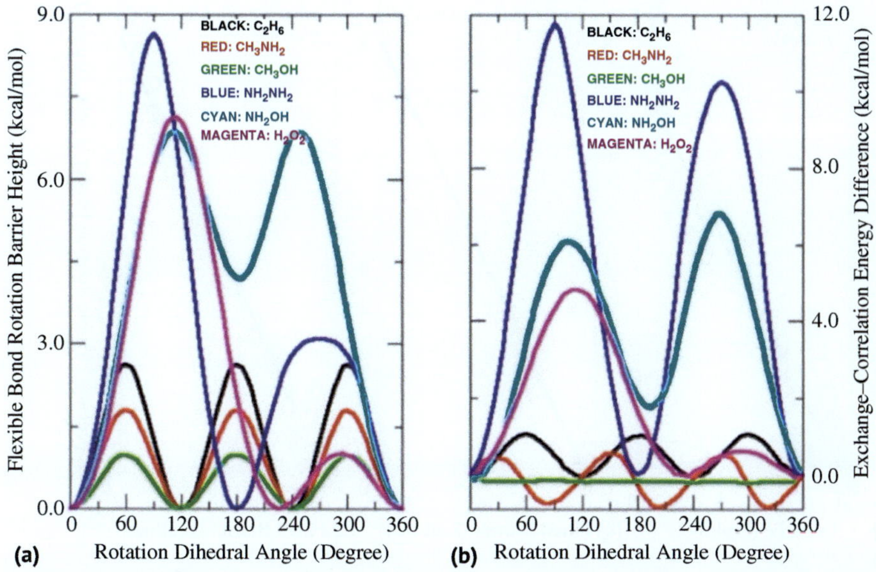

Figure 5.1 Profiles of (a) the total energy difference and (b) exchange–correlation energy difference with respect to the 360° dihedral angle rotation of six simple molecules studied in this work: C_2H_6, CH_3NH_2, CH_3OH, NH_2NH_2, NH_2OH, and H_2O_2. *Source:* Reproduced from Ref. [60] with permission of American Chemical Society.

dominance, with electrostatic terms weighted significantly higher (e.g., coefficients of 0.618 versus 0.029 for steric effects). Steric (ΔE_s) and quantum (ΔE_q) components contribute positively but less decisively, aligning with earlier debates that attributed barriers to steric clashes or hyperconjugation. This unified framework not only reconciles previously conflicting explanations based on steric and hyperconjugation viewpoints but also underscores that the net rotation barrier is determined by the subtle balance of all these effects, with the dominant role played by the electrostatics.

5.2.3 Pauli Energy, Pauli Potential, Pauli Force, and Pauli Charge

Pauli energy [29, 37] quantifies how much the *Pauli exclusion principle* contributes to the total noninteracting kinetic energy $T_S[\rho]$. It is the difference between $T_S[\rho]$ and Weizsäcker kinetic energy $T_W[\rho]$ in Eq. (5.2),

$$E_P[\rho] \equiv \int \varepsilon_P(\mathbf{r})d\mathbf{r} = T_S[\rho] - T_W[\rho] \equiv \int (\tau_s(\mathbf{r}) - \tau_w(\mathbf{r}))d\mathbf{r}, \tag{5.36}$$

where $\varepsilon_p(\mathbf{r})$ is the Pauli energy density and $\tau_s(\mathbf{r})$ and $\tau_w(\mathbf{r})$ stand for the noninteracting and Weizsäcker kinetic energy densities, respectively. In DFT, the analytical forms of T_S and τ_s as explicit density functionals are still unknown. However, in

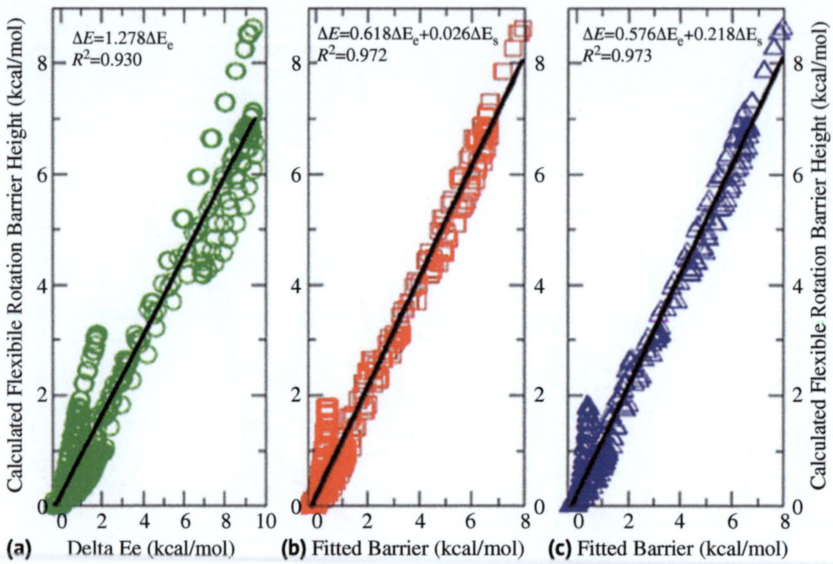

Figure 5.2 Least-square fittings with one and two energy components for the rotation barrier of all six molecules. *Source:* Reproduced from Ref. [60] with permission of American Chemical Society.

the Kohn–Sham scheme, these quantities can be expressed in terms of Kohn–Sham orbitals in two equivalent ways. The first is the Hamiltonian form:

$$\tau_s(\mathbf{r}) = -\frac{1}{2}\sum_i \varphi_i^*(\mathbf{r})\nabla^2 \varphi_i(\mathbf{r}), \tag{5.37}$$

where $\{\varphi_i(\mathbf{r})\}$ are the Kohn–Sham orbitals and the summation is over all occupied orbitals. The second is the Lagrangian form,

$$\tau_s(\mathbf{r}) = -\frac{1}{2}\sum_i |\nabla \varphi_i(\mathbf{r})|^2 = -\frac{1}{2}\sum_i \varphi_i^*(\mathbf{r})\nabla^2 \varphi_i(\mathbf{r}) + \frac{1}{4}\nabla^2\rho(\mathbf{r}). \tag{5.38}$$

Although both forms yield the same total kinetic energy T_S, the Lagrangian expression is nowhere negative. For this reason, it is the preferred form in our formulation.

In regions of molecules where either strong or weak interactions are present, such as covalent bonding and van der Waals interactions, there should be electron density redistributions. These density redistributions will induce discernible effects from the Pauli principle. In this regard, Pauli energy and its derived quantities, that is, Pauli potential, Pauli force, and Pauli charge, can be a reliable descriptor to simultaneously identify and describe both covalent bonding and noncovalent interactions [11, 31, 33, 35].

The Pauli potential is defined as the functional derivative of the Pauli energy with respect to the electron density via Eqs. (5.19) and (5.21).

$$\upsilon_P(\mathbf{r}) = \frac{\delta E_P[\rho]}{\delta\rho(\mathbf{r})} = \frac{\delta T_S[\rho]}{\delta\rho(\mathbf{r})} - \frac{\delta T_W[\rho]}{\delta\rho(\mathbf{r})}, \tag{5.39}$$

with $\frac{\delta T_W[\rho]}{\delta \rho(\mathbf{r})}$ given in Eq. (5.22). Since the exact form of $T_S[\rho]$ and $\tau_s(\mathbf{r})$ is unknown, there are two ways to approximate Eq. (5.39). The first one to take approximate forms of $T_S[\rho]$ in the generalized gradient approximation, such as this one,

$$\tau_s(\mathbf{r}) = c_{TF}\rho^{5/3}(\mathbf{r}) + \frac{1}{72}\frac{|\nabla\rho(\mathbf{r})|^2}{\rho(\mathbf{r})}, \tag{5.40}$$

one thus has

$$\frac{\delta T_S[\rho]}{\delta \rho(\mathbf{r})} = \frac{5}{3}c_{TF}\rho^{2/3}(\mathbf{r}) + \frac{1}{9}\left\{\frac{1}{8}\left[\frac{\nabla\rho(\mathbf{r})}{\rho(\mathbf{r})}\right]^2 - \frac{1}{4}\frac{\nabla^2\rho(\mathbf{r})}{\rho(\mathbf{r})}\right\} = \frac{5}{3}c_{TF}\rho^{2/3}(\mathbf{r}) + \frac{1}{9}\upsilon_s(\mathbf{r}), \tag{5.41}$$

so

$$\upsilon_P(\mathbf{r}) = \frac{5}{3}c_{TF}\rho^{2/3}(\mathbf{r}) - \frac{8}{9}\upsilon_s(\mathbf{r}). \tag{5.42}$$

Here $\upsilon_s(\mathbf{r})$ is the steric potential,

$$\upsilon_s(\mathbf{r}) = \frac{\delta T_W[\rho]}{\delta \rho(\mathbf{r})} = \frac{1}{8}\left(\frac{\nabla\rho(\mathbf{r})}{\rho(\mathbf{r})}\right)^2 - \frac{1}{4}\frac{\nabla^2\rho(\mathbf{r})}{\rho(\mathbf{r})}, \tag{5.43}$$

which will be further examined in the next section. The other way to obtain the Pauli potential is to use the Euler–Lagrange equation in Eq. (5.23) through chemical potential μ, Kohn–Sham potential, Eq. (5.16), and steric potential, Eq. (5.43),

$$\upsilon_P(\mathbf{r}) = \mu - \upsilon_{KS}(\mathbf{r}) - \upsilon_s(\mathbf{r}). \tag{5.44}$$

In this way, no kinetic energy functional is needed, but an approximate exchange–correlation functional is required to evaluate the Kohn–Sham potential. The Pauli force is defined as the negative gradient of the Pauli potential,

$$\mathbf{F}_P(\mathbf{r}) = -\nabla\upsilon_P(\mathbf{r}), \tag{5.45}$$

and Pauli charge by means of the Poisson equation of the Pauli potential,

$$\nabla^2\upsilon_P(\mathbf{r}) = -4\pi q_p(\mathbf{r}). \tag{5.46}$$

Since $E_{xc}[\rho]$ is much easier to be approximated than $T_S[\rho]$, it is anticipated that Eq. (5.44) should be more accurate than Eq. (5.42).

Shown in Figure 5.3 are Pauli potentials based on Eq. (5.44) for crystals (a) diamond, (b) BN, and (c) MgB_2, and the Pauli charge for MgB_2 [63, 64]. As can be seen from the Figure, Pauli potential and Pauli charge reproduce the atomic shell structure near the nucleus, and in the bonding region, they are able to demonstrate different features for different bonding systems. This confirms their significance in determining strong covalent bonding and weak van der Waals interactions that will be further discussed and examined later.

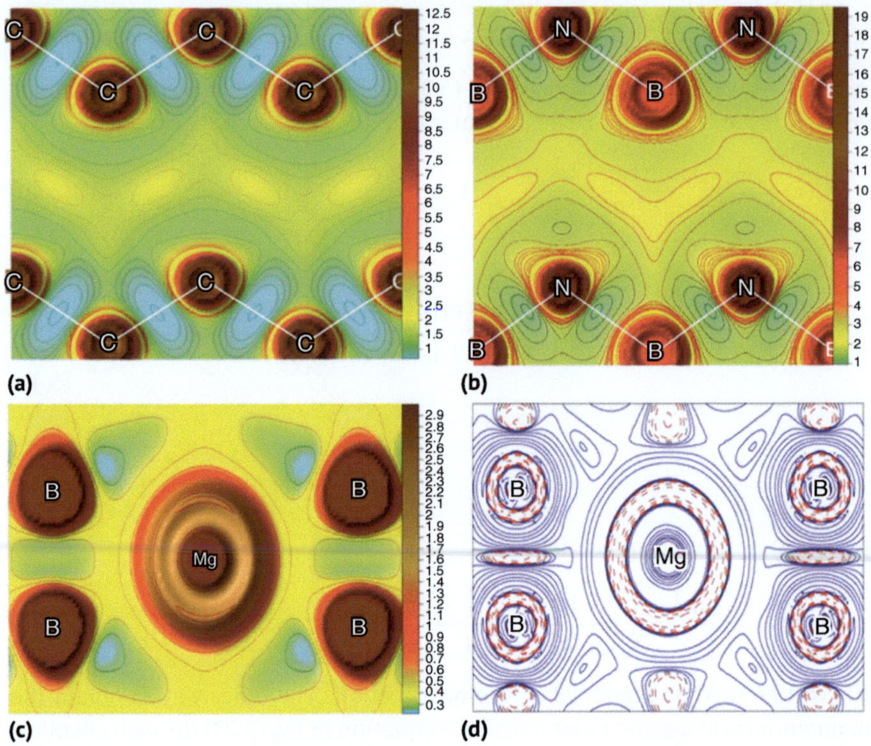

Figure 5.3 (a) Pauli potential for diamond; (b) Pauli potential for BN crystal; (c) Pauli potential for MgB_2 crystal; and (d) Pauli charge for MgB_2 crystal. *Source:* Reproduced from Ref. [63] with permission of Elsevier.

5.2.4 Steric Energy, Steric Potential, Steric Force, and Steric Charge

Our 2007 proposal [38] of quantifying steric effect using the Weizsäcker kinetic energy, $T_W[\rho]$, aligns with Weisskopf's concept of "kinetic pressure" [42] as the origin of the effect. Additionally, Fisher information I_F, differing from $T_W[\rho]$ by a factor of 1/8, effectively measures steric effect by capturing electron density nonuniformity.

$$E_s[\rho] \equiv T_W[\rho] = \frac{1}{8} I_F[\rho] = \frac{1}{8} \int \frac{|\nabla \rho(\mathbf{r})|^2}{\rho(\mathbf{r})} d\mathbf{r}. \tag{5.47}$$

This formulation of steric energy possesses key properties: it is always nonnegative, scales extensively with system size, vanishes for a homogeneous electron gas, and equals the total energy for one-electron systems. Bader's zero-flux criterion links steric energy to atomic partitioning of molecules, revealing a balance in steric repulsion for atoms in molecules. Conceptually, if electrons behaved as bosons, steric energy would represent the system's intrinsic spatial extent, independent of quantum and electrostatic effects, reinforcing the validity of the assumed energy partitioning in Eq. (5.28).

The density-based quantification of the steric effect in Eq. (5.47) is an explicit electron density functional. This steric energy can be further expanded using three key physical concepts: (i) the steric potential $v_s(\mathbf{r})$, defined as the functional derivative of the steric energy with respect to electron density, given by [38]

$$v_s(\mathbf{r}) = \frac{\delta E_s[\rho]}{\delta \rho(\mathbf{r})} = \frac{1}{8}\left(\frac{\nabla \rho(\mathbf{r})}{\rho(\mathbf{r})}\right)^2 - \frac{1}{4}\frac{\nabla^2 \rho(\mathbf{r})}{\rho(\mathbf{r})}, \tag{5.48}$$

(ii) the steric force $\mathbf{F}_s(\mathbf{r})$, defined as the negative gradient of the steric potential:

$$\mathbf{F}_s(\mathbf{r}) = -\nabla v_s(\mathbf{r}), \tag{5.49}$$

and (iii) the steric charge $q_s(\mathbf{r})$, expressed as:

$$q_s(\mathbf{r}) = -\frac{1}{4\pi}\nabla^2 v_s(\mathbf{r}) = -\frac{1}{4\pi}\nabla^2\left(\frac{\delta E_s[\rho]}{\delta \rho(\mathbf{r})}\right). \tag{5.50}$$

These quantities can be analyzed at molecular, group, and atomic levels using partitioning schemes such as Hirshfeld, Becke, and AIM. Together, steric energy, steric potential, steric force, and steric charge—defined in Eqs. (5.47)–(5.50)—constitute the density-based framework for quantifying the steric effect in DFT from the theoretical perspective.

Shown in Figure 5.4 are these steric effect–related quantities, together with the deformation electron density and Laplacian, for the (110) plane of the diamond crystal [64, 65]. The steric energy density exhibits cusps at nuclear positions and

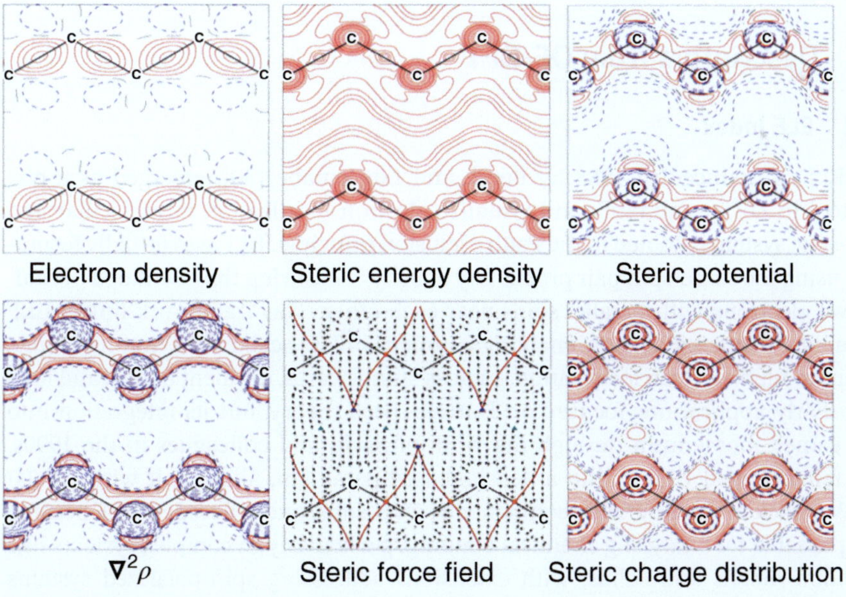

Figure 5.4 Steric energy density, steric potential, steric force, and steric charge for diamond in the (110) plane. *Source:* Reproduced from Ref. [64] with permission of American Physical Society.

vanishes at bond critical points, particularly at the midpoint of C—C bonds. In this region, deformation electron density indicates electron accumulation, while the Laplacian of electron density highlights areas of concentration. A slight increase in steric energy density is observed beyond the C atoms along the C—C axis, where deformation electron density reveals a density deficit, and the Laplacian confirms density depletion. The steric potential distributions display a shell-like structure, closely resembling the patterns of the Laplacian of electron density. Within the inner-electron shell (core) region, steric forces consistently point toward the nuclear cusp. This aligns with the DFT-based quantification of steric effect, which assumes a hypothetical bosonic electron state where electrons cluster into the lowest ground-state orbital. As illustrated in the Figure, the steric force reflects this behavior by exerting a dragging effect toward the nuclear cusp. Similarly, steric charge density exhibits a shell structure, with positive regions surrounding atomic positions.

Systematic studies and applications of the previous formulation of steric effect have been conducted by us, including investigating its impact on molecular stability and reactivity, determining stereoselectivity, and correlating with the experimental scale of Taft's steric parameters. These findings provide a deeper understanding of the steric effect and offer a quantitative tool for studying the effect from a DFT perspective. We will come back to this topic later in Chapter 11.

Collectively, these formulations show how OF-DFT turns density-based quantities—especially Pauli and steric potentials—into chemically interpretable descriptors, offering direct handles on bonding and reactivity while retaining computational efficiency.

5.3 Descriptors from OF-DFT

5.3.1 ELF Index

The ELF, introduced in 1988 by Becke and Edgecombe [30], emerged as a pivotal tool in DFT to quantify and visualize electron localization in molecular and solid-state systems. Rooted in the Pauli exclusion principle, ELF was initially formulated using the same-spin pair probability density, measuring the likelihood of finding two electrons with parallel spins near each other. This framework linked ELF to the kinetic energy density, enabling a dimensionless scalar field that highlights regions of high electron localization, such as atomic cores, covalent bonds, and lone pairs. Early applications focused on Hartree–Fock theory, but its integration into DFT gained momentum through the work of Savin and colleagues in the 1990s, who reinterpreted ELF within a density-based paradigm, aligning it with the TF and Weizsäcker kinetic energy components. Over subsequent decades, ELF became indispensable for analyzing chemical bonding, particularly in complex systems like transition metals and solids, with extensions addressing spin-polarized systems and periodic boundary conditions. Its intuitive graphical representation—via isosurfaces in molecular visualization software—revolutionized the interpretation of electron density, bridging quantum mechanics with classical chemical concepts.

ELF's evolution reflects ongoing efforts to reconcile quantum mechanical rigor with chemical intuition, solidifying its role in electronic structure theory and materials science.

In formulation, ELF is defined through the scaled Pauli energy density $\zeta(\mathbf{r})$[31],

$$\zeta(\mathbf{r}) \equiv \frac{\varepsilon_P(\mathbf{r})}{\tau_{TF}(\mathbf{r})}, \tag{5.51}$$

where $\tau_{TF}(\mathbf{r})$ is the TF kinetic energy in Eq. (5.1) and $\varepsilon_P(\mathbf{r})$ is the Pauli energy density in Eq. (5.36). This scaled Pauli energy density converts the original Pauli energy density to be dimensionless. For atoms and molecules with one and two electrons, one has $\tau_s(\mathbf{r}) = \tau_W(\mathbf{r})$ in Eq. (5.36) so $\zeta(\mathbf{r}) = 0$. When the nuclear charge Z goes to infinity, the system becomes a homogeneous electron gas, so $\tau_s(\mathbf{r}) = \tau_{TF}(\mathbf{r})$ and $\tau_W(\mathbf{r}) = 0$; thus $\zeta(\mathbf{r}) = 1$. One downside of $\zeta(\mathbf{r})$ is that because $\tau_{TF}(\mathbf{r})$ decays much faster than $\varepsilon_P(\mathbf{r})$, $\zeta(\mathbf{r})$ is asymptotically divergent for atoms and molecules. To fix this issue, ELF is defined as the following:

$$\text{ELF} = \frac{1}{1 + \zeta^2(\mathbf{r})}. \tag{5.52}$$

The advantage of the ELF index is that its value falls in the range between 0 and 1. Figure 5.5 demonstrates the ELF isosurface for water and fluorine dimer as an example. ELF can locate the lone electron pairs for O and F atoms but cannot distinguish the O—H and F—F single bonds. We will show later that (i) ELF can identify strong covalent bonds such as double, triple, and quadruple bonds and (ii) to identify single bonds we need reformulate Eq. (5.52).

5.3.2 Strong Covalent Interaction Index

To craft a density-based index to identify SCI [31, 32, 35], our hypothesis is that regions exhibiting SCI should be characterized by strong electron correlations arising from the Pauli exclusion principle. In other words, in areas where SCI exists, significant repulsion, stemming from the principle, should be present. In DFT, most of these electron correlations are primarily captured by Pauli energy, which represents the contribution to the total kinetic energy from the exchange interaction

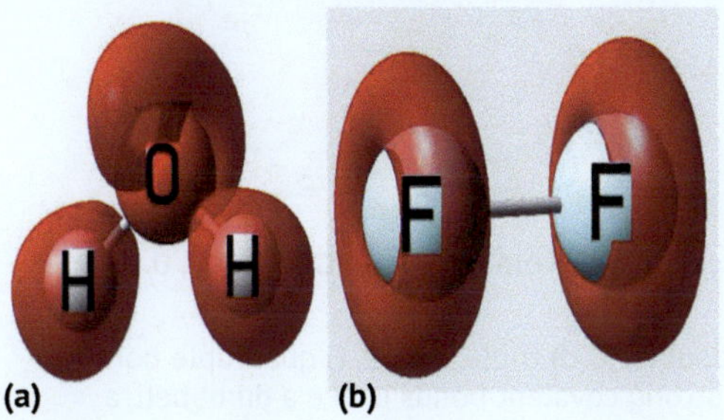

Figure 5.5 The ELF isosurface for (a) H_2O and (b) F_2 molecules with the isovalue of 0.85.

among same-spin electrons from the principle. Since the scaled Pauli energy density $\zeta(\mathbf{r})$ in Eq. (5.51) is dimensionless, it should be a decent descriptor by itself for SCI. However, as mentioned earlier, $\tau_{TF}(\mathbf{r})$ decays much faster than $\varepsilon_P(\mathbf{r})$, so $\zeta(\mathbf{r})$ is asymptotically divergent. To avoid this divergence problem, we define the SCI index as the reciprocal of the scaled Pauli energy density $\zeta(\mathbf{r})$,

$$\mathrm{SCI} = \frac{1}{\zeta(\mathbf{r})}. \tag{5.53}$$

As illustrative examples, Figure 5.6 displays the ELF and SCI distributions used to identify SCIs in (a, b) the double bond in ethylene, (c, d) the triple bond in acetylene, and (e, f) the quadruple bond in chromium dimer cations. The ELF and SCI signature surface between the two bonding atoms shows strikingly similar patterns. For double bonds, a dumbbell-like shape appears at the center of the bond; for triple bonds, the shape transforms into a donut-like structure, and for quadruple bonds, four "bean"-shaped regions are observed between the two chromium atoms. These distinct isosurface features can be effectively used to identify different types of strong covalent bonds in other systems.

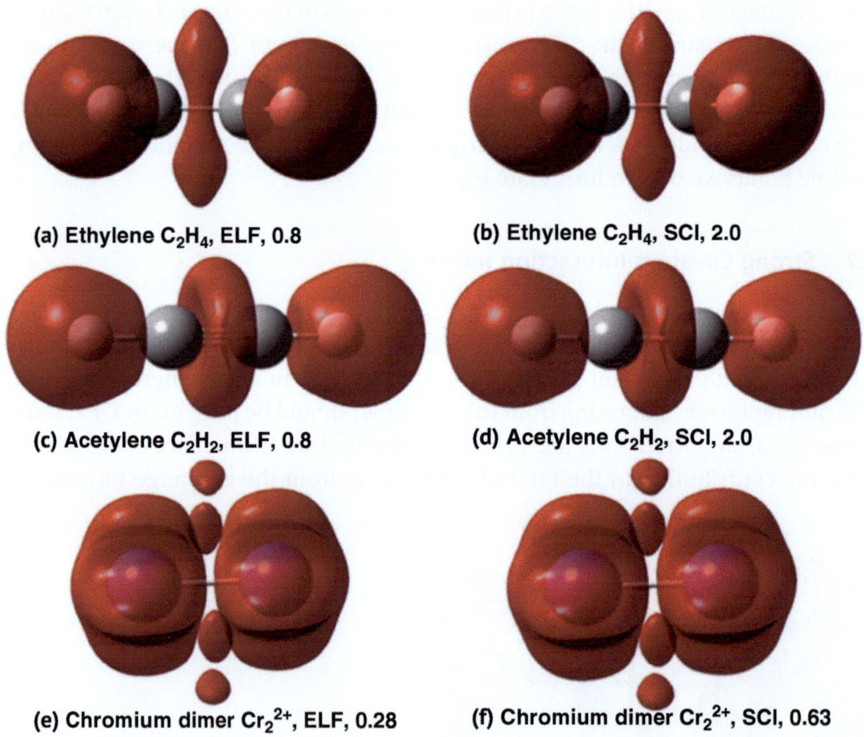

Figure 5.6 ELF and SCI isosurfaces: (a, b) double, (c, d) triple, and (e, f) quadruple bonds, whose signature shape in the center of the strong covalent bonds is like a dumbbell, a donut, and four beans, respectively. *Source:* Reproduced from Ref. [31] with permission of American Chemical Society.

5.3.3 BNI Index

While ELF and SCI indices given previously can reliably identify SCIs, they are unable to distinguish single bonds and other interaction types such as metallic bonds, hydrogen bonding, and van der Waals interactions. The other limitation of the SCI index in Eq. (5.53) is that for one- and two-electron systems, where $\tau_s(\mathbf{r})$ and $\tau_W(\mathbf{r})$ are identical, $\zeta(\mathbf{r}) = 0$, causing Eq. (5.53) to diverge. While this divergence is not a major issue for SCIs, where two or more electron pairs are shared, it becomes problematic when extending the index to other types of chemical bonding and noncovalent interactions. To address this issue, we propose a new scaled Pauli energy measure, referred to as the BNI index [35], which overcomes the divergent problem and broadens the applicability of the method.

$$\text{BNI} = \frac{\varepsilon_P(\mathbf{r})}{\tau_W(\mathbf{r})} = \frac{\tau_s(\mathbf{r}) - \tau_W(\mathbf{r})}{\tau_W(\mathbf{r})}. \tag{5.54}$$

Instead of using the TF kinetic energy density as a reference, we utilize the Weizsäcker kinetic energy density. We refer to this new function as the BNI index because it effectively identifies both covalent and noncovalent interactions. Why is this new formulation preferable to Eq. (5.53)? Replacing $\tau_{TF}(\mathbf{r})$ with $\tau_W(\mathbf{r})$ in the denominator not only ensures that the quantity remains dimensionless but also eliminates the divergence issue, as $\tau_W(\mathbf{r})$ is well-behaved asymptotically. Additionally, expressing Eq. (5.54) as the relative change of the noninteracting kinetic energy density with respect to the Weizsäcker kinetic energy density provides a mathematically sound framework. Our previous studies have shown that the Weizsäcker kinetic energy is a reliable measure of the steric effect. In this context, Eq. (5.54) separates steric effect from bonding interactions, which aligns well with physical chemistry principles. Moreover, since $\tau_W(\mathbf{r})$ is a well-behaved local function, there is no need to use its reciprocal form, as seen in the SCI index of Eq. (5.53). The only drawback of Eq. (5.54) is that for one- and two-electron systems, the BNI index vanishes, making it inapplicable to simple systems like H_2^+ and H_2. This limitation stems fundamentally from the Pauli principle and the nature of Pauli energy. Figure 5.7 summarizes the current status of using four density-based descriptors to identify both weak and strong interactions [35].

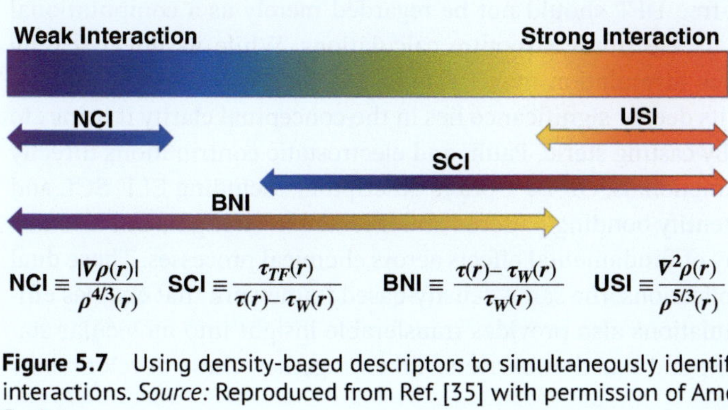

Figure 5.7 Using density-based descriptors to simultaneously identify strong and weak interactions. *Source:* Reproduced from Ref. [35] with permission of American Chemical Society.

5.4 Applications of OF-DFT

Recent literature has renewed interest in OF-DFT [1, 2, 4, 24, 25], particularly as artificial intelligence and machine learning provide new routes to develop more accurate KEDFs. These approaches can bridge the accuracy gap with Kohn–Sham methods by improving functional forms and enabling transferable, large-scale simulations, thereby broadening the scope of OF-DFT in materials and biological systems.

In this book, our focus on OF-DFT is a little different. We are interested in making use of quantities from OF-DFT to appreciate chemical reactivity. For example, we proposed to use the Weizsäcker kinetic energy functional to quantify steric effects and Pauli energy-related quantities to simultaneously identify both strong and weak interactions. By integrating these concepts, we develop density-based descriptors—such as the BNI index—that disentangle steric, electrostatic, and quantum contributions to molecular behaviors. This framework enables the analysis of reaction mechanisms, stereoselectivity, and activation barriers, as demonstrated in studies of bond rotation barriers where electrostatic effects dominate [27, 66]. More applications will be presented in later chapters of this book on specific topics such as acidity, basicity, aromaticity, and antiaromaticity [27, 28, 67]. Our approach not only provides a deeper theoretical understanding of steric and electronic effects for a variety of chemical phenomena and processes but also lays the groundwork for predictive models that can guide molecular design, catalysis, and materials development. In this way, OF-DFT becomes more than a vehicle for computational efficiency: it serves as a framework for exploring chemical reactivity through density-based descriptors. These applications demonstrate how OF-DFT unites efficiency with understanding: the same framework that allows linear-scaling simulations also delivers transferable insight into how steric, electrostatic, and quantum effects govern molecular processes. In this sense, OF-DFT does not merely replicate Kohn–Sham results more cheaply; it reframes reactivity analysis by elevating density-based descriptors to the level of chemical concepts.

5.5 Concluding Remarks

In summary, orbital-free DFT should not be regarded merely as a computational shortcut for scaling up electronic structure calculations. While its linear-scaling nature and orbital-free formulation make it indispensable for treating systems of unprecedented size, its deeper significance lies in the conceptual clarity it brings to chemical reactivity. By casting steric, Pauli, and electrostatic contributions directly in terms of density functionals, OF-DFT offers descriptors, including ELF, SCI, and BNI, that not only identify bonding patterns and interaction strengths but also disentangle the interplay of fundamental effects across chemical processes. These dual roles converge in applications: the same density-based framework that enables efficient large-scale simulations also provides transferable insight into molecular stability, stereoselectivity, and reaction barriers. Taken together, these developments demonstrate that OF-DFT unites computational efficiency with conceptual clarity, securing its place as a central pillar of density functional theory. Looking ahead, advances in machine learning are poised to refine kinetic energy density functionals and expand the reach of OF-DFT, further reinforcing its dual role as both a scalable method and a framework for chemical understanding.

References

1 Mi W, Luo K, Trickey SB, Pavanello M. Orbital-free density functional theory: an attractive electronic structure method for large-scale first-principles simulations. Chem Rev. 2023 Nov 8;123(21):12039–104.

2 Xu Q, Ma C, Mi W, Wang Y, Ma Y. Recent advancements and challenges in orbital-free density functional theory. WIREs Comput Mol Sci. 2024 May;14(3):e1724.

3 Wesolowski TA, Wang YA. Recent Progress in Orbital-Free Density Functional Theory. World Scientific; 2013.

4 Ryley MS, Withnall M, Irons TJP, Helgaker T, Teale AM. Robust all-electron optimization in orbital-free density-functional theory using the trust-region image method. J Phys Chem A. 2021 Jan 14;125(1):459–75.

5 Majumdar S, Shi Z, Vignale G. Orbital-free density functional theory for periodic solids: construction of the Pauli potential. J Chem Theory Comput. 2025 May 30;21(12):6007–22.

6 Remme R, Kaczun T, Ebert T, Gehrig CA, Geng D, Gerhartz G, et al. Stable and accurate orbital-free density functional theory powered by machine learning. J Am Chem Soc. 2025 Aug 1;147(32):28851–9.

7 Parr RG, Yang W. Density-Functional Theory of Atoms and Molecules. Oxford University Press; 1989.

8 Kohn W, Sham LJ. Self-consistent equations including exchange and correlation effects. Phys Rev. 1965 Nov 15;140(4A):A1133–8.

9 García-Aldea D, Alvarellos JE. Kinetic energy density study of some representative semilocal kinetic energy functionals. J Chem Phys. 2007 Oct 14;127(14):144109.

10 Della Sala F, Fabiano E, Constantin LA. Kinetic-energy-density dependent semilocal exchange-correlation functionals. Int J Quantum Chem. 2016 Nov 15;116(22):1641–94.

11 Liu S, Zhao D, Rong C, Lu T, Liu S. Using Pauli energy to appraise the quality of approximate semilocal non-interacting kinetic energy density functionals. J Chem Phys. 2019 May 28;150(20):204106.

12 Thakkar AJ. Comparison of kinetic-energy density functionals. Phys Rev A. 1992 Dec 1;46(11):6920–4.

13 Liu S, Karasiev V, Lopez-Boada R, De Proft F. Polynomial and Padé representations for the kinetic component $T_c[\rho]$ of the correlation energy density functional. Int J Quantum Chem. 1998;69(4):513–22.

14 Luo K, Karasiev VV, Trickey SB. A simple generalized gradient approximation forhe noninteracting kinetic energy density functional. Phys Rev B. 2018 Jul 26;98(4):041111.

15 García-González P, Alvarellos JE, Chacón E. Nonlocal kinetic-energy-density functionals. Phys Rev B Condens Matter. 1996 Apr 15;53(15):9509–12.
16 Sim E, Larkin J, Burke K, Bock CW. Testing the kinetic energy functional: kinetic energy density as a density functional. J Chem Phys. 2003 May 8;118(18):8140–8.
17 Thomas LH. The calculation of atomic fields. Proc Camb Phil Soc. 1927;23:542–48.
18 Fermi E. Eine statistische Methode zur Bestimmung einiger Eigenschaften des Atoms und ihre Anwendung auf die Theorie des periodischen Systems der Elemente. Z Phys. 1928 Jan;48(1–2):73–9.
19 Weizsäcker CF. Zur Theorie der Kernmassen. Z. Phys. 1935;96:431–58; Bethe HA, Bacher RF. Nuclear physics A. Stationary states of nuclei. Rev Mod Phys. 1935;8(2):82.
20 DePristo AE, Kress JD. Kinetic-energy functionals via Padé approximations. Phys Rev A. 1987 Jan 1;35(1):438–41.
21 Wang YA, Govind N, Carter EA. Orbital-free kinetic-energy density functionals with a density-dependent kernel. Phys Rev B. 1999 Dec 15;60(24):16350–8.
22 Mi W, Genova A, Pavanello M. Nonlocal kinetic energy functionals by functional integration. J Chem Phys. 2018 May 14;148(18):184107.
23 Moldabekov ZA, Shao X, Pavanello M, Vorberger J, Dornheim T. Nonlocal vs local pseudopotentials affect kinetic energy kernels in orbital-free DFT. Electron Struct. 2025 Mar 21;7:015006.
24 Hutcheon MJ, Wibowo-Teale AM. Exact nuclear cusps in all-electron orbital-free density functional theory. Phys Rev B. 2024 Nov 22;110(19):195146.
25 Kumar S, Jing X, Pask JE, Medford AJ, Suryanarayana P. Kohn-Sham accuracy from orbital-free density functional theory via Δ-machine learning. J Chem Phys. 2023 Dec 28;159(24):244106.
26 Zhang H, Liu S, You J, Liu C, Zheng S, Lu Z, et al. Overcoming the barrier of orbital-free density functional theory for molecular systems using deep learning. Nat Comput Sci. 2024 Mar 11;4(3):210–23.
27 Rong C, Zhao D, He X, Liu S. Development and applications of the density-based theory of chemical reactivity. J Phys Chem Lett. 2022 Dec 8;13(48):11191–200.
28 He X, Li M, Rong C, Zhao D, Liu W, Ayers PW, et al. Some recent advances in density-based reactivity theory. J Phys Chem A. 2024 Feb 22;128(7):1183–96.
29 Holas A, March NH. Construction of the Pauli potential, Pauli energy, and effective potential from the electron density. Phys Rev A. 1991 Nov 1;44(9):5521–36.
30 Becke AD, Edgecombe KE. A simple measure of electron localization in atomic and molecular systems. J Chem Phys. 1990 May;92(9):5397–403.
31 Liu S, Rong C, Lu T, Hu H. Identifying strong covalent interactions with Pauli energy. J Phys Chem A. 2018 Mar 22;122(11):3087–95.
32 Huang Y, Liu L, Rong C, Lu T, Ayers PW, Liu S. SCI: a robust and reliable density-based descriptor to determine multiple covalent bond orders. J Mol Model. 2018 Jul 21;24(8):213.

33 Zhang W, He X, Li M, Zhang J, Zhao D, Liu S, et al. Simultaneous identification of strong and weak interactions with Pauli energy, Pauli potential, Pauli force, and Pauli charge. J Chem Phys. 2023 Nov 14;159(18):184104.

34 He X, Liu W, Liu S. Using energetic information quantities from density functional theory to simultaneously identify both covalent and noncovalent interactions. ChemPhysChem. 2025 Apr 1;26(7):e202401068.

35 Zhong S, He X, Liu S, Wang B, Lu T, Rong C, et al. Toward density-based and simultaneous description of chemical bonding and noncovalent interactions with Pauli energy. J Phys Chem A. 2022 Apr 21;126(15):2437–44.

36 March NH. Some pointers towards a future orbital-free density functional theory of inhomogeneous electron liquids. Phys Chem Liq. 2010 Apr;48(2):141–55.

37 Levy M, Hui O-Y. Exact properties of the Pauli potential for the square root of the electron density and the kinetic energy functional. Phys Rev A Gen Phys. 1988 Jul 15;38(2):625–9.

38 Liu S. Steric effect: a quantitative description from density functional theory. J Chem Phys. 2007 Jun 28;126(24):244103.

39 Liu S, Govind N. Toward understanding the nature of internal rotation barriers with a new energy partition scheme: ethane and n-butane. J Phys Chem A. 2008 Jul 24;112(29):6690–9.

40 Esquivel RO, Liu S, Angulo JC, Dehesa JS, Antolín J, Molina-Espíritu M. Fisher information and steric effect: study of the internal rotation barrier of ethane. J Phys Chem A. 2011 May 5;115(17):4406–15.

41 Liu S, Govind N, Pedersen LG. Exploring the origin of the internal rotational barrier for molecules with one rotatable dihedral angle. J Chem Phys. 2008 Sep 7;129(9):094104.

42 Weisskopf VF. Of atoms, mountains, and stars: a study in qualitative physics. Science. 1975 Feb 21;187(4177):605–12.

43 Rong C, Yu D, Liu S. Information-theoretic approach. In: Liu S, editor. Conceptual Density Functional Theory: Towards a New Chemical Reactivity Theory. Wiley; 2022. pp. 281–99.

44 Rong C, Wang B, Zhao D, Liu S. Information-theoretic approach in density functional theory and its recent applications to chemical problems. WIREs Comput Mol Sci. 2020 Jul;10(4):e1461.

45 Liu S. Information-theoretic approach in density functional reactivity theory. Acta Phys-Chim Sin. 2016;32(1):98–118.

46 Liu S, Parr RG. Homogeneities in density of various LDA energy functionals. J Mol Struct: THEOCHEM. 2000 Apr;501–502:29–34.

47 Parr RG, Liu S. Consequences for finite electronic systems of homogeneity properties of density-functional energy components. Chem Phys Lett. 1997 Dec;280(3–4):159–66.

48 Liu S, Parr RG. Expansions of density functionals in terms of homogeneous functionals: justification and nonlocal representation of the kinetic energy, exchange energy, and classical Coulomb repulsion energy for atoms. Phys Rev A. 1997 Mar 1;55(3):1792–8.

49 Liu S. Expansions of the pair distribution function and the second-order density matrix in terms of homogeneous functionals. Phys Rev A. 1996 Dec;54(6):4863–7.

50 Bader RFW. Atoms in Molecules: A Quantum Theory. Oxford University Press; 1990.

51 Phipps MJS, Fox T, Tautermann CS, Skylaris C-K. Energy decomposition analysis approaches and their evaluation on prototypical protein-drug interaction patterns. Chem Soc Rev. 2015 May 21;44(10):3177–211.

52 Fang D, Piquemal J-P, Liu S, Cisneros GA. DFT-steric-based energy decomposition analysis of intermolecular interactions. Theor Chem Acc. 2014 May;133(5):1484.

53 Su P, Tang Z, Wu W. Generalized Kohn-Sham energy decomposition analysis and its applications. WIREs Comput Mol Sci. 2020 Sep;10(5):e1460.

54 Kitaura K, Morokuma K. A new energy decomposition scheme for molecular interactions within the Hartree-Fock approximation. Int J Quantum Chem. 1976 Mar;10(2):325–40.

55 Hopffgarten M von, Frenking G. Energy decomposition analysis. WIREs Comput Mol Sci. 2012 Jan;2(1):43–62.

56 Huang Y, Zhong A G, Yang Q, Liu S. Origin of anomeric effect: a density functional steric analysis. J Chem Phys. 2011 Feb 28;134(8):084103.

57 Liu S, Hu H, Pedersen LG. Steric, quantum, and electrostatic effects on S(N)2 reaction barriers in gas phase. J Phys Chem A. 2010 May 13;114(18):5913–8.

58 Ess DH, Liu S, De Proft F. Density functional steric analysis of linear and branched alkanes. J Phys Chem A. 2010 Dec 16;114(49):12952–7.

59 Torrent-Sucarrat M, Liu S, De Proft F. Steric effect: partitioning in atomic and functional group contributions. J Phys Chem A. 2009 Apr 16;113(15):3698–702.

60 Liu S. Origin and nature of bond rotation barriers: a unified view. J Phys Chem A. 2013 Feb 7;117(5):962–5.

61 Wang B, Liu S, Lei M, De Proft F. Steric effect and intrinsic electrophilicity and nucleophilicity from a conceptual density functional theory and information-theoretic approach as quantitative probes of chemical reactions. Chem Eur J. 2024 Jul 3;e202401295.

62 Zhang J, He X, Wang B, Rong C, Zhao D, Liu S. Density-based quantification of steric effect: validation by Taft steric parameters from acid-catalyzed hydrolysis of esters. Phys Chem Chem Phys. 2024;26:23447–56.

63 Tsirelson VG, Stash AI, Karasiev VV, Liu S. Pauli potential and Pauli charge from experimental electron density. Comput Theor Chem. 2013 Feb; 1006:92–9.

64 Tsirelson VG, Stash AI, Liu S. Quantifying steric effect with experimental electron density. J Chem Phys. 2010 Sep 21;133(11):114110.

65 Liu S, Liu L, Yu D, Rong C, Lu T. Steric charge. Phys Chem Chem Phys. 2018 Jan 17;20(3):1408–20.

66 An X, Zhang W, He X, Li M, Rong C, Liu S. Density-based reactivity theory applied to excited states. AAPPS Bull. 2024 Feb 2;34(1):8.

67 Zhao D, He X, Rong C, Liu S. Chemical concepts from density-based approaches in density functional theory. In: Liu S, editor. Exploring Chemical Concepts Through Theory and Computation. Wiley; 2024. pp. 71–100.

6

Recent Advances in Density-based Frameworks

TABLE OF CONTENTS

Density-based reactivity theory has evolved into a multifaceted paradigm centered on four complementary frameworks: conceptual density functional theory (CDFT), density-associated quantities (DAQs), information-theoretic approach (ITA), and orbital-free density functional theory (OF-DFT). Although developed independently, all share a reliance on the electron density as the fundamental variable and a common aim to express total energy and reactivity in chemically meaningful terms. Section 6.1 clarifies the interconnections among these frameworks, showing how variational principles, functional expansions, and energetic information provide conceptual bridges that unify them. Building on this foundation, Section 6.2 introduces the topological analysis of ITA quantities, while Section 6.3 formalizes energetic information by treating scaled energy densities as probability distributions, enabling entropy- and information-based measures of chemical processes.

The later sections broaden the scope of density-based reactivity theory through three major extensions. Section 6.4 generalizes ITA to the pair density, moving beyond one-electron measures to capture nonlocal correlations. Section 6.5 applies density-based descriptors to excited states, uncovering qualitative changes in bonding, stability, and reactivity. Section 6.6 merges ITA with machine learning, demonstrating predictive power across molecules, macromolecules, and materials. Together, these developments establish density-based frameworks as a coherent, versatile methodology that consolidates conceptual foundations and expands applicability from ground to excited states and from theory-driven analysis to data-driven prediction.

6.1 Relationship Among Four Frameworks

The foundational theorem of DFT [1, 2] establishes that the ground-state electron density uniquely determines all system properties. While the four density-based frameworks [3, 4]—CDFT, DAQ, ITA, and OF-DFT—employ distinct strategies, their shared reliance on electron density inherently embeds redundant information, creating interdependencies. This overlap is evidenced by our prior work on relationships among ITA quantities [5, 6], prompting further inquiry: Can conceptual bridges be built between these frameworks by integrating these methodologies? For instance, can CDFT descriptors directly express the total energy or its components in OF-DFT? Can ITA quantify entropy for energetic densities? Recent studies address these questions, revealing the relationships among these frameworks.

6.1.1 Bridging OF-DFT and ITA via Variational Principle and Energetic Information

In OF-DFT [7, 8], the variational principle yields the following Euler–Lagrange equation

$$\left(-\frac{1}{2}\nabla^2 + v_P(\mathbf{r}) + v_{KS}(\mathbf{r})\right)\rho^{1/2}(\mathbf{r}) = \mu\rho^{1/2}(\mathbf{r}), \tag{6.1}$$

where $v_P(\mathbf{r})$ is the Pauli potential, $v_{KS}(\mathbf{r})$ is the Kohn–Sham potential, and μ chemical potential. It is obtained by minimizing the total energy density functional $E[\rho]$ with respect to the electron density $\rho(\mathbf{r})$ under the constraint of particle conservation,

$$\frac{\delta}{\delta\rho(\mathbf{r})}\left\{E[\rho] - \mu\left[\int\rho(\mathbf{r})d\mathbf{r} - N\right]\right\} = 0, \tag{6.2}$$

with N as the total number of electrons in the system.

Recently, Nagy [9] proposed the *principle of extreme physical information*, which asserts that the *physical information* (K) of a system—defined as the difference between its Fisher information (I_F) and a collection of constraints (J)—must attain an extremum:

$$K = I_F - J = \text{extreme}. \tag{6.3}$$

The collection of constraints encapsulates three critical constraints governing the system, including (i) the antisymmetric property of the total wavefunction, which gives rise to the Pauli potential $v_P(\mathbf{r})$, (ii) electron density equivalence—the electron density of the noninteracting reference system must match that of the interacting system—which determines the Kohn–Sham potential $v_{KS}(\mathbf{r})$, and (iii) the normalization of electron density, which introduces the chemical potential μ. This variational principle recovers the Euler–Lagrange equation for OF-DFT in Eq. (6.2).

The second way to make a connection between OF-DFT and ITA is through energetic information [10–12]. We will detail this connection in the next section. Basically, energy densities from OF-DFT can be employed as the probability distribution function in ITA. For example, for an energy component $E = \int e(\mathbf{r})d\mathbf{r}$, where $e(\mathbf{r})$ is the energy density, we introduce a new distribution function called scaled energy density $\varepsilon(\mathbf{r}) \equiv e(\mathbf{r})/E$ so that $\varepsilon(\mathbf{r})$ is normalized to 1, $\int \varepsilon(\mathbf{r})d\mathbf{r} = 1$, which can be seen as a probability function. This formalism enables the application of ITA tools—Shannon entropy, Fisher information, relative entropy, and others—to energetic densities, termed *energetic information*. Minimizing the information gain in the atoms in molecules representations further yields a Hirshfeld-like partitioning scheme based on energy densities. Notably, the first-order approximation of information gain for an energetic density corresponds to its contribution to the total interaction energy, underscoring the utility of this approach. To illustrate this concept, Figure 6.1 presents the radial distributions of electron density, electrostatic energy density, and Shannon entropy of the Lagrangian kinetic energy density for an argon atom. The figure demonstrates that these distributions exhibit tremendously different local behaviors. These variations likely contain valuable information for quantifying different physicochemical properties.

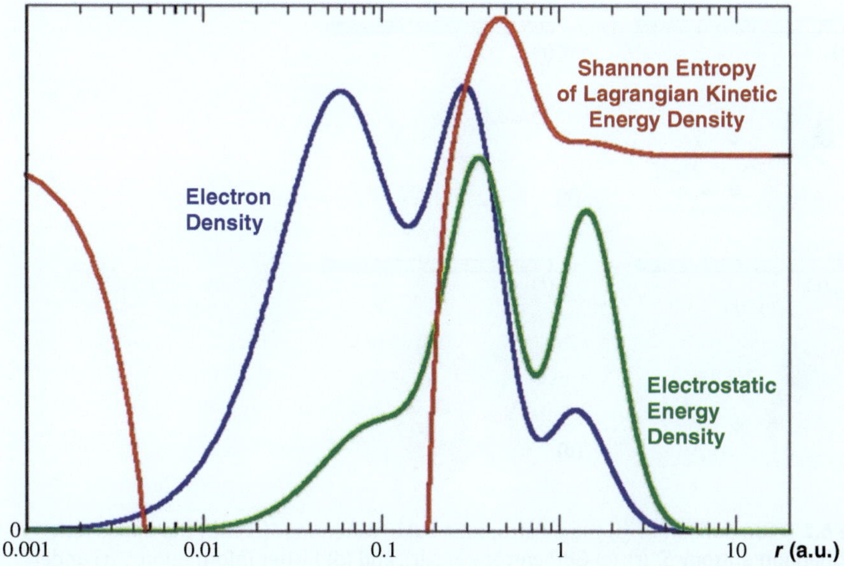

Figure 6.1 The radial distribution of the electron density (blue color), electrostatic energy density (green color), and Shannon entropy of the Lagrangian kinetic energy density (red color) for the argon atom. Atomic units (a.u.).

6.1.2 Linking ITA and CDFT Through Reactivity Descriptors

In CDFT [13–16], directional derivatives (e.g., electrophilic/nucleophilic Fukui functions) encode reactivity. To explore analogous ITA-based reactivity descriptors, we derived explicit partial derivatives of ITA quantities (e.g., Shannon entropy, Fisher information, and Ghosh–Berkowitz–Parr entropy) with respect to electron number [17]. These derivatives correlate analytically with Fukui functions, offering parallel regioselectivity predictions. Figure 6.2 demonstrates this point for the electrophilic propensity of 2-propenylbenzene, where ITA-derived regioselectivity descriptors through Shannon entropy, Fisher information, and Ghosh–Berkowitz–Parr (GBP) entropy accurately reproduce the result from the electrophilic Fukui function from CDFT. Their precise relationship with the Fukui function can be analytically derived. Moreover, we can extend the idea to other directional derivatives in CDFT. For example, similar derivatives of ITA quantities with respect to the external potential $v(\mathbf{r}')$ can be readily achieved, which further align ITA quantities with the response function $\varpi(\mathbf{r}, \mathbf{r}')$ in CDFT,

$$\varpi(\mathbf{r}, \mathbf{r}') = \left(\frac{\delta\rho(\mathbf{r})}{\delta v(\mathbf{r}')}\right)_N. \tag{6.4}$$

6.1.3 Connecting CDFT with OF-DFT via Functional Expansions

The functional expansion approach [18–21]—which expresses the total energy E of an electronic system using the reactivity descriptors from the four ensembles of CDFT—provides a direct link between OF-DFT and CDFT. This approach was

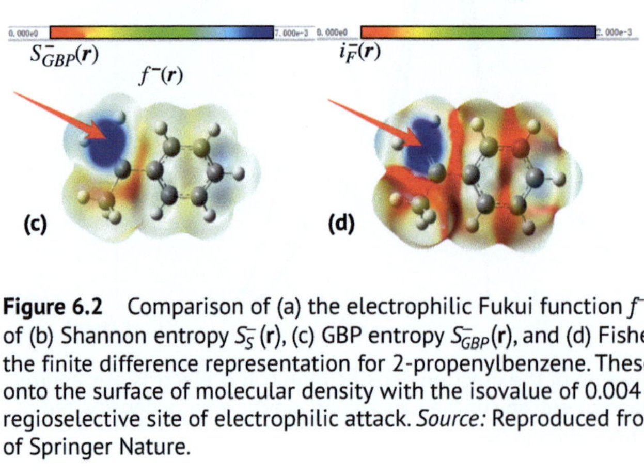

Figure 6.2 Comparison of (a) the electrophilic Fukui function $f^-(\mathbf{r})$ with the minus form of (b) Shannon entropy $S_S^-(\mathbf{r})$, (c) GBP entropy $S_{GBP}^-(\mathbf{r})$, and (d) Fisher information $i_F^-(\mathbf{r})$ under the finite difference representation for 2-propenylbenzene. These 3D functions are mapped onto the surface of molecular density with the isovalue of 0.004 au. Red arrows mark the regioselective site of electrophilic attack. *Source:* Reproduced from Ref. [17] with permission of Springer Nature.

originally developed by the present author in collaboration with Robert G. Parr. The mathematical background of this approach can be found elsewhere [22–24]. Basically, in CDFT, the total energy change ΔE due to the change of electron number ΔN and external potential $\Delta v(\mathbf{r})$ in a chemical process can be expressed. The key idea of the functional expansion approach is to make use of a re-summation technique so all contributions from the reference can be dumped into a constant C and the total energy E can be expressed in terms of CDFT descriptors without resorting to any reference. For example, in the canonical ensemble of CDFT, we have the following functional expansion expression [21]:

$$E[N,v] = C + N\mu - \frac{1}{2}N^2\eta + \int [\rho(\mathbf{r}) - Nf(\mathbf{r})]v(\mathbf{r})d\mathbf{r} \\ -\frac{1}{2}\iint v(\mathbf{r})v(\mathbf{r}')\varpi(\mathbf{r},\mathbf{r}')d\mathbf{r}d\mathbf{r}', \tag{6.5}$$

where C is a constant, N is the total number of electrons, and the involved CDFT descriptors include chemical potential μ, hardness η, Fukui function $f(\mathbf{r})$, and response function $\varpi(\mathbf{r},\mathbf{r}')$. This expression should be contrasted with what we know in OF-DFT, where the total energy can be partitioned in terms of energetic components (e.g., kinetic, electron–nuclear attraction, classical Coulomb repulsion, exchange–correlation, and others) or physicochemical effects (i.e., steric, electrostatic, and quantum). From the grand canonical ensemble of CDFT, we have

$$E[N,v] = C + \frac{1}{2}\mu^2 S + \int [\rho(\mathbf{r}) - \mu s(\mathbf{r})]v(\mathbf{r})d\mathbf{r} + \frac{1}{2}\iint v(\mathbf{r})v(\mathbf{r}')s(\mathbf{r},\mathbf{r}')d\mathbf{r}d\mathbf{r}', \tag{6.6}$$

where CDFT descriptors contained are softness S, local softness $s(\mathbf{r})$, and softness kernel $s(\mathbf{r},\mathbf{r}')$. From the grand isomorphic ensemble,

$$E[N,v] = C + N\mu - \frac{1}{2}\iint \rho(\mathbf{r})\rho(\mathbf{r}')\eta(\mathbf{r},\mathbf{r}')d\mathbf{r}d\mathbf{r}', \tag{6.7}$$

where $\eta(\mathbf{r},\mathbf{r}')$ is the hardness kernel. We can derive a similar total energy expression in the isomorphic ensemble using other CDFT descriptors. The constant C in these equations could be different. These formulas, Eqs. (6.5)–(6.7), directly connect the total energy E with different descriptors in CDFT.

6.1.4 Unifying the Four Density-based Frameworks

Even though the four density-based frameworks employ completely different strategies in deriving reactivity descriptors, their main purpose is the same: to express the total energy E in terms of physicochemically meaningful variables. For example, OF-DFT uses the electron density, $E \equiv E[\rho]$, and CDFT employs the number of electrons N and external potential $v(\mathbf{r})$, $E \equiv E[N,v]$, or their conjugate variables μ and $\rho(\mathbf{r})$. In ITA, according to the information functional theory work [25] that we proposed earlier, we have $E \equiv E[\{\text{IT}\}]$, where $\{\text{IT}\}$ stands for the collection of information-theoretic quantities. It can be proved that {IT} should determine the electron density and, consequently, all ground-state properties, including the total energy, as dictated by the basic theorems of DFT. In DAQ, even though local functions such as density gradient $\nabla\rho(\mathbf{r})$, Laplacian $\nabla^2\rho(\mathbf{r})$, and their combinations are the focus, according to Bader's quantum theory of atoms in molecules [26], the total energy can be expressed as the sum of atomic contributions, $E \equiv E\left[\{\rho_{\Omega_A}\}\right]$, where ρ_{Ω_A} denotes the atomic density in the Ω_A basin. With this, we can unify the four density-based frameworks in the following manner,

$$E \equiv E_{\text{OF-DFT}}[\rho] = E_{\text{CDFT}}[N, v] = E_{\text{ITA}}[\{\text{IT}\}] = E_{\text{DAQ}}\left[\{\rho_{\Omega_A}\}\right]. \quad (6.8)$$

where the total energy E is formulated in terms of different variables in the four different frameworks. Under this unification scheme of the four density-based reactivity frameworks, a holistic and consistent approach emerges for describing chemical reactivity from the DFT viewpoint, allowing for a more cohesive interpretation of structure, stability, reactivity, and other physicochemical properties.

Because the electron density $\rho(\mathbf{r})$ uniquely determines the external potential $v_{ext}(\mathbf{r})$ and the ground-state wavefunction (Hohenberg–Kohn theorem), all physically admissible descriptors that are functionals of $\rho(\mathbf{r})$—or are informationally equivalent to $\rho(\mathbf{r})$—must also determine the total energy. Therefore, the four frameworks differ only in the types of variables used, not in the underlying physical content. The mappings among $\{\rho\}$, $\{N, v\}$, {IT}, and $\{\rho_{\Omega_A}\}$ are conceptually invertible in the DFT sense, which justifies the unified expression in Eq. (6.8).

6.2 Topological Analysis of ITA Quantities

The mathematical foundations of topological analysis, including Hessian-based classification of critical points, were introduced in Chapter 3. Here, we extend that analysis to information-theoretic quantities. Unlike electron density or its derivatives, ITA descriptors such as Shannon entropy, Fisher information, and information gain distinct and often nonintuitive topological patterns. These differences enable new ways to interpret bonding, aromaticity, and acidity.

Chapter 3 introduced the full topological analysis of the electron density and its derivatives $(\nabla\rho, \nabla^2\rho)$, including their critical points and spatial patterns (Figure 3.4). Here, in Figure 6.3, we revisit these DAQ-based topologies together with electron localization function (ELF) and localized orbital locator (LOL), two additional localization descriptors not shown previously, to establish a baseline for comparison. This allows us to highlight the qualitative differences that emerge when the same topological machinery is applied to information-theoretic quantities, as shown in Figure 6.4. For electron density in Figure 6.3a, ncp is on the nucleus, bcp at the middle point of the bonds, and rcp at the center of the ring. But for other density-associated local functions, multiple incidences of bcp, ccp, ncp, and rcp are found in this system.

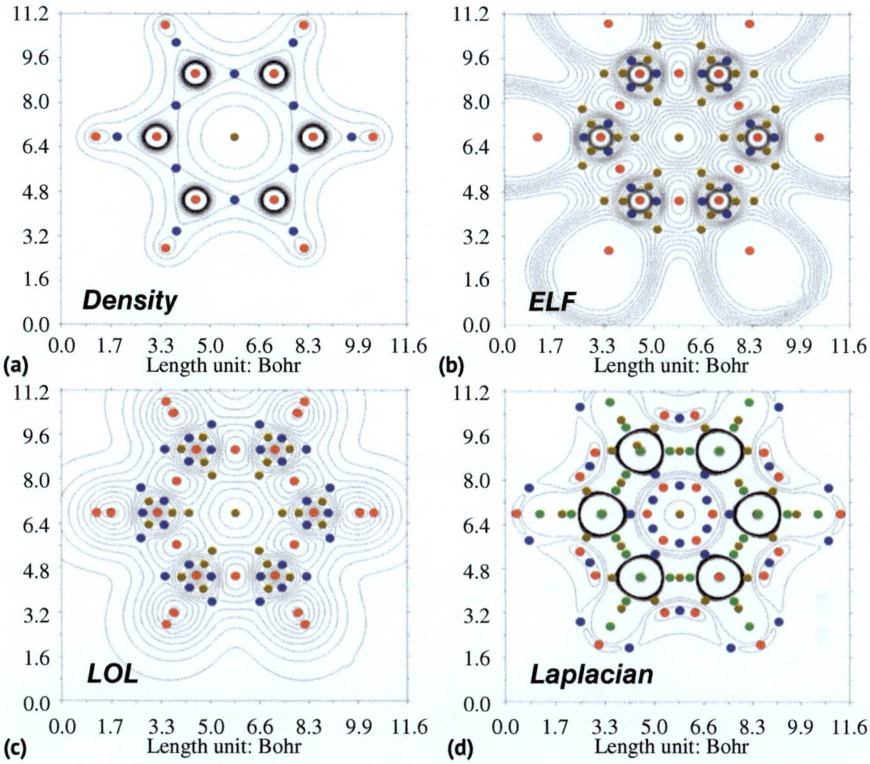

Figure 6.3 The two-dimensional representation of four kinds of cps for four electron density-associated quantities: (a) electron density itself $\rho(\mathbf{r})$; (b) ELF; (c) LOL; and (d) Laplacian, for the benzene molecule. The color code is the following for the four cps: (i) red is ncp, (3,−3); (ii) blue is bcp, (3,−1); (iii) orange is rcp, (3,+1); and green is ccp, (3,+3). *Source:* Reproduced from Ref. [27] with permission of AIP Publishing.

Shown in Figure 6.4 is the two-dimensional *cps* representation of the four ITA quantities, Shannon entropy S_S, Fisher information I_F, information gain I_G, and alternative Fisher information I'_F, for the benzene molecule, where nuclear (ncp, in red), bond (bcp, in blue), ring (rcp, in orange), and cage (ccp, in green) *cps* are displayed. As can be seen from Figure 6.4, both the number and the location of critical points differ markedly among the four ITA quantities and from those of the DAQ functions in Figure 6.3. These results can be applied to correlate with physicochemical properties such as aromaticity, acidity, and electrophilicity [27]. These ITA-based topological descriptors complement the DAQ-based topology developed in Chapter 3 and together provide a multifaceted real-space picture of chemical structure.

In a recent study [27], we analyzed a diverse set of molecular species, including seven with various interaction types (covalent, ionic, metallic, hydrogen bonding, and van der Waals), 23 benzene-fused fulvene derivatives to assess aromaticity, 33 benzoic acids (both singly and doubly substituted) to evaluate acidity, 21 carbocations to probe electrophilicity, and 19 conjugated molecules to examine nucleophilicity. Our findings demonstrated that local information-theoretic measures can yield valuable insights into

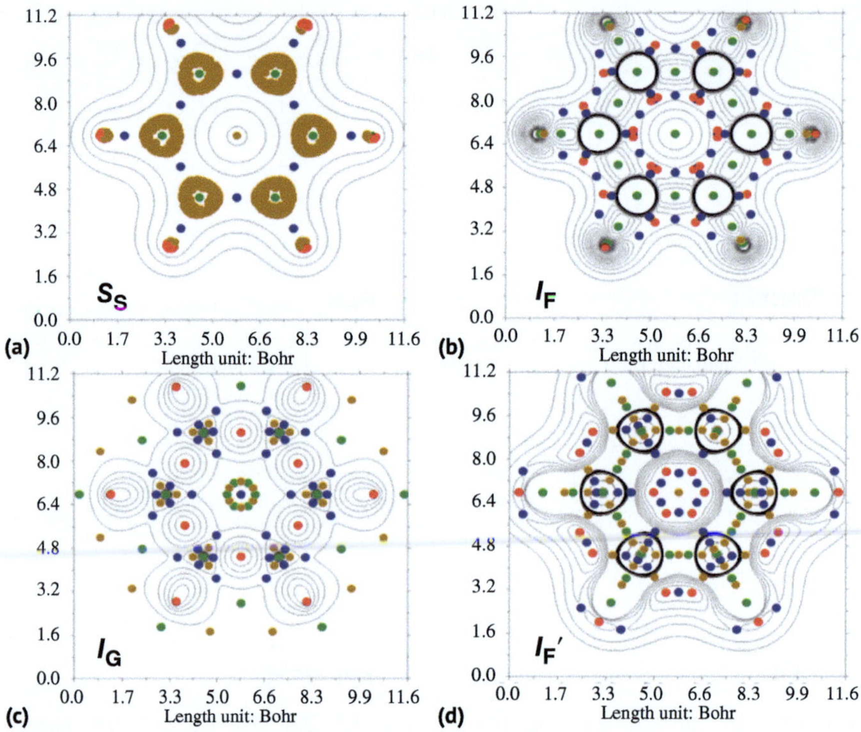

Figure 6.4 The two-dimensional representation of four kinds of cps for four ITA quantities: (a) Shannon entropy S_S; (b) Fisher information I_F; (c) information gain I_G; and (d) alternative Fisher information I'_F, for the benzene molecule. The color code is the same as Figure 6.3 for the four cps. Carbon atoms are denoted by solid circles. Also shown in the figure in grey color for each ITA quantity is its contour surface mapped onto the molecular plane. Reproduced from Ref. [27] with permission of AIP Publishing.

both strong and weak interactions. Specifically, our topological analysis revealed that while Shannon entropy, Fisher information, and information gain adequately characterized several systems, the alternative Fisher information outperformed the others, consistently achieving correlation coefficients greater than 0.90 across all species. Moreover, this quantity effectively differentiated between various bonding interaction types and more reliably captured bonding characteristics and physicochemical properties. Among the four density-related quantities evaluated in Figure 6.3, the electron density emerged as the most effective descriptor. Beyond topology, energy densities themselves can also be cast into an information-theoretic form, as discussed next.

6.3 Energetic Information

Building on the local analysis above, we next ask whether energy densities themselves can be treated as probability distributions [10–12]. We recently explored using the scaled energy density as a probability distribution function to define

ITA quantities. The concept is straightforward. Let the energy E (which may be the total energy or one of its components) be given by

$$E = \int e(\mathbf{r})d\mathbf{r}, \tag{6.9}$$

where $e(\mathbf{r})$ is the energy density. Normalizing this expression by the total energy E, we obtain the normalization condition

$$\int \varepsilon(\mathbf{r})d\mathbf{r} \equiv \int \frac{e(\mathbf{r})}{E}\, d\mathbf{r} = 1. \tag{6.10}$$

with $\varepsilon(\mathbf{r}) \equiv \dfrac{e(\mathbf{r})}{E}$ defined as the scaled energy density. This scaled energy density can now serve as a probability distribution function to construct a new set of ITA quantities. For example, we define the Shannon entropy and two forms of Fisher information as follows:

$$S_S^{\mathcal{E}} = \int s_S^{\mathcal{E}}(\mathbf{r})d\mathbf{r} = -\int \varepsilon(\mathbf{r})\ln \varepsilon(\mathbf{r})d\mathbf{r}, \tag{6.11}$$

$$I_F^{\mathcal{E}} = \int i_F^{\mathcal{E}}(\mathbf{r})d\mathbf{r} = \int \frac{|\nabla \varepsilon(\mathbf{r})|^2}{\varepsilon(\mathbf{r})} d\mathbf{r}, \tag{6.12}$$

$$I_F^{\prime\mathcal{E}} = \int i_F^{\prime\mathcal{E}}(\mathbf{r})d\mathbf{r} = -\int \nabla^2 \varepsilon(\mathbf{r})\ln \varepsilon(\mathbf{r})d\mathbf{r}, \tag{6.13}$$

and their relative forms,

$$I_G^{\mathcal{E}} = \int i_G^{\mathcal{E}}(\mathbf{r})d\mathbf{r} = \int \varepsilon(\mathbf{r})\ln \frac{\varepsilon(\mathbf{r})}{\varepsilon_0(\mathbf{r})} d\mathbf{r}, \tag{6.14}$$

$${}_F^rI^{\mathcal{E}} = \int {}_F^ri^{\mathcal{E}}(\mathbf{r})d\mathbf{r} = \int \varepsilon(\mathbf{r}) \left[\nabla \ln \frac{\varepsilon(\mathbf{r})}{\varepsilon_0(\mathbf{r})}\right]^2 d\mathbf{r}, \tag{6.15}$$

$${}_F^rI^{\prime\mathcal{E}} = \int {}_F^ri^{\prime\mathcal{E}}(\mathbf{r})d\mathbf{r} = \int \nabla^2 \varepsilon(\mathbf{r})\ln \frac{\varepsilon(\mathbf{r})}{\varepsilon_0(\mathbf{r})} d\mathbf{r}. \tag{6.16}$$

Other ITA quantities can be defined accordingly. Here, the superscript $\mathcal{E}$ indicates that these ITA quantities are defined using the scaled energy density $\varepsilon(\mathbf{r})$ as the probability distribution function. As illustrative examples, shown in Figure 6.5 are radial distributions of three Shannon entropy functions, Eq. (6.11), using (a) the kinetic energy T_S, (b) Weizsäcker kinetic energy T_W, and (c) classical Coulombic repulsion J energy density as the probability distribution function.

Figure 6.6 displays the six ITA quantities obtained by the Pauli energy from OF-DFT for the benzene molecule. As illustrated, each quantity exhibits a distinct behavior, highlighting the diversity of these energetic information measures. Although the Fisher information in Figure 6.6b and the alternative Fisher information in Figure 6.6d share a similar visual appearance, their magnitudes differ considerably, as indicated by the color bars. This variation supports our hypothesis that such energetic information quantities can be employed to characterize and, in some cases, quantitatively describe various physicochemical properties across different molecular systems.

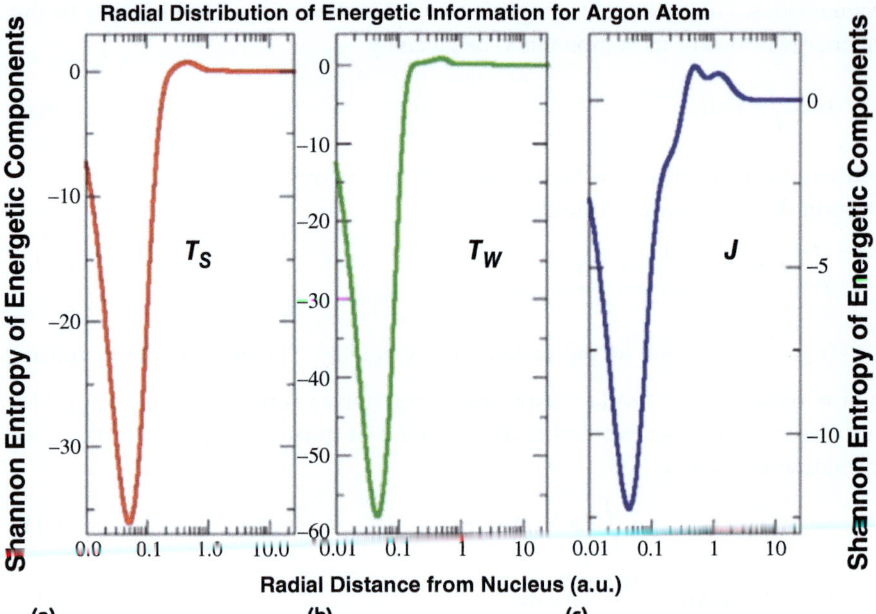

Figure 6.5 Radial distributions of Shannon entropy S_S, Eq. (6.11), for energetic components of the argon atom: (a) kinetic energy component T_S; (b) Weizsäcker kinetic energy T_W; and (c) classical Coulombic repulsion J. Reproduced from Ref. [10] with permission of AIP Publishing.

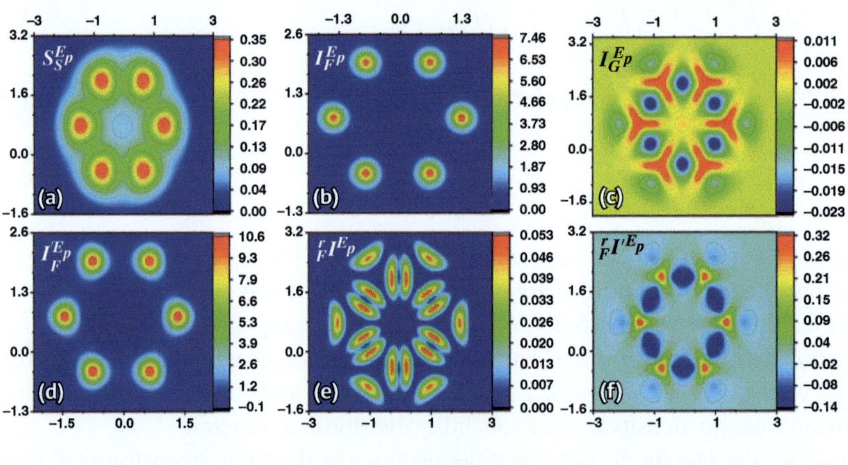

Figure 6.6 The 2D representation for six energetic information quantities of the Pauli energy for the benzene molecule: (a) Shannon entropy; (b) Fisher information; (c) information gain; (d) alternative Fisher information; (e) relative Fisher information; and (f) relative alternative Fisher information.

For example, we recently explored the use of energetic information measures to analyze bonding, conformational stability, intermolecular interactions, acidity, aromaticity, cooperativity, electrophilicity, nucleophilicity, and reactivity [11]. Our results indicate that (i) within a given system, strong linear correlations emerge from these energetic metrics; (ii) different systems display unique patterns when viewed through the lens of energetic information; and (iii) certain energetic measures are more effective than others for specific types of processes.

We found [11] that the Shannon entropy of the total kinetic energy density, $S_S^{T_s}$, emerged as the most effective descriptor for bond stretching, bond-angle bending, dihedral rotations, and overall conformational stability. Meanwhile, the information gain of the Pauli energy, $I_G^{E_p}$, played a key role in quantifying both intermolecular interaction energies and reaction barrier heights. Additionally, for physicochemical properties such as acidity, aromaticity, electrophilicity, nucleophilicity, and cooperativity, the alternative Fisher information of the steric energy, $I_F'^{E_s}$, proved to be a robust descriptor across all systems.

Our recent study also demonstrated that energetic information quantities can be used to distinguish between various types of interactions [12]. For instance, by examining ${}_F^rI'^{E_s}$ and $I_G^{E_s}$, we were able to identify unique signature isosurfaces corresponding to single, double, and multiple covalent bonds, as well as ionic, metallic, and noncovalent interactions. Interactions of the same type display qualitatively similar signature patterns, whereas interactions of different types exhibit distinct structural signatures. Figure 6.7, which highlights a double-bonded covalent system, serves as an example in comparison with other density-based interaction descriptors, such as bonding and nonbonding interaction (BNI), strong covalent interaction (SCI), and Pauli potential. These results further confirm that both covalent and noncovalent interactions can be characterized without resorting to orbital-based approaches. So far, these quantities are all local; extending ITA to the *pair density* supplies nonlocal correlation descriptors in the following Section 6.4.

Because scaled energy densities define legitimate probability distributions, energetic information becomes directly linked to variational properties of DFT. In particular, Fisher information plays an analogous role to kinetic-energy curvature, while information gain relates to interaction contributions. This provides a natural

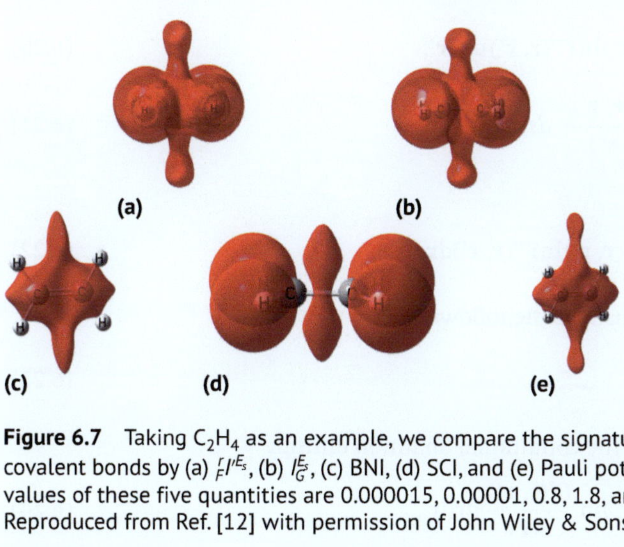

Figure 6.7 Taking C_2H_4 as an example, we compare the signature isosurfaces of double covalent bonds by (a) ${}_F^rI'^{E_s}$, (b) $I_G^{E_s}$, (c) BNI, (d) SCI, and (e) Pauli potential. The isosurface values of these five quantities are 0.000015, 0.00001, 0.8, 1.8, and 0.75 a.u., respectively. Reproduced from Ref. [12] with permission of John Wiley & Sons.

extension of the principle of extreme physical information discussed in Section 6.1.1, thereby grounding energetic information in a well-defined variational framework.

6.4 ITA Extended to Pair Density

Thus far, ITA quantities have predominantly focused on global and local measures with the probability distribution function having one variable, and nonlocal descriptors involving two variables remain largely unexplored. We recently extended ITA to the pair density [28], with which joint, conditional, and mutual information can be obtained. Pair density is defined as,

$$\Gamma(\mathbf{r},\mathbf{r}') = N(N-1)\iint\cdots\int|\Upsilon(\mathbf{r},\mathbf{r}',\mathbf{r}_3,\mathbf{r}_4,\ldots,\mathbf{r}_N)|^2\,d\mathbf{r}_3\,d\mathbf{r}_4\cdots d\mathbf{r}_N, \tag{6.17}$$

where Ψ is the system total wavefunction and $\mathbf{r}$ is the space-spin coordinate. Pair density denotes the probability of finding an electron at $\mathbf{r}$ and another electron at $\mathbf{r}'$. If we perform double-integration for pair density over the whole space, we will get,

$$\iint\Gamma(\mathbf{r},\mathbf{r}')d\mathbf{r}d\mathbf{r}' = N(N-1), \tag{6.18}$$

reflecting the fact that there are $N(N-1)$ electron pairs in the system. Following the shape function $\sigma(\mathbf{r})$ representation of ITA in Chapter 4, we introduced the unit-normalized pair density

$$\Gamma^{\sigma}(\mathbf{r},\mathbf{r}') = \frac{\Gamma(\mathbf{r},\mathbf{r}')}{N(N-1)}, \tag{6.19}$$

and define joint ITA quantities using this new function. For instance, the joint Shannon entropy, joint Fisher information, and joint alternative Fisher information are defined, respectively, as

$$S_{S(\mathbf{X},\mathbf{Y})} = -\iint\Gamma^{\sigma}(\mathbf{r},\mathbf{r}')\ln\Gamma^{\sigma}(\mathbf{r},\mathbf{r}')\,d\mathbf{r}d\mathbf{r}', \tag{6.20}$$

$$I_{F(\mathbf{X},\mathbf{Y})} = \iint\frac{|\nabla_{\mathbf{rr}'}\Gamma^{\sigma}(\mathbf{r},\mathbf{r}')|^2}{\Gamma^{\sigma}(\mathbf{r},\mathbf{r}')}d\mathbf{r}d\mathbf{r}', \tag{6.21}$$

and

$$I'_{F(\mathbf{X},\mathbf{Y})} = -\iint\nabla^2_{\mathbf{rr}'}\Gamma^{\sigma}(\mathbf{r},\mathbf{r}')\ln\Gamma^{\sigma}(\mathbf{r},\mathbf{r}')d\mathbf{r}d\mathbf{r}'. \tag{6.22}$$

The conditional pair density is the following,

$$\Gamma^{\sigma}(\mathbf{r}|\mathbf{r}') = \frac{\Gamma^{\sigma}(\mathbf{r},\mathbf{r}')}{\sigma(\mathbf{r})}, \tag{6.23}$$

with which we can define the conditional Shannon entropy

$$S_{S(\mathbf{X}|\mathbf{Y})} = -\iint\Gamma^{\sigma}(\mathbf{r},\mathbf{r}')\ln\frac{\Gamma^{\sigma}(\mathbf{r},\mathbf{r}')}{\sigma(\mathbf{r})}\,d\mathbf{r}d\mathbf{r}', \tag{6.24}$$

and mutual Shannon entropy as

$$S_{S(\mathbf{X};\mathbf{Y})} = -\iint \Gamma^{\sigma}(\mathbf{r}, \mathbf{r}')\ln \frac{\Gamma^{\sigma}(\mathbf{r}, \mathbf{r}')}{\sigma(\mathbf{r})\sigma(\mathbf{r}')} d\mathbf{r}d\mathbf{r}', \tag{6.25}$$

Notice the difference in the subscript for Shannon entropy in Eqs. (6.20), (6.24), and (6.25). These subscripts are used to differentiate joint, condition, and mutual Shannon entropies. Conditional and mutual forms of the two Fisher information can be defined accordingly:

$$I_{F(\mathbf{X}|\mathbf{Y})} = \int \sigma(\mathbf{r}')d\mathbf{r}' \int \Gamma^{\sigma}(\mathbf{r}|\mathbf{r}')| \nabla_{\mathbf{rr}'} \Gamma^{\sigma}(\mathbf{r}|\mathbf{r}')|^2 \, d\mathbf{r}, \tag{6.26}$$

$$I_{F(\mathbf{X};\mathbf{Y})} = -\iint \Gamma^{\sigma}(\mathbf{r}, \mathbf{r}') \left| \nabla_{\mathbf{rr}'} \ln \frac{\Gamma^{\sigma}(\mathbf{r}, \mathbf{r}')}{\sigma(\mathbf{r})\sigma(\mathbf{r}')} \right|^2 d\mathbf{r}d\mathbf{r}', \tag{6.27}$$

$$I'_{F(\mathbf{X}|\mathbf{Y})} = \int \sigma(\mathbf{r}')d\mathbf{r}' \int \left[\nabla^2_{\mathbf{rr}'} \Gamma^{\sigma}(\mathbf{r}|\mathbf{r}')\right] \ln \Gamma^{\sigma}(\mathbf{r}|\mathbf{r}')d\mathbf{r}, \tag{6.28}$$

and

$$I'_{F(\mathbf{X};\mathbf{Y})} = \iint \nabla^2_{\mathbf{rr}'} \Gamma^{\sigma}(\mathbf{r}, \mathbf{r}') \ln \frac{\Gamma^{\sigma}(\mathbf{r}, \mathbf{r}')}{\sigma(\mathbf{r})\sigma(\mathbf{r}')} d\mathbf{r}d\mathbf{r}'. \tag{6.29}$$

Figure 6.8 presents radial distributions of joint, conditional, and mutual forms of Shannon entropy and Fisher information for Kr, serving as illustrative examples. The figure reveals that these quantities offer particularly rich insights into shell structure. Furthermore, the results highlight how the pair-electron framework enhances the interpretation of electronic correlations and their relationship to spatial localization in atomic systems.

This work emphasizes the significance of extending beyond traditional one-electron frameworks to incorporate pair density. Future applications to molecular

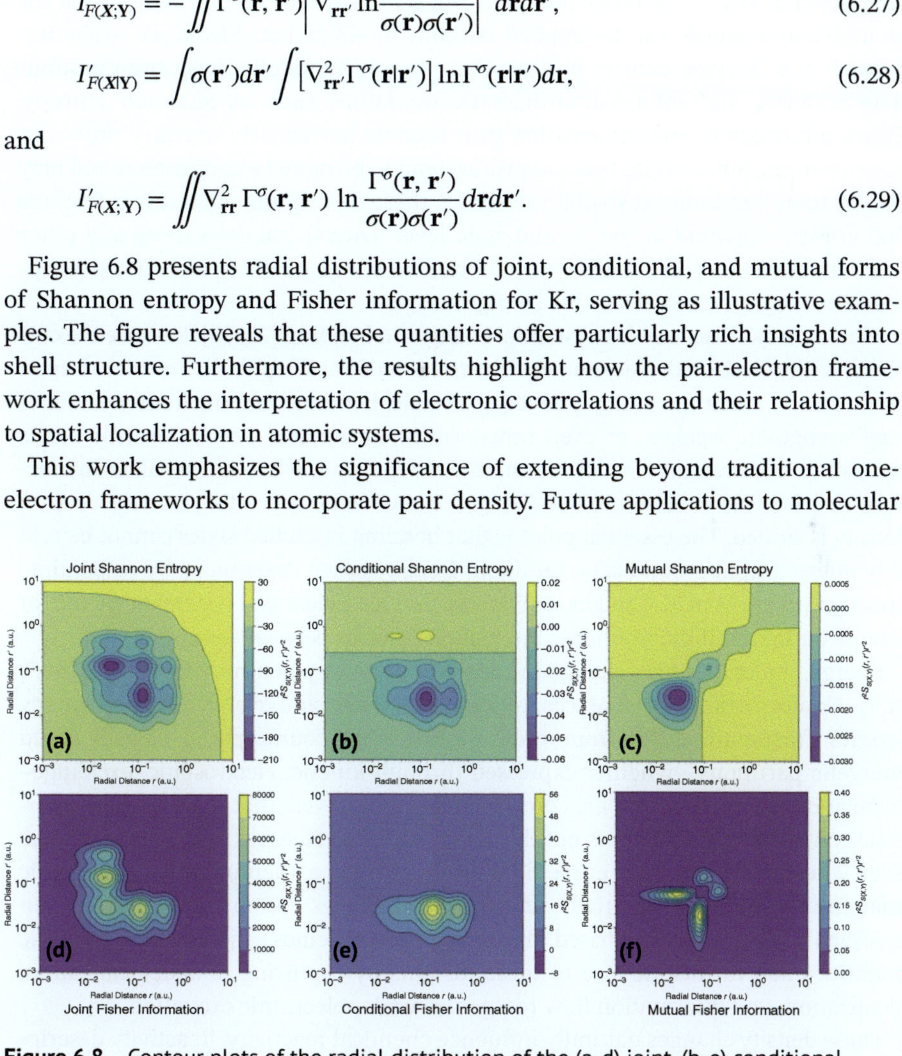

Figure 6.8 Contour plots of the radial distribution of the (a, d) joint, (b, e) conditional, and (c, f) mutual forms for (a–c) Shannon entropy and (d, f) Fisher information for the Kr atom. *Source:* Reproduced from Ref. [28] with permission of AIP Publishing.

systems will allow for a more comprehensive understanding of electron interactions, opening the door to innovative uses of nonlocal ITA descriptors for deeper insights into electronic structure and reactivity.

6.5 Extension to Excited States

Although DFT was originally formulated for ground states, density-based reactivity frameworks extend naturally to excited states because all descriptors—whether from ITA, DAQ, CDFT, or OF-DFT—ultimately depend on the electron density or on functions derived from it. Once an excited-state density is available from time-dependent DFT or other methods, the same conceptual machinery used for ground-state analysis can be applied without modification. Electronic excitation redistributes electron density in ways that often differ qualitatively from ground-state patterns, and information-theoretic quantities such as Shannon entropy, Fisher information, and information gain become particularly sensitive probes of these changes. Since excited-state densities tend to be more heterogeneous and may display more pronounced spatial variations, these descriptors often reveal features that are less apparent at the ground-state level. Directional derivatives and other reactivity indicators also remain valid in excited states, enabling a unified description of reactivity across electronic configurations.

Excitation can also lead to substantial changes in bonding [29, 30]. Because the curvature of the electron density and the distribution of Pauli repulsion are altered upon promotion of electrons to higher-energy orbitals, bonding patterns in excited states may strengthen, weaken, or even transform in unexpected ways. Covalent bonds can exhibit reduced localization or enhanced repulsion, while noncovalent interactions may become either more stabilizing or more fragile depending on how electron density is shifted. The essential point is that bonding in excited states cannot be reliably inferred from ground-state intuition. Density-based descriptors, by responding directly to changes in $\rho(\mathrm{r})$ and its derivatives, provide a clear and systematic picture of how electronic excitation reshapes interaction patterns between atoms.

Energy landscapes also change in meaningful ways when moving from S_0 to S_1 or T_1. Conformational preferences, rotational barriers, and overall potential energy profiles frequently differ from their ground-state counterparts. Density-based energetic partitions—whether expressed through kinetic, electrostatic, exchange–correlation, steric, or fermionic components—often reveal that the dominant energetic contributor to stability is not the same in excited states as in the ground state. Even when potential energy curves look similar in shape, their underlying energetic anatomy can differ substantially, and these differences become visible only when analyzed through density-based decomposition schemes. Information-theoretic quantities add further nuance to these insights by capturing how delocalization, localization, and information flow reorganize under electronic excitation.

These density changes naturally influence chemical reactivity. Reactivity descriptors derived from the electron density, such as Fukui functions, local temperature, and information gain, continue to provide reliable measures of electrophilic and nucleophilic tendencies in excited states. However, the directional preferences of reactions may shift. Substituents that direct electrophilic attack toward one position in the ground state may direct it toward a different position in the excited state, and regions of elevated or suppressed reactivity may relocate as density redistributes. Density-based reactivity theory is therefore especially valuable in excited-state chemistry, as it ties these altered patterns directly to physically interpretable features of the electron density.

In summary, the extension of density-based reactivity theory to excited states is conceptually seamless and provides a unified interpretive framework for understanding how electronic excitation affects density redistribution, bonding, stability, and chemical reactivity. Excited-state chemistry cannot be reduced to ground-state reasoning, because new density patterns and new energetic balances arise upon excitation. Information-theoretic and density-based descriptors offer a clear and coherent means of capturing these changes while maintaining continuity with the broader density-based paradigm. A full, systematic, and example-rich treatment—including detailed difference maps, case studies, energetic partitions, and reactivity analyses—is presented in Chapter 15, which serves as the dedicated chapter for excited-state density-based reactivity theory.

6.6 Merging with Machine Learning

Beyond ground and excited states, density-based descriptors also lend themselves naturally to data-driven prediction. In particular, machine learning techniques have been used to predict molecular response properties such as polarizabilities and NMR shielding [31–37]. We introduced a novel computational protocol that integrates the generalized energy-based fragmentation (GEBF) method with deep learning (DL), leveraging both molecular and atomic ITA descriptors.

This GEBF-DL(ITA) approach [31–33] utilizes molecular wavefunctions to compute electron density-based ITA descriptors, which are then used as inputs for DL models. These models demonstrated high accuracy in predicting molecular polarizabilities and NMR shielding constants for large macromolecules and their aggregates, all while maintaining relatively low computational costs. Notably, this method enabled the first-ever collection of isotropic polarizability data for large proteins and protein aggregates (as shown in Figure 6.9). The approach achieved a mean absolute percentage error of less than 1%, highlighting its exceptional predictive accuracy and efficiency for macromolecular systems containing thousands of atoms.

Building on the previous work on polarizabilities, we developed an accurate and efficient methodology for predicting molecular polarizabilities in excited states for condensed-phase systems. This approach combined post-Hartree–Fock methods with ITA quantities within the PBC–GEBF–ITA framework, where molecular polarizabilities for large periodic systems were determined through linear relationships with ITA descriptors derived from electron density.

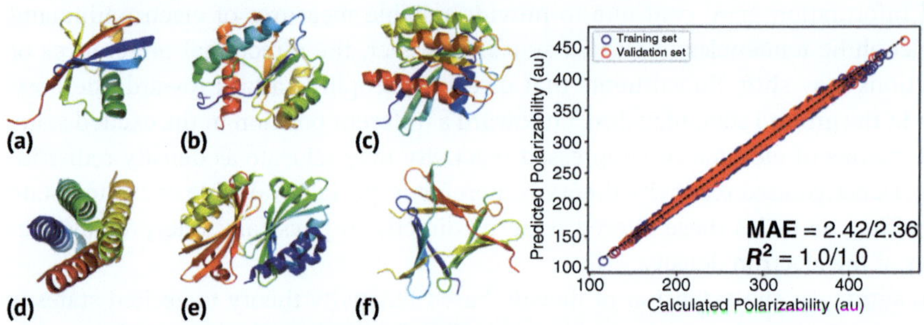

Figure 6.9 (a) 1UBQ (1240 atoms), human erythrocytic ubiquitin, and (b) 3LUC (2188 atoms), the MID (middle) domain from human protein argonaute-2, (c) wild-type KRAS (2692 atoms); 8AZY (2697 atoms), mutant KRAS^{G12D}; 8AZZ (2701 atoms), mutant KRAS^{G12V}, (d) four-helix bundled 6C52 (1792 atoms), (e) 5TPH (4084 atoms), a homodimer with cavities formed by curved β-sheets, and (f) 5XZK (1821 atoms), a homotrimer of *Pholiota squarrosa* lectin. *Source:* Reproduced from Ref. [33] with permission of American Chemical Society.

This method significantly enhanced computational efficiency while maintaining accuracy comparable to the original PBC–GEBF approach [35]. We demonstrated strong scalability and transferability, as the model reliably predicted polarizabilities for large periodic systems containing several hundred atoms per unit cell. This research paves the way for further exploration of molecular response and other physicochemical properties in condensed-phase materials using ITA descriptors.

These successes highlight why ITA descriptors are particularly suitable for machine learning: they provide compact, physically meaningful summaries of electron-density distributions, allowing neural networks to operate in an information-space that preserves chemical interpretability. Thus, machine learning (ML) does not replace physical reasoning; rather, ITA quantities provide an interpretable bridge between first-principles electron density and data-driven prediction.

6.7 Concluding Remarks and Outlook

Density-based reactivity theory has undergone rapid and profound growth [3]. By clarifying the relationships among the four frameworks—CDFT, DAQ, ITA, and OF-DFT—this chapter demonstrates that they can be unified under a single perspective where total energy and reactivity are expressed through density-derived descriptors. This unification not only consolidates the conceptual foundation of density-based approaches but also enriches their interpretive power across chemical domains. These foundations underwrite the extensions and the applications discussed earlier, which together chart a path for scalable, conceptually transparent reactivity analysis.

New developments, such as topological analysis of ITA quantities and the concept of energetic information, expand the descriptive and predictive capabilities of density-based methods. The extension of ITA to pair density introduces nonlocal measures that capture electron correlation more comprehensively. Applications to excited states illustrate that density-based descriptors can reveal qualitative and quantitative changes in bonding, stability, and reactivity—domains traditionally considered beyond the reach of ground-state DFT. Finally, the synergy between ITA and machine learning underscores the potential for predictive modeling at scales ranging from small molecules to proteins and materials, paving the way for future integration of physics-based descriptors with data-driven methods.

The take-home message is the following. The recent advances discussed in this chapter consolidate density-based reactivity theory as a coherent and versatile paradigm. By bridging frameworks, extending descriptors, and merging with machine learning, density-based approaches are poised to provide a unified, scalable, and conceptually transparent path toward deeper chemical understanding in the new era of the quantum age.

Looking ahead, all four density-based frameworks are poised for further conceptual and methodological expansion. OF-DFT is expected to advance through improved Pauli potentials and machine-learned kinetic energy functionals, enabling routine large-scale and excited-state simulations. CDFT will likely deepen its influence through incorporating density matrix and higher-order derivatives, multicomponent reactivity descriptors, and direct connections to free-energy landscapes. DAQ methods may evolve toward examining other dimensionless quantities, supported by multiresolution real-space analysis and unified topological descriptors that integrate density, energy, and information. ITA, with its rapidly growing scope, is uniquely positioned to become a crosscutting backbone for hierarchical modeling, pair-density analysis, and interpretable ML representations. Together, these anticipated developments suggest that the next decade will see an increasingly seamless and unified density-based paradigm capable of addressing complex chemical behavior across scales and states of matter.

On the application side, the continued development of these density-based frameworks is expected to broaden their reach across diverse chemical domains. In excited-state and time-dependent processes, density-derived descriptors may provide real-time, chemically intuitive insights into ultrafast charge migration, photochemical reactivity, and nonadiabatic pathways. In materials and energy research, density-based analysis, particularly energetic information and pair-density ITA, holds promise for the rational design of catalysts, photovoltaic materials, and solar-energy conversion systems. In the biological and medicinal arenas, the combination of density-based descriptors with machine learning is positioned to accelerate molecular recognition, protein–ligand interactions, and drug-design workflows through physically interpretable models. As these frameworks mature, their integration with dynamics, large-scale simulations, and data-driven pipelines will likely establish density-based reactivity theory as an essential tool across chemical, materials, and biological sciences.

References

1 Hohenberg P, Kohn W. Inhomogeneous electron gas. Phys Rev. 1964 Nov 9; 136(3B):B864–71.

2 Parr RG, Yang W. Density-Functional Theory of Atoms and Molecules. Oxford University Press; 1989.

3 Rong C, Zhao D, He X, Liu S. Development and applications of the density-based theory of chemical reactivity. J Phys Chem Lett. 2022 Dec 8;13(48): 11191–200.

4 An X, Zhang W, He X, Li M, Rong C, Liu S. Density-based reactivity theory applied to excited states. AAPPS Bull. 2024 Feb 2;34(1): 8.

5 Liu S. On the relationship between densities of Shannon entropy and Fisher information for atoms and molecules. J Chem Phys. 2007 May 21;126(19): 191107.

6 Liu S. Identity for Kullback-Leibler divergence in density functional reactivity theory. J Chem Phys. 2019 Oct 14;151(14): 141103.

7 Mi W, Luo K, Trickey SB, Pavanello M. Orbital-free density functional theory: an attractive electronic structure method for large-scale first-principles simulations. Chem Rev. 2023 Nov 8;123(21): 12039–104.

8 Xu Q, Ma C, Mi W, Wang Y, Ma Y. Recent advancements and challenges in orbital-free density functional theory. WIREs Comput Mol Sci. 2024 May;14(3):e1724.

9 Nagy Á. Fisher information and density functional theory. Int J Quantum Chem. 2022 Apr 15;122(8):e26679.

10 Liu S. Quantifying energetic information in density functional theory. J Chem Phys. 2022 Sep 14;157(10): 101103.

11 He X, Lu T, Rong C, Liu W, Ayers PW, Liu S. Energetic information from information-theoretic approach in density functional theory as quantitative measures of physicochemical properties. J Chem Theory Comput. 2024 Jul 12; 20(14): 6049–61.

12 He X, Liu W, Liu S. Using energetic information quantities from density functional theory to simultaneously identify both covalent and noncovalent interactions. ChemPhysChem. 2025 Apr 1;26(7):e202401068.

13 Liu S. Conceptual density functional theory and some recent developments. Acta Phys-Chim Sin. 2009;25(3):590–600.

14 Liu S, editor. Conceptual Density Functional Theory: Towards a New Chemical Reactivity Theory. 1st ed. Wiley-VCH GmbH: Wiley; 2022.

15 Geerlings P, De Proft F, Langenaeker W. Conceptual density functional theory. Chem Rev. 2003 May;103(5): 1793–873.

16 Geerlings P, Chamorro E, Chattaraj PK, De Proft F, Gázquez JL, Liu S, et al. Conceptual density functional theory: status, prospects, issues. Theor Chem Acc. 2020 Feb;139(2): 36.

17 Guo C, Li M, Rong C, Liu S. Finite difference representation of information-theoretic approach in density functional theory. Theor Chem Acc. 2023 Sep;142(9): 83.

18 Liu S. Local-density approximation, hierarchy of equations, functional expansion, and adiabatic connection in current-density-functional theory. Phys Rev A. 1996 Aug;54(2): 1328–36.

19 Liu S, Parr RG. Expansions of density functionals in terms of homogeneous functionals: justification and nonlocal representation of the kinetic energy, exchange energy, and classical Coulomb repulsion energy for atoms. Phys Rev A. 1997 Mar 1;55(3): 1792–8.

20 Liu S. Expansions of the pair distribution function and the second-order density matrix in terms of homogeneous functionals. Phys Rev A. 1996 Dec;54(6): 4863–7.

21 Liu S, Parr RG. Second-order density-functional description of molecules and chemical changes. J Chem Phys. 1997 Apr 1;106(13): 5578–86.

22 Parr RG, Liu S, Kugler AA, Nagy Á. Some identities in density-functional theory. Phys Rev A. 1995 Aug;52(2): 969–76.

23 Liu S, Nagy Á, Parr RG. Expansion of the density-functional energy components E_c and T_c in terms of moments of the electron density. Phys Rev A. 1999 Feb 1;59(2): 1131–4.

24 Liu S, Parr RG. Expansions of the correlation-energy density functional Ec. Phys Rev A. 1996 Apr;53(4): 2211–9.

25 Zhou X-Y, Rong C, Lu T, Zhou P, Liu S. Information functional theory: electronic properties as functionals of information for atoms and molecules. J Phys Chem A. 2016 May 26;120(20): 3634–42.

26 Bader RFW. Atoms in Molecules. Oxford University Press; 1989.

27 He X, Lu T, Rong C, Liu S, Ayers PW, Liu W. Topological analysis of information-theoretic quantities in density functional theory. J Chem Phys. 2023 Aug 7;159(5): 054112.

28 Zhao Y, Zhao D, Rong C, Liu S, Ayers PW. Extending the information-theoretic approach from the (one) electron density to the pair density. J Chem Phys. 2025 Jun 28;162(24): 244108.

29 Liu S, Rong C, Lu T, Hu H. Identifying strong covalent interactions with Pauli energy. J Phys Chem A. 2018 Mar 22;122(11): 3087–95.

30 Zhong S, He X, Liu S, Wang B, Lu T, Rong C, et al. Toward density-based and simultaneous description of chemical bonding and noncovalent interactions with Pauli energy. J Phys Chem A. 2022 Apr 21;126(15): 2437–44.

31 Zhao D, Zhao Y, He X, Ayers PW, Liu S. Efficient and accurate density-based prediction of macromolecular polarizabilities. Phys Chem Chem Phys. 2023 Jan 18;25(3): 2131–41.

32 Zhao D, Zhao Y, He X, Li Y, Ayers PW, Liu S. Accurate and efficient prediction of post-Hartree-Fock polarizabilities of condensed-phase systems. J Chem Theory Comput. 2023 Sep 26;19(18): 6461–70.

33 Zhao D, Zhao Y, Xu E, Liu W, Ayers PW, Liu S, et al. Fragment-based deep learning for simultaneous prediction of polarizabilities and NMR shieldings of macromolecules and their aggregates. J Chem Theory Comput. 2024 Mar 26;20(6): 2655–65.

34 Zhao Y, Zhao D, Liu S, Rong C, Ayers PW. Why are information-theoretic descriptors powerful predictors of atomic and molecular polarizabilities. J Mol Model. 2024 Oct 3;30(11): 361.

35 Yuan Y, Zhao Y, Lu L, Wang J, Chen J, Liu S, et al. Multiproperty deep learning of the correlation energy of electrons and the physicochemical properties of molecules. J Chem Theory Comput. 2025 Jun 24;21(12): 5997–6006.

36 Zhao D, He X, Ayers PW, Liu S. Excited-state polarizabilities: a combined density functional theory and information-theoretic approach study. Molecules. 2023;28(6):2576.

37 Zhao D, Liu S, Chen D. A density functional theory and information-theoretic approach study of interaction energy and polarizability for base pairs and peptides. Pharmaceuticals. 2022;15(8):938.

Part II

Applications

7

Covalent and Noncovalent Interactions

TABLE OF CONTENTS

7.1 Introduction

Chemistry focuses on the behavior and interactions of molecules, driving transformations in their structure and properties—such as stability and reactivity [1, 2]. Unraveling the origins of these changes hinges on grasping the fundamental forces governing atomic interactions within molecules. What mechanisms enable atoms to bond and assemble into molecules or materials? How do their electronic configurations evolve during bonding processes, altering their chemical properties? What defines the nature of these bonds, and how can stability and reactivity be quantified at the atomic or molecular level? Questions like these have challenged and inspired chemists for centuries.

Atoms interact through forces that span a wide spectrum—from strong covalent bonds to weak van der Waals attractions [3–6]. Covalent bonds arise from electron sharing and define molecular structure, while ionic, metallic, hydrogen-bonding, and dispersion interactions shape stability, supramolecular organization, and reactivity. Despite their diversity, all interactions originate from the redistribution of electron density when atoms assemble into molecules, reflecting the underlying quantum mechanical principles that govern chemical structure and behavior.

Chemical bonding is one of the few traditional chemical concepts that electronic structure theories, including valence bond theory (VBT) [7–12] and molecular orbital theory (MOT) [13–15], were immediately applied to provide deep understanding and novel insights. VBT, pioneered by Walter Heitler, Fritz London, and Linus Pauling, emphasized the role of overlapping atomic orbitals and hybridization in forming localized bonds, offering an intuitive picture of covalent interactions. In contrast, MOT, developed by Robert Mulliken, Friedrich Hund, and others, introduced the concept of delocalized orbitals encompassing entire molecules, enabling the explanation of aromaticity, bonding in diatomic species like O_2, and the quantum mechanical nature of electron distribution. These deeper insights have enabled the explanation and prediction of the structure, stability, and reactivity properties of molecules. Moreover, by incorporating phenomena such as resonance, electron delocalization, feedback bonding, and multicenter bonding, these approaches continue to shape chemical research, bridging the gap between empirical observations and theoretical understandings from the quantum mechanical perspective.

Density functional theory (DFT) [16], a groundbreaking electronic structure theory, has reshaped computational studies of molecular and material systems over the past few decades due to its remarkable balance of accuracy and efficiency. Yet, its application to analyze atomic interactions—such as covalent bonding and noncovalent interactions (NCIs) like hydrogen bonding or van der Waals forces—remained constrained until very recently. This limitation stemmed from the traditional reliance on orbital-based VBT and MOT frameworks (atomic or molecular orbitals) to describe bonding, which contrasts with DFT's density-centric approach that inherently lacks explicit orbital descriptions. Recently, we circumvented this orbital dependency by leveraging density-derived quantities to identify and quantify both covalent bonding and NCIs [1, 2]. Remarkably, a density-based formulation was found to capture a broad spectrum of interactions, from strong covalent bonds to delicate noncovalent forces, within a unified conceptual framework.

In this chapter, we begin by reviewing the orbital-based approach from VBT and MOT to use atomic and molecular orbitals to appreciate chemical bonding. We then shift to the density-based approach offered by DFT, which allows for the description of both covalent bonding and NCIs. In the orbital-based picture, covalent bonds arise directly from orbital overlap, whereas ionic, metallic, hydrogen-bonding, and dispersion interactions rely on electrostatic or correlation-driven explanations rather than orbital mixing. By contrast, the density-based approach treats all interaction types uniformly through electron density and its derivatives. This terminology—distinguishing *covalent bonding* from NCIs—follows conventional orbital-based usage and is adopted here solely for conceptual consistency with earlier chapters.

7.2 Orbital-based Approaches

7.2.1 Valence Bond Theory

VBT, the earliest quantum-mechanical description of chemical bonding, depicts a covalent bond as the overlap of localized atomic orbitals that share electron pairs. This localized perspective naturally accounts for bond directionality and the structural features of molecules.

Orbital Overlap. At the heart of VBT is the concept of orbital overlap. Consider the hydrogen molecule, H_2, as an example. Each hydrogen atom contributes a 1s atomic orbital. The overlap of these orbitals leads to the formation of a covalent bond between two H atoms. The VB wave function for H_2 can be written as:

$$\Psi_{H_2} = N\left[\phi_{1s}(1_A)\phi_{1s}(2_B) + \phi_{1s}(2_A)\phi_{1s}(1_B)\right], \tag{7.1}$$

where N is a normalization constant ensuring that the total wave function is normalized, and the indices *1* and *2* denote the electrons assigned to atoms A and B. This symmetric combination emphasizes that each electron is shared between the two atoms. The stabilization resulting from orbital overlap can be quantitatively assessed by calculating the energy expectation value, which shows a lower energy for the combined system compared with isolated atoms. The strength of the bond is directly related to the extent of orbital overlap. When the orbitals have a significant spatial overlap, the probability of finding electron density between the two nuclei increases, thereby stabilizing the bond. In contrast, minimal or poor overlap results in weaker bonding interactions. Also, VBT can distinguish between σ bonds (head-on overlap) and π bonds (sideways overlap).

Hybridization and Bond Directionality. A powerful extension of VBT is the concept of hybridization, which explains molecular geometry by allowing atoms to mix s and p orbitals to form oriented hybrid orbitals optimized for overlap. For example, in methane (CH_4), the 2s and 2p orbitals of the carbon atom hybridize to form four equivalent sp^3 orbitals,

$$\phi_{sp^3} = \frac{1}{2}\left(\phi_{2s} + \phi_{2p_x} + \phi_{2p_y} + \phi_{2p_z}\right), \tag{7.2}$$

with similar expressions for the other three sp^3 orbitals differing in phase and orientation. This hybridization results in a tetrahedral geometry with bond angles of approximately 109.5°, matching experimental observations. The directional characteristics of these hybrid orbitals are crucial for predicting the shapes of more complex molecules and understanding their reactivity.

Resonance and Electron Delocalization. While VBT primarily describes localized bonds, it also provides a framework for understanding resonance in molecules where electron density is delocalized over several atoms. A classic example is benzene (C_6H_6), where the bonding cannot be depicted by a single Lewis structure. For cases like this, VBT invokes the concept of resonance, the superposition of multiple Lewis structures. The actual wavefunction Ψ is a weighted average:

$$\Psi = c_1\Psi_1 + c_2\Psi_2 + c_3\Psi_3 + \cdots \tag{7.3}$$

This superposition captures electron delocalization over the carbon ring in benzene, leading to bond lengths that are intermediate between typical single and double bonds. The energy decrease due to the resonance between Kekulé structures can be quantified by the *resonance energy* (ΔE_{res}),

$$\Delta E_{\text{res}} = E_{\text{localized}} - E_{\text{delocalized}}. \tag{7.4}$$

With this effect of electron delocalization, VBT can be used to explain conjugation, hyperconjugation, and aromaticity. Based on this, Shaik and Hiberty et al. [6, 17] proposed a new type of chemical bond, the charge-shift bond, in which the resonance energy contributes over 50% to the total bond dissociation energy.

7.2.2 Molecular Orbital Theory

MOT describes bonding through delocalized orbitals formed from linear combinations of atomic orbitals. Electrons occupy these molecular orbitals according to their energies, providing a complementary view to VBT's localized picture and enabling intuitive explanations of bond order, electron delocalization, aromaticity, and magnetism.

Linear Combination of Atomic Orbitals (LCAOs). The core principle of MO theory is the LCAO method. When two atoms approach each other, their atomic orbitals combine to form bonding (lower energy) and antibonding (higher energy) molecular orbitals. Again, using H_2 as an example, the 1s atomic orbital, ϕ_A and ϕ_B, of the two H atoms combines to form two molecular orbitals: a *bonding* orbital and an *antibonding* orbital. The bonding molecular orbital (lower in energy) is given by:

$$\psi_+ = \frac{1}{\sqrt{2(1+S)}}(\phi_A + \phi_B), \tag{7.5}$$

and the antibonding molecular orbital (higher in energy) is:

$$\psi_- = \frac{1}{\sqrt{2(1-S)}}(\phi_A - \phi_B), \tag{7.6}$$

with S being the overlap integral between ϕ_A and ϕ_B. Electrons fill these orbitals according to the Pauli exclusion principle and Hund's rule, prioritizing lower energy bonding orbitals. In the bonding orbital ψ_+, electron density is enhanced between the nuclei, yielding a net attractive force that holds the two atoms together. In contrast, the antibonding orbital ψ_- contains a node between the nuclei and generally leads to destabilization if occupied. The *bond order* quantifies bond strength and is calculated as:

$$\text{Bond order} = \frac{\text{Bonding Electrons} - \text{Antibonding Electrons}}{2}. \tag{7.7}$$

With this, one has single, double, triple, quadruple, and even stronger covalent bonds.

Localized Molecular Orbitals (LMOs). Although MOT yields delocalized orbitals, unitary transformations can produce LMOs that resemble traditional σ bonds, π bonds, and lone pairs. Localization schemes such as Boys or Edmiston–Ruedenberg methods maximize spatial separation of electron density, bridging quantum-mechanical descriptions with the chemical intuition of localized bonding. The ability to localize molecular orbitals underscores that electrons are often concentrated in specific regions of a molecule, aligning MOT with Lewis structures. In polyatomic systems, LMOs neatly describe bonds between adjacent atoms and lone pairs not directly involved in bonding interactions.

Feedback (Back) Bonding. Back bonding occurs when a ligand donates a σ lone pair to a metal center, while the metal simultaneously donates d-electron density into the ligand's π* orbital. This two-way interaction strengthens the metal–ligand bond, stabilizes the complex, and often shortens bond lengths relative to systems lacking such feedback donation. A simplified depiction of back bonding can be written as:

$$\sigma: \ L \rightarrow M(\text{donation}) \tag{7.8}$$

$$\pi*: \ M \rightarrow L(\text{back} - \text{donation}) \tag{7.9}$$

where L represents the ligand and M is the metal center.

MOT can also account for conjugation, hyperconjugation, aromaticity, multicenter bonding, and other nonclassical interactions by describing electrons as delocalized over a portion or even the entirety of a molecular structure. This unifying framework bridges localized bonding models with quantum-mechanical descriptions and enables the prediction of unconventional bonding patterns relevant to materials science and molecular design.

7.3 Density-based Approaches

DFT has emerged as the dominant computational tool in theoretical and computational chemistry over recent decades. However, its reliance on electron density—a three-dimensional local function—as the foundational variable inherently limits its ability to provide an orbital-based perspective on chemical bonding. This constraint complicates the interpretation of quintessential chemical concepts, such as bond formation, resonance, and reactivity, which traditionally rely on molecular orbitals and localized bonding interactions. Kohn–Sham orbitals and energy decomposition analysis [18–22] can help reconcile these viewpoints by offering orbital-based interpretations, but they lie outside the density-based framework. In contrast, the approaches reviewed in this section focus on how bonding and nonbonding interactions (BNIs) can be described purely from density-derived quantities. Recent developments demonstrate that these interactions can be understood in a coherent and conceptually transparent way directly from the density itself [1, 2, 23, 24]. In what follows, we summarize these advances.

7.3.1 NCI Index

NCIs—such as hydrogen bonding, van der Waals forces, and π–π stacking—govern molecular recognition, protein folding, and material assembly, playing a pivotal role in chemistry, biology, and materials science [25–27]. Despite their ubiquity, these weak interactions are notoriously challenging to characterize due to their subtle energetics and sensitivity to environmental fluctuations. DFT excels in modeling electron density distributions but struggles to provide intuitive, orbital-based frameworks because the basic variable is the density, a three-dimensional scalar field that lacks explicit information about orbitals. The NCI [28–30] was the first DFT-based descriptor to emerge as a transformative tool and enabled us to visualize and quantify weak interactions directly from electron density.

The NCI index, pioneered by Johnson, Yang, and collaborators [28, 29], leverages two key properties derived from the electron density (ρ): the *reduced density gradient* ($\mathbf{s}$) (RDG) and the eigenvalues of the density Hessian matrix. The RDG is defined as:

$$s(\mathbf{r}) = \frac{|\nabla\rho(\mathbf{r})|}{2(3\pi)^{1/3}\rho^{4/3}(\mathrm{r})}. \tag{7.10}$$

which measures deviations from a homogeneous electron distribution. Regions with small $\mathbf{s}$ and ρ values—such as hydrogen bonds, steric clashes, or van der Waals contacts—correspond to NCIs.

By analyzing the second eigenvalue (λ_2) of the Hessian matrix of the electron density, the NCI index distinguishes between attractive ($\lambda_2 < 0$) and repulsive ($\lambda_2 > 0$) interactions, with interaction strength proportional to $\rho(\mathbf{r})$. When plotted, the function $s(\mathbf{r})$ versus $\mathrm{sign}(\lambda_2)\rho(\mathbf{r})$ exhibits characteristic spikes whose positions correlate with the type and strength of the interaction. In practice, isosurfaces of $s(\mathbf{r})$ can be rendered and color-coded according to $\mathrm{sign}(\lambda_2)\rho(\mathbf{r})$, yielding a clear, three-dimensional map of NCIs in the molecular system. For example, as shown in Figure 7.1, hydrogen bonds appear as blue isosurfaces (strong attraction), while steric repulsions are red, and van der Waals interactions are green.

This approach is inherently transferable because it relies solely on electron density, which can be obtained from either quantum-mechanical calculations or experimental X-ray diffraction data. Its independence from molecular orbital transformations makes it particularly suited for large systems, including proteins and

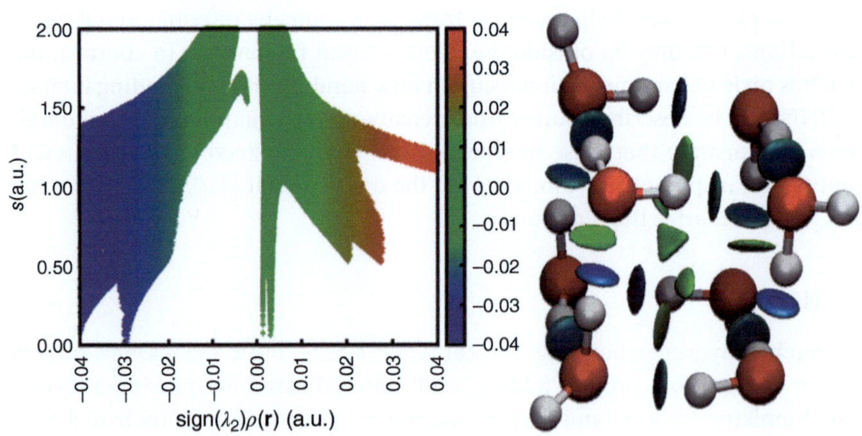

Figure 7.1 NCI representation in 3D and 2D of an eight water molecule cluster.

nanomaterials. Moreover, the visualization achieved through the NCI index not only offers a qualitative picture but can also be accompanied by quantitative measures that help in comparing interaction strengths across different systems. The method has been implemented in various computational chemistry packages and stands as an accessible tool seeking to interpret DFT results in terms of physically intuitive interactions.

7.3.2 Strong Covalent Interaction Index

Although the NCI index is highly effective for mapping weak NCIs, it is inherently ill-suited for detecting or characterizing strong covalent bonds. This shortcoming is rooted in its original design, which utilizes the RDG to accentuate subtle changes in electron density in low-density regions characteristic of van der Waals forces, hydrogen bonds, and π–π interactions. In contrast, covalent bonds feature sharply localized electron density accumulations and steep density gradients between nuclei, phenomena that exceed the sensitivity range of the RDG. These regions of intense electron concentration, often marked by large positive Laplacian values $(\nabla^2\rho)$ and significant Pauli repulsion, embody the charge distributions essential for covalent bonding yet remain obscured by the NCI index's emphasis on smooth, low-density gradients. Therefore, to achieve a thorough characterization of strong covalent interactions (SCIs), additional descriptors, such as the electron localization function (ELF) [31–33], the SCI index [34, 35], or other density-based measures, must be employed to complement the NCI approach and provide a complete depiction of the bonding landscape in molecular systems.

We introduced the SCI index [34, 35] in 2018 to map out the regions of significant electron repulsion that define the nature of SCIs. It is based on the hypothesis that in the interatomic regions where strong covalent bonding occurs, the Pauli repulsion—stemming from the Pauli exclusion principle and quantified by the Pauli energy—becomes dominant. This premise suggests that the Pauli energy can serve as a definitive marker for identifying and characterizing strong covalent bonds. Furthermore, we posited that covalent bonds sharing the same multiplicity (e.g., both are double bonds) should exhibit similar Pauli repulsion characteristics. Building on these ideas and using the distinctive isosurface patterns observed in bonding regions as indicators, we developed a qualitative framework that correlates local Pauli energy contributions with covalent bond multiplicities (e.g., double, triple, and quadruple bonds). This approach enables us to map out the regions of significant electron repulsion that define the nature of covalent bonding interactions.

In DFT, Pauli energy is defined as

$$E_P[\rho] \equiv \int e_P(\mathbf{r})d\mathbf{r} = T_S[\rho] - T_W[\rho] \equiv \int [\tau_S(\mathbf{r}) - \tau_W(\mathbf{r})]\, d\mathbf{r}, \tag{7.11}$$

where E_P, T_S, and T_W stand for the Pauli energy, total noninteracting kinetic energy, and Weizsäcker kinetic energy, respectively, and e_P, τ_S, and τ_W represent their corresponding local energy density. The analytical expression for the Weizsäcker kinetic energy is the following,

$$T_W[\rho] = \int \tau_W(\mathbf{r})d\mathbf{r} \equiv E_s = \int e_s(\mathbf{r})d\mathbf{r} = \frac{1}{8}\int \frac{|\nabla\rho(\mathbf{r})|^2}{\rho(\mathbf{r})} d\mathbf{r}, \tag{7.12}$$

which is also the steric energy E_s from our earlier study. Pauli energy in Eq. (7.11) is the energetic contribution to the total kinetic energy from the Pauli exclusion principle. The principle also contributes to the potential energy, called the exchange–correlation energy, but its magnitude is much smaller, so it could be neglected.

Based on the above assumption, SCIs should be gauged by E_P in Eq. (7.11). However, it is a kinetic energy per se, so we need to convert it to be dimensionless. To do that, we can use the Thomas–Fermi kinetic energy T_{TF} for a homogeneous electron gas as the reference,

$$T_{TF}[\rho] = \int \tau_{TF}(\mathbf{r})d\mathbf{r} = c_{TF} \int \rho^{5/3}(\mathbf{r})d\mathbf{r}, \tag{7.13}$$

with $c_{TF} = \frac{3}{10}(3\pi^2)^{2/3}$, yields the following scaled Pauli energy $\zeta(\mathbf{r})$ as a descriptor of SCIs,

$$\zeta(\mathbf{r}) = \frac{\tau_S(\mathbf{r}) - \tau_W(\mathbf{r})}{\tau_{TF}(\mathbf{r})}. \tag{7.14}$$

For atoms and molecules with one and two electrons, one has $\tau_S(\mathbf{r}) = \tau_W(\mathbf{r})$, so $\zeta(\mathbf{r}) = 0$. On the other hand, when the nuclear charge Z goes to infinity, the system becomes a homogeneous electron gas, so $\tau(\mathbf{r}) = \tau_{TF}(\mathbf{r})$, $\tau_W(\mathbf{r}) = 0$, and $\zeta(\mathbf{r}) = 1$. However, $\tau_{TF}(\mathbf{r})$ decays much faster than $e_P(\mathbf{r})$, so $\zeta(\mathbf{r})$ is asymptotically divergent. To avoid this divergence problem, we define the SCI index as its reciprocal to appreciate SCIs,

$$\text{SCI} = \frac{1}{\zeta(\mathbf{r})}. \tag{7.15}$$

Although Eq. (7.15) circumvents divergence for large-Z cases, it will diverge in the one- and two-electron cases (where $\zeta(\mathrm{r}) = 0$). Fortunately, since most systems with SCIs involve more than two electrons, this divergence is typically not a significant issue. Nonetheless, this limitation indicates that the SCI index cannot effectively identify covalent bonding in systems containing only one or two electrons.

Closely related to SCI is the ELF [31], originally proposed by Becke and Edgecombe to measure the likelihood of finding an electron with the same spin in the neighborhood space of a reference electron located at a given point $\mathbf{r}$. Its definition is as follows,

$$\text{ELF} = \frac{1}{1 + \zeta^2(\mathbf{r})}. \tag{7.16}$$

As can be seen from the two definitions, both SCI and ELF indices are defined using the same scaled Pauli energy. Although the ELF index is confined within a range of 0 to 1, thus eliminating any divergence, the SCI index preserves the clear physical interpretation that Pauli energy measures SCIs. As demonstrated previously in Figure 5.6, both indices generate nearly identical isosurface patterns to indicate SCIs. Specifically, double bonds are characterized by a dumbbell-like isosurface at the bond center, triple bonds display a toroidal (donut-like) shape, and quadruple bonds reveal four bean-like features between the two transition metals.

These unique and characteristic isosurface shapes serve as effective markers for identifying and classifying different types of strong covalent bonds across various systems.

To confirm if different systems with the same bond multiplicity should possess the same characteristic isosurface [34], shown in Figure 7.2 are six metallic and organometallic systems with known metal–metal triple covalent bonds. They are (a) Cr_2^{2-}, (b) W_2, (c) Th_2H_2, (d) $W_2(NH_2)_6$, (e) Cr_2Ar_2, and (f) $Os_2Cl_8^{2-}$. As can be seen from the isosurface between the metals for each of the six systems, the donut shape isosurface between metal atoms is verified. We also confirmed this point for six inorganic and organometallic systems with known quadruple bonds between two transition metal atoms, as shown in Figure 7.3. As can be seen from the figure, the signature four-bean-shaped isosurface is unambiguously discernible for each of these systems, confirming the validity of this signature isosurface as the indicator of quadruple covalent bonds.

These studies yield two main conclusions. First, different classes of covalent bonds exhibit unique signature isosurfaces. Second, compounds featuring the same type of covalent bond multiplicity display an identical signature isosurface. We extended our studies to systems containing other varieties of covalent bonds. For example, two complexes with a quintuple metal–metal bond were observed to have the same isosurface signature, and we even identified a bond stronger than a quintuple bond [34]. Together, these findings demonstrate that SCI and ELF provide a robust, density-based framework for identifying strong covalent bonds. Moreover, this work reinforces that covalent bonding can be successfully *described* within a density-based framework, even without explicit reference to molecular orbitals.

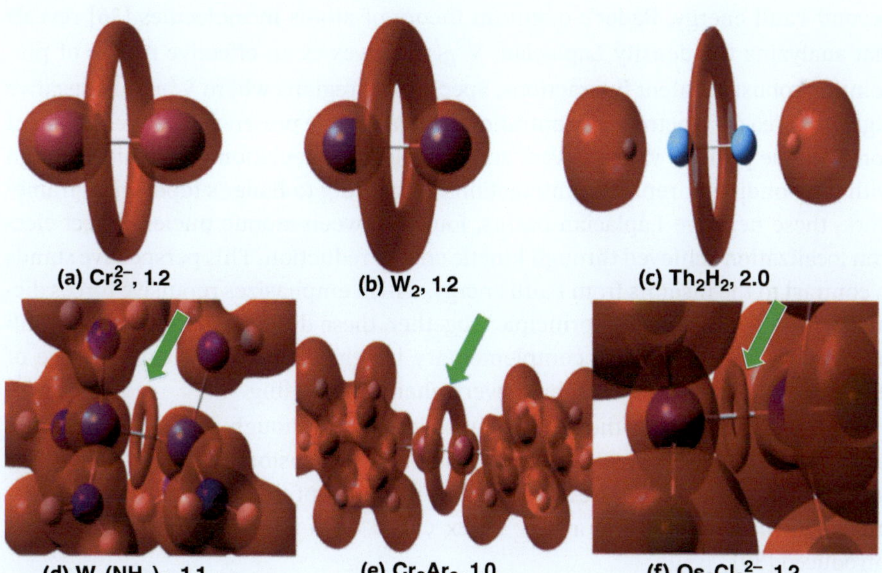

Figure 7.2 SCI distributions for six metallic and organometallic systems with known metal–metal triple bonds whose signature isosurface is donut-like. *Source*: Reproduced from Ref. [34] with permission of American Chemical Society.

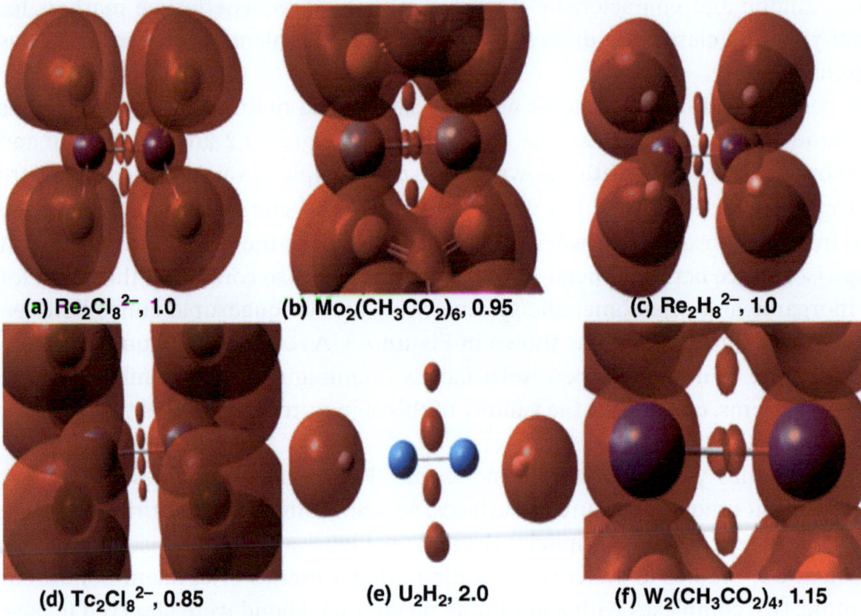

Figure 7.3 SCI distributions for six inorganic and organometallic systems with known metal–metal quadruple bonds, which can be characterized by the four beans shown between the two metal atoms. *Source*: Reproduced from Ref. [34] with permission of American Chemical Society.

7.3.3 Ultra-strong Interaction Index

Beyond Pauli energy, Bader's quantum theory of atoms in molecules [36] reveals that analyzing the density Laplacian, $\nabla^2\rho(\mathbf{r})$, serves as an effective means of pinpointing robust covalent interactions. Specifically, regions where $\nabla^2\rho(\mathbf{r})$ is negative highlight areas of electron concentration, indicating the presence of shared covalent bonds, while positive values reveal zones of electron depletion that typically align with nonbonding or repulsive interactions. According to Bader's topological framework, these negative Laplacian basins, found between atomic nuclei, reflect electron localization achieved through kinetic energy reduction. This perspective stands in contrast to the insights from Pauli energy, which emphasizes repulsive forces dictated by the Pauli exclusion principle. Together, these dual perspectives from Pauli energy and QTAIM provide complementary insights into the delicate balance of attractive and repulsive forces that govern chemical bonding.

There is one issue with the density Laplacian $\nabla^2\rho(\mathbf{r})$ though. It is not dimensionless. In coordinate scaling [37, 38], it has the same dimensionality as kinetic energy. Following what we did for the SCI index, we make use of the Thomas–Fermi kinetic energy as the reference, so a new index called ultra-strong interaction (USI) is introduced [39],

$$\mathrm{USI} \equiv \frac{\nabla^2\rho(\mathbf{r})}{\rho^{5/3}(\mathbf{r})}. \tag{7.17}$$

We refer to this quantity as the USI index because it provides a highly robust descriptor for systems featuring USIs, including quadruple, quintuple, and sextuple covalent bonds. This USI index stands in sharp contrast to the NCI index presented in Eq. (7.10), where the RDG is used to highlight the subtle density gradient features in low-density regions—a signature of weak NCIs such as van der Waals forces.

Figure 7.4 presents examples of compounds with quadruple bonds (Figure 7.4a–f), quintuple bonds (Figure 7.4g, h), and sextuple bonds (Figure 7.4i, j). Notably, we observed that (i) compounds with the same bond multiplicity share an identical signature isosurface, and (ii) these signature shapes match those obtained from the SCI index. Specifically, quadruple bonds exhibit a four-bead isosurface, quintuple bonds display a "double corn-bean" pattern, and sextuple bonds form two parallel tori, or donuts. These findings demonstrate that the USI index produces results consistent with the SCI index.

7.3.4 BNI Index

The aforementioned density-based indices—NCI, SCI, and USI—each selectively capture NCIs (e.g., weak van der Waals forces) or strong covalent bonds. Specifically, the NCI index relies on the RDG to target noncovalent forces, while the SCI and USI indices leverage Pauli energy and the density Laplacian, respectively, to characterize SCIs. Yet, a critical question remains: Is it possible to use a single density-based descriptor to identify *both* covalent bonding and NCIs? Unfortunately, limitations in the density gradient and Laplacian prevent NCI and USI from achieving this goal,

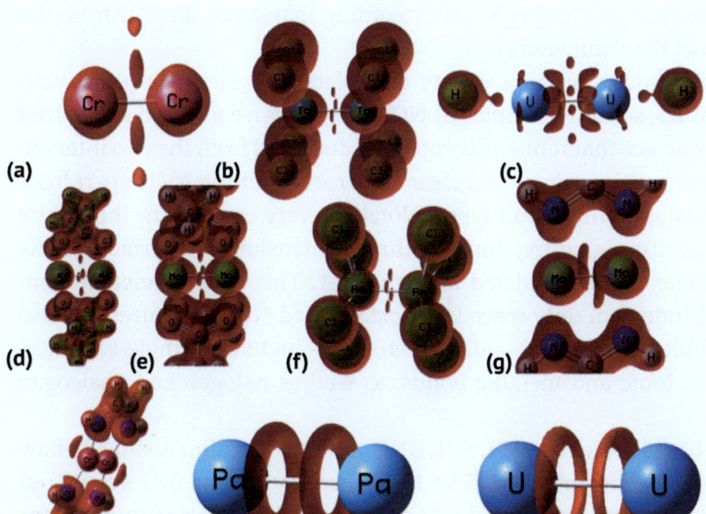

Figure 7.4 Signature isosurfaces from the USI index for systems with quadruple (a–f), quintuple (g, h), and sextuple (i, j) covalent bonds. *Source*: Reproduced from Ref. [39] with permission of American Chemical Society.

and the SCI index is hindered by divergence issues in systems or regions containing only one or two electrons, which are often the case for weakly bonded interactions. In response to these challenges, we have recently introduced a new descriptor, the BNI index [39], which aims to overcome these shortcomings. Rather than employing the Thomas–Fermi energy as the reference, we use the Weizsäcker kinetic energy directly:

$$\mathrm{BNI} = \frac{\tau_S(\mathbf{r}) - \tau_W(\mathbf{r})}{\tau_W(\mathbf{r})}. \tag{7.18}$$

We refer to it as the BNI index because it is capable of detecting both BNI signatures within a single formulation.

A key advantage of this new index is that substituting $\tau_{TF}(\mathbf{r})$ with $\tau_W(\mathbf{r})$ in the denominator not only preserves the dimensionless character but also circumvents divergence issues, thanks to the well-behaved asymptotic properties of $\tau_W(\mathbf{r})$. Recasting Eq. (7.18) as the relative change in noninteracting kinetic energy density with respect to the Weizsäcker kinetic energy provides a more mathematically robust framework. Moreover, our earlier work demonstrated that $\tau_W(\mathbf{r})$ effectively quantifies steric effects, which are crucial in understanding various chemical processes and reactivity phenomena—including stereoselectivity. By excluding the steric contribution in the analysis of bonding interactions, BNI aligns more closely with the principles of physical chemistry. Additionally, because $\tau_W(\mathbf{r})$ is a well-defined local function, there is no need for its reciprocal form, unlike what is required in the SCI index. However, a notable limitation of this index is that it vanishes in one- and two-electron systems (e.g., H_2^+ and H_2), a shortcoming that stems directly from the inherent properties of the Pauli energy.

We applied the BNI index across a variety of molecular systems to probe both BNIs [39]. In Figure 7.5, several examples of NCIs are shown: each is characterized by a disc-shaped isosurface that remains completely detached from the two interacting atoms. This disconnection serves as a clear fingerprint, setting NCIs apart from single covalent bonds and other bond types. Notably, every example in the Figure displays the identical disc signature, and this three-dimensional isosurface closely parallels the NCI index results depicted in Figure 7.1. This concordance demonstrates that the BNI index not only reproduces established NCI signatures but also extends its applicability to diverse bonding scenarios—including single and multiple covalent bonds, ionic and metallic bonds, as well as halogen and chalcogen bonds.

The four density-based indices—NCI, SCI, BNI, and USI—exhibit distinct behaviors due to their unique formulations. The key takeaway is that BNI stands out as the only index capable of simultaneously detecting both covalent interactions and NCIs. However, our findings reveal a critical limitation: no single index comprehensively spans the full spectrum of interaction strengths, from the weakest van der Waals forces to the strongest covalent bonds. This necessitates a context-dependent selection of indices, where specific interactions dictate the choice of analytical tool. Figure 7.6 synthesizes these insights, providing a visual framework that maps the strengths, limitations, and optimal applications of each index within diverse chemical systems. Figure 7.6 builds upon the general descriptor map introduced earlier in Figure 5.7, but here it is expanded and reinterpreted in the context of specific chemical interactions—covalent, ionic, metallic, hydrogen-bonding, and dispersion—reflecting the practical bonding scenarios discussed in this chapter.

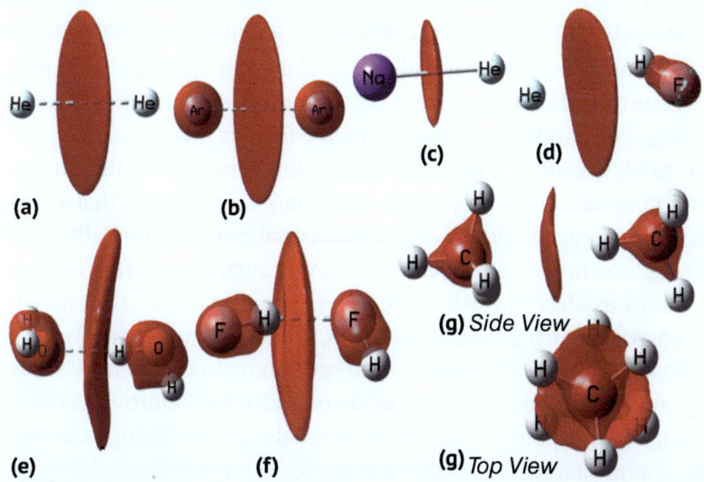

Figure 7.5 Using BNI to describe noncovalent interactions, including dispersion, charge, and dipole interactions and hydrogen bonding for (a) He dimer, (b) Ar dimer, (c) He and Na^+ diatom, (d) HF and He complex, (e) water dimer, (f) HF dimer, and (g) methane dimer. *Source*: Reproduced from Ref. [39] with permission of American Chemical Society.

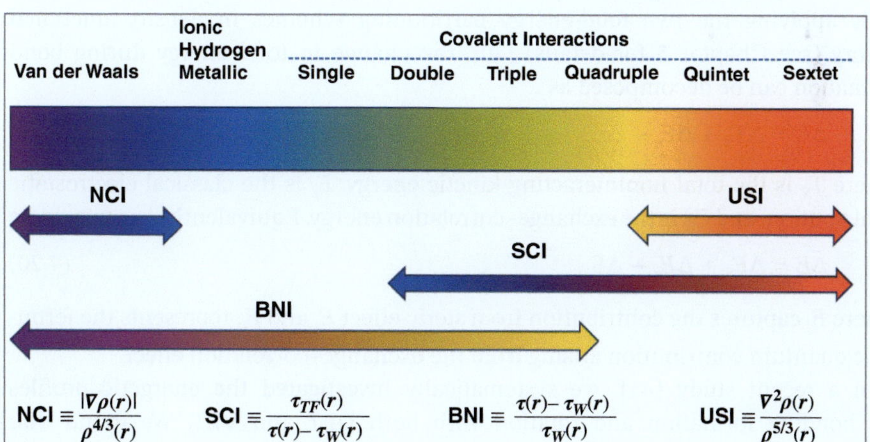

Figure 7.6 Density-based descriptors (NCI, BNI, SCI, and USI) organized according to the types and strengths of chemical interactions. This figure extends the conceptual map of Figure 5.7 by specifying where covalent, ionic, metallic, hydrogen-bonding, and dispersion interactions fall within the descriptor landscape.

7.4 Energetics of Bonding

Before examining how individual energy components contribute to bond formation, it is useful to recall the decomposition schemes introduced in Chapter 5. In the Kohn–Sham formalism, the total energy change is expressed in terms of the noninteracting kinetic energy T_s, the classical electrostatic component E_e, and the exchange–correlation contribution E_{xc}. An equivalent density-based reformulation separates the total change into the steric contribution E_s and the fermionic quantum contribution E_q. These two decompositions are mathematically equivalent but highlight different physical perspectives, one rooted in conventional DFT, the other emphasizing density-derived chemical effects. The parallel presentation below follows this structure.

Bond formation invariably leads to a net reduction in total energy. But the precise balance among kinetic, potential, electrostatic, and other energy components that drives this stabilization has been debated for decades. One influential camp, led by Hellmann, Ruedenberg, Kutzelnigg, and others, attributes the primary stabilizing influence to a lowering of kinetic energy as electrons delocalize across bonding atoms. In contrast, another school, with advocates such as Slater, Feynman, and Bader, maintains that the key factor is the decrease in electrostatic potential energy when electron density shifts into the bonding region between nuclei. The virial theorem further intertwines these components, stipulating that the potential energy decrease dominates (being twice the magnitude of kinetic energy changes) in stable bonds. Yet, interpretations diverge—some argue this theorem underscores potential energy's primacy, while others contend that kinetic effects, such as electron correlation and exchange interactions, underpin the quantum mechanical stability. Despite extensive quantum chemical studies on even the simplest systems like H_2^+, no consensus has yet emerged, highlighting the subtle interplay of energetic contributions in chemical bond formation.

By applying the two total-energy partitioning schemes in density-functional theory (see Chapter 5 for details) [40], the change in total energy during bond formation can be decomposed as

$$\Delta E = \Delta T_S + \Delta E_e + \Delta E_{xc} \tag{7.19}$$

where T_S is the total noninteracting kinetic energy, E_e is the classical electrostatic contribution, and E_{xc} is the exchange–correlation energy. Equivalently, one may write

$$\Delta E = \Delta E_S + \Delta E_e + \Delta E_q, \tag{7.20}$$

where E_s captures the contribution from steric effect E_s and E_q represents the fermionic quantum contribution arising from the exchange–correlation effect.

In a recent study [41], we systematically investigated the energetic profiles for bonding formation and rotation with both DFT and VBT. We found that (i) electrostatic energy is the dominant contributor in most cases; (ii) within the electrostatic terms, nuclear–electron attraction is often, but not always, the leading contributor; (iii) during bond formation, different interactions dominate at different stages; and (iv) in multi-electron systems, steric effects consistently contribute positively to the total energy decrease due to the spatial constraints imposed by the Pauli exclusion principle. These findings are consistently supported by both theories.

Figure 7.7 illustrates how the interaction energy ΔE varies with the F—F bond length in the fluorine dimer [41]. At short separations, the electrostatic term ΔE_e is the main driver of the attractive $\Delta E < 0$, whereas the kinetic-energy contribution ΔT_s works against bonding ($\Delta T_s > 0$). In contrast, at longer distances (see inset), ΔT_s becomes the dominant term, and ΔE_e actually turns positive. This demonstrates that no single energy component can be singled out as "responsible" for bond formation—different terms take on leading or opposing roles depending on the bond distance.

To further illustrate this behavior, Figure 7.8 displays the energy-decomposition profile for an ionic bond. At short separations, the electrostatic term ΔE_e remains dominant, whereas at longer distances (see inset), the kinetic energy term ΔT_s becomes the more favorable contributor. In both cases, the steric effect contribution ΔE_s is attractive, but the fermionic quantum contribution ΔE_q is only favorable at long range—becoming repulsive at short distances. Finally, the exchange-correlation term ΔE_{xc} can switch sign in each case, and for the ionic system ΔE_e stays negative throughout. These results directly mirror the steric and fermionic partitioning introduced in Chapter 5, reaffirming that density-based energetics captures the same balance of forces long debated in orbital-based frameworks.

These results suggest that both schools of thought—those emphasizing one particular energy component over all others—are at once justified and incomplete: each captures an aspect of bond formation at certain given ranges, yet no single

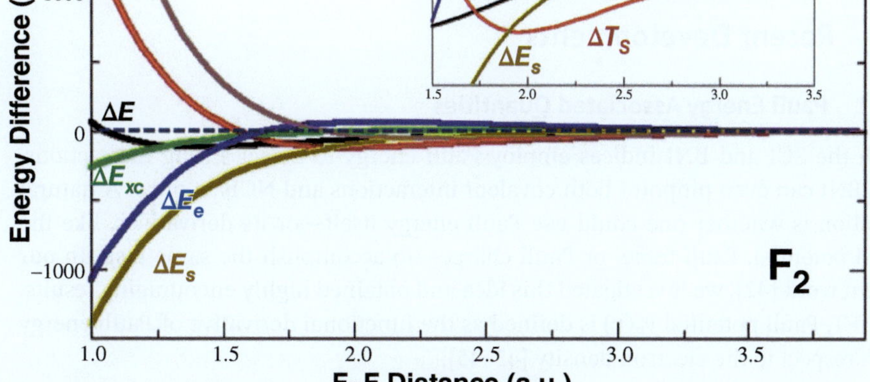

Figure 7.7 Total energy change and its components as a function of the F—F covalent bond distance for F_2. The inset is the same figure within the range between 1.5 and 3.5 Å. *Source*: Reproduced from Ref. [41] with permission of AIP Publishing.

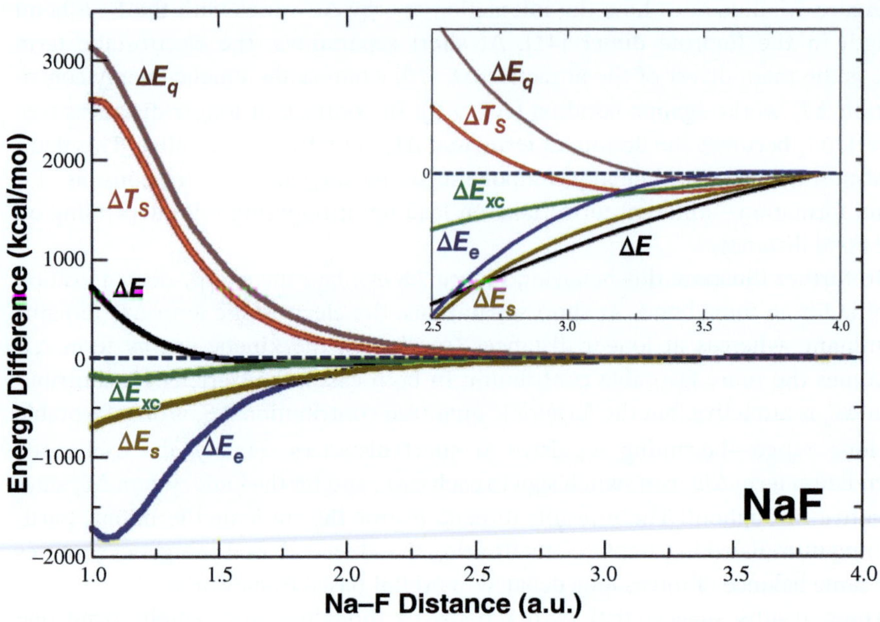

Figure 7.8 The total energy change and its components as a function of the Na—F ionic bond distance for NaF. The inset shows the same figure within the range between 2.5 and 4.0 Å. *Source*: Reproduced from Ref. [41] with permission of AIP Publishing.

component universally governs the entire process. Rather, bond formation emerges from a dynamic interplay among electrostatic attraction, kinetic-energy effects, and exchange–correlation interactions, with each term taking the leading or opposing role at different interatomic distances. A complete picture, therefore, must recognize how these contributions shift in balance as atoms approach or recede, rather than privileging one component in isolation [41].

7.5 Recent Developments

7.5.1 Pauli Energy Associated Quantities

Both the SCI and BNI indices employ Pauli energy to detect strong interactions, and BNI can even pinpoint both covalent interactions and NCIs at once. A natural question is whether one could use Pauli energy itself—or its derivatives, like the Pauli potential, Pauli force, or Pauli charge—to accomplish the same task. In our recent work [42], we investigated this idea and obtained highly encouraging results. In DFT, Pauli potential $v_P(\mathbf{r})$ is defined as the functional derivative of Pauli energy with respect to the electron density [43–45],

$$v_P(\mathbf{r}) = \frac{\delta E_P[\rho]}{\delta\rho(\mathbf{r})} = \frac{\delta T_s[\rho]}{\delta\rho(\mathbf{r})} - \frac{\delta T_w[\rho]}{\delta\rho(\mathbf{r})}. \tag{7.21}$$

with

$$\frac{\delta T_w[\rho]}{\delta \rho(\mathbf{r})} = \frac{1}{8}\frac{|\nabla \rho(\mathbf{r})|^2}{\rho^2(\mathbf{r})} - \frac{1}{4}\frac{\nabla^2 \rho(\mathbf{r})}{\rho(\mathbf{r})}. \tag{7.22}$$

To obtain $\delta T_s[\rho]/\delta\rho(\mathbf{r})$, one can use either Kohn–Sham potential or approximate kinetic energy density functionals. Taking the partial derivative of the Pauli potential in Eq. (7.21) with respect to the spatial variable $\mathbf{r}$, we yield the Pauli force F_P as

$$\boldsymbol{F}_P = -\nabla_{\mathbf{r}} v_P(\mathbf{r}). \tag{7.23}$$

Since the Pauli force is a vector, for the purpose of convenient presentation, we only plot and visualize its magnitudes $|\boldsymbol{F}_P|$. Following our earlier work on the exchange–correlation charge, we have defined the Pauli charge $q_P(\mathbf{r})$ by means of the Poisson equation,

$$q_P(\mathbf{r}) = \nabla^2 v_P(\mathbf{r})/(-4\pi). \tag{7.24}$$

In contrast to the Pauli force, the Pauli charge is a scalar property, satisfying the following normalization condition,

$$\int q_P(\mathbf{r})d\mathbf{r} = 0. \tag{7.25}$$

Applying these quantities to different bonding systems, we found [42] that Pauli potential faithfully reproduces the characteristic isosurface of the BNI index, while Pauli force and Pauli charge each generate distinct yet uniquely shaped signature isosurfaces for different interaction types; in contrast, Pauli energy on its own does not offer definitive insight into interaction classification—consistent with prior SCI and BNI findings—and, importantly, all three Pauli energy-derived measures exhibit strong correlations with computed bond orders, underscoring their viability for quantitative evaluation of chemical interactions. Shown in Figure 7.9 is an illustrative example to demonstrate the point for the double covalent bond in ethylene.

7.5.2 Energetic Information

Using local energy densities as the probability function to define different information-theoretic approach (ITA) quantities, we recently extended the scope of ITA and then applied the newly defined energetic information quantities to identify covalent interactions and NCIs [46–48]. For example, with the steric energy density in Eq. (7.12), we can define the *corresponding* information gain

$$I_G^{E_s} = \int i_G^{E_s}(\mathbf{r})d\mathbf{r} = \int e_s(\mathbf{r})\ln\frac{e_s(\mathbf{r})}{e_{s_0}(\mathbf{r})}d\mathbf{r}, \tag{7.26}$$

and relative alternative Fisher information,

$$ {}^{r}_{F}I'^{E_s} = \int {}^{r}_{F}i'^{E_s}(\mathbf{r})d\mathbf{r} = \int \nabla^2 e_s(\mathbf{r}) \ln \frac{e_s(\mathbf{r})}{e_{s_0}(\mathbf{r})} d\mathbf{r}. \tag{7.27} $$

Our studies show that these quantities generate unique signature isosurfaces for single, double, and multiple covalent bonds as well as for ionic interactions, metallic interactions, and NCIs. Moreover, for any given interaction type, ${}^{r}_{F}I'^{E_s}$ and I_G^{Es} produce qualitatively similar isosurface shapes. These density-based energetic information descriptors thus reaffirm that both covalent interactions and NCIs can be characterized without invoking molecular orbitals. Figure 7.10 presents a selection of systems featuring triple covalent bonds.

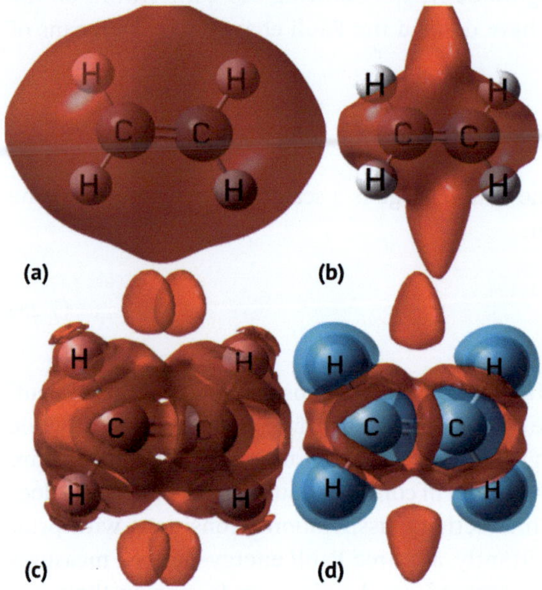

Figure 7.9 The signature isosurface shape of (a) Pauli energy, (b) Pauli potential, (c) Pauli force, (d) Pauli charge for the double C=C covalent bond in C_2H_4. *Source*: Reproduced from Ref. [42] with permission of AIP Publishing.

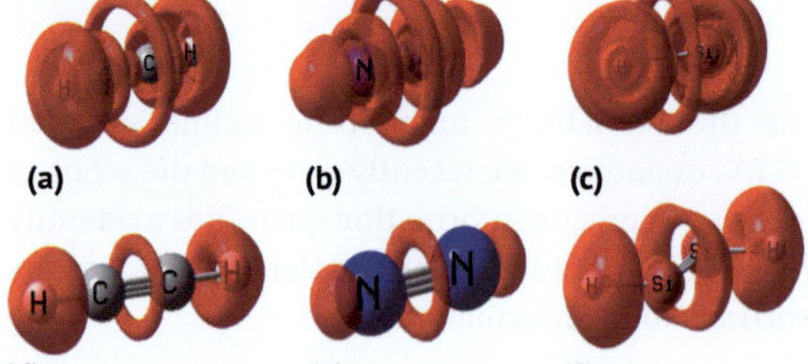

Figure 7.10 Using ${}^{r}_{F}I'^{E_s}$ (top panel) and $I_G^{E_s}$ (bottom panel) to identify the triple covalent bonds for (a) and (d) C_2H_2, (b) and (e) N_2, (c) and (f) Si_2H_2. *Source*: Reproduced from Ref. [48] with permission of John Wiley & Sons.

7.6 Concluding Remarks

Chemical bonding has traditionally been interpreted through orbital-based concepts such as overlap, hybridization, bond order, and resonance. These frameworks remain indispensable for developing intuition about covalent structures and for rationalizing aromaticity, multicenter interactions, and metal–ligand bonding. Yet, as this chapter demonstrates, the electron density—the foundational variable in DFT—offers an alternative and equally powerful means of characterizing both covalent interactions and NCIs.

A central insight emerging from recent work is that *different density-derived descriptors excel in different interaction regimes.* The NCI index captures delicate, weak noncovalent forces; SCI and USI identify strong and ultra-strong covalent bonds with characteristic isosurface signatures tied to bond multiplicity; and the BNI index bridges these extremes by detecting both bonded and nonbonded interactions within a single formulation. These descriptors, though diverse in origin, collectively reveal a coherent picture: chemical interactions of every strength leave identifiable fingerprints in the electron density and its derivatives.

The energetic analysis presented in this chapter further reinforces this perspective. No single energetic component—kinetic, electrostatic, steric, or exchange–correlation—dominates universally across all bonding situations. Instead, bond formation reflects a shifting and distance-dependent balance among these contributions. Density-based energy partitioning thus complements orbital-based reasoning, offering a quantitative route to disentangling stabilizing and destabilizing forces in both covalent and ionic systems.

Recent developments such as the Pauli potential, Pauli force, Pauli charge, and energetic-information measures extend this density-based framework beyond static indicators, revealing new avenues for identifying interaction signatures and correlating them with bond orders and chemical reactivity. These advances suggest that density-derived concepts can not only parallel but also deepen our understanding of bonding phenomena that have historically been interpreted through orbital language.

Taken together, the results of this chapter establish an integrated picture of covalent interactions and NCIs grounded in the electron density. Orbital-based and density-based perspectives, far from competing, offer complementary views of the same underlying quantum mechanical reality. This synthesis lays essential groundwork for subsequent chapters, where chemical concepts such as acidity and basicity, electrophilicity and nucleophilicity, and reactivity patterns will be revisited through the unified lens of density-based reactivity theory.

References

1 Rong C, Zhao D, He X, Liu S. Development and applications of the density-based theory of chemical reactivity. J Phys Chem Lett. 2022 Dec 8;13(48): 11191–200.

2 He X, Li M, Rong C, Zhao D, Liu W, Ayers PW, et al. Some recent advances in density-based reactivity theory. J Phys Chem A. 2024 Feb 22;128(7): 1183–96.

3 Pauling L. The nature of the chemical bond—1992. J Chem Educ. 1992 Jul;69(7): 519.
4 Mulliken RS. Spectroscopy, molecular orbitals, and chemical bonding. Science. 1967 Jul 7;157(3784): 13–24.
5 Roos BO, Borin AC, Gagliardi L. Reaching the maximum multiplicity of the covalent chemical bond. Angew Chem Int Ed. 2007;46(9):1469–72.
6 Shaik S, Danovich D, Galbraith JM, Braïda B, Wu W, Hiberty PC. Charge-shift bonding: a new and unique form of bonding. Angew Chem. 2020 Jan 13;132(3): 996–1013.
7 Gerratt J, Cooper DL, Karadakov PB, Raimondi M. Modern valence bond theory. Chem Soc Rev. 1997;26(2):87.
8 Zhou C, Wu X, Ying F, Wu W. Ab initio valence bond theory for strongly correlated systems. J Chem Theory Comput. 2025 Jul 8;21(13): 6243–57.
9 Dunning TH, Xu LT, Takeshita TY, Lindquist BA. Insights into the electronic structure of molecules from generalized valence bond theory. J Phys Chem A. 2016 Mar 24;120(11): 1763–78.
10 Shaik S, Hiberty PC. Valence bond theory, its history, fundamentals, and applications: a primer. In: Lipkowitz KB, Larter R, Cundari TR, editors. Reviews in Computational Chemistry. Wiley; 2004. pp. 1–100.
11 Zhou C, Ying F, Wu W. Chemical concepts from *ab initio* valence bond theory. In: Liu S, editor. Exploring Chemical Concepts Through Theory and Computation. Wiley; 2024. pp. 23–42.
12 Stuyver T, Shaik S. Bridging conceptual density functional and valence bond theories. In: Liu S, editor. Conceptual Density Functional Theory: Towards a New Chemical Reactivity Theory. Wiley; 2022. pp. 391–415.
13 Long Gu F, Yu J, Yang W. Chemical concepts from molecular orbital theory. In: Liu S, editor. Exploring Chemical Concepts Through Theory and Computation. Wiley; 2024. pp. 1–22.
14 Murrell JN. The origins and later developments of molecular orbital theory. Int J Quantum Chem. 2012 Sep 5;112(17): 2875–9.
15 Hehre WJ. Ab initio molecular orbital theory. Acc Chem Res. 1976 Nov 1;9(11): 399–406.
16 Parr RG, Yang W. Density-Functional Theory of Atoms and Molecules. Oxford University Press; 1989.
17 Shaik S, Danovich D, Wu W, Hiberty PC. Charge-shift bonding and its manifestations in chemistry. Nat Chem. 2009 Sep;1(6): 443–9.
18 von Hopffgarten M, Frenking G. Energy decomposition analysis. WIREs Comput Mol Sci. 2012 Jan;2(1): 43–62.
19 Su P, Tang Z, Wu W. Generalized Kohn-Sham energy decomposition analysis and its applications. WIREs Comput Mol Sci. 2020 Sep;10(5):e1460.
20 Fang D, Piquemal J-P, Liu S, Cisneros GA. DFT-steric-based energy decomposition analysis of intermolecular interactions. Theor Chem Acc. 2014 May;133(5): 1484.
21 Hettich CP, Zhang X, Kemper D, Zhao R, Zhou S, Lu Y, et al. Multistate energy decomposition analysis of molecular excited states. JACS Au. 2023 Jul 24;3(7): 1800–19.

22 Phipps MJS, Fox T, Tautermann CS, Skylaris C-K. Energy decomposition analysis approaches and their evaluation on prototypical protein-drug interaction patterns. Chem Soc Rev. 2015 May 21;44(10): 3177–211.

23 An X, Zhang W, He X, Li M, Rong C, Liu S. Density-based reactivity theory applied to excited states. AAPPS Bull. 2024 Feb 2;34(1): 8.

24 Zhao D, He X, Rong C, Liu S. Chemical concepts from density-based approaches in density functional theory. In: Liu S, editor. Exploring Chemical Concepts Through Theory and Computation. Wiley; 2024. pp. 71–100.

25 Sak MH, Jacobsen EN. Selective noncovalent catalysis with small molecules. Chem Rev. 2025 May 15;125(10): 8319–69.

26 Scheiner S. Understanding noncovalent bonds and their controlling forces. J Chem Phys. 2020 Oct 14;153(14): 140901.

27 Tretiakov S, Nigam A, Pollice R. Studying noncovalent interactions in molecular systems with machine learning. Chem Rev. 2025 Jun 9;125(12): 5776–5829.

28 Johnson ER, Keinan S, Mori-Sánchez P, Contreras-García J, Cohen AJ, Yang W. Revealing noncovalent interactions. J Am Chem Soc. 2010 May 12;132(18): 6498–506.

29 Contreras-García J, Johnson ER, Keinan S, Chaudret R, Piquemal J-P, Beratan DN, et al. NCIPLOT: a program for plotting non-covalent interaction regions. J Chem Theory Comput. 2011 Mar 8;7(3): 625–32.

30 Suresh CH, Anila S. Molecular electrostatic potential topology analysis of noncovalent interactions. Acc Chem Res. 2023 Jul 4;56(13): 1884–95.

31 Becke AD, Edgecombe KE. A simple measure of electron localization in atomic and molecular systems. J Chem Phys. 1990 May;92(9): 5397–403.

32 Savin A, Nesper R, Wengert S, Fässler TF. ELF: the electron localization function. Angew Chem Int Ed. 1997 Sep 17;36(17): 1808–32.

33 Silvi B, Savin A. Classification of chemical bonds based on topological analysis of electron localization functions. Nature. 1994 Oct;371(6499): 683–6.

34 Liu S, Rong C, Lu T, Hu H. Identifying strong covalent interactions with Pauli energy. J Phys Chem A. 2018 Mar 22;122(11): 3087–95.

35 Huang Y, Liu L, Rong C, Lu T, Ayers PW, Liu S. SCI: a robust and reliable density-based descriptor to determine multiple covalent bond orders. J Mol Model. 2018 Jul 21;24(8): 213.

36 Bader RFW. Atoms in Molecules. Oxford University Press; 1989.

37 Liu S. Expansions of the pair distribution function and the second-order density matrix in terms of homogeneous functionals. Phys Rev A. 1996 Dec;54(6): 4863–7.

38 Liu S, Parr RG. Expansions of density functionals in terms of homogeneous functionals: justification and nonlocal representation of the kinetic energy, exchange energy, and classical Coulomb repulsion energy for atoms. Phys Rev A. 1997 Mar 1;55(3): 1792–8.

39 Zhong S, He X, Liu S, Wang B, Lu T, Rong C, et al. Toward density-based and simultaneous description of chemical bonding and noncovalent interactions with Pauli Energy. J Phys Chem A. 2022 Apr 21;126(15): 2437–44.

40 Liu S. Steric effect: a quantitative description from density functional theory. J Chem Phys. 2007 Jun 28;126(24): 244103.

41 An X, Zhang C, Zhou C, Wu W, Liu S. Understanding energetics of bond formation and bond rotation with density functional theory and valence bond theory. J Chem Phys. 2025 Oct 7;163(13): 134106.

42 Zhang W, He X, Li M, Zhang J, Zhao D, Liu S, et al. Simultaneous identification of strong and weak interactions with Pauli energy, Pauli potential, Pauli force, and Pauli charge. J Chem Phys. 2023 Nov 14;159(18): 184104.

43 Holas A, March NH. Construction of the Pauli potential, Pauli energy, and effective potential from the electron density. Phys Rev A. 1991 Nov 1;44(9): 5521–36.

44 Liu S, Zhao D, Rong C, Lu T, Liu S. Using Pauli energy to appraise the quality of approximate semilocal non-interacting kinetic energy density functionals. J Chem Phys. 2019 May 28;150(20): 204106.

45 Rincón L, Torres FJ, Almeida R. Is the Pauli exclusion principle the origin of electron localisation? Mol Phys. 2017 Aug 16;116: 1–10.

46 Liu S. Quantifying energetic information in density functional theory. J Chem Phys. 2022 Sep 14;157(10): 101103.

47 He X, Lu T, Rong C, Liu W, Ayers PW, Liu S. Energetic information from information-theoretic approach in density functional theory as quantitative measures of physicochemical properties. J Chem Theory Comput. 2024 Jul 12;20(14): 6049–61.

48 He X, Liu W, Liu S. Using energetic information quantities from density functional theory to simultaneously identify both covalent and noncovalent interactions. ChemPhysChem. 2025 Apr 1;26(7):e202401068.

8

Cooperativity and Frustration

TABLE OF CONTENTS

8.1 Introduction

In the previous chapter, we concentrated on the atomic scale, using density-based quantities to provide a unified description of both covalent bonds and noncovalent interactions such as van der Waals forces, thereby clarifying the fundamental forces governing atomic interactions within molecules [1–3]. In this chapter, we move beyond pairwise bonding signatures to collective many-body coupling among molecules.

At this level, the focus is on intermolecular interactions, such as hydrogen bonding, dipole–dipole interactions, and dispersion forces. These interactions are ubiquitous across diverse molecular systems, spanning from biological to material sciences. Although generally weaker than the intramolecular bonds holding atoms together, they play crucial roles in determining the behavior and properties of molecular assemblies.

Central to these systems are two defining characteristics: cooperativity and frustration [4, 5]. Cooperativity refers to a phenomenon where the interaction between two molecules enhances subsequent interactions with other molecules. This creates a synergistic effect, where the collective behavior of the system becomes more pronounced than the sum of its individual parts [6–8]. For instance, the presence of one interaction might stabilize or facilitate additional connections, amplifying the system's overall structure or functionality, as seen in hemoglobin's oxygen-binding efficiency or the self-reinforcing order of crystalline materials.

Frustration [5], in contrast, emerges when interactions within the system compete with one another, preventing all of them from being optimized simultaneously. This conflict results in a complex energy landscape, where the system may adopt multiple possible configurations rather than settling into a single, ideal state. These two phenomena are interconnected, each reflecting a different facet of the collective behavior within molecular assemblies. Cooperativity emphasizes the emergent synergy of a system, where interactions between components amplify collective behavior (e.g., enhanced ligand binding in hemoglobin), while frustration reflects the competing interactions inherent to its parts, where individual elements cannot simultaneously satisfy all energetic or structural preferences. The former optimizes global function, while the latter underscores the unavoidable trade-offs for the parts required to maintain stability in the entire system.

Cooperativity and frustration originate from the intrinsic many-body and non-additive nature of intermolecular interactions. Cooperativity arises when interactions reinforce one another, creating positive feedback loops that stabilize ordered states, whereas frustration appears when competing geometric or energetic constraints prevent all interactions from being simultaneously optimized. These dual tendencies are widely observed—from hydrogen-bonded water networks to protein folding and spin-glass systems—and reflect different responses of molecular assemblies to collective coupling. Despite their broad relevance across chemical and biological systems, a unified, first-principles formulation has long been lacking—a gap this chapter addresses using density-based reactivity theory.

8.2 Traditional Theory of Cooperativity

The traditional theory of cooperativity [9] focuses on describing how the binding of a ligand to one site on a multisite molecule affects the binding affinity at other sites, leading to nonlinear ("sigmoidal") binding curves rather than the hyperbolic curves of independent sites.

Early twentieth-century experiments on hemoglobin by Christian Bohr revealed that oxygen binding at one heme facilitated further oxygen binding, a phenomenon now known as positive homotropic cooperativity. Subsequent formalizations—the Hill, Adair, Klotz, and Pauling equations—provided increasingly detailed mathematical descriptions of cooperative binding. Mechanistic models proposed by Koshland–Némethy–Filmer (KNF) [10] and Monod–Wyman–Changeux (MWC) [11] further linked these observations to conformational changes in oligomeric proteins. Modern applications extend beyond hemoglobin to enzymes, ion channels, transcription factors, and synthetic biomolecular assemblies, underscoring cooperativity's central role in biology and biotechnology.

In 1910, Hill introduced a phenomenological equation relating fractional occupancy ($\overline{Y}$) to ligand concentration $[L]$, incorporating an empirical "n" factor (the Hill coefficient) to capture the degree of cooperativity.

$$\overline{Y} = \frac{[L]^n}{k_d^n + [L]^n}, \tag{8.1}$$

where $n > 1$ indicates positive cooperativity and $n < 1$ indicates negative cooperativity. Assuming a receptor P has n identical sites binding ligand L simultaneously,

$$P + nL \leftrightharpoons PL_n \tag{8.2}$$

with the dissociation equilibrium constant k_d^n

$$k_d^n = \frac{[P][L]^n}{[PL_n]}. \tag{8.3}$$

Defining the fractional saturation $\overline{Y}$

$$\overline{Y} = \frac{\text{bound sites}}{\text{total sites}} = \frac{[PL_n]}{[P]_{tot}} = \frac{[PL_n]}{[P] + [PL_n]} = \frac{[L]^n}{k_d^n + [L]^n}. \tag{8.4}$$

The Hill plot, $\log(\overline{Y}/(1 - \overline{Y}))$ versus $\log[L]$ is linear with slope n, but this slope remains fixed across ligand concentrations.

Hill's model treated cooperativity as a fixed property (n constant across saturations) and did not provide a molecular basis for this behavior. Adair (1925) extended Hill's work by formulating a stepwise binding model that used macroscopic association constants for each intermediate occupancy state to describe how affinity changes with progressive ligand binding. The Adair equation is as follows:

$$\overline{Y} = \frac{1}{n}\frac{K_I[L] + 2K_{II}[L]^2 + \cdots + nK_n[L]^n}{1 + K_I[L] + 2K_{II}[L]^2 + \cdots + [L]^n}, \tag{8.5}$$

where each K_i is an apparent macroscopic constant for adding the i-th ligand. Unlike Hill's model, the Adair formalism allows cooperativity to vary with saturation and yields a nonlinear Hill plot.

Klotz further decomposed Adair's constants into microscopic stepwise association constants $K_1, K_2, \ldots$ governed by mass-action law, providing a clear criterion for positive cooperativity when K_i values exceed those expected for independent sites. Later, Pauling introduced subunit interaction energy (α) into the Adair framework, deriving equations for different subunit arrangements (e.g., tetrahedral for hemoglobin), thus linking phenomenological constants to underlying subunit interactions.

The 1960s marked a paradigm shift with the introduction of mechanistic models explaining cooperativity through protein conformational changes. Proposed by Koshland et al. [10], the KNF model posits that ligand binding induces a conformational change in a single subunit from inactive (T) to active (R), which then influences neighboring subunits to adopt the active state *sequentially*. Cooperativity arises through inter-subunit interaction energies that depend on relative conformations, and mixed conformations within the oligomer are permitted. Meanwhile, Monod, Wyman, and Changeux [11] introduced a *concerted* framework in which all subunits of a symmetric oligomer switch simultaneously between two states (T and R), each with distinct ligand affinities. The equilibrium constant ratio $K = T_0/R_0$ and the affinity ratio $c = K_d^R/K_d^T$ determine the shape of the binding curve:

$$\overline{Y} = \frac{\alpha(1+\alpha)^{n-1} + Kc\alpha(1+c\alpha)^{n-1}}{(1+\alpha)^n + K(1+c\alpha)^n}, \quad \text{with} \quad \alpha = \frac{[L]}{K_d^R}. \tag{8.6}$$

This model elegantly links thermodynamics to cooperativity and explains ultrasensitive responses when the slope of the Hill plot varies with saturation.

Hill, Adair, MWC, and KNF models collectively form the cornerstone of the traditional theory of cooperativity. While the Hill equation remains a tool for quantifying cooperativity, the MWC and KNF models revolutionized biochemistry by linking macroscopic behavior to molecular dynamics. However, these models have limitations: the MWC model oversimplifies systems with mixed conformations, while the KNF model's complexity can obscure predictive insights. Moreover, they are all kinetic models, and none of them captures atomistic-level details.

8.3 Quantification of Cooperativity in Density Functional Theory

8.3.1 Formulation

For a molecular system bound together by noncovalent interactions, we assume it has the following general formula AR_n, where A stands for the accessory component and R is the constituting unit with a total of n repeating pieces (i.e., building block). For example, for a water cluster consisting of 20 copies of water molecules, it is $(H_2O)_{20}$, with $R = H_2O$, $n = 20$ and no A, and for the cluster, $F^-(H_2O)_{10}$, $A = F^-$, $R = H_2O$ and $n = 10$. For these systems with noncovalent interactions, we define cooperativity κ as [4]

$$\kappa = -\left(\frac{\partial \Delta E_n}{\partial n}\right), \tag{8.7}$$

where E_n denotes the interaction energy change ΔE per building block

$$\Delta E_n = \frac{\Delta E}{n} = \frac{E - nE_R - E_A}{n}, \tag{8.8}$$

with E, E_A, and E_R as the total energy of the whole system, the accessory, and the building block, respectively. With this, we have

$$\begin{cases} \kappa > 0, & \text{positive cooperativity;} \\ \kappa = 0, & \text{no cooperativity;} \\ \kappa < 0, & \text{negative cooperativity.} \end{cases} \tag{8.9}$$

The cooperativity index (κ) quantifies how the interaction energy per building block changes as the total number of blocks in the system increases. When a new building block is added to a system containing a specific number of existing blocks, a positive κ indicates that the interaction energy per building block decreases (becomes more attractive). This implies that the addition strengthens interactions between blocks, reflecting *positive cooperativity* in the system. Conversely, a negative κ signifies that introducing a new block raises the interaction energy per unit (making it less favorable or more repulsive), pointing to *negative cooperativity*, where added units disrupt or weaken interactions within the system. Again, here ΔE is the total interaction energy of the cluster, whereas $\Delta E_n = \Delta E/n$ denotes the interaction energy *per building block*. Throughout this chapter, ΔE_n is consistently used as the size-normalized measure whose curvature with respect to n defines the cooperativity index κ.

The formulation presented above may be viewed as a first-principles analogue of the classical Hill or Adair treatments discussed in Section 8.2. In those empirical models, the Hill coefficient n measures the steepness of the binding curve and hence the degree of nonadditivity in site affinities. In the present formulation, the cooperativity index κ plays an equivalent role but is derived directly from total-energy changes upon incremental aggregation. Whereas Hill's n is an empirical slope obtained from experimental data, κ measures the curvature of the interaction-energy profile with respect to cluster size, providing a quantum-mechanical, parameter-free measure of cooperativity.

Figure 8.1 presents two examples of cooperativity profiles (ΔE_n versus n): one for a neutral water cluster and the other for a protonated water cluster, both over the range $n = 1$–20. The neutral water cluster exhibits positive cooperativity, as the interaction energy per unit (ΔE_n) decreases with the addition of more building blocks (i.e., water molecules). This trend indicates that each added unit strengthens the interactions, thereby enhancing the overall stability of the cluster. In contrast, the protonated water cluster displays negative cooperativity—adding more building blocks leads to an increase in ΔE_n, implying weaker interactions and reduced stability as the cluster grows.

From Figure 8.1, we can also see the main differences between the two cases of cooperativity. In neutral water clusters, hydrogen bonding is the main driving force, whereas in the protonated water cluster, besides short-range hydrogen bonding, there are long-range electrostatic interactions because the system is charged. This difference not only displays the difference in cooperativity propensity but also results in substantial differences in ΔE_n values, as elucidated in the Figure.

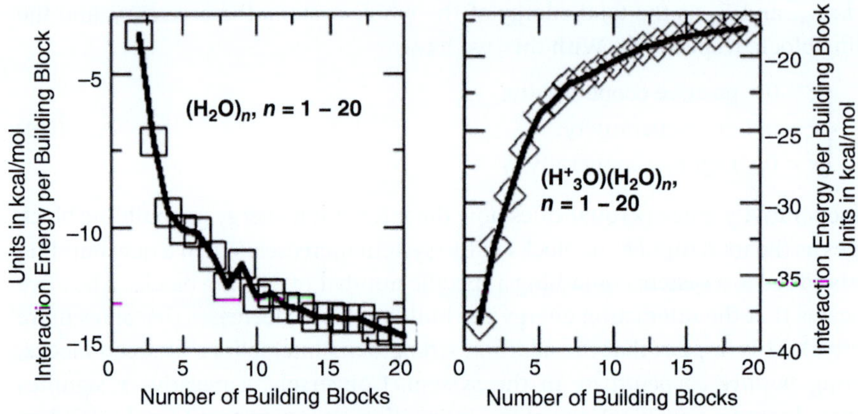

Figure 8.1 The cooperativity profile of (left) neutral and (right) protonated water clusters as a function of the number of water molecules. *Source:* Reproduced from Ref. [4] with permission of The Royal Society of Chemistry.

The sigmoidal ΔE_n profiles in Figure 8.1 resemble the well-known logistic form, which arises whenever growth is self-enhancing at small sizes but self-limiting at large sizes. This behavior can be modeled by the generalized logistic ordinary differential equation (ODE),

$$\frac{dP(t)}{dt} = rP(t)\left\{1 - \left[\frac{P(t)}{K}\right]^n\right\}, \tag{8.10}$$

where the cooperativity exponent n determines the steepness of the response. Here, the logistic equation serves purely as a phenomenological surrogate; it is *not* intended as a mechanistic model of cluster growth. By replacing the continuous variable t with the discrete cluster size n, this expression becomes a compact phenomenological surrogate for the nonlinear buildup of interaction strength during cluster growth. In this context, it captures the universal S-shaped profile without implying a mechanistic model of molecular growth.

8.3.2 Patterns of Cooperativity

Can we identify a predictive criterion to determine whether a system exhibits positive or negative cooperativity? Our recent study demonstrates that such a criterion exists [6]. We revealed a striking pattern: homogeneous molecular systems exhibit positive cooperativity, whereas charged molecular systems display negative cooperativity [6]. This distinction might be rooted in the dominant intermolecular forces at play. As illustrated in Figure 8.2 as an example for $n = 20$, we analyzed six cluster models, including four homogeneous systems and two charged systems. The interactions governing these systems differ fundamentally. For example, Cl_2Ar_n is dominated by van der Waals interactions, $(HF)_n$ is primarily governed by hydrogen bonding, and $Li^+(H_2O)_n$ involves a combination of hydrogen bonding and

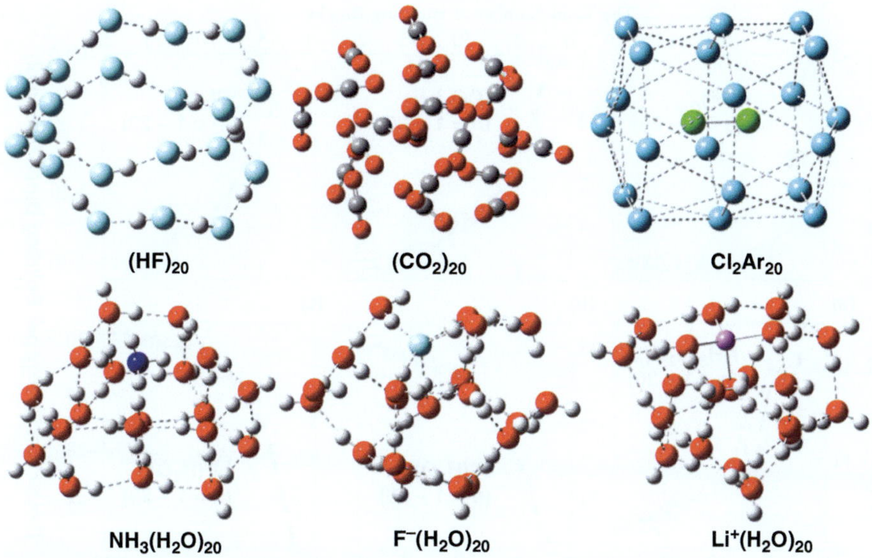

Figure 8.2 Six molecular clusters studied. Reproduced from Ref. [6] with permission of American Chemical Society.

Coulombic interactions. These findings highlight how the nature of intermolecular forces—nonpolar, hydrogen-bonded, or ionic—dictates the cooperative behavior of a system. Homogeneous systems with symmetric, weaker interactions (e.g., van der Waals) favor positive cooperativity, whereas charged systems, dominated by strong electrostatic interactions, exhibit negative cooperativity, as shown in Figure 8.3. We also found that positive cooperativity is dominated by the exchange–correlation interaction and steric effect, but the negative cooperativity is dictated by the electrostatic interaction [6]. From the information-theoretic approach (ITA) perspective, these two categories of systems also demonstrated vastly different behaviors.

8.3.3 Quantification with Density-based Quantities

Besides the total interaction energy per building block, can we use other quantities to represent or quantify cooperativity? The answer is yes. Traditionally, one employed quantities such as binding affinity or rate constants to indicate how the binding of one molecule influences subsequent binding events. Another widely recognized metric is the Hill coefficient, commonly used in biochemistry to quantify cooperativity in processes like enzyme kinetics or ligand binding.

In density-based frameworks [2, 3], we can make use of energetic components, information, or other descriptors. Different descriptors might behave differently for different cooperative propensities. Are there any density-based quantities behaving similarly to the total interaction energy? We show two examples here.

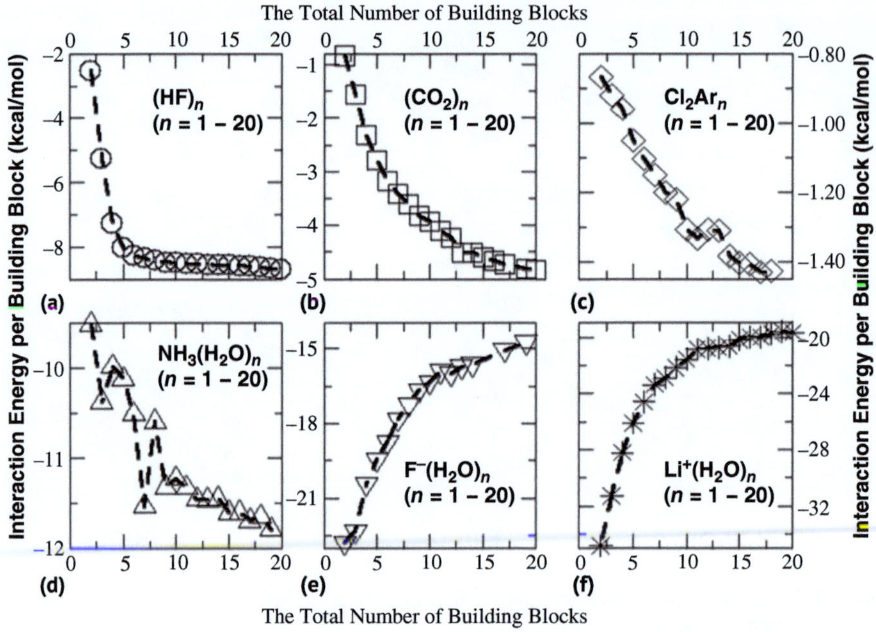

Figure 8.3 Cooperativity profiles of the six molecular clusters. *Source:* Reproduced from Ref. [6] with permission of American Chemical Society.

The first one is the electronic energy $V_e[\rho]$, one component of total energy, which is the sum of the total electrostatic energy $E_e[\rho]$ and the exchange-correlation energy $E_{xc}[\rho]$. In density functional theory (DFT) [12], it can be expressed as

$$V_e[\rho] = E_e[\rho] + E_{xc}[\rho], \tag{8.11}$$

where the electrostatic energy is the sum of the electron–nuclear attraction, classical electron–electron Coulomb repulsion, and nuclear–nuclear repulsion energies. The second quantity is the information gain, or relative Shannon entropy, or Kullback–Leibler divergence in ITA [13, 14], which is defined as

$$I_G = \int \rho(\mathbf{r}) \ln \frac{\rho(\mathbf{r})}{\rho_0(\mathbf{r})} d\mathbf{r}, \tag{8.12}$$

with $\rho_0(\mathbf{r})$ as the reference density. The reason that these quantities can be applied to represent cooperativity is based on their nonlinear nature, which is key to the cooperativity effect. Note that information gain has also been applied to quantify other molecular reactivity properties such as electrophilicity and nucleophilicity [15, 16], the capability of atoms in molecules to accept and donate the maximum number of electrons in chemical transformations. We will discuss this in more detail in Chapter 10.

Shown in Figures 8.4 and 8.5 are results using these two descriptors, electronic energy (Figure 8.4) and information gain (Figure 8.5), for four systems, two of which, water and CO_2 clusters, are positive cooperativity, and the other two, protonated and

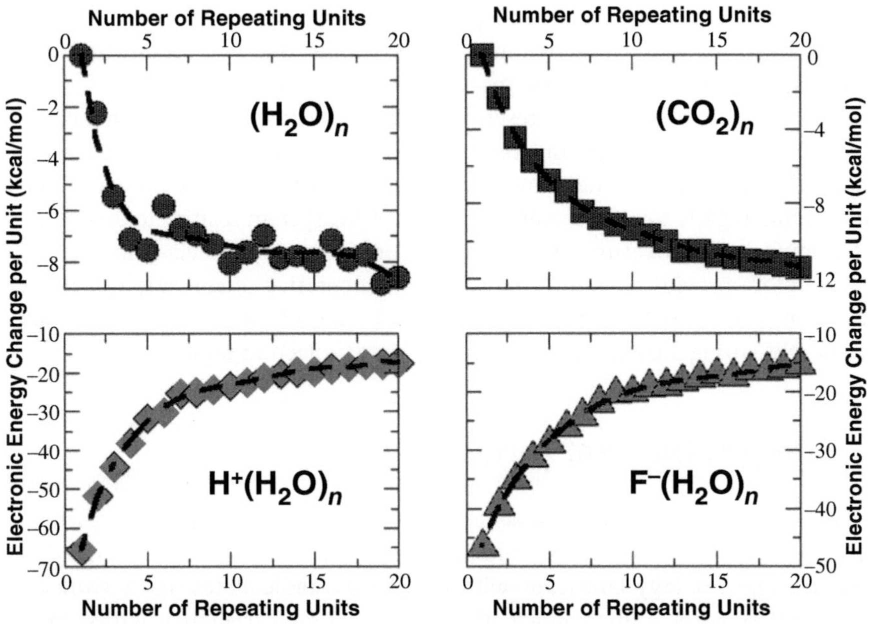

Figure 8.4 Electronic energy, Eq. (8.11), is a descriptor of cooperativity for four systems. Two are positive cooperativity, and the other two are negative cooperativity.

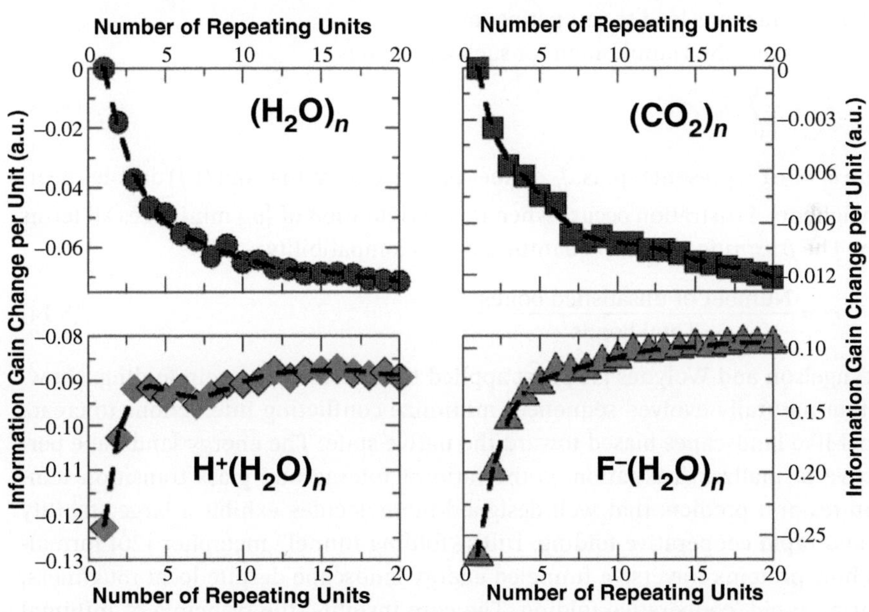

Figure 8.5 Information gain, Eq. (8.12), is a descriptor of cooperativity for four systems. Two are positive cooperativity, and the other two are negative cooperativity.

fluoride water clusters, are negative cooperativity. As can be seen from the Figures, these two descriptors can correctly represent the cooperativity propensity for these systems, same as the interaction energy.

The use of ITA quantities to represent cooperativity rests on the idea that cooperative assembly increases the statistical dependence among constituent densities. As molecules interact, their individual electron distributions become mutually constrained, reducing the total information entropy of the system relative to that of the isolated components. Measures such as information gain or Kullback–Leibler divergence therefore provide a natural indicator of cooperative organization: a larger information gain reflects greater mutual conditioning of electron densities and, consequently, stronger cooperative coupling among molecular units.

8.4 Classical Theory of Frustration

The classical theory of frustration was inspired by early work on spin glasses, where competing interactions prevent simultaneous minimization of all local energies, yielding a rugged energy landscape with many metastable states. Early work by Wannier (1950) on the antiferromagnetic Ising model for a triangular lattice demonstrated that geometric constraints could prevent a system from achieving a simple ordered state [17]. The term "frustration" was formalized by Vannimenus and Toulouse [18] in the context of magnetic systems, particularly spin glasses, where random interactions lead to competing forces.

To be specific, the Hamiltonian for such a system is

$$\mathcal{H} = -\sum_{(i,j)} J_{ij}\sigma_i\sigma_j, \tag{8.13}$$

where $\sigma_i = \pm 1$ represents spins, J_{ij} defines pairwise coupling, and (i,j) denotes nearest neighbors. Frustration occurs when no configuration of $\{\sigma_i\}$ minimizes all terms $J_{ij}\sigma_i\sigma_j$. The *frustration index* ϕ quantifies this incompatibility:

$$\phi = \frac{\text{Number of unsatisfied bonds}}{\text{Total bonds}}. \tag{8.14}$$

Bryngelson and Wolynes [19] first applied these ideas to protein folding, showing that naturally evolved sequences minimize conflicting interactions to create funnel-like landscapes biased toward the native state. The energy landscape perspective formalizes frustration as the ratio of folding and glass transition temperatures and predicts that well-designed biomolecules exhibit a large stability gap and rapid cooperative folding. Dill's "folding funnel" metaphor [20] formalized how proteins traverse a funneled energy landscape despite local roughness, ensuring rapid, cooperative folding. The core insight—the principle of minimal frustration—posits that evolution sculpts sequences to reduce energetic conflicts while retaining localized frustration necessary for function, thus balancing stability, kinetics, and dynamics.

In a nutshell, the free energy landscape describes how the free energy of a system changes as a function of its configuration or order parameters. It is related to the partition function Z by,

$$Z = -k_B T \ln Z, \tag{8.15}$$

where k_B is Boltzmann's constant, T is the absolute temperature, and Z sums over all microscopic states,

$$Z = \int e^{-\beta E(x)}\, dx \tag{8.16}$$

with $\beta = 1/k_B T$, and $E(x)$ the energy at configuration x.

Instead of considering all degrees of freedom, the free energy landscape simplifies the description to only a few collective variables, such as the reaction coordinates q. The probability density of the system having coordinated q, $P(q)$ is

$$P(q) = \frac{1}{Z}\int \delta(q - q(x))e^{-\beta E(x)}\, dx, \tag{8.17}$$

so the effective free energy as a function of q is

$$F(q) = -k_B T \ln P(q). \tag{8.18}$$

In proteins, frustration emerges from conflicting interactions between amino acids. Hydrophobic residues may favor burial in the core, while polar residues seek solvation, creating competing demands. The total frustration $\mathscr{F}$ of a protein can be modeled as:

$$\mathscr{F} = \sum_{(i<j)} \left(\epsilon_{ij}^{\text{native}} - \epsilon_{ij}^{\text{non-native}}\right), \tag{8.19}$$

where ϵ_{ij} represents interaction energies.

The *minimum frustration principle* [21] states that natural systems evolve toward minimizing frustration in their energy landscapes,

$$\mathscr{F} \approx 0. \tag{8.20}$$

and the free energy landscape becomes funnel-shaped toward the native or functional state, with few conflicting barriers. In contrast, a random heteropolymer (maximally frustrated) exhibits a rugged landscape with many local minima [22, 23]. Table 8.1 summarizes the key differences between physical and biological systems in frustration and the minimum frustration principle.

Wales and coworkers [24] constructed energy landscapes by first defining a system's potential energy surface (PES) based on atomic coordinates and an interaction model. Then, they systematically locate local minima (stable structures) and transition states (first-order saddle points) using optimization algorithms. These minima and saddles are then connected into a network, where edges represent possible transitions between states over energy barriers. To efficiently explore low-energy regions, Wales developed the basin-hopping algorithm, which simplifies

Table 8.1 Comparison of frustration and the minimum frustration principle across contexts.

	Physics (e.g., spin glasses)	Biology (e.g., protein folding)
Definition of frustration	Competing interactions or geometric constraints prevent all interactions from being satisfied.	Conflicting interactions hinder efficient folding into the native state.
Minimum frustration principle	Systems adopt configurations that minimize total frustration, balancing competing forces.	Proteins evolve sequences to minimize frustration, ensuring a funneled energy landscape.
Energy landscape	Rugged, with many local minima; minimization reduces degeneracy.	Funneled, with a clear native state minimum and few traps.
Examples	Antiferromagnetic triangular lattice, spin glasses.	Thioredoxin family proteins, fast-folding proteins.
Applications	Understanding glassy dynamics, novel material properties.	Predicting protein folding rates, designing stable proteins.

the search by focusing on minimized structures. The landscape's structure is visualized using disconnectivity graphs, which reveal whether the system has funnels (favoring efficient transitions) or frustration (many competing minima). Finally, kinetic and thermodynamic properties are extracted through rate theories and master equations, allowing prediction of dynamics like folding, melting, or glassy relaxation based on the mapped landscape.

The main issues of the classical theory of frustration are that (i) it ignored solvent effect and quantum details; (ii) real folding events occur far from equilibrium, challenging classical assumptions; (iii) more importantly, it completely neglects the other side of the same coin, cooperativity.

8.5 Quantification of Frustration in DFT

8.5.1 Formulation

Frustration is an inherent feature of molecular complexes. It emerges because individual molecules must deviate from their optimal, isolated structures to establish the noncovalent interactions—such as hydrogen bonds or van der Waals forces—that hold the complex together. This deviation is essential for the stability of the complex, requiring each molecule to sacrifice its own ideal configuration. As a result, every component within the complex exists in a frustrated state, unable to maintain its most stable individual form. Furthermore, the extent of this frustration, or how much each molecule must compromise, is a fundamental characteristic of the molecular complex itself.

With this said, a straightforward quantification of frustration [5] is to examine the energetic state of individual molecules with their respective optimal structures in an isolated state as the reference state. For a molecular system with the general formula of $AB_n(\mathbf{R}_A, \mathbf{R}_B)$, where A is the optional accessory component with the structure of $\mathbf{R}_A$, B the building block with the structure of $\mathbf{R}_B$, and n the total number of building blocks the total energy in its optimal structure is $E(AB_n)$ and the energy sum of the individual parts in the reference state is

$$E_0(\mathbf{R}_0) = E_A(\mathbf{R}_{0A}) + n\, E_B(\mathbf{R}_{0B}). \tag{8.21}$$

The energy sum of all individual components in the optimal state of the complex is the following,

$$E_0(\mathbf{R}) = E_A(\mathbf{R}_A) + \sum_{B=1}^{n} E_B(\mathbf{R}_B). \tag{8.22}$$

We define the total frustration energy E_{frust} as the difference between Eqs. (8.22) and (8.21),

$$\Delta E_{\text{frust}} = E_0(\mathbf{R}) - E_0(\mathbf{R}_0) > 0. \tag{8.23}$$

When individual components assemble into a molecular complex, the overall system becomes more stable, resulting in an energetic gain from a global perspective. However, each component incurs a structural penalty by deviating from its optimal geometry in the isolated state. The quantity defined in Eq. (8.23) corresponds to what is commonly referred to in the literature as the distortion energy, particularly in the context of transition state structures.

Following cooperativity, we use the change of the frustration energy per building block as the definition of frustrativity,

$$\zeta = \partial \Delta E_{0n} / \partial n. \tag{8.24}$$

where

$$\Delta E_{0n} = [E_0(\mathbf{R}) - E_0(\mathbf{R}_0)]/n = \Delta E_{\text{frust}}/n > 0. \tag{8.25}$$

When more building blocks are added to a system, an increase in ζ indicates positive frustrativity, meaning the system's frustration rises. Conversely, a decrease in ζ indicates negative frustrativity, suggesting the system's frustration diminishes. That is,

$$\begin{cases} \zeta > 0, & \text{positively frustrative} \\ \zeta = 0, & \text{non frustrative} \\ \zeta < 0, & \text{negatively frustrative} \end{cases} \tag{8.26}$$

Figure 8.6 presents the frustration profiles for four molecular complexes. As illustrated, frustrativity can be either positive or negative. For instance, in the case of the water cluster—which exhibits positive cooperativity—frustrativity is also positive. This indicates that as additional water molecules are incorporated into the cluster, the interaction energy becomes more negative (stronger), while each water molecule deviates further from its optimal isolated geometry. Conversely, for the protonated water cluster, which displays negative cooperativity, the frustrativity is negative as well.

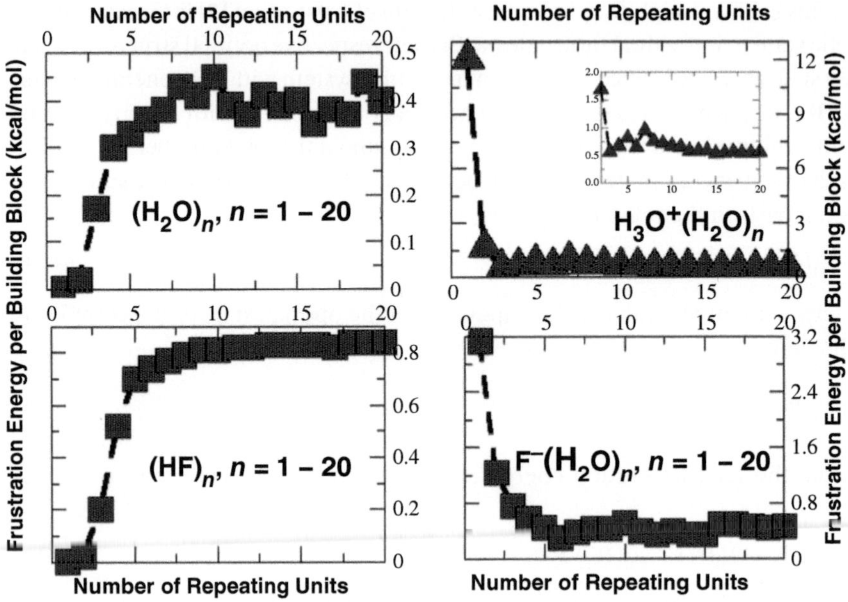

Figure 8.6 Frustration profiles (ΔE_{0n} versus n) for four molecular complexes.

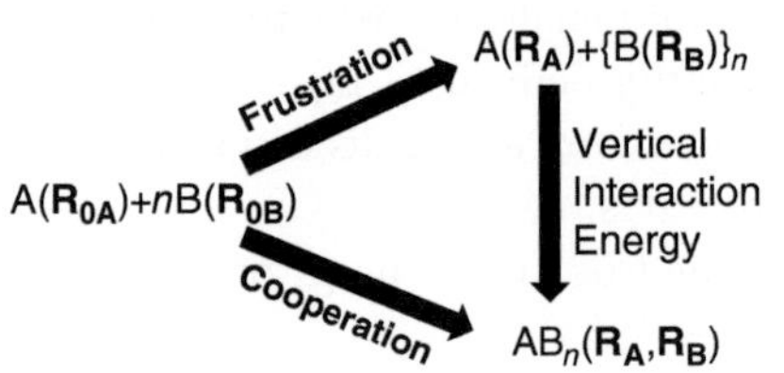

Figure 8.7 Relationship between cooperation and frustration. *Source:* Reproduced from Ref. [5] with permission of American Chemical Society.

This suggests that adding more water molecules weakens the interaction energy and results in less structural distortion of the individual water molecules. This is also the case for other systems.

The correspondence between cooperativity and frustrativity follows directly from their definitions. Positive cooperativity implies increasing stabilization with system growth, whereas negative cooperativity implies the opposite. Conversely, in negatively cooperative systems, added units weaken interactions and relax structural distortions, producing negative frustrativity. The κ–ζ relationship in Figure 8.7 therefore reflects the same underlying energetic nonadditivity, expressed once through interaction strength (κ) and once through structural distortion (ζ).

8.5.2 Minimum and Maximum Frustration

It is important to note that the "maximum" and "minimum" frustration principles refer to different descriptive levels and therefore are not contradictory. In energy landscape theory, the *principle of minimal frustration* refers to the global topology of the free-energy surface—evolutionary optimization of biomolecules toward a smooth, funneled landscape that minimizes kinetic traps. In contrast, the present density-based definition concerns *local structural frustration*, quantified as the energetic cost individual components pay by deviating from their isolated equilibrium geometries. A system can thus exhibit minimal global frustration while still possessing high local frustration within its most stable configuration, reflecting the balance between collective stability and individual compromise.

The main findings in the preceding section—demonstrating that positive cooperativity corresponds to positive frustrativity, and vice versa—suggest that even at the same level, it is possible to have different frustration propensities for different systems. Positively cooperative systems were found to exhibit maximum frustration, while negatively cooperative systems were shown to display minimal frustration [5]. This stands in contrast to the minimum frustration principle discussed in Section 8.4, even at the local level.

To validate this observation, we analyzed two positively cooperative systems—$(CH_3OH)_5$ and $(H_2O)_6$—and two negatively cooperative systems—$F^-(H_2O)_6$ and $H_3O^+(H_2O)_5$—examining the relationship between stability and frustration. For each species, multiple conformations were sampled. The results are presented in Figure 8.8, where the global minimum (lowest-energy structure) is indicated by a red star. As shown, in positively cooperative systems (Figure 8.8a and b), frustration energy is highest at the global minimum. In contrast, for negatively cooperative systems (Figure 8.8c and d), the global minimum corresponds to the lowest frustration energy. These findings reinforce the correlation between cooperativity and frustrativity described earlier.

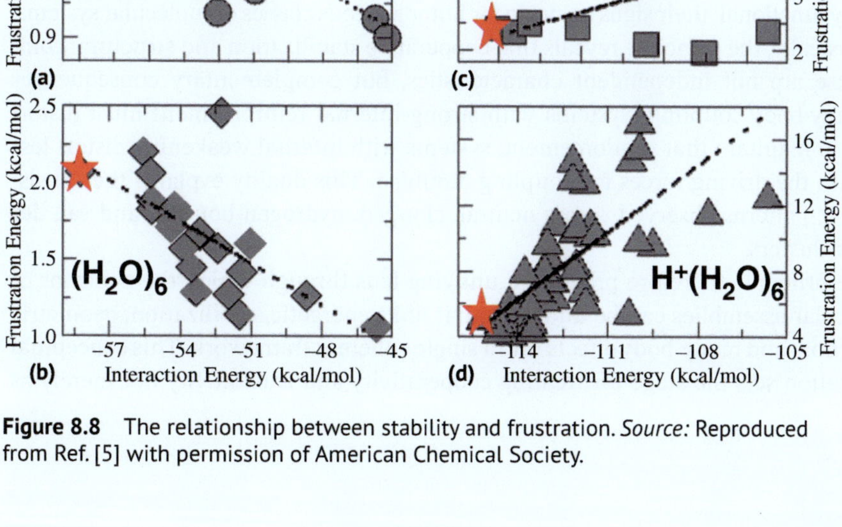

Figure 8.8 The relationship between stability and frustration. *Source:* Reproduced from Ref. [5] with permission of American Chemical Society.

8.6 Principle of Cooperativity and Frustrativity

The primary advantage of the proposed scheme for quantifying cooperativity and frustrativity lies in its ability to treat both concepts within a unified framework, unlike classical theories of cooperation and frustration, which rely on separate foundational principles. This unified approach facilitates a more consistent and holistic analysis of the complex behaviors exhibited by molecular complexes, enabling direct comparisons, exploration of their interplay, and the identification of general patterns in which cooperativity and frustrativity either coexist or counterbalance one another.

Specifically, we discovered that homogeneous molecular complexes, such as water molecular clusters and CO_2 clusters, tend to exhibit positive cooperativity, whereas charged molecular systems, for instance, water clusters containing Li^+ cation or F^- anion, are characterized by negative cooperativity. Furthermore, our analysis revealed a strong correlation between cooperativity and frustrativity: molecular systems with positive cooperativity often show positive frustrativity, while those with negative cooperativity typically display negative frustrativity. Taken together, these findings form what we refer to as the *principle of cooperativity and frustrativity*.

Formally, the *principle of cooperativity and frustrativity* states that the second-order energetic responses of a molecular assembly—those encoded in κ, Eq. (8.7), and ζ, Eq. (8.24)—are intrinsically correlated because they reflect the same many-body coupling that governs how building blocks interact collectively. Positive cooperativity ($\kappa > 0$) implies that successive units experience increasingly favorable interactions; achieving this increased stabilization requires each unit to deviate further from its isolated optimum, leading necessarily to positive frustrativity ($\zeta > 0$). Conversely, when additional units weaken intermolecular interactions ($\kappa < 0$), structural deviations become less severe, yielding negative frustrativity ($\zeta < 0$).

Mathematically, κ and ζ measure curvature in two orthogonal but related spaces: κ reflects curvature of the interaction-energy hypersurface with respect to system size, whereas ζ reflects curvature of the structural-distortion energy with respect to the same variable. Because both originate from the same nonadditive term in the total energy functional, their signs coincide for homogeneous classes of molecular systems.

Physically, the principle reveals that cooperative stabilization and structural compromise are not independent characteristics, but complementary consequences of many-body coupling. Systems with strong internal reinforcement must distort more to maintain that reinforcement; systems with internal weakening distort less because the driving forces for coupling diminish. This duality explains the consistent κ–ζ patterns observed across neutral, charged, hydrogen-bonded, and van der Waals clusters.

The principle therefore provides a unifying lens through which the behavior of molecular assemblies can be understood: it links energetic stabilization, geometric distortion, and many-body effects into a single coherent framework. This conceptual unification sets the stage for treating cooperativity and frustrativity not merely as descriptors, but as fundamental indicators of the emergence of collective chemical behavior.

The generalized logistic ODE (also known as Richards' equation) was adopted as a heuristic model to capture both positive and negative cooperativity in Section 8.3.1, given their characteristic sigmoidal (S-shaped) behavior. A similar approach can be applied to frustrativity, which, as demonstrated in the previous section, also exhibits sigmoidal patterns. This naturally leads to the following question: can these two phenomena be integrated into a unified mathematical model capable of simulating both positive and negative cooperativity and frustrativity? A straightforward answer is to construct a coupled system of ODEs, one describing cooperativity and the other frustrativity, allowing their coupling to be modeled in tandem. More studies along this is needed in the future to further explore the possibility and applicability of this coupling.

8.7 Recent Advances

We recently extended the concepts of cooperativity and frustrativity to classical and multiscale molecular simulations using both polarizable and nonpolarizable force fields [25, 26]. By design, nonpolarizable force fields model intermolecular interactions using pairwise terms, neglecting the many-body interactions among three or more particles that are crucial for capturing cooperative and frustrated behaviors. Polarizable force fields, on the other hand, incorporate many-body effects through polarization, although they still treat van der Waals and permanent electrostatic interactions in a pairwise manner. Despite these inherent limitations, we demonstrated that both types of force fields can successfully reproduce cooperation and frustration phenomena, largely due to compensating errors within their interaction models.

In another recent extension, we generalized the concept of cooperativity to examine the homochiral phenomenon in nature systems such as proteins and DNA [27–29]. It was found that the generation process of homochirality is positive cooperativity, suggesting that once a chiral preference begins to emerge, subsequent molecular interactions increasingly favor and reinforce that same chirality. This self-reinforcing mechanism may have played a critical role in driving the uniform handedness observed in biological macromolecules, pointing to cooperativity as a fundamental principle underlying molecular evolution and symmetry breaking in early biochemical systems.

Despite recent advances, our grasp of cooperativity and frustrativity remains in its infancy. These concepts, by definition, in Eqs. (8.7) and (8.24), rely on the second derivative, reflecting how the response of a system changes in relation to perturbations in its components. Their sensitivity to curvature highlights their unique ability to probe nonlinear dynamics and emergent properties in many-body systems, offering insights into how collective interactions give rise to unexpected behaviors. To fully leverage their potential, it will be essential to conduct thorough investigations into their mathematical foundations and physical implications, as well as explore their applications in chemical and biological systems. Such efforts will pave the way for embedding these concepts and associated principles into theoretical and computational models across disciplines, from chemistry to physics, biology, and beyond, ultimately enriching our toolkit for deciphering interconnected systems at different scales.

Together, these developments show that cooperativity and frustrativity are not limited to static quantum descriptions but extend across scales and representations, providing a unifying language for analyzing collective behavior in molecular, biological, and materials systems.

8.8 Outlook: From Cooperation and Frustration to Emergence

The discussions in the preceding sections reveal that cooperativity and frustrativity are second-order response phenomena, reflected in the curvature of interaction- and distortion-energy profiles and, more broadly, in the curvature of energy and information landscapes. Yet beyond their numerical characterization lies a broader conceptual perspective: together, these intertwined effects mark the threshold of *emergence* in chemical systems.

Cooperativity embodies positive feedback—interactions among components reinforce one another, producing collective stabilization—whereas frustration introduces negative feedback, as competing constraints prevent complete optimization and preserve structural diversity. Their balance generates emergent behavior: new structures, functions, and reactivities that cannot be inferred from isolated components.

From a physical standpoint, emergence arises when the potential-energy or information landscape acquires curvature and correlation beyond pairwise additivity. Positive curvature associated with reinforcing interactions drives order and coherence, whereas conflicting curvature introduces ruggedness and metastability. The sigmoidal patterns captured by logistic-type equations already signal the onset of emergent organization: growth that accelerates through self-enhancement but ultimately saturates through internal limitations.

This duality shapes reactivity across scales. In catalysis, neighboring active sites interact cooperatively through electron delocalization or polarization, while steric and electrostatic constraints introduce frustration; optimal efficiency emerges at their balance point. In allosteric regulation, cooperative conformational transitions are modulated by local frustration to achieve switch-like responses. Likewise, supramolecular assembly, nucleation, micellization, and crystallization follow cooperative growth laws, yet packing constraints generate frustration that limits domain size and permits polymorphism. Even aromaticity and resonance can be viewed through this lens: cyclic π systems exhibit cooperative stabilization, whereas ring strain or antiaromaticity introduces frustration that modulates their final character.

Within the density-based reactivity framework, the cooperativity index (κ) and the frustrativity index (ζ) provide quantitative probes of this interplay. As second-order response measures, they encode the degree of nonadditivity and thus the system's potential for collective behavior. Large $|\kappa|$ and $|\zeta|$ values indicate strong nonlinear coupling, whereas systems with $\kappa \approx \zeta \approx 0$ behave as nearly additive aggregates.

In summary, emergence in chemistry arises when cooperative reinforcement and structural frustration coexist in a dynamic equilibrium. Cooperativity imparts cohesion and order; frustration provides flexibility and adaptability. Their interplay gives rise to new reactivity patterns, catalytic mechanisms, and structural motifs. Recognizing this triad—cooperation, frustration, and emergence—extends density-based reactivity theory from describing energetic trends to explaining how new chemical behavior materializes from many-body coupling, thereby connecting naturally to the hierarchical modeling of chemical understanding developed in the next chapter.

References

1 Zhong S, He X, Liu S, Wang B, Lu T, Rong C, et al. Toward density-based and simultaneous description of chemical bonding and noncovalent interactions with Pauli energy. J Phys Chem A. 2022 Apr 21;126(15):2437–44.

2 Rong C, Zhao D, He X, Liu S. Development and applications of the density-based theory of chemical reactivity. J Phys Chem Lett. 2022 Dec 8;13(48):11191–200.

3 He X, Li M, Rong C, Zhao D, Liu W, Ayers PW, et al. Some recent advances in density-based reactivity theory. J Phys Chem A. 2024 Feb 22;128(7):1183–96.

4 Rong C, Zhao D, Yu D, Liu S. Quantification and origin of cooperativity: insights from density functional reactivity theory. Phys Chem Chem Phys. 2018 Jul 4;20(26):17990–8.

5 Liu S, Rong C. Quantifying frustrations for molecular complexes with noncovalent interactions. J Phys Chem A. 2021 Jun 10;125(22):4910–7.

6 Rong C, Zhao D, Zhou T, Liu S, Yu D, Liu S. Homogeneous molecular systems are positively cooperative, but charged molecular systems are negatively cooperative. J Phys Chem Lett. 2019 Apr 18;10(8):1716–21.

7 Zhou T, Liu S, Yu D, Zhao D, Rong C, Liu S. On the negative cooperativity of argon clusters containing one lithium cation or fluorine anion. Chem Phys Lett. 2019 Feb;716:192–8.

8 Zhang W, Zhao Y, An X, Fu J, Zhang J, Zhao D, et al. Cooperativity and reactivity properties of medium-sized boron clusters: a combined density functional theory and information-theoretic approach study. Mol Phys. 2022;e2157774.

9 Acerenza L, Mizraji E. Cooperativity: a unified view. Biochim Biophys Acta (BBA) – Protein Struct Mol Enzymol. 1997 Apr;1339(1):155–66.

10 Koshland DE, Némethy G, Filmer D. Comparison of experimental binding data and theoretical models in proteins containing subunits. Biochemistry. 1966 Jan;5(1):365–85.

11 Changeux JP. Allostery and the Monod-Wyman-Changeux model after 50 years. Annu Rev Biophys. 2012 Jan 6;41:103–33.

12 Parr RG, Yang W. Density-Functional Theory of Atoms and Molecules. Oxford University Press; 1989.

13 Liu S. Identity for Kullback-Leibler divergence in density functional reactivity theory. J Chem Phys. 2019 Oct 14;151(14):141103.

14 Nalewajski RF, Parr RG. Information theory, atoms in molecules, and molecular similarity. Proc Natl Acad Sci USA. 2000 Aug 1;97(16):8879–82.

15 Liu S, Rong C, Lu T. Information conservation principle determines electrophilicity, nucleophilicity, and regioselectivity. J Phys Chem A. 2014 May 22;118(20):3698–704.

16 Wang B, Rong C, Chattaraj PK, Liu S. A comparative study to predict regioselectivity, electrophilicity and nucleophilicity with Fukui function and Hirshfeld charge. Theor Chem Acc. 2019 Dec;138(12):124.

17 Moessner R, Sondhi SL. Ising models of quantum frustration. Phys Rev B. 2001 May 4;63(22):224401.

18 Vannimenus J, Toulouse G. Theory of the frustration effect. II. Ising spins on a square lattice. J Phys C: Solid State Phys. 1977 Sep 28; 10:L537.

19 Bryngelson JD, Wolynes PG. Intermediates and barrier crossing in a random energy model (with applications to protein folding). J Phys Chem. 1989 Sep;93(19):6902–15.

20 Dill KA. Polymer principles and protein folding. Protein Sci. 1999 Jun;8(6):1166–80.

21 Tzul FO, Vasilchuk D, Makhatadze GI. Evidence for the principle of minimal frustration in the evolution of protein folding landscapes. Proc Natl Acad Sci USA. 2017 Feb 28;114(9):E1627–32.

22 Onuchic JN, Luthey-Schulten Z, Wolynes PG. Theory of protein folding: the energy landscape perspective. Annu Rev Phys Chem. 1997;48:545–600.

23 Schön JC. Energy landscapes-past, present, and future: a perspective. J Chem Phys. 2024 Aug 7;161(5):050901.

24 Wales DJ, Doye JPK, Miller MA, Mortenson PN, Walsh TR. Energy landscapes: from clusters to biomolecules. In: Prigogine I, Rice SA, editors. Advances in Chemical Physics. Wiley; 2000. pp. 1–111.

25 Nochebuena J, Piquemal J-P, Liu S, Cisneros GA. Cooperativity and frustration effects (or lack thereof) in polarizable and non-polarizable force fields. J Chem Theory Comput. 2023 Nov 14;19(21):7715–30.

26 Nochebuena J, Liu S, Cisneros GA. Relative cooperativity in neutral and charged molecular clusters using QM/MM calculations. J Chem Phys. 2024 Apr 7;160(13): 134301.

27 Liu S. Homochirality originates from the handedness of helices. J Phys Chem Lett. 2020 Oct 15;11(20):8690–6.

28 Liu S. Principle of chirality hierarchy in three-blade propeller systems. J Phys Chem Lett. 2021 Sep 16;12(36):8720–5.

29 Li M, He X, Chen J, Wang B, Liu S, Rong C. Density functional theory and information-theoretic approach study on the origin of homochirality in helical structures. J Phys Chem A. 2021 Feb 11;125(5):1269–78.

9

Homochirality and Principle of Chirality Hierarchy

TABLE OF CONTENTS

The previous chapters have progressively built density-based frameworks for understanding chemical structure and reactivity, emphasizing how electrostatics, information-theoretic quantities, and cooperativity shape molecular structures across scales. Central ideas—such as collective stabilization, hierarchical coupling, and energetic asymmetry—naturally lead to the concept of homochirality, one of the most striking examples of molecular order in chemistry and biology. In this chapter, we extend the principles of cooperativity and hierarchical modeling to chirality, showing that handedness at different dimensional levels (0D, 1D, 2D, 3D) interacts through long-range electrostatic forces and density-derived quantities. Homochirality, therefore, is not an isolated stereochemical curiosity but a direct manifestation of the density-based principles developed earlier in this book.

9.1 What Is Chirality?

Chirality—from the Greek χεἰρ ("hand")—refers to the geometric property whereby an object is nonsuperposable on its mirror image. In chemistry, chirality underpins stereochemistry and governs how molecules interact with other chiral environments (e.g., biological receptors). At the heart of chirality lies the absence of mirror-plane symmetry (σ), inversion center (i), or improper rotation axes (S_n); a chiral molecule belongs to a point group that contains only proper rotations (C_n) and the identity (E).

In chemistry, chirality most often arises when a molecule contains a carbon atom bonded to four *different* groups or atoms. Such a carbon atom is called a chiral center or *stereocenter*. For example, consider bromochlorofluoromethane (BrClFCH), where the carbon atom is attached to four different atoms: bromine, chlorine, fluorine, and hydrogen. This molecule is chiral because its mirror image cannot be rotated or manipulated to match the original structure. In contrast, dichloromethane (CH_2Cl_2), where the carbon atom is bonded to two hydrogen atoms and two chlorine atoms, has a plane of symmetry and is therefore achiral.

Two nonsuperposable mirror images are called *enantiomers*, which are mirror-image stereoisomers that share identical physical properties but rotate plane-polarized light in opposite directions. *Diastereomers* are stereoisomers that are not mirror images and thus exhibit different physical and chemical behaviors. A *stereogenic* (or chiral) center is most often an sp^3-hybridized carbon bearing four different substituents, while a prochiral center can be converted into a stereocenter by replacing one of two identical groups. In an achiral environment, a 1:1 mixture of enantiomers forms a *racemate*, which shows no net optical activity, and the predominance of one enantiomer over the other is measured by *enantiomeric excess* (*ee*). Enantiomeric purity is critical in pharmaceuticals: often only one enantiomer is biologically active or may exhibit reduced toxicity.

The Cahn–Ingold–Prelog (CIP) priority rules [1] assign an absolute configuration (*R* or *S*) to stereogenic centers. Priorities are assigned by atomic number, with higher atomic numbers receiving higher priority; in case of ties, one moves outward along substituent chains. The arrangement of substituents around the stereocenter—viewed such that the lowest-priority substituent points away—determines if the sequence $1 \rightarrow 2 \rightarrow 3$ is clockwise (*R*) or counterclockwise (*S*). *Stereoselectivity* and *stereospecificity* distinguish whether a reaction favors one stereoisomer or whether different stereoisomeric reactants necessarily yield different stereoisomeric products. The stereocontrolled construction of chiral centers is central to organic synthesis. *Asymmetric catalysis* uses chiral catalysts—organocatalysts or metal complexes bearing chiral ligands (e.g., BINAP, salen, proline)—to induce enantioselectivity in reactions such as hydrogenation, epoxidation, and C—C bond formation. *Stereoconvergence* is a process in which two or more stereoisomeric substrates (e.g., *E*/*Z* alkenes or enantiomers) are converted under the same conditions into a single, identical stereoisomeric product. Conversely, *stereodivergence* uses a single stereoisomeric substrate and diverges it into different stereoisomeric products (e.g., *R* versus *S* alcohols) simply by altering catalysts or reaction conditions.

The concept of chirality [2] dates to 1815, when Jean-Baptiste Biot first observed that certain organic substances rotate the plane of polarized light, inaugurating the field of optical activity. In 1848, Louis Pasteur famously resolved the enantiomers of tartaric acid by manually sorting their crystallographic "hands," demonstrating that mirror-image molecules could be physically separated. The theoretical foundation of stereochemistry was laid independently by Jacobus Henricus van't Hoff and Joseph Le Bel in 1874, who proposed that the tetrahedral arrangement of four different substituents around a carbon atom gives rise to no superposable mirror images. In 1894, Hermann Emil Fischer introduced the *D*/*L* notation for sugars, linking molecular asymmetry to biological function. The mid-twentieth century saw the formalization of absolute configuration with the CIP priority rules (1956) and the advent of modern enantioselective synthesis, marked by William Knowles's asymmetric hydrogenations (Nobel Prize, 2001) [3] and Barry Sharpless's catalytic epoxidations (Nobel Prize, 2001) [4]. Together, these milestones trace the evolution of chirality from curious optical phenomena to a central principle in chemical theory, synthesis, and pharmacology.

9.2 Homochirality and Theories of Homochirality

Molecules can possess multiple stereogenic centers. If a molecule has n stereogenic centers, it can exist as a maximum of 2^n stereoisomers. These stereoisomers can be related as enantiomers (nonsuperimposable mirror images) or diastereomers (stereoisomers that are not mirror images). Homochirality is a special case of such systems when every chiral center adopts the same absolute configuration.

Homochirality is unique to the biopolymers of life [5], most notably proteins and nucleic acids, which are composed exclusively of one enantiomeric form: L-amino acids in proteins and D-sugars in nucleic acids and carbohydrates. This uniform chirality is remarkable because abiotic syntheses typically produce racemic mixtures (50:50 enantiomeric ratios). Homochirality underlies the high fidelity of enzyme–substrate interactions, the regular secondary and tertiary structures of biomolecules (α-helices, β-sheets, double helices), and, ultimately, the emergence of self-replicating systems. If non-biological chemical processes tend to produce racemic mixtures, but life is profoundly homochiral, how did this transition occur? The origin of biological homochirality is one of the most significant and challenging unanswered questions related to the origin of life (abiogenesis). How did prebiotic chemical systems on early Earth select L-amino acids and D-sugars from what was likely a near-racemic primordial soup? Understanding how homochirality arose and is maintained connects fundamental questions in stereochemistry, origin-of-life research, and materials science.

Homochirality became a focal point only starting from the mid-twentieth century, with the realization that biological macromolecules display absolute enantiomeric purity. However, the emergence of homochirality poses a "chicken-and-egg" dilemma: Did homochirality arise before life, or was it a product of early biological evolution? Theories of homochirality [6–8] have evolved through Frank's seminal 1953 kinetic model—demonstrating that autocatalysis plus mutual antagonism can amplify an infinitesimal enantiomeric imbalance—into concrete chemical systems like the Soai reaction (1995) [9], which achieves >99% *ee* via self-propagating autocatalysis. Parallel work [10] on Viedma ripening in the early 2000s showed that simple grinding and recrystallization of racemic conglomerates can yield a single enantiomorph purely through physical cycling, while studies of circularly polarized light and chiral mineral surfaces have provided plausible environmental "seed" biases. More recently, spatially resolved and stochastic computational models have explored how diffusion, compartmentalization, and random fluctuations can lock in and propagate chirality in confined prebiotic settings, integrating chemical, physical, and environmental pathways into a cohesive picture of how life's uniform handedness may have arisen.

Theories of homochirality at the present time can be categorized into the following five types. (i) *Deterministic models* suggest external chiral influences, such as circularly polarized light in interstellar regions or parity-violating weak nuclear forces, created an initial enantiomeric excess, which was then amplified by processes like autocatalysis—exemplified by the Frank model and the Soai reaction, where chiral molecules catalyze their own production while inhibiting their mirror image. (ii) *Extraterrestrial delivery theories* posit that meteorites like the Murchison meteorite, which contained slight L-amino acid excesses, seeded Earth with chiral bias, though critics argue racemization would erase such imbalances over time. (iii) *Stochastic models* propose that random fluctuations in racemic mixtures, combined with autocatalytic amplification, led to homochirality, as seen in computational RNA-world simulations where template-directed polymerization favored uniform chirality at the polymer level despite racemic monomers. (iv) *Crystallization-based mechanisms*, such as Viedma ripening, demonstrate how grinding or temperature cycling can drive symmetry breaking in solids, though their prebiotic relevance is debated. Lastly, (v) *thermodynamic and evolutionary models* argue homochirality emerged from functional advantages, such as stable polymer folding or enzymatic specificity, with recent studies highlighting chiral selection in tRNA-amino acid interactions during early translation systems. While no consensus exists, these theories collectively emphasize the interplay of chance, physical forces, and evolutionary pressure in resolving biology's chiral puzzle.

9.3 Types of Chirality and Chirality Hierarchy

The chirality we discussed earlier was focused on *point chirality* arising from the tetrahedral stereogenic center of carbon atoms. Beyond carbon, other atoms such as phosphorus (in phosphines), sulfur (in sulfoxides), and even metals in coordination complexes can serve as centers of point chirality. In organometallic chemistry, octahedral and trigonal bipyramidal complexes with appropriate ligands can exhibit chirality if the arrangement of ligands breaks mirror symmetry. The CIP rules systematize the assignment of their absolute configuration (R/S). Because this type of chirality is localized at individual atoms, it is called *point chirality* or *0D (zero-dimensional) chirality*.

Chirality can also arise from the arrangement of substituents around an axis of chirality [11, 12]. This occurs when rotation around a bond is sufficiently restricted due to steric hindrance or incorporation into a rigid structure (atropisomerism), and the substituents at the ends of the axis meet specific symmetry requirements. The molecule lacks S_n axes and exists as nonsuperimposable mirror images. Conceptually, this can be visualized as an elongated tetrahedron. For example, in biphenyls, when ortho-substituents on each aromatic ring are bulky enough to restrict free rotation around the central C—C bond, the two rings adopt a twisted conformation. If the molecule lacks a plane of symmetry, it can exist as enantiomers due to axial chirality. Similarly, allenes ($R_1 R_2C{=}C{=}CR_3 R_4$) exhibit axial chirality when the substituents at each terminal carbon are distinct, resulting in a cumulative double-bond system with a linear, chiral geometry. BINAP (2,2′-*bis*(diphenylphosphino)-1,1′-binaphthyl) is a widely used atropisomeric ligand in asymmetric catalysis. One example is the design of the anticancer agent navitoclax. Certain spiro compounds, where two rings share a single common atom, can also exhibit axial chirality if appropriately substituted.

Axial chirality [13] arises when four substituents are arranged in two pairs around a rigid molecular axis such that the molecule and its mirror image are nonsuperposable. Since this type of chirality lies along a single molecular axis, it is commonly referred to as one-dimensional (*1D*) chirality. To assign its absolute configuration, the extended CIP rules rank the two "near" substituents above the two "far" ones when viewing end-on along the axis, and a clockwise sequence of priorities #1→#2→#3 defines R_a while a counterclockwise sequence defines S_a. Alternatively, the helical sense of the substituent arrangement can be described by *P* (right-handed screw, also **Δ**) or *M* (left-handed screw, also **Λ**). These conventions and nomenclatures were endorsed by IUPAC and widely applied to atropisomeric biaryls, allenes, and helicenes to unambiguously name axially chiral compounds.

Helical chirality [14] is a special case of *1D chirality*. Helical chirality (helicity) arises from a screw-shaped, propeller-like arrangement of atoms in a molecule. Its stereoisomerism arises solely due to a helical or screw-shaped molecular backbone, independent of any stereogenic center or substituent pairs, though it can be assigned by the *P*/*M*(**Δ**/**Λ**) nomenclature. Biomolecules such as DNA (typically right-handed B-form) and proteins (α-helices are right-handed) and synthetic helicenes, which are fused aromatic compounds forming a rigid, helical shape, are classical examples of molecules with helical chirality. Its key difference with axial chirality is that helical chirality is an inherent property of a helical shape itself, whereas axial chirality depends on substituent arrangements around a defined axis.

Planar chirality, or *2D chirality* [15], refers to systems in which a two-dimensional arrangement (usually a plane of atoms or groups) generates chirality due to asymmetry above and below the plane. This form is less intuitive than point (0D) or axial (1D) chirality and often appears in metallacycles, macrocycles, or substituted ferrocenes. For instance, in ferrocenes, when one or both cyclopentadienyl rings are substituted in an unsymmetrical fashion, the resulting molecule exhibits planar chirality. The iron atom lies between the two rings, and the relative orientation of substituents breaks mirror symmetry. Other examples of planar chirality exist in substituted paracyclophanes, *ansa* compounds (molecules with an "ansa" bridge spanning an aromatic ring), and *E*-cyclooctene. The nomenclature used in planar chirality is based on the absolute configuration of the planar chiral unit, designated as either R_p or S_p. The process involves identifying a "pilot atom" (the highest priority atom directly attached to the chiral plane but not in it), tracing a path along the atoms of the chiral plane (from highest to second-highest priority atom bonded to the pilot atom), and determining the direction (clockwise *R*, counterclockwise *S*) relative to the pilot atom's position above or below the plane.

3D chirality [16] transcends point, axial, or planar asymmetry by referring to chirality arising from the three-dimensional arrangement of atoms in molecules or molecules in molecular assembly. Within this category, we have three specific classes of chirality. *Topological chirality* [17] is related to the molecule's overall topology, independent of local stereogenic units. It arises in molecules like knots (e.g., trefoil knot) and links (catenanes, rotaxanes). Interconversion between enantiomers requires breaking and reforming covalent bonds. A trefoil knot, for instance, is chiral regardless of the atoms composing the loop. *Inherent chirality* [18] refers to molecules possessing curved structures where chirality is intrinsic to the framework itself. Examples include certain functionalized fullerenes or saddle-shaped ("warped") aromatic molecules. *Supramolecular chirality* [19] emerges from the ordered, noncovalent assembly of achiral or chiral molecules into larger superstructures. Examples are liquid crystals, self-assembled monolayers, aggregates, and gels. This type of chirality can be transferred or amplified through these assemblies through, for example, the "sergeants-and-soldiers" principle and the "majority rules" principle.

Chirality hierarchy describes how different forms of chirality—0D (point), 1D (axial or helical), 2D (planar), and 3D (topological or supramolecular)—are organized and interrelated within a system. These chiral elements are not isolated; rather, they coexist and often influence one another. For example, the 0D chirality of L-amino acids biases the peptide backbone toward a right-handed α-helix (1D), while the axial chirality of BINAP (1D) can impose planar chirality (2D) on its metal complexes. A local stereocenter can thus steer the formation of a preferred helical handedness, and higher-level chiral frameworks can, in turn, stabilize or destabilize stereocenters. Chirality hierarchy emphasizes that stereochemical preferences are distributed across levels and can reinforce, oppose, or reshape each other.

This multilevel coupling of chirality arises because different chiral elements interact through the underlying electrostatic, steric, and exchange–correlation effects that shape molecular conformations. When these interactions are aligned, lower-dimensional chirality reinforces higher-dimensional structure. Conversely, when geometric or electronic constraints oppose each other, local and global chirality preferences may compete, producing subtle energetic balances or unexpected stereochemical outcomes.

Importantly, this interaction can be bidirectional. While 0D chirality often guides the formation of 1D or 2D chirality, supramolecular or macroscopic chirality may also imprint handedness onto local (i.e., lower level) units, demonstrating that higher levels can influence or even override the preferences of lower ones. In this sense, chirality hierarchy reflects a dynamic network of mutual stereochemical influence, rather than a one-way cascade, and provides a conceptual bridge to understanding how handedness propagates through molecular architectures.

9.4 Chirality Transmission

Chirality transmission [20, 21] refers to the process by which chiral information is relayed from one part of a molecular or supramolecular system to another. In systems containing only one type of chirality, transmission typically occurs when a stereocenter influences the conformation, reactivity, or physical properties of a remote site. For example, a single chiral center can dictate the configuration of a newly formed stereocenter or direct the helical preference of a polymer backbone.

In systems where multiple types of chirality coexist, transmission becomes richer and more complex. Different chiral elements—such as 0D stereocenters, 1D helices, and 2D planar units—can reinforce or compete, leading to nonlinear propagation of chiral bias across the molecular framework. This hierarchical coupling is central to biological systems: the 0D chirality of amino acids and nucleotides imposes the 1D handedness of protein and DNA helices, which in turn determines higher-order structure and function.

Chirality transmission also resembles allosteric communication in proteins [22, 23]: a perturbation or bias at one site can induce a structural response at a distant location through conformational or electronic changes. Such long-range coupling enables chiral information to be conveyed across multiple length scales and can even induce chirality at previously achiral sites. External factors such as temperature, solvent, light, or co-assembling molecules can modulate this transmission pathway, allowing chiral properties to be switched or amplified.

Examples span both natural and synthetic systems. Proteins and nucleic acids are classical cases where local stereochemistry gives rise to global chirality with functional consequences. Synthetic foldamers, metal–organic frameworks, supramolecular assemblies, and mechanically interlocked molecules likewise exhibit chirality transmission, often engineered to achieve enantioselective catalysis, chiral recognition, or tunable chiroptical properties.

In essence, chirality transmission describes how chiral information is conveyed, amplified, or transformed across interconnected elements of a molecular system. This cooperative relay provides the mechanistic basis for the chirality hierarchies analyzed quantitatively in the following sections.

9.5 Chirality Hierarchy Case Study: Helices

To better understand the intricacy of chirality hierarchy, we recently examined a few right-handed helical systems that incorporate stereogenic centers (either *R* or *S* configuration) within their backbone [24–27]. Helices are intrinsically 1D chirality, and stereogenic centers are 0D chirality, so these systems naturally exhibit a hierarchy of two chirality types. The systems we investigated are (a) hexahelicene with six outer-edged carbon on the three middle benzene rings, each added by one methyl group, (b) 3_{10}-helix with alanine attached to the C_α carbon, (c) α-helix with alanine residue attached to the C_α carbon, and two single-stranded DNA deoxyribose

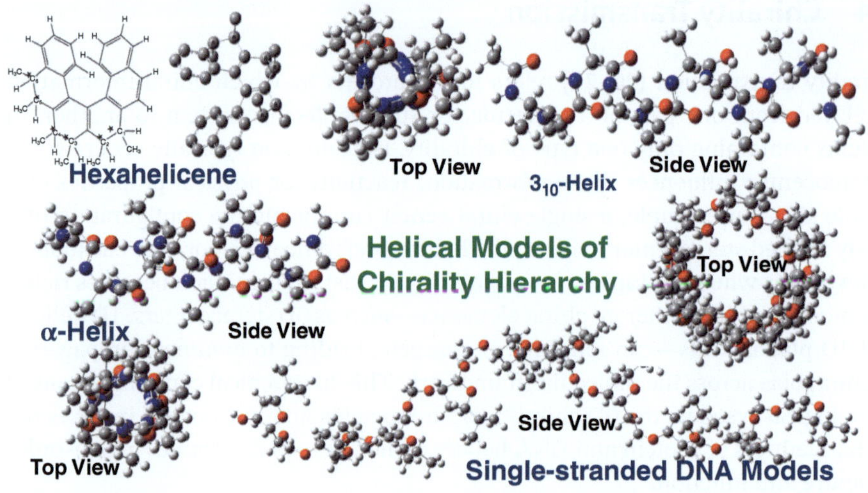

Figure 9.1 Helical models of chirality hierarchy examined in our study.

models capped with (d) iso-propyl (iPr) and (e) dimethylamine (NMe_2). Shown in Figure 9.1 are these models.

We first examined right-handed helical models in which all stereogenic centers adopt the same chiral configuration—either all-*R* or all-*S* (i.e., homochiral)—for both 3_{10}-helices and α-helices. Our results reveal a stability difference of up to 15 kcal/mol (depending on the helical type and number of residues), favoring the all-*S* configuration. In contrast, no energetic difference is observed for single- and double-stranded β-sheets. These findings indicate that, for the two helical structures, the homochiral *S* enantiomers are consistently more stable than their homochiral *R* counterparts. Furthermore, when comparing 3_{10}-helix with α-helix, we found that the α-helix generally exhibits a greater *S–R* energy difference, indicating a stronger preference for the all-*S* configuration over the 3_{10}-helix.

Does this result depend on the identity of the side chain at the C_α atom? To answer this, we examined nine additional amino acids—isoleucine (Ile), leucine (Leu), cysteine (Cys), valine (Val), methionine (Met), serine (Ser), phenylalanine (Phe), lysine (Lys), and histidine (His)—alongside alanine (Ala). For each amino acid, we constructed six-residue homochiral models in both 3_{10}-helix and α-helix conformations, where all six C_α positions adopted either all-*R* or all-*S* configurations. The total energy differences (*S-R*) are negative for all amino acids except cysteine, indicating that the all-*S* form is energetically more favorable. In the case of cysteine, the result aligns with the well-established natural preference for the *R* configuration, consistent with experimental observations. This inversion relative to other amino acids arises because the sulfur substituent changes CIP priorities, making the natural L-form of cysteine correspond to the R configuration.

A similar preference for one homochiral configuration over the other is also observed in models of hexahelicene and DNA. However, unlike the peptide helices, hexahelicene favors the homochiral *R*-form over the *S*-form. In the case of deoxyribose units in single- and double-stranded DNA helices, experimental evidence confirms a natural preference for the *R* configuration. Our computational results for the two single-stranded DNA models are consistent with this, showing that the all-*R* enantiomer is more stable than the all-*S* form.

Does there exist a cooperativity effect [28–31] in these homochiral systems with two types of chirality? To quantify their cooperativity effect, we extend the definition of cooperativity previously used for molecular complexes in Chapter 8 to homochiral systems by using the same expression [24],

$$k = -(\partial \Delta E_n / \partial n). \tag{9.1}$$

where n is the number of repeating units but ΔE_n is the total energy difference of the two homochiral configurations per repeating unit, defined by

$$\Delta E_n = \frac{E_{\text{favored}} - E_{\text{unfavored}}}{n} \tag{9.2}$$

with E_{favored} and $E_{\text{unfavored}}$ represent the total energy of the energetically preferred and less preferred homochiral configurations, respectively. By convention, $\Delta E_n < 0$ indicates the favored homochiral configuration is lower in energy. If $k > 0$, cooperativity is positive, meaning that adding an additional unit enhances the stabilization of the homochiral configuration. Conversely, a negative k implies negative cooperativity, where additional units weaken the overall stability. If $k = 0$, no cooperativity is present. The minus sign in Eq. (9.1) ensures that positive k corresponds to enhanced stabilization as the system grows. Because ΔE_n becomes more negative for stronger homochiral preference, $-\partial \Delta E_n / \partial n$ provides a positive measure of cooperativity.

Figure 9.2 presents the cooperativity profiles (ΔE_n versus n) for four model systems: the 3_{10}-helix, α-helix, and two single-stranded DNA models studied in Figure 9.1. All systems exhibit positive cooperativity except for the 3_{10}-helix. This result supports the assertion that cooperativity should exist and thus play an important role in natural systems. The contrasting cooperativity behavior between two helical secondary structures (3_{10}-helix and α-helix) may help explain why α-helices are far more prevalent than 3_{10}-helices in naturally occurring proteins.

Is homochirality favored over heterochirality in systems exhibiting chirality hierarchy? To answer this question, we constructed all 64 possible stereoisomers (2^6 combinations) for both 3_{10}- and α-helices composed of six alanine residues, with each C_α center adopting either the *R* or *S* configuration. Our results show that the all-*S* configuration has the lowest energy in both helical types, indicating it is the most stable configuration. This finding affirms that homochirality is indeed favored in these systems.

Finally, what interaction governs homochirality and the preference for one enantiomer over another? We employ the total energy partition scheme in density functional theory (DFT) [32, 33],

$$\Delta E = \Delta T_S + \Delta E_{xc} + \Delta E_e, \tag{9.3}$$

where the total energy difference ΔE comes from three independent contributions, noninteracting kinetic energy ΔT_S, exchange–correlation energy ΔE_{xc}, and

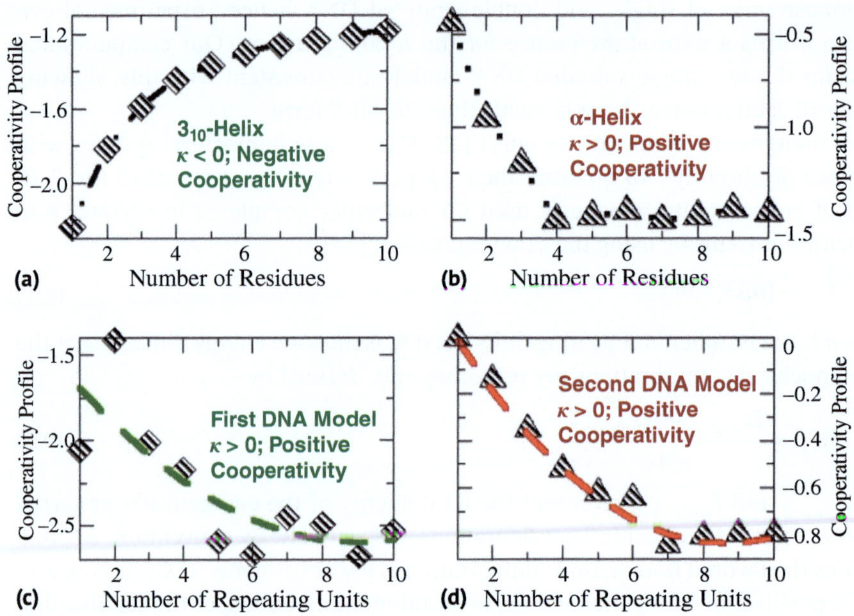

Figure 9.2 The cooperativity profile of (a) 3_{10}-helix, (b) a-helix, (c) and (d) two DNA models up to 10 units. For (a) and (b), the corresponding all-R enantiomer was employed as the reference, but for (c) and (d), all-S enantiomer was the reference because they are less favored. *Source*: Reproduced from Ref. [24] with permission of American Chemical Society.

electrostatic energy ΔE_e. For the 64 stereoisomers analyzed above, we find that the negative electrostatic term ΔE_e is primarily responsible for the overall negative ΔE, indicating that the greater stability of the favored form arises from stronger electrostatic interaction. In contrast, the kinetic and exchange–correlation components (ΔT_S and ΔE_{xc}) are consistently positive and therefore oppose the stabilization. The dominant role of ΔE_e suggests that stronger electrostatic interactions in the all-*S* configuration underlie its enhanced stability over other stereoisomers.

9.6 Chirality Hierarchy Case Study: Propellers

To validate the findings from the previous right-handed helical systems, we investigated a left-handed three-blade propeller molecule (Figure 9.3) [25]. As illustrated, the propeller structure consists of three key components: a central atom (X), a six-membered ring (6MR), and chiral substituents attached to the 6MR at the *meta*, *para*, or both positions. We constructed a series of propeller models by varying each of these components: six options for the central atom (X = Al, B, C^+, C^-, N, and P), eight alternatives for the six-membered ring (6MR = phenyl, pyridine, pyrimidine, pyrazine, triazine, and borazine), and ten types of chiral groups, $C^*R_1R_2R_3$, with $R_1R_2R_3$ combinations drawn from H, F, Cl, OH, and Me.

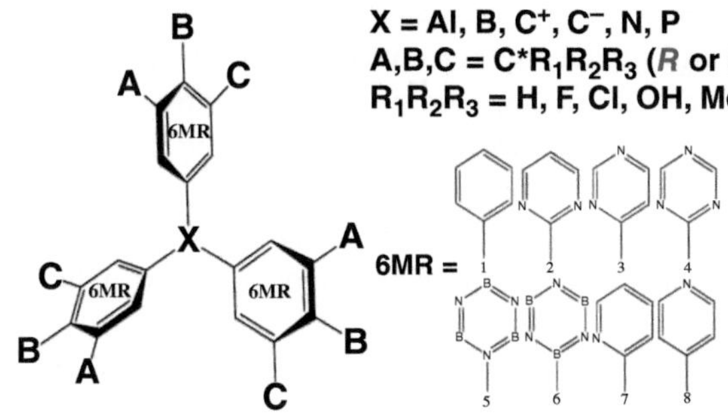

Figure 9.3 Propeller molecular models as a chirality hierarchy. *Source*: Reproduced from Ref. [25] with permission of American Chemical Society.

For the model system with X = B and a phenyl 6MR, where the 0D chiral group CHFMe is attached at the *meta*, *para*, or two *meta* positions on each blade, we observed the following [25]: (i) the stereoisomeric energy difference is consistently nonzero, indicating that the presence of chiral centers on the 6MR induces a thermodynamic preference for one enantiomer over the other; (ii) the magnitude of this energy difference varies depending on the number and positions of the chiral centers; and (iii) the stereoisomeric energy differences are relatively insensitive to the choice of basis sets and exchange–correlation functionals used in the DFT calculations.

The stereoisomeric energy difference depends on the specific nature of the three components that make up each propeller blade. For instance, when the central atom of the propeller chosen from Al, B, C^+, C^-, N, and P combined with 10 different chiral groups placed at a single *meta* position on each blade, we observe that the S–R energy difference can be either negative or positive depending on the chiral group, even with the same central atom. Similarly, for a fixed chiral group, the energy difference varies with different central atoms. These findings highlight the complexity of chirality hierarchy, revealing how axial chirality (embodied by the propeller structure) interplays intricately with point chirality (the *R*/*S* configuration preference and associated stability of the chiral substituent).

We also explored cases where the same chiral group is placed on different 6MRs and paired with various propeller centers, as well as scenarios where the same chiral group is positioned at different locations on the 6MR, with varying numbers of chiral substitutions. Across all these variations, the *S*–*R* energy difference is found to be nonzero yet can be either positive or negative, further reinforcing the intricacy and complexity of the chirality hierarchy described earlier.

Does cooperativity effect also exist in these propeller systems? To investigate this, we studied a model with boron as the propeller center and phenyl rings as the blades, systematically increasing the number of chiral groups at the *meta* positions from one to six, all in either the R or S configuration. The resulting cooperativity profile shows an unambiguous downward trend, indicating positive cooperativity. In other words, as more chiral groups are added, the stereoisomeric energy difference per unit increases in magnitude, reinforcing the preference for one homochiral configuration.

What drives the preference for homochirality in these systems? To address this, again, we constructed 64 heterochiral models by permuting all six chiral centers (*R* or *S*) on the blades of a propeller system composed of a boron center (X = B), phenyl rings (6MR), and CHFMe groups at all *meta* positions. Among these, the all-*S* homochiral isomer was found to have the lowest total energy, confirming it as the most stable configuration. This reinforces that homochirality is a fundamental feature of the propeller system, consistent with the behavior observed in helical structures [24].

Figure 9.4 presents energetic analyses of the 64 stereoisomers using Eq. (9.3) and another total energy partition scheme, proposed recently by this author [32],

$$\Delta E = \Delta E_s + \Delta E_e + \Delta E_q, \tag{9.4}$$

where E_s and E_q are contributions from the steric and fermionic quantum effects. Details of this partition are available in Chapter 5. In Figure 9.4a, a strong linear correlation is observed between the stereoisomeric stability (referenced to the all-R isomer for consistency; note that the all-S form is more stable) and the electrostatic energy component. This indicates that enhanced electrostatic interactions in the all-*S* form are the dominant factor stabilizing the system. This trend mirrors the findings of the helical systems. Furthermore, Figure 9.4b and c shows that incorporating exchange–correlation or steric energy terms alongside the electrostatic component

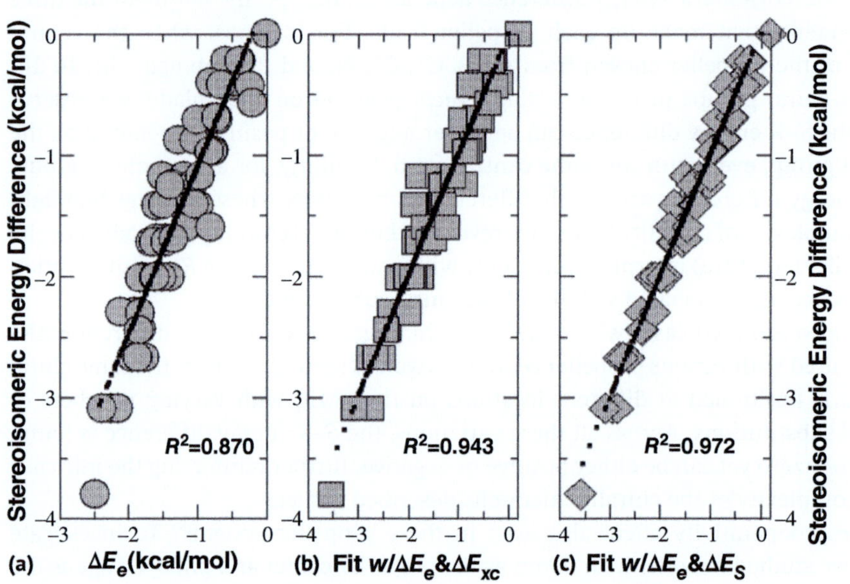

Figure 9.4 Energetic analyses for 64 propeller stereoisomers. The strong linear correlation of the stereoisomeric energy difference (with all-*R* as the reference) with (a) the electrostatic component, (b) two-variable fitting with electrostatic and exchange-correlation energy components, and (c) two-variable fitting with electrostatic and steric components. *Source*: Reproduced from Ref. [25] with permission of American Chemical Society.

improves the correlation, suggesting that while these factors play secondary roles, they are still essential contributors to overall stereoisomeric stability. This result is also consistent with that from the helical systems in Section 9.5, as well as with other systems that we will extensively discuss in Chapter 14.

9.7 Principle of Chirality Hierarchy

The principle of chirality hierarchy states that in systems containing multiple levels of chirality—0D stereocenters, 1D helices, 2D planar units, and 3D supramolecular arrangements—the preferred handedness emerges from cooperative coupling among chiral elements across these levels. Lower- and higher-dimensional chiralities do not act independently: lower-dimensional chirality biases the formation of higher-dimensional structure, while higher-dimensional configuration can also reinforce or reshape stereocenter preferences. The resulting handedness reflects a collective energetic optimum, not merely the sum of isolated chiral units.

Our results from helices and propellers provide a clear empirical foundation for this principle. First, all systems examined consistently favor one homochiral configuration over the other, establishing homochirality as an intrinsic energetic preference. Second, the magnitude of this preference grows with system size, indicating positive cooperativity. Third, energy decomposition shows that this stabilization is dominated by electrostatic interactions, while exchange–correlation and kinetic contributions oppose it. Thus, the preferred handedness is determined primarily by how extended chiral structures optimize long-range electrostatic interactions.

A minimal energetic formulation captures this insight. Let χ_i represent the handedness variable (R/S, P/M, etc.) associated with the i-th chiral element and E_{chir}, the total chiral stabilization energy. The coupling among chiral levels may then be expressed schematically as

$$\frac{\partial E_{\mathrm{chir}}}{\partial \chi_i} = \sum_i f_{ij}\left(\chi_i, \chi_j, r_{ij}\right), \tag{9.5}$$

where f_{ij} quantifies cooperative or antagonistic interactions between chiral elements i and j at separation r_{ij}.

This formulation is the chirality-specific instance of the general hierarchical coupling model introduced in Eq. (8.10). There, the variable χ_i captured any local degree of freedom subject to cooperative or frustrative interactions. By identifying χ_i with χ_i, Eq. (9.5) shows how chiral propagation emerges from the same multilevel coupling architecture: positive f_{ij} reinforces coherent handedness across units, while negative f_{ij} produces antagonistic or frustrative patterns. Strong positive coupling drives convergence toward a single collective handedness, whereas weaker or competing interactions lead to partial, mixed, or scale-dependent chirality.

At the same time, hierarchical coupling is not strictly unidirectional. Although 0D chirality commonly directs the emergence of 1D or 2D structure, higher-dimensional chirality can influence or even invert local preferences. Examples include cysteine in helices, where the favored configuration corresponds to the experimentally established R form, and the left-handed propeller systems, where local R/S preferences sometimes oppose the global M helicity. These mismatches illustrate that chirality hierarchy reflects a balance of short-range steric interactions and long-range electrostatic interactions, not a simple deterministic rule.

From a density-based perspective, the principle of chirality hierarchy follows naturally from how electron density organizes under asymmetric environments. Chirality at any level corresponds to an uneven distribution of $\rho(\mathbf{r})$ and its derivatives, shaping the electrostatic potential and modulating the forces that stabilize molecular and supramolecular structure. When these asymmetric effects align across levels, the resulting stabilization propagates through the structure; when they misalign, steric and electrostatic frustration arises. In this view, global handedness emerges from the distribution of electron density, linking stereochemical propensity to the density-based concepts developed earlier in the book without reintroducing system-specific energetic details.

The implications of this principle extend beyond molecular structure. In origin-of-life chemistry, it complements kinetic symmetry-breaking models by revealing an intrinsic energetic bias toward homochirality. In catalysis, it provides a unifying rationale for how local stereogenic asymmetry controls global selectivity through cooperative effects. In supramolecular systems, it explains "majority-rules" and "sergeants-and-soldiers" phenomena within a density-based energetic framework. Viewed through the broader perspective of hierarchical modeling [34, 35], chirality hierarchy offers a clear molecular prototype of how local interactions propagate across levels through recursive coupling, giving rise to the global stereochemical structure. In this sense, the principle of chirality hierarchy serves not only as a stereochemical insight but also as a general paradigm of chemical emergence and multilevel organization.

9.8 Concluding Remarks

Chirality, though traditionally treated as a geometric and stereochemical concept, emerges in this chapter as a profoundly energetic and hierarchical phenomenon. By integrating insights from density-based reactivity theory, we showed that homochirality and chirality hierarchy are not accidental outcomes of molecular geometry but natural consequences of long-range electrostatic coherence, positive cooperativity, and the organized distribution of electron density. Across diverse systems—from peptide helices to helicenes, DNA fragments, and three-blade propellers—the same pattern recurs: one homochiral arrangement is consistently stabilized, and the stabilization grows cooperatively with system size, and electrostatics plays the dominant role in enforcing a single handedness.

These findings reveal that chirality hierarchy reflects a deep coupling between local asymmetry and global order. Lower-dimensional chirality can bias higher-dimensional structure, yet higher-dimensional chirality can also reshape local stereochemical preferences, producing nontrivial mismatches and cooperative networks across scales. This bidirectionality links chirality tightly to the broader themes of emergence and hierarchical modeling developed throughout this book.

From the perspective of chemical understanding, the chirality hierarchy demonstrates how subtle variations in electron density propagate through cooperative interactions to yield macroscopic stereochemical structure. This connects stereochemistry with the density-based view established in earlier chapters—particularly, the interplay of electrostatics, information-theoretic measures, and collective stabilization. Moreover, the energetic principles uncovered here complement kinetic models of symmetry breaking and enrich our understanding of life's homochirality by providing a foundational density-based rationale. Thus, chirality hierarchy illustrates how density-based principles and cooperative, hierarchical interactions shape molecular structure and, eventually, chemical understanding, a theme that will be further expanded in later chapters.

References

1 Hanson RM, Musacchio S, Mayfield JW, Vainio MJ, Yerin A, Redkin D. Algorithmic analysis of Cahn-Ingold-Prelog rules of stereochemistry: proposals for revised rules and a guide for machine implementation. J Chem Inf Model. 2018 Sep 24;58(9):1755–65.

2 Mauskopf S. A history of chirality. In: Busch KW, Busch MA, editors. Chiral Analysis. Elsevier; 2006. pp. 3–24.

3 Knowles WS. Asymmetric hydrogenations (Nobel lecture). Angew Chem Int Ed. 2002 Jun 17;41(12):1998–2007.

4 Katsuki T, Sharpless KB. The first practical method for asymmetric epoxidation. J Am Chem Soc. 1980 Aug;102(18):5974–6.

5 Fujii N, Saito T. Homochirality and life. Chem Rec. 2004;4(5):267–78.

6 Keszthelyi L. Origin of the homochirality of biomolecules. Q Rev Biophys. 1995 Nov;28(4):473–507.

7 Janoschek R. Theories on the origin of biomolecular homochirality. In: Janoschek R, editor. Chirality. Berlin, Heidelberg: Springer Berlin Heidelberg; 1991. pp. 18–33.

8 Fiore M. Homochirality emergence: a scientific enigma with profound implications in origins of life studies. Symmetry. 2025 Mar 20;17(3):473.

9 Soai K, Kawasaki T, Matsumoto A. Asymmetric autocatalysis: the Soai reaction, an overview. In: Soai K, Kawasaki T, Matsumoto A, editors. Asymmetric Autocatalysis. The Royal Society of Chemistry; 2022. pp. 1–32.

10 Sögütoglu L-C, Steendam RRE, Meekes H, Vlieg E, Rutjes FPJT. Viedma ripening: a reliable crystallisation method to reach single chirality. Chem Soc Rev. 2015 Oct 7;44(19):6723–32.

11 Prelog V. Chirality in chemistry. Science. 1976 Jul 2;193(4247):17–24.

12 Barron LD. Symmetry and molecular chirality. Chem Soc Rev. 1986;15(2):189.

13 LaPlante SR, Edwards PJ, Fader LD, Jakalian A, Hucke O. Revealing atropisomer axial chirality in drug discovery. ChemMedChem. 2011 Mar 7;6(3):505–13.

14 Rickhaus M, Mayor M, Juríček M. Strain-induced helical chirality in polyaromatic systems. Chem Soc Rev. 2016 Mar 21;45(6):1542–56.

15 López R, Palomo C. Planar chirality: a mine for catalysis and structure discovery. Angew Chem Int Ed. 2022 Mar 21;61(13):e202113504.

16 Zhang J, Kürti L. Multi-layer 3D chirality: its enantioselective synthesis and aggregation-induced emission. Natl Sci Rev. 2021 Jan; 8(1):nwaa205.

17 Schalley CA, Beizai K, Vögtle F. On the way to rotaxane-based molecular motors: studies in molecular mobility and topological chirality. Acc Chem Res. 2001 Jun;34(6):465–76.

18 Szumna A. Inherently chiral concave molecules—from synthesis to applications. Chem Soc Rev. 2010 Nov;39(11):4274–85.

19 Liu M, Zhang L, Wang T. Supramolecular chirality in self-assembled systems. Chem Rev. 2015 Aug 12;115(15):7304–97.

20 Pandey S, Mandal S, Danielsen MB, Brown A, Hu C, Christensen NJ, et al. Chirality transmission in macromolecular domains. Nat Commun. 2022 Jan 10;13(1):76.

21 Puigcerver J, Marin-Luna M, Iglesias-Sigüenza J, Alajarin M, Martinez-Cuezva A, Berna J. Mechanically planar-to-point chirality transmission in [2]rotaxanes. J Am Chem Soc. 2024 Feb 7;146(5):2882–7.

22 Morrow SM, Bissette AJ, Fletcher SP. Transmission of chirality through space and across length scales. Nat Nanotechnol. 2017 May 5;12(5):410–9.

23 Kumar M, George SJ. Homotropic and heterotropic allosteric regulation of supramolecular chirality. Chem Sci. 2014;5(8):3025–30.

24 Liu S. Homochirality originates from the handedness of helices. J Phys Chem Lett. 2020 Oct 15;11(20):8690–6.

25 Liu S. Principle of chirality hierarchy in three-blade propeller systems. J Phys Chem Lett. 2021 Sep 16;12(36):8720–5.

26 Li M, He X, Chen J, Wang B, Liu S, Rong C. Density functional theory and information-theoretic approach study on the origin of homochirality in helical structures. J Phys Chem A. 2021 Feb 11;125(5):1269–78.

27 Wang B, Zhao D, Lu T, Liu S, Rong C. Quantifications and applications of relative fisher information in density functional theory. J Phys Chem A. 2021 May 6; 125(17):3802–11.

28 Rong C, Zhao D, Yu D, Liu S. Quantification and origin of cooperativity: insights from density functional reactivity theory. Phys Chem Chem Phys. 2018 Jul 4; 20(26):17990–8.

29 Rong C, Zhao D, Zhou T, Liu S, Yu D, Liu S. Homogeneous molecular systems are positively cooperative, but charged molecular systems are negatively cooperative. J Phys Chem Lett. 2019 Apr 18;10(8):1716–21.

30 Nochebuena J, Liu S, Cisneros GA. Relative cooperativity in neutral and charged molecular clusters using QM/MM calculations. J Chem Phys. 2024 Apr 7; 160(13):134301.

31 Zhang W, Zhao Y, An X, Fu J, Zhang J, Zhao D, et al. Cooperativity and reactivity properties of medium-sized boron clusters: a combined density functional theory and information-theoretic approach study. Mol Phys. 2022 Dec 22:e2157774.

32 Liu S. Steric effect: a quantitative description from density functional theory. J Chem Phys. 2007 Jun 28;126(24):244103.
33 Parr RG, Yang W. Density-Functional Theory of Atoms and Molecules. Oxford University Press; 1989.
34 Liu S. Harvesting chemical understanding with machine learning and quantum computers. ACS Phys Chem Au. 2024 Mar 27;4(2):135–42.
35 Liu S. The evolving quest for chemical understanding in the quantum age. J Chem Theory Comput. 2025 Oct 9;21(20):10068–79.

10

Electrophilicity and Nucleophilicity

TABLE OF CONTENTS

10.1 Introduction

The previous chapter examined how subtle imbalances in electron density—manifested through chirality hierarchy, cooperative interactions, and density-driven amplification—govern the emergence of handedness in molecular systems. Although homochirality represents a global organizational phenomenon, its origin ultimately rests on local differences in electron density. These same local principles also govern how molecules donate or accept electrons, making electrophilicity and nucleophilicity natural continuations of the conceptual themes established in the previous chapter. This chapter extends that perspective by examining electronic asymmetry—how local regions of a molecule tend to donate or accept electron density. Electrophilicity and nucleophilicity, as complementary density-based reactivity properties, describe the intrinsic tendency for atoms in molecules to donate or accept electrons. In this sense, the same underlying principles of local density asymmetry also underpin the reactivity concepts explored in the present chapter.

At their core, nucleophilicity and electrophilicity express how local regions of a molecule respond to changes in electron density. Nucleophiles possess excess electron density and readily donate it, while electrophiles are electron-deficient and readily accept it. This donor–acceptor interplay dictates the course of countless chemical reactions—substitutions, additions, pericyclic reactions, redox steps, and many enzymatic processes—and forms one of the most intuitive and widely used conceptual frameworks in chemical reactivity. Representative *nucleophiles* include anions and neutral lone-pair donors, whereas *electrophiles* include cations and electron-poor functional groups.

The concepts of nucleophilicity and electrophilicity have their roots in nineteenth-century theories of chemical affinity [1–3]. Jöns Jacob Berzelius proposed the idea of "dualism," classifying compounds based on opposing charges. In the 1920s, Gilbert N. Lewis [4] shifted the focus to electron-pair interactions, introducing the concepts of Lewis acids and Lewis bases. The specific terms "nucleophilic" and "electrophilic" were coined in the 1930s by Christopher Kelk Ingold [5], a founder of physical organic chemistry. Ingold used these terms to describe ionic reagents and linked them to kinetic and thermodynamic studies, laying the foundation for understanding organic reaction mechanisms in terms of electron transfer. Starting from the 1960s, Swain and Scott first [6], Ritchie later [7], and then Herbert Mayr and coworkers [1, 8–13] developed quantitative experimental scales for nucleophilicity and electrophilicity based on reaction rate constants. Later advancements, including Robert Parr's introduction of the Fukui function (1986) [14–17] and electrophilicity index (1999) [18–20], along with our own recent work (2014) using information-theoretic approach (ITA) [21] from the density functional theory (DFT) framework, have further bridged empirical data with quantum mechanical theory, creating a more comprehensive and predictive framework for these concepts.

From a molecular orbital standpoint, nucleophilicity is associated with highest-occupied molecular orbital (HOMO), while electrophilicity correlates with lowest-unoccupied molecular orbital (LUMO). A good nucleophile has a high-energy HOMO, while a good electrophile has a low-energy LUMO. The interaction between these frontier orbitals is critical in pericyclic reactions, such as the Diels–Alder reaction [22–24], where HOMO of the diene (nucleophile) interacts with LUMO of the dienophile (electrophile). While frontier orbitals provide an intuitive orbital-based picture, density-based reactivity theory recasts these tendencies directly in terms of density-based quantities.

Two conceptual pairs—electrophilicity/nucleophilicity and Lewis acidity/basicity—are closely related but fundamentally distinct. Both involve the interaction of electron pairs, yet they differ in scope and emphasis. Lewis acidity and basicity describe a species' thermodynamic tendency to accept or donate an electron pair, independent of specific reaction pathways. In contrast, electrophilicity and nucleophilicity are kinetic concepts, focusing on how rapidly and effectively a species engages in electron transfer during a particular reaction. While Lewis theory broadly classifies species based on their inherent ability to accept or donate electrons (e.g., BF_3 as a Lewis acid, NH_3 as a Lewis base), electrophilicity and nucleophilicity depend on factors such as reaction conditions, solvent effects, and the nature of the reaction partner. For example, BF_3 can function as an electrophile, but its reactivity may be limited in certain contexts, whereas highly reactive electrophiles like carbocations do not always qualify as classical Lewis acids. Similarly, while many nucleophiles are also Lewis bases, their reactivity can be influenced by steric hindrance, solvation, and polarizability, which may not affect basicity to the same extent. This distinction is essential: understanding chemical reactivity requires both a thermodynamic perspective (via Lewis theory) and a kinetic one (via electrophilicity/nucleophilicity), as they illuminate different yet complementary aspects of how and why reactions proceed.

This chapter focuses on electrophilicity and nucleophilicity, while Lewis acidity and basicity, including the hard–soft acid–base [25–27] principle, are treated in Chapter 12. From a density-based perspective, electrophilicity and nucleophilicity capture the fundamental duality of molecular reactivity: the intrinsic tendency for atoms in molecules to accept or donate electrons. They describe complementary modes of information flow with the electron density and naturally bridge the energy-based descriptors of conceptual DFT with the information-based quantities of ITA. Together, they provide a unified foundation for understanding how electron density reorganizes during chemical transformations.

10.2 Experimental Scales

In 1953, Swain and Scott [6] first quantified nucleophilicity by comparing reaction rates of various nucleophiles with methyl bromide relative to water:

$$\log_{10}\left(\frac{k}{k_0}\right) = sn \tag{10.1}$$

where k is the pseudo-first-order rate constant for a given nucleophile, k_0 that for water, n the nucleophilicity constant, and s a substrate sensitivity parameter (1.00 for methyl bromide). Typical values include $n = 4.2(OH^-)$, $5.0\,(I^-)$, and $6.4\,(S_2O_3^{2-})$ in water at 25 °C, illustrating a predictive ordering over several orders of magnitude.

Another early attempt was the Edwards equation (1954), which related nucleophilicity to the basicity and polarizability of a nucleophile [28]:

$$\log_{10}\left(\frac{k}{k_0}\right) = \alpha E_n + \beta H \tag{10.2}$$

where E_n is a parameter related to the oxidation potential (polarizability) of a nucleophile, H is a parameter related to the basicity of the nucleophile, and α/β are constants that depend on the electrophile and reaction conditions. While incorporating electronic properties, the Edwards equation suffered from dependence on the specific reaction system.

Ritchie's 1972 equation removed substrate dependence entirely, proposing [7]:

$$\log_{10}\left(\frac{k}{k_0}\right) = N^+ \tag{10.3}$$

with N^+ as the nucleophile's intrinsic parameter. While elegant, the assumption of constant selectivity (i.e., that relative reactivities are invariant across electrophiles) conflicts with the reactivity–selectivity principle and limits applicability primarily to cationic systems such as tropylium or diazonium ions.

Legon and Millen [29] determined intrinsic nucleophilicity based on the intermolecular stretching force constant k, derived from the rotational and infrared spectra of dimers of the form B$\cdots$HX. These dimers are formed between a nucleophile B and a series of HX molecules (where X is a halogen) or other neutral electrophiles. The nucleophilicity value is obtained using the following empirical relationship:

$$k = \mathrm{c}NE \tag{10.4}$$

where c is a constant, N represents the nucleophilicity of B, and E denotes the electrophilicity of HX. In this model, reference values were assigned as follows: a nucleophilicity number $N = 10$ for H_2O and an electrophilicity number $E = 10$ for HF.

Mayr and Patz [1] introduced in 1994 a two-parameter equation that re-incorporates substrate effects via a nucleophile-specific slope s:

$$\log k = s(N + E) \tag{10.5}$$

where k is the second-order rate constant (20 °C), N the nucleophilicity parameter (relative to 2-methyl-1-pentene), and E the electrophilicity parameter (relative to a reference benzhydrylium ion). This formalism allows direct comparison across diverse reaction classes—S_N1, S_N2, electrophilic additions, and aromatic substitutions. These parameters are derived by measuring rate constants for a series of benchmark reactions. For nucleophiles, reactions with standardized electrophiles (e.g., benzhydrylium ions) yield N and s, whereas electrophile parameters E emerge from reactions with reference nucleophiles such as 2-methyl-1-pentene. High-accuracy kinetic techniques (stopped-flow UV–vis spectroscopy, NMR) ensure reliable data across 1 M to 10^{-6} M concentrations, with rate constants ranging from 10^{-5} to 10^{8} $M^{-1}\,s^{-1}$.

The Mayr database currently documents over 1302 nucleophiles and 359 electrophiles, covering $-8.80 \leq N \leq 31.92$ and $-29.60 \leq E \leq 8.02$. Reactants are classified into molecular classes (carbon, heteroatom, radical, organometallic) and ranked by star ratings indicating data precision (five stars are the best). Shown in Figures 10.1

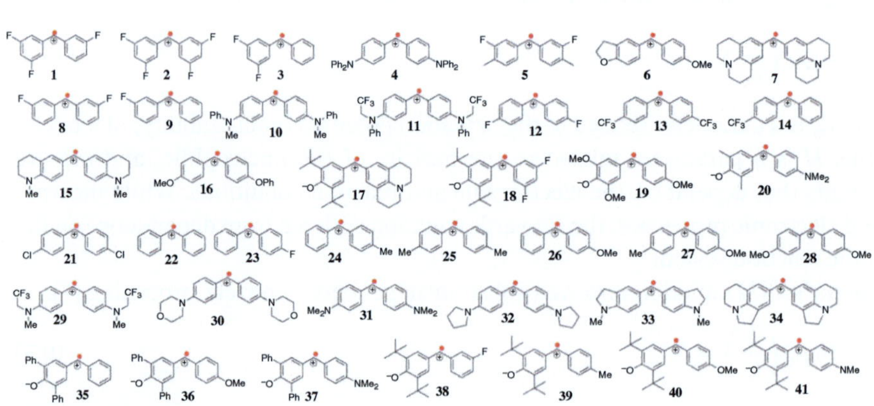

Figure 10.1 Electrophilic systems studied whose electrophilicity scales are experimentally known. The reactive centers are marked by red stars. The solvent is CH_2Cl_2. *Source*: Reproduced from Ref. [30] with permission of Springer Nature.

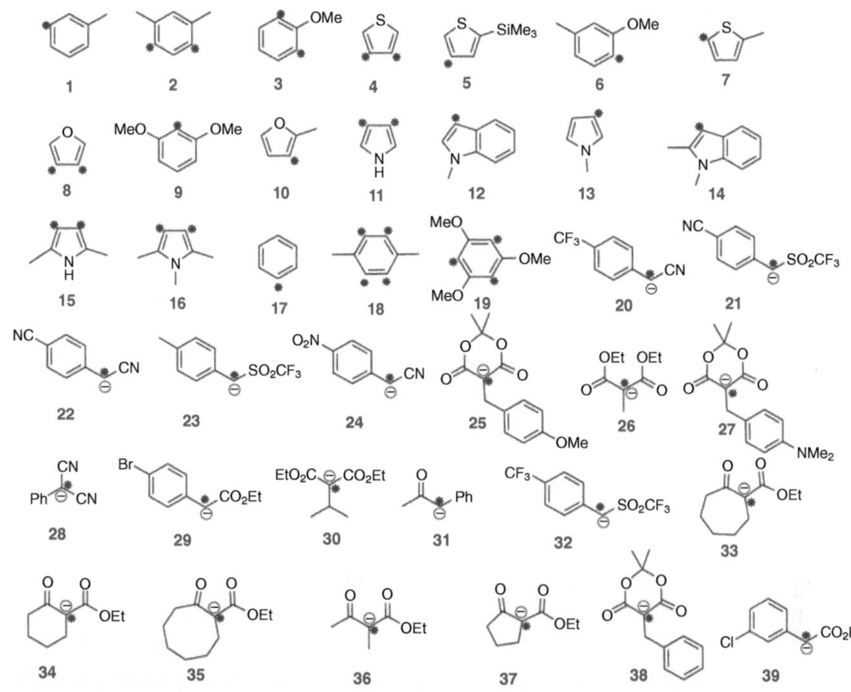

Figure 10.2 Nucleophilic systems studied whose nucleophilicity scales are experimentally available in the literature. The first 19 species are in CH_2Cl_2 solvent, and the last 20 species are in dimethyl sulfoxide solvent. The reactive centers are marked by blue stars. Reproduced from Ref. [30] with permission of Springer Nature.

and 10.2 are two sets of systems, one electrophilic (Figure 10.1) and the other nucleophilic (Figure 10.2), whose respective experimental scale is known.

10.3 Quantification in Conceptual Density Functional Theory

As discussed in Chapter 2, conceptual density functional theory (CDFT) [20, 31–34] derives chemically meaningful concepts and reactivity descriptors directly from the electron density and its derivatives with respect to variables such as the number of electrons N and the external potential $v_{\text{ext}}(\mathbf{r})$, which depends on the positions and charges of the nuclei. These descriptors, including chemical potential, hardness/softness, Fukui function, and electrophilicity/nucleophilicity, provide quantitative measures of intrinsic reactivity, offering a theoretical foundation for understanding where and why chemical reactions take place. In CDFT, electrophilicity and nucleophilicity can be quantified in two ways: globally, using indices like the electrophilicity index and nucleophilicity index and, locally, through functions such as the Fukui function and dual descriptor.

10.3.1 Electrophilicity Index

In 1999, building on an earlier proposal by Maynard et al. [35], the concept of the electrophilicity index was quantified by Parr, Szentpaly, and this author [18]. This index represents the stabilization energy associated with the acquisition of additional electronic charges by atoms or molecules in their ground states from the surrounding environment. The underlying idea is the extent to which partial electron transfer contributes to the decrease of the total energy through the maximal flow of electrons.

Consider an electrophile immersed in an idealized free electron sea at zero temperature—an approximation that may resemble its binding environment in systems such as proteins, DNA helices, or surfaces. In this environment, the electrophile absorbs electrons until its chemical potential matches that of the electron sea. To second order, according to CDFT, the resulting energy change ΔE associated with the electron transfer ΔN is given by the expression:

$$\Delta E = \mu \Delta N + \frac{1}{2}\eta(\Delta N)^2, \tag{10.6}$$

where μ is chemical potential (the negative of the electronegativity) and η represents the chemical hardness, whose definitions are available in Chapter 2.

According to Mulliken, using the finite difference approximation, chemical potential and chemical hardness can be expressed through the following working equations:

$$\mu = -\chi = -\frac{1}{2}(I + A) \tag{10.7}$$

and

$$\eta = I - A, \tag{10.8}$$

where I is the first ionization potential (IP) and A is the electron affinity (EA). Under Koopmans' theorem, that is, frozen-orbital approximation, which applies to closed-shell molecules, these quantities can be approximated using frontier molecular orbital energies:

$$I \approx -\varepsilon_{\text{HOMO}};\ A \approx -\varepsilon_{\text{LUMO}}, \tag{10.9}$$

where $\varepsilon_{\text{HOMO}}$ and $\varepsilon_{\text{LUMO}}$ are the energies of the highest-occupied and lowest-unoccupied molecular orbitals, respectively.

If the electron sea supplies enough electrons, the electrophile becomes saturated with the following condition met:

$$\frac{d\Delta E}{d\Delta N} = 0, \tag{10.10}$$

Solving for the maximum electron transfer yields

$$\Delta N_{\text{max}} = -\frac{\mu}{\eta}, \tag{10.11}$$

and the corresponding total energy decrease due to this charge transfer is then:

$$\Delta E_{\min} = -\frac{\mu^2}{2\eta}. \tag{10.12}$$

Since $\eta > 0$, it follows that $\Delta E < 0$, indicating that the charge transfer process is always energetically favorable. We defined this quantity as *electrophilicity index* ω

$$\omega = -\frac{\mu^2}{2\eta}, \tag{10.13}$$

which serves as a quantitative measure of an electrophile's electrophilicity. This definition is justified by analogy to classical electrostatics, where the power is given by V^2/R; in this context, μ plays the role of the electrostatic potential V, and η corresponds to the resistance R.

Since its introduction, the electrophilicity index has garnered significant attention as a theoretical measure of an electrophile's capacity to accept the maximal number of electrons. A comprehensive review on this topic was published by Chattaraj, Sarkar, and Roy in *Chemical Reviews* [19]. One interesting extension is the proposal of a nucleophilicity index in the same spirit. In this context, the grand canonical ensemble must be resorted to. Similar to Eq. (10.6), one has [2, 36]

$$\Delta\Omega = -N\Delta\mu - \tfrac{1}{2}S(\Delta\mu)^2 \tag{10.14}$$

where S is the softness, with $S = 1/\eta$. Using the charge transfer in the grand canonical ensemble obtained by the electronegativity equalization principle in the literature,

$$N = \frac{\mu_A - \mu_B}{\eta_A + \eta_B}, \tag{10.15}$$

one can define the nucleophilicity index ω^- as

$$\omega^- = \frac{1}{2}\left(\frac{\mu_A - \mu_B}{\eta_A + \eta_B}\right)^2 \eta_A. \tag{10.16}$$

This index is, however, limited by its electrophile-dependence, meaning that there is no universal nucleophilicity scale. This dependence arises because nucleophilicity involves electron donation into an acceptor with specific hardness and chemical potential, making the descriptor inherently two-body. Other ways to quantify nucleophilicity are available in the literature. A major limitation of this global approach in Eqs. (10.13) and (10.16), to quantify electrophilicity and nucleophilicity, is its lack of regioselective resolution, because it does not indicate which specific atoms within a molecule are the preferred sites for electrophilic or nucleophilic attack.

10.3.2 Fukui Function

Fukui function, introduced by Yang and Parr [16, 17], is the first local descriptor to quantify electrophilicity and nucleophilicity. It quantifies the change in electron density $\rho(\mathbf{r})$ at a specific point $\mathbf{r}$ upon the addition or removal of an infinitesimal amount of charge. Mathematically, it is defined as the derivative of the electron density with respect to the number of electrons N, at a fixed external potential $v(\mathbf{r})$:

$$f(\mathbf{r}) = \left(\frac{\partial \rho(\mathbf{r})}{\partial N}\right)_{v} = \left(\frac{\delta \mu}{\delta v(\mathbf{r})}\right)_{N}, \tag{10.17}$$

Alternatively, the Fukui function can equally be defined as the derivative of chemical potential μ with respect to the external potential $v(\mathbf{r})$ with the total number of electrons N fixed. The Fukui function identifies regions within a molecule that are most responsive to changes in the total number of electrons, highlighting the sites most likely to participate in chemical reactions involving electron transfer or sharing. If an atom exhibits electrophilic or nucleophilic behavior, a significant change in its electron density should be observed upon the addition or removal of an electron. This sensitivity makes the Fukui function a powerful tool not only for quantifying electrophilicity and nucleophilicity but also for identifying regioselective sites within a molecule.

Perdew et al. [37] showed that the exact ground-state energy of an open system is strictly linear in N between integers, with derivative discontinuities at integer electron numbers—an essential feature for correctly describing processes like molecular dissociation and band gaps in solids. Fukui function experiences jumps at integer N, yielding ambiguous or even undefined values exactly at those points. This discontinuity makes a straightforward analytic derivative impossible and undermines the reliability of $f(\mathbf{r})$ as a smooth local reactivity descriptor. To resolve this discontinuity issue, directional derivatives are employed,

$$f^{\pm}(\mathbf{r}) = \left(\frac{\partial \rho(\mathbf{r})}{\partial N}\right)_{v}^{\pm}. \tag{10.18}$$

The "+" symbol denotes the right-hand derivative corresponding to electron addition to the system, reflecting the propensity for nucleophilic attack, while the "−" symbol stands for the derivative associated with electron removal, quantifying the reactivity tendency of electrophilic attack.

To numerically evaluate this function, two approaches, the finite-difference approximation (the first equality below) and frozen-orbital approximation using frontier orbitals, are available:

$$f^{+}(\mathbf{r}) = \rho_{N+1}(\mathbf{r}) - \rho_{N}(\mathbf{r}) \approx |\psi_{\mathrm{LUMO}}(\mathbf{r})|^2, \tag{10.19}$$

$$f^{-}(\mathbf{r}) = \rho_{N}(\mathbf{r}) - \rho_{N-1}(\mathbf{r}) \approx |\psi_{\mathrm{HOMO}}(\mathbf{r})|^2. \tag{10.20}$$

where $\rho_{N+1}(\mathbf{r})$, $\rho_{N}(\mathbf{r})$, and $\rho_{N-1}(\mathbf{r})$ denote the electron density with $N+1$, N, and $N-1$ electrons, and $|\psi_{\mathrm{LUMO}}(\mathbf{r})|^2$ and $|\psi_{\mathrm{HOMO}}(\mathbf{r})|^2$ stand for LUMO and HOMO densities, respectively. Under the frozen-orbital approximation, the frontier molecular orbital form of the Fukui function is recovered. This is the precise reason why this local function is called the Fukui function.

In practical applications of predicting regioselectivity, we often look at the Fukui function condensed to individual atoms (f_k). These atomic Fukui indices are

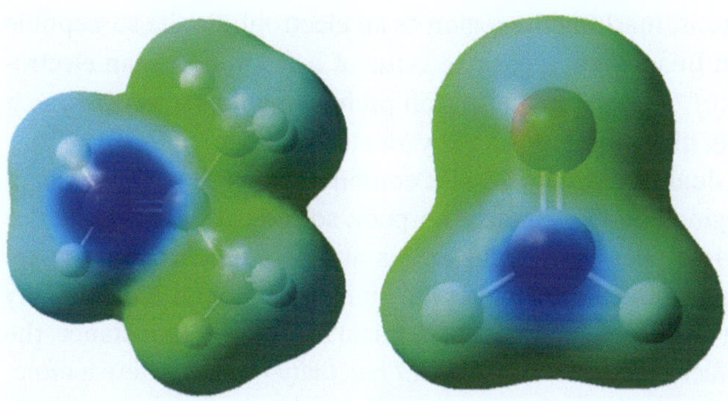

Figure 10.3 Illustrative examples of (a) nucleophilic and (b) electrophilic Fukui functions mapped onto the van der Waals surface of the molecules. The blue regions are indicative of the reaction center for the respective reaction types.

calculated by integrating the Fukui function over the volume associated with atom k, yielding

$$f_k^+ = q_k(N) - q_{k+1}(N+1), \tag{10.21}$$

$$f_k^- = q_k(N-1) - q_{k+1}(N), \tag{10.22}$$

where $q_k(N)$ represents the partial charge on atom k in the N-electron system.

Shown in Figure 10.3 are illustrative examples of (a) electrophilic Fukui function for a system with C═C double bond and (b) nucleophilic Fukui function for formaldehyde. As can be seen from the figure, the Fukui function can correctly predict the reaction center (regioselectivity) in both cases.

10.3.3 Dual Descriptor

Reactivity descriptors from CDFT are often limited to the first- and second-order derivatives. One exception is the dual descriptor denoted by $\Delta f(\mathbf{r})$ or $f^{(2)}(\mathbf{r})$, which was introduced by Toro-Labbé, Morell, and so on [38], defined by

$$\Delta f(\mathbf{r}) = \left(\frac{\partial^2 \rho(\mathbf{r})}{\partial N^2}\right)_v . \tag{10.23}$$

Within the finite-difference approximation, it becomes

$$\Delta f(\mathbf{r}) = f^+(\mathbf{r}) - f^-(\mathbf{r}) = \rho_{N+1}(\mathbf{r}) - 2\rho_N(\mathbf{r}) + \rho_{N-1}(\mathbf{r}), \tag{10.24}$$

and in the frozen-orbital approximation, we have

$$\Delta f(\mathbf{r}) \approx |\psi_{\text{LUMO}}(\mathbf{r})|^2 - |\psi_{\text{HOMO}}(\mathbf{r})|^2 . \tag{10.25}$$

The physical meaning and interpretation of a dual descriptor can be directly derived from its definition as the difference between two Fukui functions in Eq. (10.24). If $\Delta f(\mathbf{r}) > 0$, that is, $f^+(\mathbf{r}) > f^-(\mathbf{r})$, the region experiences a larger increase in electron density upon the addition of an electron, so the site is preferentially prone to accepting electrons, marking the region as an electrophilic site susceptible to nucleophilic attack. In this regard, a positive value of $\Delta f(\mathbf{r})$ indicates an electrophilic region. If $\Delta f(\mathbf{r}) < 0, f^{-}(\mathbf{r}) > f^{+}(\mathbf{r})$, the region prefers to donate electrons, so it acts as a nucleophilic site, likely to be attacked by an electrophile.

One advantage of the dual descriptor is that it combines two Fukui functions in one formula. Positive regions/atoms are electron-poor, so they are likely to be targets for nucleophiles, while negative regions/atoms are electron-rich, so they are likely to be targets for electrophiles. This makes the dual descriptor an exceptionally useful tool for predicting the regioselectivity of chemical reactions. For instance, the dual descriptor can satisfactorily describe *ambident reactivity*—cases where a molecule exhibits competing sites for both electron donation and acceptance. For thiocyanate (SCN^-), for instance, $\Delta f(\mathbf{r})$ reveals that sulfur is the preferred nucleophilic site with $\Delta f < 0$ and nitrogen is the preferred electrophilic site with $\Delta f > 0$, explaining its context-dependent dual reactivity.

We applied the dual descriptor to predict the PCET (proton-coupled electron transfer) mechanism [39, 40]. Shown in Figure 10.4 is a QM/MM model system of the photosystem II reaction center (PDB ID 2axt) with dual descriptor plotted. This system is well known for its widely accepted PCET mechanism. As the proton is transferred from tyrosine 160 (Tyr160) to histidine 190 (His190), a long-range (ca. 10 Å) electron transfer occurs from Tyr160 to P680. The plotted dual descriptor contour surface shows that the electron that originates from the nucleophilic tyrosine π system will be transferred to the electrophilic region on P680, consistent with the proposed PCET mechanism.

Moreover, dual descriptor is particularly insightful for analyzing pericyclic reactions, providing a complementary DFT perspective on the Woodward–Hoffmann rules [22, 41]. Tracking the evolution of $\Delta f(\mathbf{r})$ along the reaction coordinate reveals

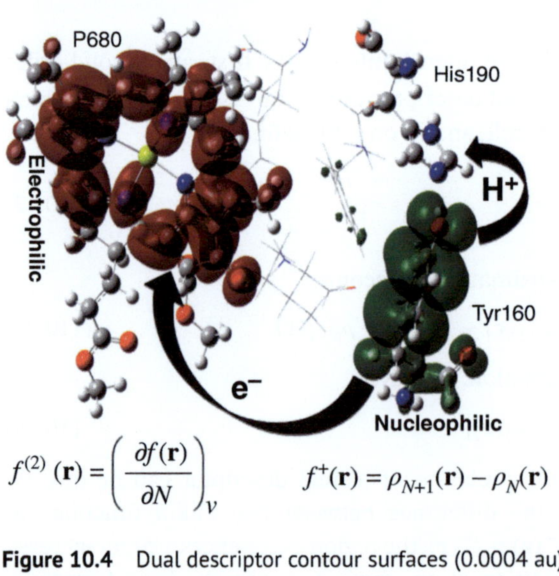

Figure 10.4 Dual descriptor contour surfaces (0.0004 au) for the photosynthesis II protein model. *Source*: Reproduced from Ref. [39] with permission of American Chemical Society.

how electron-donating (nucleophilic) and electron-accepting (electrophilic) regions develop and interact. This dynamic analysis directly informs the prediction of allowedness for concerted reactions (like cycloadditions and electrocyclizations). In molecular orbital theory, these predictions hinge on maintaining orbital symmetry and the phase matching of frontier molecular orbitals throughout the reaction. The power of the dual descriptor here is that its sign and spatial distribution of $\Delta f(\mathbf{r})$ correlate directly with the phase continuity characteristics of the relevant frontier orbitals, offering an unambiguous density-based interpretation and validation of the symmetry conservation principle underpinning the Woodward–Hoffmann rules.

10.4 Quantification in ITA

The preceding quantification of electrophilicity and nucleophilicity was based on energetic and electron density considerations. In contrast, ITA [21, 42–44] offers an entirely different perspective, which is rooted in information content. Rather than focusing on energy or density changes, ITA characterizes chemical reactivity in terms of information uncertainty and complexity within the electron density. Whereas CDFT describes reactivity through energy derivatives with respect to N and $v(\mathbf{r})$, ITA complements this view by examining how information is redistributed with the density under perturbation. Measures such as Kullback–Leibler divergence (also called information gain) and relative Rényi entropy are employed to capture how the electronic structure responds to perturbations in chemical processes, providing insights into reactivity trends, selectivity, and stability from a fundamentally different angle.

This scheme was proposed in 2014 [45], when we first introduced the *information conservation principle*, which posits that, under the first-order approximation, the total information content remains conserved before and after the formation of a molecular system. The formulation starts with the atoms-in-molecules representation of information gain,

$$I_G = \sum_A \int \rho_A(\mathbf{r}) \ln \frac{\rho_A(\mathbf{r})}{\rho_A^0(\mathbf{r})} d\mathbf{r}, \tag{10.26}$$

where $\rho_A(\mathbf{r})$ is the electron density on atom (or group) A in a molecule, whose total electron density is $\rho(\mathbf{r})$ and the total number of electrons is N, and $\rho_A^0(\mathbf{r})$ is the counterpart of atom (or group) A in the reference state, which can be a neutral atom, or an ion, or a group, and so on, and satisfies the same normalization condition as $\rho_A(\mathbf{r})$ and $\rho(\mathbf{r})$,

$$\int \rho(\mathbf{r}) d\mathbf{r} = \sum_A \int \rho_A(\mathbf{r}) d\mathbf{r} = \sum_A \int \rho_A^0(\mathbf{r}) d\mathbf{r} = N. \tag{10.27}$$

Minimizing information gain, Eq. (10.26), subject to the normalization condition, Eq. (10.27),

$$\delta\left\{I_G - \lambda\left[\sum_A \int \rho_A(\mathbf{r})d\mathbf{r} - N\right]\right\} = 0, \tag{10.28}$$

one obtains the Hirshfeld partition of the electron density

$$\rho_A(\mathbf{r}) = \frac{\rho_A^0(\mathbf{r})}{\sum_A \rho_A^0(\mathbf{r})}\rho(\mathbf{r}). \tag{10.29}$$

Since ρ_A and ρ_A^0 are similar, we introduce a new variable, $x = (\rho_A - \rho_A^0)/\rho_A$, which is expected to be small, so information gain in Eq. (10.26) becomes

$$I_G = \sum_A \int \rho_A \ln\frac{1}{1-x} d\mathbf{r}. \tag{10.30}$$

Expanding $\ln\frac{1}{1-x} \approx x$ as the first-order approximation, we have

$$I_G \approx \sum_A \int \left(\rho_A - \rho_A^0\right) d\mathbf{r} = -\sum_A q_A = 0, \tag{10.31}$$

where q_A is the Hirshfeld charge on atom (or group) A. We assumed the system is neutral in the last equality. If it is not, the net information gain in Eq. (10.31) becomes constant, equal to the negative of the total charge of the system.

The above formulation underpins the *information conservation principle.* According to this principle, when a new molecular system is formed from its constituent components, the identity of each component is preserved—at least to the first-order approximation. In other words, a species that is electrophilic or nucleophilic in nature will retain that character within the newly formed system. To maintain this identity, the components must adjust themselves such that their electron densities redistribute according to Hirshfeld's stockholder partitioning, resulting in each fragment acquiring partial charges in proportion to its original contribution. With this, we proposed to employ information gain and its first-order approximation, Hirshfeld charge, to simultaneously quantify electrophilicity and nucleophilicity [45, 46].

Figure 10.5 exhibits the strong correlations of experimental electrophilicity and nucleophilicity scales with information gain and Hirshfeld charge for the 41 electrophilic systems in Figure 10.1 and 39 nucleophilic species in Figure 10.2 [30]. The validity of using information gain and Hirshfeld charge to quantify this pair of reactivity is confirmed. For electrophilicity, the larger the experimental scale, the smaller the information gain, the larger the Hirshfeld charge, and thus the stronger the capability for the reactive center to accept electrons. In the nucleophilicity case, the larger the experimental nucleophilicity scale, the larger the information gain, the smaller (more negative) the Hirshfeld charge, and thus the stronger the capability for the reactive center to donate electrons.

Now, let us consider an alternative variable y in Eq. (10.26), with $y = \frac{\rho_A(\mathbf{r}) - \rho_A^0(\mathbf{r})}{\rho_A^0(\mathbf{r})}$, which is the relative difference of the atomic density with respect to that of an isolated atom before a molecule is formed, and then Eq. (10.26) can be rewritten as

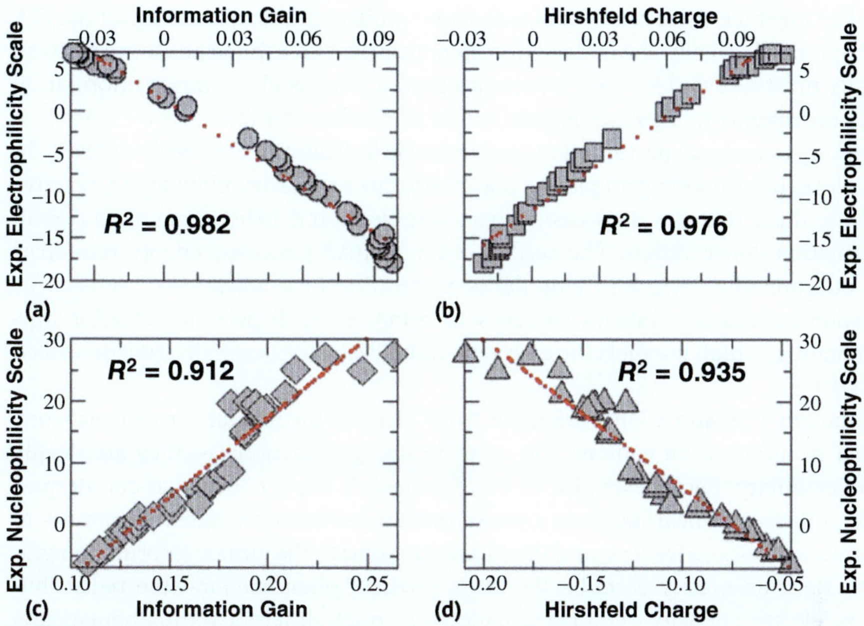

Figure 10.5 Strong correlations of experimental electrophilicity and nucleophilicity scales with information gain and Hirshfeld charge for 41 electrophilic species and 39 nucleophilic species. *Source*: Reproduced from Ref. [30] with permission of Springer Nature.

$$I_G = \sum_A \int \rho_A(\mathbf{r}) \ln(1+y)d\mathbf{r}. \tag{10.32}$$

Again, since y should be small, with $\ln(1+y) \approx y$ as the first-order approximation, we have

$$I_G \approx \sum_A \int \left\{ \frac{\rho_A^2(\mathbf{r})}{\rho_A^0(\mathbf{r})} - \rho_A(\mathbf{r}) \right\} d\mathbf{r} = E_2^r - N, \tag{10.33}$$

where E_2^r is the second-order relative Onicescu information energy in the atoms-in-molecules representation, defined as follows:

$$E_2^r = \sum_A \int \frac{\rho_A^2(\mathbf{r})}{\rho_A^0(\mathbf{r})} d\mathbf{r}. \tag{10.34}$$

Higher-order expressions can be derived accordingly. Eq. (10.33) signifies that the relative Onicescu information energy can also be employed to quantify electrophilicity and nucleophilicity, the same as Kullback–Leibler divergence and Hirshfeld charge. We have proved that Kullback–Leibler divergence should satisfy an identity associated with two forms of the relative Fisher information [47, 48]. These formulas confirm again that ITA quantities are intrinsically interrelated with each other due to the existence of two basic DFT theorems of Hohenberg–Kohn. These relationships also prove useful and insightful when other ITA quantities, besides information gain, can be employed to quantify reactivity propensities such as electrophilicity and nucleophilicity.

Taken together, these relationships provide intuitive physical meaning for the ITA descriptors. Physically, the information gain measures how much an atom's electron density must reorganize when it becomes part of a molecule. A larger information gain corresponds to a greater redistribution of density and therefore a higher susceptibility to external perturbation—an intrinsic indicator of chemical reactivity. In this sense, information gain plays a role analogous to polarizability: highly reactive sites are those that can most easily alter their electron density to maintain overall information conservation. The second-order relative Onicescu energy reinforces this interpretation by quantifying the concentration or localization of density; its variation reflects the system's tendency to compress or disperse information during bonding, which parallels the energetic stabilization or destabilization described in CDFT.

Although ITA and CDFT originate from distinct theoretical formalisms—one rooted in information content, the other in energy functionals—they are deeply complementary. Each descriptor in one framework has a conceptual counterpart in the other: chemical hardness corresponds to information stability, softness to information responsiveness, and the Fukui function to the atomic information gradient. Both ultimately describe the same physical phenomenon—the redistribution of electron density upon perturbation—through different yet mathematically consistent languages. This dual representation unifies energetic and informational views of chemical reactivity, confirming that a molecule's ability to donate or accept electrons can be viewed either as energetic relaxation or as informational reorganization of the density distribution.

The key advantage of using information gain to simultaneously quantify electrophilicity and nucleophilicity lies in its unified, system-independent framework, firmly rooted in the principles of information theory. Notably, it aligns with Hirshfeld charges under the first-order approximation. This connection is particularly significant, as it parallels the conceptual foundation of Fukui function, which also relies on charge distribution. However, while Fukui function requires evaluating changes in electron density upon the addition or removal of an electron, Hirshfeld charges can be directly employed to assess reactivity. Specifically, if the Hirshfeld charge of an atom in a molecule is negative, the atom is nucleophilic; if positive, it is electrophilic.

10.5 Regioselectivity

Regioselectivity is a fundamental concept in chemical reactivity, describing the preference for a reaction to occur at one specific atom or region within a molecule over another. This selective behavior is crucial in shaping the structure and properties of the resulting products, particularly when multiple reactive sites are present. The origin of regioselectivity lies in the unequal reactivity of different molecular sites, governed by a complex interplay of electronic, steric, orbital, and environmental factors. Among these influences, electronic factors, particularly how local electron density responds to perturbation, dominate in most systems and form the basis of density-based reactivity analysis. Electron-donating groups (EDGs) or

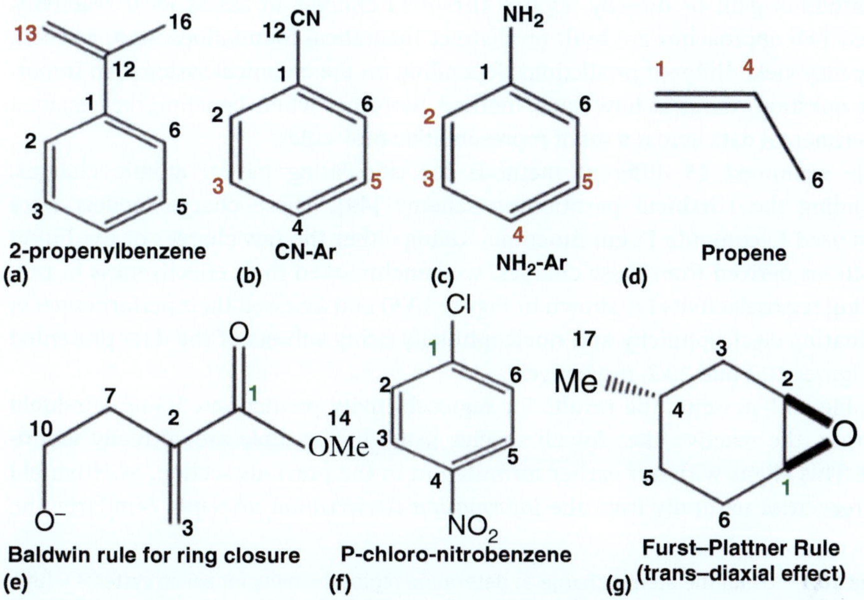

Figure 10.6 Regioselective systems to illustrate how to use Hirshfeld charge and Fukui function to determine regioselectivity for both (a)–(d) electrophilic and (e)–(g) nucleophilic attacks. Red integers are used to mark the regioselective site for electrophilic attack, and green integers are regioselective sites for nucleophilic attack. *Source*: Reproduced from Ref. [45] with permission American Chemical Society.

electron-withdrawing groups (EWGs), resonance effects, and frontier molecular orbital distributions influence the most favorable sites for electrophilic or nucleophilic attack. Additionally, steric hindrance can impede access to certain regions, while reaction conditions and catalytic systems—such as enzymes or transition-metal complexes—can further direct the reaction pathway.

We can use CDFT and ITA to determine regioselectivity. In ITA, regioselectivity is ruled by information gain and Hirshfeld charge. For electrophilic attacks, the site with the largest information gain and thus the most negative charge is preferred, and for nucleophilic attacks, the site with the smallest information gain and thus the most positive charge is selected. We showcased this rule for seven systems in Figure 10.6 [45], the first 4 being electrophilic and the last 3 being nucleophilic. In every case, our computational result agrees well with the experimental evidence. It is particularly noteworthy that Hirshfeld charges provide the most consistent correct prediction across all systems.

10.6 Benchmark CDFT and ITA Quantifications

Both CDFT and ITA offer practical tools for predicting regioselectivity, electrophilicity, and nucleophilicity. CDFT relies on the Fukui function, which evaluates changes in partial charges upon electron addition or removal, while ITA uses information gain or directly applies Hirshfeld charges to assess local reactivity. These two approaches are built on distinct theoretical foundations, and, as such, they may yield different predictions depending on the chemical system. An important question, then, is how each method performs when benchmarked against experimental data across a set of representative molecules.

We examined 15 different methods for calculating partial atomic charges, including the Hirshfeld partitioning scheme [49]. These charge models were then used to compute Fukui functions. Using either the raw charges or the Fukui functions derived from these charges, we benchmarked their effectiveness in predicting regioselectivity (as shown in Figure 10.6) and assessed their performance in evaluating electrophilicity and nucleophilicity using subsets of the data presented in Figures 10.1 and 10.2, respectively.

Table 10.1 presents the results for regioselectivity predictions. Using Hirshfeld charges, the reactive sites for all species listed in the table are correctly identified. This aligns with our earlier formulation in the previous section, as Hirshfeld charges arise naturally from the *information conservation principle*. Similarly, the

Table 10.1 Using the atomic charge to determine regioselectivity for seven systems with both electrophilic and nucleophilic attacks in Figure 10.6. The integer shown in the Table is the serial number of the atom in the molecule shown in Figure 10.6. The red fonts indicate correct predictions.

	Electrophilic attack				Nucleophilic attack		
	a	**b**	**c**	**d**	**e**	**f**	**g**
Hirshfeld	13	3	4	1	1	1	1
VDD	13	3	2	1	1	1	1
Mulliken	16	12	1	6	2	1	4
Löwdin	12	1	4	6	14	3	1
Ros–Schuit	1	1	2	4	14	1	17
Stout–Politer	6	12	3	6	2	1	4
Bickelhaupt	13	12	1	6	2	1	4
Becke	16	3	2	1	1	1	2
Chelpg ESP	16	3	2	1	1	4	2
Merz–Kollmann	13	2	6	1	10	4	4
ADCH	13	2	2	1	1	4	4
AIM	13	4	4	1	1	4	2
Hirshfeld-I	16	6	6	6	1	1	2
CM5	16	3	4	6	1	4	2
EEM	16	2	2	6	1	4	2

Table 10.2 Using the Fukui function obtained from different partial charges to determine regioselectivity for seven systems with both electrophilic and nucleophilic attacks in Figure 10.6. See caption in Table 10.1.

	Electrophilic attack				Nucleophilic attack		
	a	**b**	**c**	**d**	**e**	**f**	**g**
Hirshfeld	13	4	4	1	14	1	5
VDD	13	4	4	1	10	1	1
Mulliken	12	1	4	1	7	3	3
Löwdin	13	4	4	1	14	1	5
Ros–Schuit	3	4	3	6	7	2	5
Stout–Politzer	12	1	4	1	3	2	4
Bickelhaupt	12	12	2	1	10	3	4
Becke	13	4	4	1	10	3	2
Chelpg ESP	13	4	4	1	3	3	2
Merz–Kollmann	13	4	4	1	10	5	2
ADCH	13	4	4	1	10	3	2
AIM	13	5	4	1	10	5	1
Hirshfeld-I	13	4	4	1	10	3	6
CM5	13	4	4	1	14	1	5
EEM	1	3	3	1	2	2	3

Voronoi Deformation Density (VDD) charges also yield accurate predictions across all cases. In contrast, the remaining 13 charge models fail to consistently identify the correct reaction sites, with none achieving complete accuracy across the dataset. As shown in Table 10.2, the regioselectivity predictions based on Fukui functions are less accurate overall. None of the 15 charge schemes successfully predicts the correct reactive sites for all seven species. Comparatively, the predictions for electrophilic attacks show better performance than those for nucleophilic attacks, as reflected in the data.

Table 10.3 presents the correlation coefficients (R) between experimentally derived electrophilicity/nucleophilicity scales and calculated values using either Hirshfeld charges or Fukui functions, using 15 different ways of calculating the partial charge. For electrophilic systems from Figure 10.1, using atomic charges directly as a measure of electrophilicity yields reliable correlations with several methods, including Hirshfeld, VDD, Löwdin, Becke, ADCH, Chelpg ESP, AIM, Hirshfeld-I, and CM5 charges. When Fukui functions are computed from these charges, strong agreement is still observed for Hirshfeld, VDD, Löwdin, and CM5. However, for nucleophilic

Table 10.3 Correlation coefficients (*R*) between experimental results of electrophilicity/nucleophilicity and results from two theoretical approaches, Fukui function and Hirshfeld charge, using 15 different ways to calculate the partial charge. Red fonts indicate strong correlations with the magnitude of the correlation coefficient *R* larger than 0.90.

	Electrophilicity		**Nucleophilicity**	
Different charges	**Atomic charge**	**Fukui function**	**Atomic charge**	**Fukui function**
Hirshfeld	0.995	0.905	−0.988	−0.344
VDD	0.993	0.948	−0.961	−0.297
Mulliken	−0.047	−0.006	0.020	−0.448
Löwdin	0.994	0.925	−0.140	−0.338
Ros–Schuit	0.115	0.346	−0.095	0.182
Stout–Polizter	0.172	0.201	0.147	−0.117
Bickelhaupt	0.098	0.036	0.036	−0.333
Becke	0.923	0.882	−0.683	−0.433
ADCH	0.977	0.832	−0.890	−0.421
Chelpg ESP	0.956	0.758	−0.431	−0.239
Merz–Kollmann	0.881	0.563	−0.423	−0.264
AIM	0.988	−0.291	−0.005	−0.692
Hirshfeld-I	0.929	−0.375	−0.423	−0.271
CM5	0.995	0.905	−0.964	−0.344
EEM	0.376	0.666	−0.469	0.020

species from Figure 10.2, the results differ significantly. Only Hirshfeld, VDD, and CM5 charges provide reasonable predictions when atomic charges are used directly, while none of the charge-derived Fukui functions show strong correlation with the experimental nucleophilicity scale.

These benchmark results highlight that electrophilicity is generally easier to model than nucleophilicity, as several partial charge schemes can closely reproduce the experimental electrophilicity scale. In contrast, nucleophilicity proves more challenging to capture accurately—particularly when using Fukui functions. The findings suggest that using atomic charges directly offers a more reliable approach than relying on Fukui functions for quantifying both electrophilicity and nucleophilicity. Notably, the Hirshfeld charge consistently performs well, accurately predicting regioselectivity and simultaneously providing dependable estimates for both electrophilicity and nucleophilicity. While the Fukui function performs reasonably in predicting electrophilic behavior, it falls short for nucleophilic systems. Among the 14 other charge models tested (excluding Hirshfeld), the VDD charge shows the strongest agreement with the Hirshfeld charge.

The higher predictability of electrophilicity compared with nucleophilicity can be rationalized in physical terms. Electron acceptance involves delocalization into low-lying vacant orbitals and is dominated by long-range electrostatic and polarization effects that are relatively well captured by static electron density descriptors. In contrast, electron donation depends on high-energy frontier orbitals whose behavior is more sensitive to dynamic correlation, solvation, and conformational flexibility—factors that are inherently system- and environment-dependent. This intrinsic asymmetry highlights that electrophilicity represents a collective, global response of the density distribution, whereas nucleophilicity is a context-dependent property of the donor site. Together, these benchmarks establish Hirshfeld charge, which is the first-order approximation of information gain, as the most reliable and broadly applicable density-based descriptor for electrophilicity, nucleophilicity, and regioselectivity.

10.7 *Ortho/Para* and *Meta* Group Directing Effect

While many other applications of the above schemes for quantifying electrophilicity and nucleophilicity are possible, this section highlights their use in predicting *ortho*/*para* versus *meta* directing effects in electrophilic aromatic substitution (EAS) reactions.

EAS is a fundamental class of reactions in organic chemistry, where an atom—typically hydrogen—on an aromatic ring is replaced by an electrophile. Due to the stability of the aromatic system, these reactions proceed via a mechanism that temporarily disrupts aromaticity, forming a high-energy σ-complex (also known as an arenium ion), before restoring aromaticity through the loss of a proton. Common electrophiles in EAS include halogens (in halogenation), nitronium ions (in nitration), and acylium ions (in Friedel–Crafts acylation), among others. These reactions are widely used in the synthesis of aromatic compounds in pharmaceuticals, agrochemicals, and dyes due to their regioselective nature.

The regioselectivity of EAS is strongly influenced by substituents already present on the aromatic ring. These substituents can direct incoming electrophiles to specific positions relative to themselves—typically classified as *ortho* (adjacent), *meta* (one carbon removed), or *para* (opposite). EDGs, such as $-OH$, $-OCH_3$, $-NH_2$, typically activate the ring and direct electrophiles to the *ortho* and *para* positions, while EWGs, such as $-NO_2$, $-CN$, $-COOH$, tend to deactivate the ring and direct substitution to the *meta* position. It is commonly believed that EDGs stabilize the carbocation intermediate formed by attack at these *ortho/para* positions through resonance or inductive effects, and EWGs destabilize the carbocation intermediates formed by *ortho/para* attack more significantly than those formed by *meta* attack, making the *meta* position the "least unfavorable" option for electrophilic substitution. However, this traditional explanation provides only a partial picture.

Our approach to quantifying nucleophilicity using information gain and Hirshfeld charge offers a new perspective on the origin of group directing effects in EAS [50, 51]. In a study of 85 aromatic species, we found that the regioselectivity of electrophilic attack is closely linked to the nucleophilicity of specific carbon atoms on the aromatic ring, which can be reliably assessed by the Hirshfeld charge. Specifically, the more negative the Hirshfeld charge at a given site, the more nucleophilic—and thus more reactive—it is toward electrophilic substitution. *Ortho*/*para* directing groups tend to induce the most negative charges at the *ortho* and *para* positions, making these sites favored for substitution. In contrast, *meta* directing groups typically result in the highest nucleophilicity—and therefore greatest negative charge—at the *meta* position, explaining their directing behavior.

We challenged the conventional notion that *ortho*/*para* directing groups are inherently electron-donating while *meta* directing groups are electron-withdrawing. Our analysis reveals that, in fact, most neutral substituents exhibit electron-withdrawing character, regardless of their directing behavior. Anionic substituents consistently act as electron donors, yet this does not necessarily correlate with their position-directing tendencies because directing effects depend on where charge is redistributed on the ring, not on the overall donor/acceptor character of the substituent. This observation highlights that directing effects arise from *relative charge redistribution* within the ring, not the absolute donor/acceptor nature of the substituent. Notably, some EDGs, such as $-GeH_3$ and $-SiH_3$, are actually *meta* directing, contradicting traditional expectations. These findings indicate that the electron-donating or electron-withdrawing nature of a substituent alone does not determine its directing effect. Instead, it is the redistribution of electron density within the aromatic ring upon substitution that governs regioselectivity in EAS. Using this approach, we predicted the *ortho*/*para* or *meta* directing behavior of a range of substituent groups whose regioselectivity had not been previously documented in the literature.

In a follow-up study [51], we investigated the transition-state structures and activation energies of identity EAS reactions involving a series of monosubstituted benzene derivatives reacting with hydrogen fluoride, catalyzed by BF_3 in the gas phase. The analysis included 18 different substituent groups—nine known to be *ortho*/*para* directing and nine *meta* directing. Our results revealed a strong correlation between the reaction barrier height and the Hirshfeld charge at the regioselective site, with correlation coefficients (R^2) exceeding 0.96 for both directing groups. While a weaker correlation was also observed between barrier height and HOMO energy, the consistency and strength of the Hirshfeld charge correlation underscore its value as a reliable predictor of both reactivity and regioselectivity in EAS reactions. This further validates the use of Hirshfeld charge as a powerful tool for understanding and forecasting outcomes in this fundamental class of transformations.

10.8 Concluding Remarks

Electrophilicity and nucleophilicity form one of the most enduring conceptual pairs in chemistry. What began as qualitative notions of “electron-loving” and “nucleus-loving” behavior has evolved into a fully quantitative framework grounded in both energetics and information. Through CDFT, reactivity can be expressed in terms of global quantities such as the electrophilicity index, together with local descriptors like the Fukui function and dual descriptor. The complementary ITA reframes the same phenomena in terms of information gain and charge redistribution. Together, these perspectives reveal that reactivity is an intrinsic property of the electron density—measurable, interpretable, and directly comparable to experiment.

The ability of these density-based descriptors to reproduce kinetic scales, regioselectivity, and substituent effects demonstrates that classical reactivity concepts retain their explanatory power when cast into rigorous theoretical form. By bridging experimental kinetics with first-principles theory, this integration transforms intuition into prediction. The vocabulary of electrophiles and nucleophiles thus remains not a relic of pre-quantum language, but a living framework that continues to deepen in precision and scope. In a broader sense, the success of this synthesis exemplifies how theoretical chemistry can unify empirical observation and quantum theory without sacrificing the conceptual clarity that defines the discipline.

References

1 Mayr H, Patz M. Scales of nucleophilicity and electrophilicity: a system for ordering polar organic and organometallic reactions. Angew Chem Int Ed Engl. 1994 May 18;33(9):938–57.

2 Jaramillo P, Pérez P, Contreras R, Tiznado W, Fuentealba P. Definition of a nucleophilicity scale. J Phys Chem A. 2006 Jul 6;110(26):8181–7.

3 Pratihar S. Electrophilicity and nucleophilicity of commonly used aldehydes. Org Biomol Chem. 2014 Aug 14;12(30):5781–8.

4 Lewis GN. Acids and bases. J Franklin Inst. 1938 Sep;226(3):293–313.

5 Shoppee CW. Christopher Kelk Ingold, 1893–1970. Biogr Mem Fellows R. Soc. 1972 Nov;18:348–411.

6 Tsuji Y, Richard JP. Swain-Scott relationships for nucleophile addition to ring-substituted phenonium ions. Can J Chem. 2015;93(4):428–34.

7 Ritchie CD. Nucleophilic reactivities toward cations. Acc Chem Res. 1972 Oct 1;5(10):348–54.

8 Gross C, Eitzinger A, Hampel N, Mayer P, Ofial AR. Defining the synthetic scope of ortho-quinone methides by quantifying their electrophilicity. Chem Eur J. 2025 Feb 3;31(7):e202403785.

9 Mayr H, Ammer J, Baidya M, Maji B, Nigst TA, Ofial AR, et al. Scales of Lewis basicities toward C-centered Lewis acids (carbocations). J Am Chem Soc. 2015 Feb 25;137(7):2580–99.

10 Jangra H, Chen Q, Fuks E, Zenz I, Mayer P, Ofial AR, et al. Nucleophilicity and electrophilicity parameters for predicting absolute rate constants of highly asynchronous 1,3-dipolar cycloadditions of aryldiazomethanes. J Am Chem Soc. 2018 Dec 5;140(48):16758–72.

11 Timofeeva DS, Ofial AR, Mayr H. Kinetics of electrophilic fluorinations of enamines and carbanions: comparison of the fluorinating power of N–F reagents. J Am Chem Soc. 2018 Sep 12;140(36):11474–86.

12 An F, Maji B, Min E, Ofial AR, Mayr H. Basicities and nucleophilicities of pyrrolidines and imidazolidinones used as organocatalysts. J Am Chem Soc. 2020 Jan 22;142(3):1526–47.

13 Li L, Mayer RJ, Ofial AR, Mayr H. One-bond-nucleophilicity and -electrophilicity parameters: an efficient ordering system for 1,3-dipolar cycloadditions. J Am Chem Soc. 2023 Apr 5;145(13):7416–34.

14 Peng D, Yang W. Fukui function and response function for nonlocal and fractional systems. J Chem Phys. 2013 May 14;138(18):184108.

15 Yang W, Cohen AJ, De Proft F, Geerlings P. Analytical evaluation of Fukui functions and real-space linear response function. J Chem Phys. 2012 Apr 14;136(14):144110.

16 Yang W, Parr RG. Hardness, softness, and the Fukui function in the electronic theory of metals and catalysis. Proc Natl Acad Sci USA. 1985 Oct;82(20):6723–6.

17 Yang W, Parr RG, Pucci R. Electron density, Kohn–Sham frontier orbitals, and Fukui functions. J Chem Phys. 1984;81(6):2862.

18 Parr RG, Szentpály LV, Liu S. Electrophilicity index. J Am Chem Soc. 1999 Mar 1;121(9):1922–4.

19 Chattaraj PK, Sarkar U, Roy DR. Electrophilicity index. Chem Rev. 2006 Jun;106(6):2065–91.

20 Liu S. Conceptual density functional theory and some recent developments. Acta Phys-Chim Sin. 2009;25(3):590–600.

21 Rong C, Wang B, Zhao D, Liu S. Information-theoretic approach in density functional theory and its recent applications to chemical problems. WIREs Comput Mol Sci. 2020 Jul;10(4):e1461.

22 Geerlings P, Ayers PW, Toro-Labbé A, Chattaraj PK, De Proft F. The Woodward-Hoffmann rules reinterpreted by conceptual density functional theory. Acc Chem Res. 2012 May 15;45(5):683–95.

23 Ayers PW, Morell C, De Proft F, Geerlings P. Understanding the Woodward-Hoffmann rules by using changes in electron density. Chem Eur J. 2007;13(29):8240–7.

24 Woodward RB, Hoffmann R. The conservation of orbital symmetry. Angew Chem Int Ed. 1969 Nov;8(11):781–853.

25 Chattaraj PK, Lee H, Parr RG. HSAB principle. J Am Chem Soc. 1991 Feb;113(5):1855–6.

26 Ho T-L. Hard soft acids bases (HSAB) principle and organic chemistry. Chem Rev. 1975 Feb 1;75(1):1–20.

27 Ayers PW, Parr RG, Pearson RG. Elucidating the hard/soft acid/base principle: a perspective based on half-reactions. J Chem Phys. 2006 May 21;124(19):194107.

28 Edwards JO. Correlation of relative rates and equilibria with a double basicity scale. J Am Chem Soc. 1954 Mar;76(6):1540–7.

29 Legon AC, Millen DJ. Hydrogen bonding as a probe of electron densities: limiting gas-phase nucleophilicities and electrophilicities of B and HX. J Am Chem Soc. 1987 Jan;109(2):356–8.

30 Fu J, Li M, Rong C, Zhao D, Liu S. Information-theoretic quantities as effective descriptors of electrophilicity and nucleophilicity in density functional theory. J Mol Model. 2024 Oct;30:341.

31 Parr RG, Yang W. Density-Functional Theory of Atoms and Molecules. Oxford University Press; 1989.

32 Ayers PW, Liu S. Basic formalism. In: Liu S, editor. Conceptual Density Functional Theory: Towards a New Chemical Reactivity Theory. Wiley; 2022. pp. 47–60.

33 Geerlings P, De Proft F, Langenaeker W. Conceptual density functional theory. Chem Rev. 2003 May;103(5):1793–873.

34 Geerlings P, Chamorro E, Chattaraj PK, De Proft F, Gázquez JL, Liu S, et al. Conceptual density functional theory: status, prospects, issues. Theor Chem Acc. 2020 Feb;139(2):36.

35 Maynard AT, Huang M, Rice WG, Covell DG. Reactivity of the HIV-1 nucleocapsid protein p7 zinc finger domains from the perspective of density-functional theory. Proc Natl Acad Sci USA. 1998 Sep 29;95(20):11578–83.

36 Hoffmann G, Chermette H, Morell C. Revisiting nucleophilicity: an index for chemical reactivity from a CDFT approach. J Mol Model. 2024 Jun 28;30(7):232.

37 Perdew JP, Parr RG, Levy M, Balduz JL. Density-functional theory for fractional particle number: derivative discontinuities of the energy. Phys Rev Lett. 1982 Dec 6;49(23):1691–4.

38 Morell C, Grand A, Toro-Labbé A. New dual descriptor for chemical reactivity. J Phys Chem A. 2005 Jan 13;109(1):205–12.

39 Liu S, Ess DH, Schauer CK. Density functional reactivity theory characterizes charge separation propensity in proton-coupled electron transfer reactions. J Phys Chem A. 2011 May 12;115(18):4738–42.

40 Kumar N, Liu S, Kozlowski PM. Charge separation propensity of the coenzyme B_{12}–tyrosine complex in adenosylcobalamin-dependent methylmalonyl–CoA mutase enzyme. J Phys Chem Lett. 2012 Apr 19;3(8):1035–8.

41 De Proft F, Forquet V, Ourri B, Chermette H, Geerlings P, Morell C. Investigation of electron density changes at the onset of a chemical reaction using the state-specific dual descriptor from conceptual density functional theory. Phys Chem Chem Phys. 2015 Apr 14;17(14):9359–68.

42 Nalewajski RF, Parr RG. Information theory, atoms in molecules, and molecular similarity. Proc Natl Acad Sci USA. 2000 Aug 1;97(16):8879–82.

43 Rong C, Yu D, Liu S. Information-theoretic approach. In: Liu S, editor. Conceptual Density Functional Theory: Towards a New Chemical Reactivity Theory. Wiley; 2022. pp. 281–99.

44 Liu S. Information-theoretic approach in density functional reactivity theory. Acta Phys-Chim Sin. 2016;32(1):98–118.

45 Liu S, Rong C, Lu T. Information conservation principle determines electrophilicity, nucleophilicity, and regioselectivity. J Phys Chem A. 2014 May 22;118(20):3698–704.

46 Zhou X, Rong C, Lu T, Liu S. Hirshfeld charge as a quantitative measure of electrophilicity and nucleophilicity: nitrogen-containing systems. Acta Phys-Chim Sin. 2014;30(11):2055–62.

47 Liu S. Identity for Kullback-Leibler divergence in density functional reactivity theory. J Chem Phys. 2019 Oct 14;151(14):141103.

48 Wang B, Zhao D, Lu T, Liu S, Rong C. Quantifications and applications of relative fisher information in density functional theory. J Phys Chem A. 2021 May 6;125(17):3802–11.

49 Wang B, Rong C, Chattaraj PK, Liu S. A comparative study to predict regioselectivity, electrophilicity and nucleophilicity with Fukui function and Hirshfeld charge. Theor Chem Acc. 2019 Dec;138(12):124.

50 Liu S. Where does the electron go? The nature of ortho/para and meta group directing in electrophilic aromatic substitution. J Chem Phys. 2014 Nov 21;141(19):194109.

51 Liu S. Quantifying reactivity for electrophilic aromatic substitution reactions with Hirshfeld charge. J Phys Chem A. 2015 Mar 26;119(12):3107–11.

11

Steric Effect and Stereoselectivity

TABLE OF CONTENTS

11.1 Steric Effect

In the preceding chapters, we have examined how electron density responds to perturbations in charge, geometry, and symmetry, from bonding and cooperativity to chirality and charge-transfer reactivity. These phenomena describe how electron density reorganizes within the molecular framework. In this chapter, we shift gears to the complementary question: how the *spatial distribution* of electron density itself governs molecular structure, reactivity, and selectivity. Steric effect and stereoselectivity emerge from this spatial aspect of electron density, reflecting how molecules resist compression and differentiate between competing reaction pathways.

11.1.1 What Is Steric Effect?

Steric effect [1–4] is one of the most fundamental and widely applied concepts in chemistry. It arises from the fact that atoms and molecular groups occupy finite regions of space, and their spatial presence influences molecular structure, reactivity, and physical properties. In contrast to electronic effects—such as inductive or resonance influences—which originate from charge redistribution or orbital interactions, steric effects are inherently nonbonded in nature and stem purely from spatial crowding.

At the microscopic level, steric effects manifest when electron densities associated with nearby atoms begin to overlap. This overlap produces repulsive forces that prevent atoms from approaching too closely, often leading to geometric distortions, altered conformational preferences, and modified reactivity. In essence, steric effect reflects the spatial constraints imposed by a molecular framework: molecular behavior is shaped not only by how electrons are distributed or delocalized but also by the simple fact that multiple atoms cannot occupy the same region of space.

The consequences of steric effects permeate nearly every area of chemistry. They influence reaction rates and mechanisms, control conformational equilibria, regulate catalytic selectivity, shape biomolecular recognition, and determine the physical and structural properties of materials. Whether in organic synthesis, transition-metal catalysis, supramolecular assembly, or biological function, steric considerations provide a unifying language for understanding how spatial demand and molecular crowding direct chemical behavior.

11.1.2 Types of Steric Effect

The steric effect encompasses a range of spatial phenomena arising from the physical presence and size of atoms or molecular groups. The most familiar manifestation is steric hindrance [5, 6], where bulky substituents block or slow down chemical reactions by limiting access to reactive sites. This is especially significant in nucleophilic substitution (e.g., S_N2 reactions), coordination chemistry, and catalysis, where crowded environments prevent effective bonding or transition state formation. Steric hindrance also influences conformational preferences, as molecules adopt shapes that minimize unfavorable atomic crowding, such as chair conformation in cyclohexane.

Beyond hindrance, steric effects manifest in various other forms. Steric relief occurs when removing or repositioning bulky groups reduces internal strain, stabilizing the molecule [7]. Steric strain [8] refers to the internal repulsion within a molecule due to the close spatial proximity of large groups, often observed in small rings or eclipsed conformations. Steric steering uses bulk to control the directionality of reactions, guiding regio- or stereoselectivity. In steric protection, large groups shield reactive sites from unwanted interactions, common in organometallic and polymer chemistry. Additional manifestations include conformational locking, where bulky substituents restrict rotational freedom, and packing effects in solids, where steric bulk influences crystallinity and material properties. Together, these diverse expressions highlight the pervasive role of steric effects in dictating molecular behavior and reactivity.

11.1.3 Origin of Steric Effect

The origin of the steric effect has been the subject of ongoing discussion and refinement in the literature for more than a century [2, 9, 10]. While its empirical consequences, such as slowed reaction rates, conformational preferences, and geometric distortions, are well documented, its theoretical foundation has been approached from multiple perspectives, shaped by evolving theoretical frameworks and available computational tools.

Because the steric effect does not correspond to a physical observable, there is no unique way for its determination. As a result, different schemes of quantification have been developed to reflect the understanding of its origin and capture its influence, each emphasizing different aspects such as spatial volume, electron density, or energy contributions.

The earliest view of steric effects treats atoms as hard spheres that physically occupy space. This mechanistic interpretation, inspired by van der Waals and early molecular models, considers atoms and groups as rigid entities whose interaction is largely a matter of geometrical crowding. It ignores electronic structure, treating atoms like billiard balls, so it is unable to explain favorable steric effects, such as steric relief, conformational preference, or steric steering.

With the rise of quantum mechanics, steric effects began to be understood in terms of electron cloud overlap and the Pauli exclusion principle. In this view, steric repulsion arises because electrons with the same spin cannot occupy the same region of space. The introduction of energy decomposition analysis (EDA) methods [11–13] dissects molecular interactions into electrostatics, orbital interactions, and steric repulsion, providing quantitative insights into the source of steric energy. In EDA, steric repulsion is obtained by the exchange repulsion or Pauli repulsion term. In this approach, the steric effect is regarded as part of the quantum effect.

According to Weisskopf [14], a renowned physicist, the steric effect stems from the Pauli exclusion principle. When two atoms or molecular groups are brought together, their overlapping electron clouds result in their occupying higher energy states, as they can no longer share the same low-energy orbitals due to the exclusion principle. Nevertheless, the kinetic energy of the electrons also increases, creating a repulsive force that pushes the atoms apart. Weisskopf described this phenomenon as "kinetic energy pressure," emphasizing that the origin of steric repulsion is of a kinetic nature. With this in mind, we recently quantified the steric effect in the density functional theory (DFT) framework using the Weizsäcker kinetic energy density functional [15]. More details about this approach will be introduced in the following sections.

11.1.4 Brief History of Steric Effect

The recognition of steric effects as a distinct chemical phenomenon has evolved over centuries, from early qualitative observations to modern quantitative treatments. The term "steric hindrance" emerged in the late nineteenth and early twentieth centuries as chemists like Victor Meyer observed that bulky groups could prevent or slow down certain reactions [16]. Meyer, in the 1890s, studied the esterification of carboxylic acids and noticed that introducing large substituents near the carboxyl group significantly inhibited the reaction. He attributed this to "steric hindrance," a physical blocking effect. This was one of the first explicit recognitions of spatial effects on chemical reactivity.

Christopher K. Ingold [6, 17] and his school in the 1930s and 1940s made significant contributions to the understanding of reaction mechanisms, including S_N1 and S_N2 reactions. Their systematic kinetic studies provided clear experimental evidence for the profound impact of steric bulk on reaction rates, particularly highlighting the dramatic decrease in S_N2 reactivity, with increasing substitution at the reacting carbon due to steric hindrance to backside attack.

A major step forward came with Robert W. Taft's dual-parameter approach to reaction rates [18], which separated steric and electronic contributions in substituent effects. This approach provided one of the first systematic ways to isolate and quantify steric effects, leading to widespread application in physical organic chemistry.

Marvin Charton [19, 20] developed a more general steric parameter, which is based on the minimum van der Waals radius of a substituent measured in a specific direction. This approach attempts to provide a more fundamental measure of the size of a substituent and its ability to interact sterically with other parts of the molecule, independent of a specific reaction series. It is often considered more theoretically sound than E_s for general steric bulk.

The 1970s saw the introduction of the cone angle by Chadwick A. Tolman [21], further expanding the quantitative assessment of steric effects, particularly in the rapidly growing field of organometallic chemistry. This parameter became indispensable for rationalizing the reactivity and selectivity of transition metal catalysts based on the steric bulk of their ligands.

11.2 Stereoselectivity

11.2.1 What Is Stereoselectivity?

Stereoselectivity [22] refers to the preference of a chemical reaction to yield one stereoisomer over another when multiple isomers are possible. It is a central concept in stereochemistry, influencing the outcome of countless organic, inorganic, and biochemical reactions in pharmaceuticals, natural product synthesis, and catalysis. In a stereoselective reaction, the reactants can give rise to two or more stereoisomeric products, but the reaction pathway favors one over the others, often due to differences in energy, steric interactions, or electronic effects.

Stereoselectivity and stereospecificity are related but fundamentally different concepts in stereochemistry. Stereoselectivity refers to a reaction's preference for forming one stereoisomer over another, without necessarily being dictated by the stereochemistry of the starting material. In contrast, stereospecificity applies to reactions in which the stereochemical configuration of the reactant directly determines the stereochemistry of the product, meaning that different stereoisomers of the starting material yield different stereoisomeric products. While every stereospecific reaction is inherently stereoselective, the reverse is not always true.

Stereoselectivity can be further categorized into two main types: *enantioselectivity* [23], which is the preferential formation of one enantiomer over its mirror image in a chiral product, and *diastereoselectivity* [24], which refers to favoring one diastereomer over others. Two related concepts deepen our understanding of stereoselective outcomes [25]: *stereoconvergence*, where different stereoisomeric substrates (such as *E*/*Z* alkenes or enantiomers) are transformed into the same stereoisomeric product under identical conditions, and *stereodivergence*, where a single stereoisomeric substrate can lead to different stereoisomeric products depending on the choice of catalyst or reaction conditions.

The concept of stereoselectivity in chemistry emerged from the early recognition of the steric effect as a key influence on reactivity and selectivity. This understanding evolved significantly in the mid-twentieth century with the development of theoretical models such as *Cram's rule* [26, 27] and the *Felkin–Anh Model* [28], which provided predictive frameworks for diastereoselective transformations. A major leap occurred in the late twentieth century with the rise of enantioselective catalysis, particularly in the realm of transition-metal catalysis [29, 30] led by pioneers like Knowles, Noyori, and Sharpless, enabling precise synthesis of chiral molecules. In the early twenty-first century, the advent of *organocatalysis*, driven by the work of List [31] and MacMillan [32], offered metal-free approaches to stereocontrol using simple organic molecules, expanding the scope and sustainability of stereoselective synthesis across pharmaceuticals, agrochemicals, and materials science.

11.2.2 Examples of Stereoselectivity

The Sharpless epoxidation [30] is a classic example of an enantioselective reaction, where a prochiral allylic alcohol is oxidized to give an epoxide with high enantiomeric excess. Using titanium isopropoxide, diethyl tartrate, and tert-butyl hydroperoxide, one enantiomer is formed preferentially, controlled by the chirality of the tartrate ligand. This has been widely used in the synthesis of chiral epoxides for pharmaceuticals and natural products.

The addition of bromine to an alkene exemplifies stereoselectivity through its *anti* addition mechanism. The reaction proceeds via a cyclic bromonium intermediate, where bromine bridges the double bond, forcing the second bromine atom to attack from the opposite side. For a symmetrical alkene like trans-2-butene, this yields a *meso* compound, while an unsymmetrical alkene produces a racemic mixture of enantiomers. In cyclic alkenes, such as cyclohexene, the ring structure further dictates the stereochemistry, resulting in a trans-dibromide product with high specificity.

In enantioselective hydrogenation [32, 33], a prochiral alkene or ketone is reduced using a chiral metal catalyst, such as Rh or Ru complexes with chiral ligands (e.g., BINAP). The spatial environment around the metal center favors one face of the molecule over the other, leading to the selective formation of one enantiomer—a critical process in the synthesis of optically active drugs.

The Diels–Alder reaction, a [4+2] cycloaddition between a diene and a dienophile, is another renowned example of its stereoselectivity, often favoring the *endo* product, even though the *exo* product might be thermodynamically more stable. This preference arises from stabilizing secondary orbital interactions in the endo transition state, where the diene's π-system interacts with substituents on the dienophile, stabilizing the *endo* pathway over the *exo* alternative. For instance, the reaction of cyclopentadiene with maleic anhydride predominantly forms the *endo* adduct, enabling the predictable synthesis of complex cyclic molecules with defined stereochemistry.

11.2.3 Origin of Stereoselectivity

The origin of stereoselectivity lies in the differential energy landscapes available to competing transition states leading to the various stereoisomeric products. For a reaction to be stereoselective, the activation energy (E_a) leading to one stereoisomer must be lower than the activation energy leading to another. This energy difference, often subtle, biases the reaction toward the formation of the more favored product.

Steric effect is the *primary driver* of stereoselectivity. It is the most common and often dominant factor. Bulky substituents can obstruct certain reaction pathways, favoring less hindered routes. In the bromination of alkenes, for example, the bromonium ion shields one face, directing the nucleophile to the opposite side, minimizing steric clashes. Other factors, such as electronic effects from dipole–dipole interactions, frontier orbital overlapping, or hydrogen bonding and other noncovalent interactions, can also play important roles. In the Diels–Alder reaction, for instance, secondary orbital interactions enhance the *endo* transition state's stability, lowering its energy relative to the *exo* pathway. Catalysts, especially chiral ones, can create asymmetric environments that favor one stereoisomer. In Sharpless epoxidation, for example, the chiral titanium complex differentiates the enantiotopic faces of the alkene, reducing the energy barrier for one epoxide's formation. Across these different scenarios, steric interactions remain the dominant origin of transition-state discrimination, while electronic factors modulate but rarely override the steric preference.

Closely related to steric effect and stereoselectivity is chirality, a topic that we discussed in Chapter 9. Chirality profoundly influences how steric effects operate and plays a pivotal role in determining the outcome of stereoselective reactions. The presence of chiral centers gives rise to distinct three-dimensional arrangements, such as enantiomers and diastereomers, which experience different steric environments during a reaction. As a result, a chiral catalyst or reactant can interact more favorably with one face or conformer of a substrate, directing the reaction toward a specific stereoisomer. Together, chirality, steric effect, and stereoselectivity form an integrated framework that underpins much of asymmetric synthesis and molecular design.

11.2.4 Models of Stereoselectivity

Predicting and rationalizing stereoselectivity in complex reactions requires conceptual models that simplify the interactions occurring in the transition state. These models aim to identify the most favorable pathway by considering mostly steric effects and, in some cases, electronic factors.

Cram's Rule. For nucleophilic additions to carbonyls adjacent to a chiral center, Cram [26, 27] proposed that the most stable conformation for the transition state would have the largest group on the chiral center anti-periplanar to the incoming nucleophile. The carbonyl oxygen is then oriented to minimize steric hindrance with the other substituents on the chiral center.

Felkin–Anh Model. This is a more refined and widely accepted model for nucleophilic additions to carbonyls adjacent to a chiral center. It acknowledges that the transition state for nucleophilic addition is an early, product-like one [27, 28]. It postulates that the incoming nucleophile approaches the carbonyl from an angle (typically about 107° from the C=O bond, known as the Bürgi–Dunitz angle) to achieve optimal orbital overlap [34, 35]. The key prediction is that the largest group on the adjacent chiral center will orient itself *anti* to the incoming nucleophile to minimize steric interactions. This allows the nucleophile to approach from the least sterically hindered face of the carbonyl. The medium group is then oriented perpendicular to the developing bond, and the smallest group is placed near the carbonyl oxygen.

Anti-Felkin–Anh Model (also known as *chelation-controlled model*) [28, 36, 37]. When a metal ion is present and capable of chelating (forming multiple bonds) with the carbonyl oxygen and an atom (e.g., an oxygen or nitrogen) on the chiral center, a rigid five- or six-membered ring can form. This chelation fixes the conformation of the substrate, and the nucleophile then attacks the carbonyl from the less sterically hindered face of this chelated complex. This overrides the predictions of non-chelation models like *Felkin–Anh model.*

Zimmerman–Traxler Model. It is a foundational transition-state model used to explain diastereoselectivity observed in aldol reactions and related carbon–carbon bond-forming processes [38]. It proposes that the reaction proceeds through a six-membered, chair-like cyclic transition state in which the enolate nucleophile and the electrophilic carbonyl compound are aligned in a specific geometry mediated by a metal cation, such as lithium or boron. This conformation allows for predictable stereochemical outcomes, as substituents prefer to occupy equatorial positions to minimize 1,3-diaxial interactions, favoring the formation of either *syn* or *anti* diastereomers depending on the configuration of the starting materials.

Electronic Models. Not all stereoselective models are based on the steric effect. For example, the *endo* selectivity in the Diels–Alder reaction is attributed to a stabilizing interaction between the π-orbitals of the dienophile's substituents and the developing π-system of the diene in the transition state. In these stereoselective reactions, the specific symmetry and spatial orientation of these interacting frontier orbitals (highest-occupied molecular orbital and lowest-unoccupied molecular orbital) in the transition state can dictate which stereoisomer is formed.

11.3 Experimental Scales of Steric Effect

Several experimental scales have been developed to quantify steric effects in chemistry, particularly in organic reactions and physical organic studies. These scales vary in their basis—some are derived from thermodynamic data, others from kinetic measurements, and some from structural or empirical considerations. Here are a few most widely recognized experimental steric scales.

11.3.1 Taft's Steric Parameter (E_S)

Introduced by R. W. Taft in 1956, the E_S parameter is one of the earliest and most widely used experimental measures of steric effect [1, 18, 39–41]. It is derived from the rate of ester hydrolysis of para-substituted alkyl esters compared to methyl esters. As part of his work on linear free energy relationships (LFERs), Taft extended the Hammett equation to aliphatic systems. The core idea behind E_S is to isolate the steric contribution to a reaction rate by comparing the rates of acid-catalyzed ester hydrolysis. Taft assumed that under strongly acidic conditions, electronic substituent effects are minimized, allowing the remaining variation in rate to be attributed *primarily* to steric hindrance.

The E_S value for a substituent R is defined relative to a reference substituent (methyl for which E_S is set to 0):

$$E_S = \log\left(\frac{k_X}{k_{Me}}\right)_{\text{acid-catalyzed}}, \tag{11.1}$$

where k_X is the rate constant for the acid-catalyzed hydrolysis of an ester with the substituent X (e.g., $XCOOCH_3$) and k_{Me} is the rate constant for the acid-catalyzed hydrolysis of the methyl ester (i.e., CH_3COOCH_3). A more negative E_S value indicates a larger steric hindrance. For instance, the E_S value (−1.54) for a tert-butyl group is significantly more negative than that for a methyl group (0.0), reflecting its larger size. For H, $E_S = +1.24$. Taft's parameters are often used in the Taft equation to separate polar, resonance, and steric effects in reaction series:

$$\log(k_X/k_{Me}) = \rho\sigma + \delta\, E_S, \tag{11.2}$$

where ρ and δ are reaction constants and σ is the Hammett substituent constant for X.

11.3.2 Charton's Steric Parameter (ν)

Developed by Marvin Charton [9, 19, 20], the Charton steric parameter (ν) is based on van der Waals radii. Unlike the Taft parameter, which is derived from kinetic data, Charton's parameters are more fundamentally related to the physical size of the substituent. They are calculated based on the minimum van der Waals radius of a symmetrical substituent or a characteristic radius for unsymmetrical substituents. For symmetrical groups like XH_n, the ν value can be related to the van der Waals radius (r_v) of the atom X via

$$v = r_v - r_{v,H}, \tag{11.3}$$

where $r_{v,H}$ is the van der Waals radius of hydrogen (typically 1.20 Å). For more complex or unsymmetrical groups, empirical relationships are used to derive ν values. Charton parameters are often used in multiparameter LFERs alongside electronic parameters to understand the contribution of steric bulk to various chemical and physical properties. Larger ν values indicate bulkier substituents. For example, for H, $\nu = 1.20$, for CH_3, $\nu = 1.52$, and for t-Bu, $\nu = 2.78$. This scale is advantageous because it relies on physical measurements rather than reaction-based data, making it widely applicable.

11.3.3 A-values

A-values quantify the preference of axial versus equatorial positioning of substituents in monosubstituted cyclohexanes, reflecting steric strain [42]. This scale is derived from the equilibrium constants of cyclohexane derivatives, where the A-value represents the difference in free energy between the two conformations.

$$A = \Delta G^{\circ} = G_{\text{axial}} - G_{\text{equatorial}} = -RT\ln K. \tag{11.4}$$

Here, A is the A-value (kcal/mol), K is the equilibrium constant for the axial–equatorial interconversion, and R and T have their usual meanings. The difference in Gibbs free energy ($\Delta G°$) between the chair conformers where the substituent (X) is equatorial versus axial. It is typically measured by NMR or IR spectroscopy. A higher A-value indicates a larger, more sterically demanding substituent that destabilizes the axial position more. For instance, for H, it is 0.0 kcal/mol, for CH_3, it is 1.74, and for t-Bu, it is 4.9.

11.3.4 Tolman Cone Angle

Used primarily in organometallic chemistry, Tolman's cone angle [21] quantifies the steric bulk of ligands, particularly phosphines (PR_3). Determined from X-ray crystallographic data of nickel complexes, $NiL(CO)_3$, the Tolman cone angle θ is defined with the metal (Ni) at the apex and the van der Waals radii of the outermost atoms of the PR_3 ligand forming the base of a cone.

$$\theta = 2 * \arcsin(r_v/d), \tag{11.5}$$

where r_v is the van der Waals radius of the outermost ligand atoms and d is the Ni—P distance. Larger cone angles indicate bulkier ligands, which can affect metal complex stability and reactivity. For example, for PH_3, $\theta = 87°$; for PMe_3, $\theta = 118°$; for PPh_3, $\theta = 145°$; and for $P(t\text{-}Bu)_3$, $\theta = 182°$. Measuring a ligand's *spatial occupation* around the metal center, this scale is highly specific to phosphine ligands in coordination chemistry but extremely influential in organometallic catalysis.

11.3.5 Verloop STERIMOL Parameters

Designed for quantitative structure–activity relationship (QSAR) studies to describe substituent shape, the STERIMOL parameters [9, 43], developed by Arie Verloop and coworkers, offer a more sophisticated and multidimensional description of a substituent's steric bulk compared to the single-value parameters like E_S or ν. These parameters are calculated from the three-dimensional geometry of a substituent and describe its dimensions along different axes relative to the point of attachment to the rest of the molecule. The main STERIMOL parameters are (i) L (length), the length of the substituent along the bond axis to the main molecule; (ii) B1 (minimum width), the minimum width of the substituent perpendicular to the bond axis, and (iii) B5 (maximum width), the maximum width of the substituent perpendicular to the bond axis.

These parameters are derived from molecular models (originally Corey–Pauling–Koltun, CPK, models, now often computed from optimized geometries) and account for the van der Waals radii of all atoms in the substituent. The multidimensional nature of STERIMOL parameters makes them particularly useful in QSAR and rational catalyst design, as they can differentiate the steric effects of substituents in different directions around the reaction center. For instance, a substituent might be long but narrow, or short and bulky, and STERIMOL parameters can capture these nuances, which a single steric constant cannot.

11.4 Steric Effect: A Density-based Quantification

The qualitative origin of the steric effect described in Section 11.1.3—repulsion arising from Pauli exclusion and electron-density crowding—can now be formulated quantitatively. The "kinetic-energy pressure" envisaged by Weisskopf finds an exact analogue in the Weizsäcker kinetic-energy functional in DFT [15]. In this way, the historical concept of steric hindrance acquires a rigorous density-based foundation, linking classical intuition with quantum mechanical expression.

While it is true that there is no unique way to quantify the steric effect, different theoretical frameworks offer different approaches to capture its essence. Regardless of the methodology, any meaningful quantification must reflect two fundamental characteristics: the steric effect should be inherently repulsive and spatially extensive. Its repulsive nature stems from the physical occupation of space. When atoms or groups approach too closely, they experience mutual exclusion due to overlapping electron clouds. Its extensiveness lies in its dependence on size. The larger or bulkier the substituent or molecular fragments, the more significant the steric interference becomes.

Also, although the steric effect concerns the spatial demand of atoms or molecules, it does not act in isolation. Other fundamental interactions, particularly quantum mechanical and electrostatic effects, work together with the steric effect to shape the molecular framework. Quantum contributions stem from the Pauli exclusion principle (manifested as the Fermi hole), which prevents same-spin electrons from occupying the same region, and from dynamic electron correlation (the Coulomb hole), which discourages proximity between opposite-spin electrons. Electrostatic effects, including electron–electron and nucleus–nucleus Coulomb repulsions, are long-range interactions, and they further stabilize molecular geometry by maintaining appropriate distances among charged particles.

Therefore, to quantify the steric effect meaningfully and exclusively, it is essential to isolate it from these quantum and electrostatic contributions. A rigorous depiction should distinguish steric repulsion as a spatial phenomenon, separate from the energy contributions arising from electrostatic and quantum effects.

Based on this argument, in early work [15], we assume that the total energy E of an electronic system comes from the contribution of three independent physicochemical effects, steric E_s, electrostatic E_e, and quantum E_q,

$$E \equiv E_s + E_e + E_q\,. \tag{11.6}$$

This contrasts with the conventional partition of the total energy in DFT [44], which is the following,

$$E = T_S + V_{ne} + J + V_{nn} + E_{xc}, \tag{11.7}$$

where T_s, V_{ne}, J, V_{nn}, and E_{xc} stand for the total noninteracting kinetic energy, nuclear–electron attraction, classical electron–electron Coulomb repulsion, nuclear–nuclear repulsion, and exchange–correlation energy, respectively. Since V_{ne}, V_{nn}, and J are all electrostatic, they can be put together, yielding the total electrostatic energy

$$E_e = V_{ne} + J + V_{nn}, \tag{11.8}$$

and thus we have

$$E = T_s + E_e + E_{xc}\,. \tag{11.9}$$

On the other hand, the quantum contribution E_q has two ingredients,

$$E_q = E_P + E_{xc} = T_s - E_s + E_{xc}, \tag{11.10}$$

one from the exchange–correlation energy E_{xc}, and the other from the kinetic energy E_P, called Pauli energy, which is the contribution of the Pauli exclusion principle to the total kinetic energy, expressed by the difference between T_s and the Weizsäcker kinetic energy T_w

$$T_W = \frac{1}{8}\int \frac{|\mathrm{N}\rho(\mathbf{r})|^2}{\rho(\mathbf{r})} d\mathbf{r}. \tag{11.11}$$

With Eqs. (11.6) and (11.9), we can readily prove that

$$E_s \equiv T_W. \tag{11.12}$$

That is, the contribution from the steric effect can be quantified by the Weizsäcker kinetic energy [45]. The latter measures the energetic penalty associated with local inhomogeneity of the electron density. Regions with steep density gradients reflect strong localization, which increases the kinetic "pressure" [14] required by the Pauli exclusion principle to keep electrons apart. Hence, T_W naturally represents the energy cost of spatial crowding—the quantum mechanical foundation of steric repulsion.

The Weizsäcker kinetic energy also possesses a deep information-theoretic meaning [46–48]. Its integrand, $|\nabla\rho|^2/\rho$, is mathematically identical (up to a constant factor) to Fisher information [49–53] of the electron density, which quantifies the degree of localization or spatial order in a probability distribution. In this sense, the steric energy $E_s = T_W$ measures the information stored in the spatial inhomogeneity of the electronic cloud—the greater the localization, the larger both the Fisher information and the kinetic energy pressure. This equivalence establishes a natural bridge between information-theoretic approach (ITA) and OF-DFT, showing that both frameworks describe the same underlying physics of how electrons distribute and resist compression in space.

This quantification scheme has several outstanding features. It is always nonnegative, scales extensively with the size of the system, satisfying the two conditions set out at the beginning of this section. It vanishes in the case of a homogeneous electron gas and reduces to the total energy in one-electron systems. It is also consistent with Weisskopf's attribution of the steric effect to the "kinetic energy pressure" [14] because T_W is one kinetic energy. Through Bader's zero-flux condition [54], steric energy is connected to the atomic partitioning of molecules, highlighting how steric repulsion is balanced among constituent atoms.

From a physical standpoint, if the electrons in an atom or molecule were bosons, they would all collapse into the lowest-energy orbital in the ground state. If one constructs a hypothetical bosonic reference state constrained to reproduce the same electron density as the physical fermionic system, the total wavefunction would simply be $\sqrt{[\rho(\mathbf{r})/N)}$ (with N being the number of electrons). In this scenario, excluding quantum and electrostatic interactions, the total energy of the system would be purely steric energy. The spatial extent occupied by this bosonic state, denoted by a steric energy term, thus reflects an intrinsic property of the system. In this context, the steric energy quantifies the inherent spatial demand of the electron cloud, entirely isolated from quantum and electrostatic contributions. This property reinforces the conceptual soundness of the energy partitioning expressed in Eq. (11.6).

In Chapter 5, we have applied these two total energy partition schemes to study single-bond rotations and come up with a unified view about the origin of bond rotation barriers [53, 55–62]. Here, we examine the relative contribution of the three independent effects, steric, electrostatic, and quantum, to the S_N2 barrier heights in the gas phase for a series of S_N2 reactions (Figure 11.1). As mentioned in Section 11.1.4, Ingold and coworkers in the 1930s and 1940s discovered clear experimental evidence that bulky substituting groups on the reaction carbon center dramatically decrease S_N2 reactivity because of steric hindrance to backside attack [58].

We found that the steric effect dominates the barrier height, in accordance with the experimental findings of Ingold and coworkers. However, it is largely compensated by the opposing, stabilizing contribution from the quantum effect (Figure 11.2b). Moreover, we found that it is the positive contribution from the electrostatic effect that is best correlated with the S_N2 barrier height for all the systems that we investigated (Figure 11.2a). This latter result is consistent with what we discovered for the bond rotation barrier (Figure 5.2a).

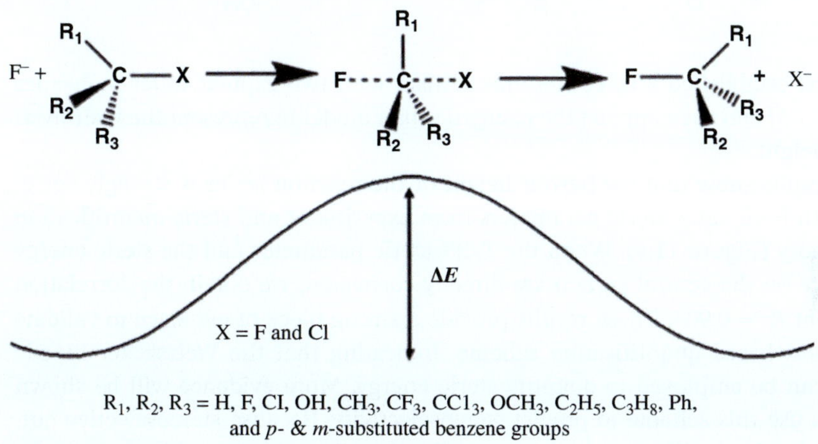

Figure 11.1 S_N2 systems studied to examine the impact of steric and other effects. *Source*: Reproduced from Ref. [58] with permission of American Chemical Society.

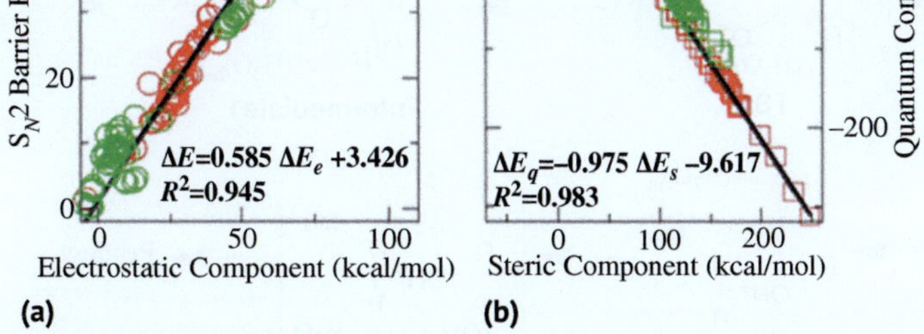

Figure 11.2 Strong linear correlations (a) between the transition state barrier height, ΔE, and the total electrostatic energy difference, ΔE_e, and (b) between the steric energy difference, ΔE_s, and the quantum energy difference, ΔE_q, for the S_N2 reactions in Figure 11.1. $CR_1R_2R_3F$ (red) and $CR_1R_2R_3Cl$ (green). *Source*: Reproduced from Ref. [58] with permission of American Chemical Society.

11.5 Validation by Taft's Steric Parameters

A more stringent test for the validity of our density-based quantification of steric effect is to correlate with Taft's steric parameters. This is what we did in a recent study [63]. In this work, we examined the acid-catalyzed hydrolysis of esters with 20 different substituent groups, whose experimental scale of Taft steric parameters was previously established in the literature.

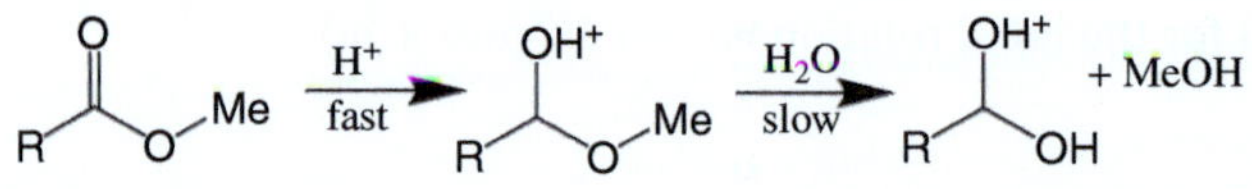

We first established a three-step mechanism with two explicit water molecules (Figure 11.3) and then applied the energetic span model to represent the hydrolysis barrier height.

Our results show that the barrier height of the reaction series is strongly correlated with both Taft's steric parameters from experiment and steric quantification from theory (Figure 11.4). When the Taft's steric parameter and the steric energy difference on the central carbon are directly correlated, we obtain the correlation coefficient $R^2 = 0.904$. These results provide a strong piece of evidence to validate our density-based quantification scheme, indicating that the Weizsäcker kinetic energy can be employed to quantify steric energy. More evidence will be shown below to use this scheme to predict stereoselectivity. Because stereoselective outcomes arise from differential steric crowding in competing transition states, the same density-based steric quantification naturally extends to reactions forming stereoisomeric products.

It should be noted, however, that the identification of E_s with T_W is most accurate for systems in which steric crowding is primarily determined by localized valence electron distributions. In delocalized π-systems, transition-metal complexes, or heavy-atom environments where d- and f-orbitals and relativistic effects contribute

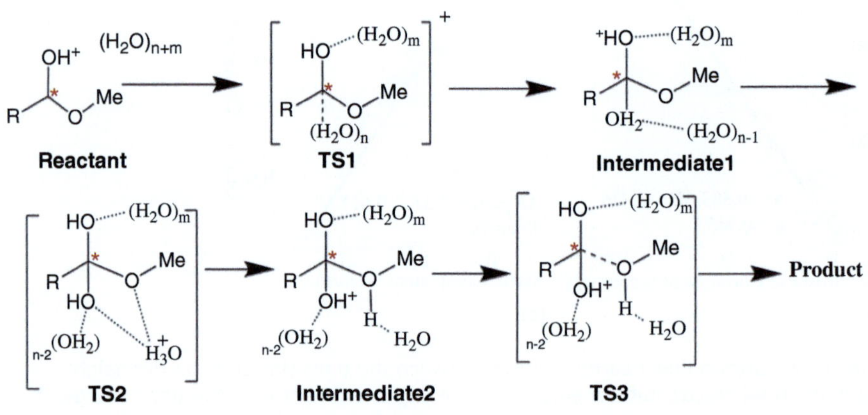

Figure 11.3 The three-step reaction mechanism of acid-catalyzed ester hydrolysis in water. *Source*: Reproduced from Ref. [63] with permission of The Royal Chemical Society.

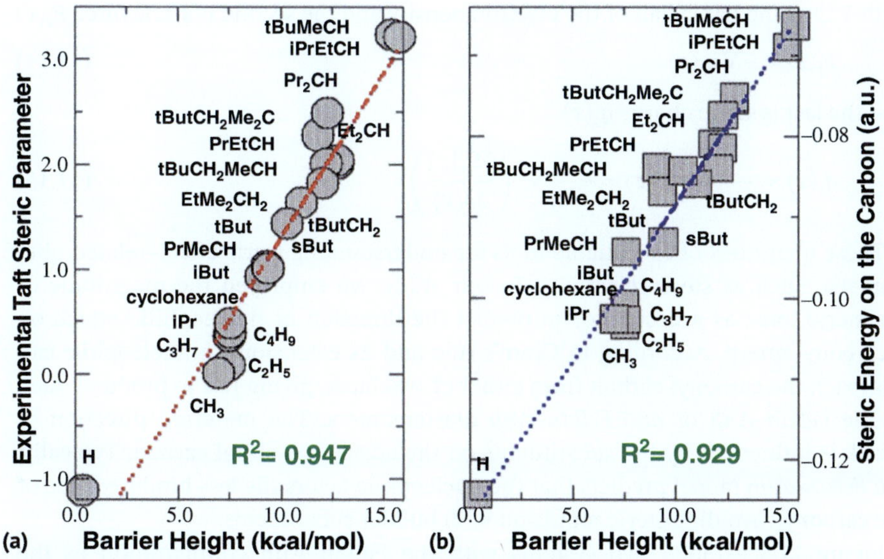

Figure 11.4 Strong linear correlations of the two-water hydrolysis barrier height with (a) experimental Taft parameters and (b) the difference of the steric energy between TS2 and the reactant in the two-water hydrolysis mechanism in Figure 11.3 on the central carbon atom for 20 alkyl groups. *Source*: Reproduced from Ref. [63] with permission of The Royal Chemical Society.

significantly, the purely gradient-based T_W form may underestimate steric contributions. Incorporating orbital deformation or nonlocal kinetic-energy corrections might improve the quantitative accuracy of this approach.

11.6 Stereoselectivity: A Density-based Description

Having validated the density-based quantification of steric energy, we now extend this concept to phenomena in which subtle differences in spatial crowding dictate distinct chemical outcomes—namely, stereoselectivity. Stereoselectivity can be viewed as the dynamic manifestation of steric asymmetry within a reaction's transition-state ensemble. By mapping these asymmetries through the steric potential, steric force, and steric charge, a direct density-based route emerges for predicting and interpreting stereochemical preferences.

From the DFT perspective, the density-based quantification of the steric effect, Eq. (11.12), is an explicit density functional, $E_s = E_s[\rho]$, suggesting that three associated physical quantities can be derived. The first one is the steric potential, which is the functional derivative of the steric energy with respect to the electron density [15, 64],

$$v_s(\mathbf{r}) = \frac{\delta E_s[\rho]}{\delta \rho(\mathbf{r})} = \frac{1}{8}\frac{|\nabla \rho(\mathbf{r})|^2}{\rho(\mathbf{r})} - \frac{1}{4}\frac{\nabla^2 \rho(\mathbf{r})}{\rho(\mathbf{r})}, \tag{11.13}$$

with $\nabla^2\rho$ as the Laplacian of the electron density and the second is steric force $\boldsymbol{F}_s(\mathbf{r})$

$$\boldsymbol{F}_s(\mathbf{r}) = -\nabla v_s(\mathbf{r}), \tag{11.14}$$

and the last is steric charge $q_s(\mathbf{r})$

$$q_s(\mathbf{r}) = -\frac{1}{4\pi}\nabla^2 v_s(\mathbf{r}) = -\frac{1}{4\pi}\nabla^2\left(\frac{\delta E_s[\rho]}{\delta\rho(\mathbf{r})}\right). \tag{11.15}$$

These quantities offer valuable tools for understanding steric effect–related phenomena, such as stereoselectivity. In our study, we employed the magnitude of the steric force as a descriptor to predict the direction of nucleophilic attack on a carbonyl group. According to Cram's rule and its extensions, a nucleophile can approach the carbonyl carbon from either of two faces, giving rise to products such as the *Felkin–Anh* or *anti-Felkin–Anh* diastereomers. The preferred direction of attack is influenced by the substituents on the adjacent α-chiral carbon. Typically, the *Felkin–Anh* model predicts that the nucleophile favors the less hindered face of the carbonyl, avoiding steric repulsion with bulkier substituents.

Figure 11.5 exhibits two systems with the Felkin–Anh configuration as the addition product. In these systems, the small group is H and the medium group is CH_3, but the large group differs. One is phenyl, and the other is chlorine. In both cases, the attack of the incoming nucleophile from the small group is much more likely than that from the large group site, as indicated by the magnitude of the steric force mapped onto the van der Waals surface in the Figure marked with the green arrows.

An even more interesting system is for the substituted norbornone (Figure 11.6), whose nucleophilic addition product with methyllithium depends on the nature

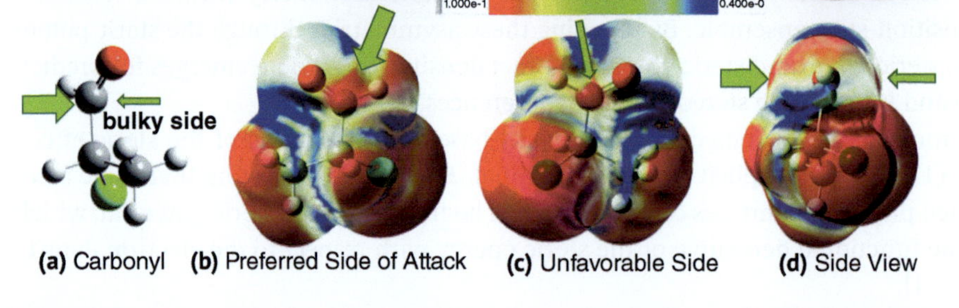

Figure 11.5 The magnitude of the steric force mapped onto the van der Waals surface for PhMeHC–CHO and ClMeHC–CHO molecules, whose nucleophilic addition product on the carbonyl group is known to be Felkin–Anh-like. *Source*: Reproduced from Ref. [65] with permission of The Royal Chemical Society.

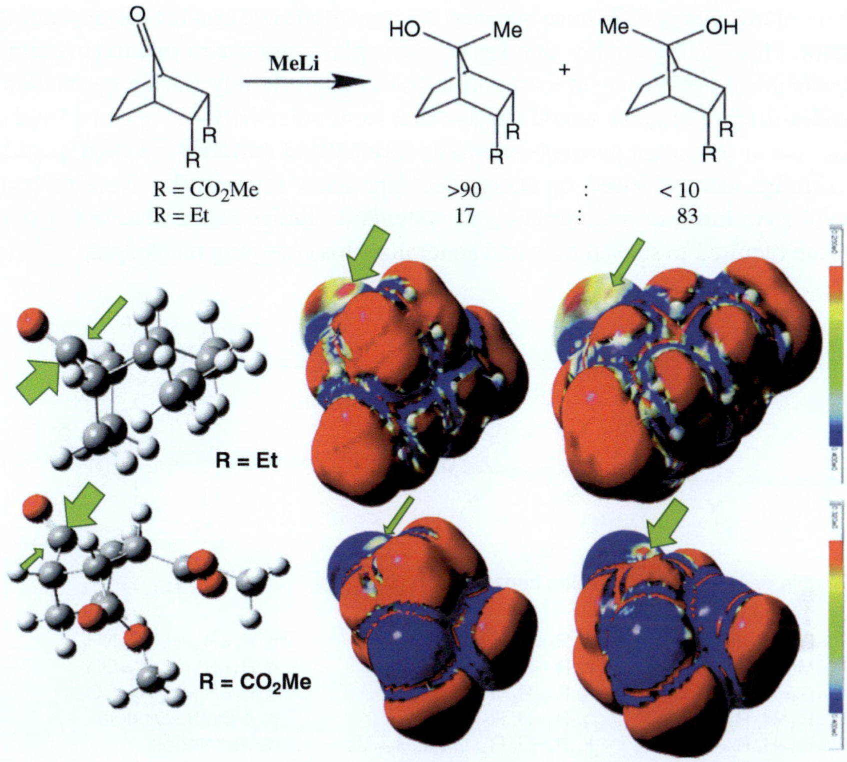

Figure 11.6 Dependence of carbonyl attack side on the substituted groups for norbornones. *Source*: Reproduced from Ref. [65] with permission of The Royal Chemical Society.

of the substituent group R at the C_β position. Using the steric force, we are able to explain the difference as exhibited in Figure 11.6. From the Figure, we find that when R = Et, the nucleophile prefers to attack from the other side of the substituent group, but when R = CO_2Me, the tendency is reversed, in consensus with the experimental evidence.

11.7 Quantification of Stereoselectivity

Is it possible to quantify stereoselectivity in the same manner as we did for electrophilicity and nucleophilicity in the previous chapter? This question remains largely unanswered, yet our preliminary findings suggest that such quantification may indeed be feasible. In an unpublished study, we investigated the stereochemical outcomes of reactions between 17 indole derivatives and 4,6-dinitrobenzofuroxan (see reaction schemes sketched below), both of which are achiral reactants that yield products containing two newly formed chiral centers. Remarkably, our results indicate that the difference in activation barriers between competing transition states—serving as a measure of stereoselectivity—correlates strongly $(R^2 \approx 0.80)$

with the steric energy difference between the transition state and the corresponding reactant. This finding implies that steric effects play a dominant role in governing stereochemical preference, at least within this reaction family. Such a correlation, if validated more broadly, would suggest that stereoselectivity—long considered a qualitative or empirical concept—could be rationalized and perhaps even quantified through density-based or energy-decomposition frameworks. Nevertheless, these observations are preliminary, and systematic studies across diverse reaction types are required to substantiate and generalize this emerging relationship.

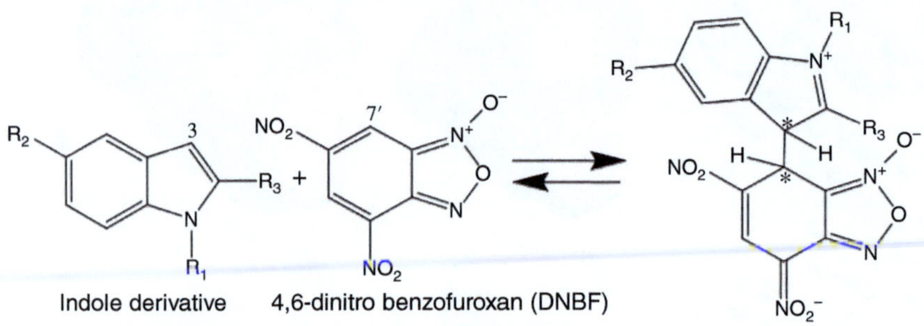

a. $R_1=R_2=R_3=H$
b. $R_1=Me$, $R_2=R_3=H$
c. $R_1=R_3=H$, $R_2=OMe$
d. $R_1=R_3=H$, $R_2=CN$
e. $R_1=R_3=H$, $R_2=Br$
f. $R_1=R_3=H$, $R_2=Cl$
g. $R_1=R_3=H$, $R_2=Me$
h. $R_1=H$, $R_2=Cl$, $R_3=Me$
i. $R_1=R_2=H$, $R_3=Me$
j. $R_1=H$, $R_2=R_3=Me$
k. $R_1=H$, $R_2=OMe$, $R_3=Me$
l. $R_1=R_3=Me$, $R_2=OMe$
m. $R_1=R_3=H$, $R_2=NH_2$
n. $R_1=R_3=H$, $R_2=OH$
o. $R_1=R_3=H$, $R_2=COOH$
p. 4-methoxyindole
q. 7-azaindole

11.8 Summary and Conclusions

Steric effect and stereoselectivity, once viewed as purely qualitative concepts, can now be understood within the same quantitative, density-based framework used previously for electrophilicity and nucleophilicity. Both originate from the electron density—its gradients, localization, and deformation—but illuminate complementary aspects of reactivity: the former governs how space is occupied, the latter how charge moves through it.

In this chapter, the steric effect is identified with the Weizsäcker kinetic energy, $E_s = T_W$, which represents the kinetic "pressure" enforcing the Pauli exclusion principle. Because the integrand of T_W is proportional to Fisher information, steric energy also measures the informational content of the density: the more localized the electrons, the higher both Fisher information and the steric repulsion. This equivalence naturally bridges the information-theoretic approach and orbital-free DFT, revealing that steric resistance and informational order describe the same underlying physics.

Correlation of T_W-derived steric energies with Taft's experimental parameters validates this formulation and shows that molecular crowding can be quantified from first principles. Extending the concept to stereoselectivity, the derived steric potential, steric force, and steric charge map how asymmetry in the electron density directs reaction pathways. Preliminary studies on indole derivatives confirm that differences in activation barriers track closely with differences in steric energy, providing a quantitative measure of stereochemical preference.

Compared with Chapter 10, the present study shifts the focus from energetic response to electron number to spatial response to density deformation. Together, these chapters define two orthogonal dimensions of chemical behavior—the charge axis and the space axis—through which ITA and OF-DFT converge. Both describe how electrons distribute, interact, and resist compression in space.

The success of this density-based approach lies in its unification of empirical reactivity notions with rigorous functionals of the electron density. By linking information content to kinetic energy and to observable chemical selectivity, it provides a coherent and predictive language for understanding how molecules organize and react. Thus, steric effects and stereoselectivity, long viewed as empirical and geometric in nature, emerge as rigorous manifestations of how electron density resists compression and asymmetrically reorganizes along reaction coordinates.

References

1 MacPhee JA, Panaye A, Dubois J-E. Steric effects—I. Tetrahedron. 1978 Jan;34 (24):3553–62.
2 Pinter B, Fievez T, Bickelhaupt FM, Geerlings P, De Proft F. On the origin of the steric effect. Phys Chem Chem Phys. 2012 Jul 28;14(28):9846–54.
3 Bott G, Field LD, Sternhell S. Steric effects. A study of a rationally designed system. J Am Chem Soc. 1980 Aug;102(17):5618–26.
4 Wagner JP, Schreiner PR. London dispersion in molecular chemistry--reconsidering steric effects. Angew Chem Int Ed. 2015 Oct 12;54(42):12274–96.
5 Hughes ED. Steric hindrance. Q Rev Chem Soc. 1948;2(2):107.
6 Ingold CK. Quantitative study of steric hindrance. Q Rev Chem Soc. 1957;11(1):1.
7 Napper DH. Steric stabilization. J Colloid Interface Sci. 1977 Feb;58(2):390–407.
8 Brown HC. Chemical effects of steric strains. J Chem Soc. 1956; 1248.
9 Harper KC, Bess EN, Sigman MS. Multidimensional steric parameters in the analysis of asymmetric catalytic reactions. Nat Chem. 2012 Mar 18;4(5):366–74.
10 Pophristic V, Goodman L. Hyperconjugation not steric repulsion leads to the staggered structure of ethane. Nature. 2001 May 31;411(6837):565–8.
11 Su P, Tang Z, Wu W. Generalized Kohn-Sham energy decomposition analysis and its applications. WIREs Comput Mol Sci. 2020 Sep;10(5):e1460.
12 Khaliullin RZ, Cobar EA, Lochan RC, Bell AT, Head-Gordon M. Unravelling the origin of intermolecular interactions using absolutely localized molecular orbitals. J Phys Chem A. 2007 Sep 13;111(36):8753–65.
13 Fang D, Piquemal J-P, Liu S, Cisneros GA. DFT-steric-based energy decomposition analysis of intermolecular interactions. Theor Chem Acc. 2014 May;133(5):1484.

14 Weisskopf VF. Of atoms, mountains, and stars: a study in qualitative physics. Science. 1975 Feb 21;187(4177):605–12.
15 Liu S. Steric effect: a quantitative description from density functional theory. J Chem Phys. 2007 Jun 28;126(24):244103.
16 Cameron MD. Victor Meyer and the thiophene compounds. J Chem Educ. 1949 Oct;26(10):521.
17 Shoppee CW. Christopher Kelk Ingold, 1893-1970. Biogr Mem Fellows R Soc. 1972 Nov;18:348–411.
18 Taft RW. Linear steric energy relationships. J Am Chem Soc. 1953 Sep;75(18): 4538–9.
19 Charton M. Steric effects. 13. Composition of the steric parameter as a function of alkyl branching. J Org Chem. 1978 Oct;43(21):3995–4001.
20 Charton M. Steric effects. I. Esterification and acid-catalyzed hydrolysis of esters. J Am Chem Soc. 1975 Mar;97(6):1552–6.
21 Tolman CA. Steric effects of phosphorus ligands in organometallic chemistry and homogeneous catalysis. Chem Rev. 1977 Jun;77(3):313–48.
22 Kellogg RM. Practical stereochemistry. Acc Chem Res. 2017 Apr 18;50(4):905–14.
23 Dalko PI, Moisan L. Enantioselective organocatalysis. Angew Chem Int Ed. 2001 Oct 15;40(20):3726–48.
24 Webb TH, Wilcox CS. Enantioselective and diastereoselective molecular recognition of neutral molecules. Chem Soc Rev. 1993;22(6):383.
25 Beletskaya IP, Nájera C, Yus M. Stereodivergent catalysis. Chem Rev. 2018 May 23;118(10):5080–200.
26 Mulzer J. Cram's rule: theme and variations. In: Waldman H, editor. Organic Synthesis Set. Wiley; 2003. pp. 3–8.
27 Smith RJ, Trzoss M, Bühl M, Bienz S. The cram rule revisited once more — revision of the Felkin–Anh model. Eur J Org Chem. 2002 Aug 1;2002(16): 2770–75.
28 Rosenberg RE, Kelly WJ. Felkin-Anh is not enough. J Phys Org Chem. 2015 Jan;28(1):47–56.
29 Knowles WS. Asymmetric hydrogenations (Nobel lecture). Angew Chem Int Ed. 2002 Jun 17;41(12):1998–2007.
30 Katsuki T, Sharpless KB. The first practical method for asymmetric epoxidation. J Am Chem Soc. 1980 Aug;102(18):5974–6.
31 List B. Introduction: organocatalysis. Chem Rev. 2007 Dec;107(12):5413–5.
32 Ouellet SG, Walji AM, MacMillan DWC. Enantioselective organocatalytic transfer hydrogenation reactions using Hantzsch esters. Acc Chem Res. 2007 Dec;40(12): 1327–39.
33 Tang W, Zhang X. New chiral phosphorus ligands for enantioselective hydrogenation. Chem Rev. 2003 Aug;103(8):3029–70.
34 Fernández I, Bickelhaupt FM, Svatunek D. Unraveling the Bürgi-Dunitz angle with precision: the power of a two-dimensional energy decomposition analysis. J Chem Theory Comput. 2023 Oct 24;19(20):7300–6.
35 Rodríguez HA, Bickelhaupt FM, Fernández I. Origin of the Bürgi-Dunitz angle. ChemPhysChem. 2023 Sep 1;24(17):e202300379.

36 Bubert C, Reiser O. Exceptionally high Felkin-Anh control for the addition of nucleophiles to a β-aminocyclopropylcarbaldehyde. Tetrahedron Lett. 1997 Jul;38 (28):4985–8.

37 Reetz MT. Structural, mechanistic, and theoretical aspects of chelation-controlled carbonyl addition reactions. Acc Chem Res. 1993 Sep;26(9):462–8.

38 Mejuch T, Gilboa N, Gayon E, Wang H, Houk KN, Marek I. Axial preferences in allylation reactions via the Zimmerman-Traxler transition state. Acc Chem Res. 2013 Jul 16;46(7):1659–69.

39 Bumbăcilă B, Putz MV. Steric Taft parameters. In: Putz MV, editor. New Frontiers in Nanochemistry: Concepts, Theories, and Trends. Includes bibliographical references and indexes. Contents: Volume 1. Structural Nanochemistry – Volume 2. Topological Nanochemistry – Volume 3. Sustainable Nanochemistry. Apple Academic Press; 2020. pp. 437–46.

40 Taft RW. Linear free energy relationships from rates of esterification and hydrolysis of aliphatic and ortho-substituted benzoate esters. J Am Chem Soc. 1952 Jun;74(11):2729–32.

41 Taft RW. Polar and steric substituent constants for aliphatic and o-benzoate groups from rates of esterification and hydrolysis of esters.[1]J Am Chem Soc. 1952 Jun;74(12):3120–8.

42 Solel E, Ruth M, Schreiner PR. London dispersion helps refine steric a-values: dispersion energy donor scales. J Am Chem Soc. 2021 Dec 15;143(49):20837–48.

43 Bumbăcilă B, Putz MV. Sterimol Verloop parameters. In: Putz MV, editor. New Frontiers in Nanochemistry: Concepts, Theories, and Trends. Includes bibliographical references and indexes. | Contents: Volume 1. Structural Nanochemistry – Volume 2. Topological Nanochemistry – Volume 3. Sustainable Nanochemistry. Apple Academic Press; 2020. pp. 447–56.

44 Parr RG, Yang W. Density-Functional Theory of Atoms and Molecules. Oxford University Press; 1989.

45 Weizsäcker CF. Zur Theorie der Kernmassen Z Phys. 1935;96:431–58; Bethe HA, Bacher RF. Nuclear physics A. Stationary states of nuclei. Rev Mod Phys. 1935;8(2):82.

46 Rong C, Wang B, Zhao D, Liu S. Information-theoretic approach in density functional theory and its recent applications to chemical problems. WIREs Comput Mol Sci. 2020 Jul;10(4):e1461.

47 Liu S. Information-theoretic approach in density functional reactivity theory. Acta Phys-Chim Sin. 2016;32(1):98–118.

48 Finley KJ, He X, Liu W, Ayers PW, Liu S. Some recent advances of information-theoretic approach in density functional theory. J Chem Sci. 2025 Sep 26;137(4):86.

49 Nagy Á. Fisher information and density functional theory. Int J Quantum Chem. 2022 Apr 15;122(8):e26679.

50 Liu S. On the relationship between densities of Shannon entropy and Fisher information for atoms and molecules. J Chem Phys. 2007 May 21;126(19):191107.

51 Nagy Á. Fisher information in density functional theory. J Chem Phys. 2003 Nov 8;119(18):9401–5.

52 Ludeña EV, Torres FJ, Becerra M, Rincón L, Liu S. Shannon entropy and Fisher information from a non-Born–Oppenheimer perspective. J Phys Chem A. 2020 Jan 16;124(2):386–94.

53 Esquivel RO, Liu S, Angulo JC, Dehesa JS, Antolín J, Molina-Espíritu M. Fisher information and steric effect: study of the internal rotation barrier of ethane. J Phys Chem A. 2011 May 5;115(17):4406–15.

54 Bader RFW. Atoms in Molecules. Oxford University Press; 1989.

55 Tsirelson VG, Stash AI, Liu S. Quantifying steric effect with experimental electron density. J Chem Phys. 2010 Sep 21;133(11):114110.

56 Liu S, Govind N. Toward understanding the nature of internal rotation barriers with a new energy partition scheme: ethane and n-butane. J Phys Chem A. 2008 Jul 24;112(29):6690–9.

57 Huang Y, Zhong A-G, Yang Q, Liu S. Origin of anomeric effect: a density functional steric analysis. J Chem Phys. 2011 Feb 28;134(8):084103.

58 Liu S, Hu H, Pedersen LG. Steric, quantum, and electrostatic effects on S(N)2 reaction barriers in gas phase. J Phys Chem A. 2010 May 13;114(18):5913–8.

59 Ess DH, Liu S, De Proft F. Density functional steric analysis of linear and branched alkanes. J Phys Chem A. 2010 Dec 16;114(49):12952–7.

60 Torrent-Sucarrat M, Liu S, De Proft F. Steric effect: partitioning in atomic and functional group contributions. J Phys Chem A. 2009 Apr 16;113(15):3698–702.

61 Liu S. Origin and nature of bond rotation barriers: a unified view. J Phys Chem A. 2013 Feb 7;117(5):962–5.

62 Wang B, Liu S, Lei M, De Proft F. Steric effect and intrinsic electrophilicity and nucleophilicity from a conceptual density functional theory and information-theoretic approach as quantitative probes of chemical reactions. Chem Eur J. 2024 Jul 3;30:e202401295.

63 Zhang J, He X, Wang B, Rong C, Zhao D, Liu S. Density-based quantification of steric effect: validation by Taft steric parameters from acid-catalyzed hydrolysis of esters. Phys Chem Chem Phys. 2024;26:13447–56.

64 Liu S, Liu L, Yu D, Rong C, Lu T. Steric charge. Phys Chem Chem Phys. 2018 Jan 17;20(3):1408–20.

65 Liu S, Rong C, Lu T. Electronic forces as descriptors of nucleophilic and electrophilic regioselectivity and stereoselectivity. Phys Chem Chem Phys. 2017 Jan 4;19(2):1496–503.

12

Acidity and Basicity

TABLE OF CONTENTS

12.1 Introduction

Building on the density-based analyses of bonding, cooperative interactions, chirality, charge-transfer reactivity, and steric effects developed in the preceding chapters, we now turn to acidity and basicity as additional manifestations of how electron density responds to proton or electron-pair transfer.

12.1.1 Types of Acids and Bases

Acidity and basicity [1] are fundamental chemical concepts that permeate virtually every branch of chemistry, including organic synthesis, catalysis, environmental science, and biochemistry. At their essence, these concepts describe the molecular propensity to donate or accept protons (H^+) or electron pairs. The most widely taught framework is the Brønsted–Lowry theory [2, 3], which defines acids as proton donors and bases as proton acceptors. This approach is especially useful in aqueous environments and is quantitatively linked to measurable properties such as pK_a, which reflects the position of the acid dissociation equilibrium.

Complementing the Brønsted–Lowry model is the Lewis theory [4], which offers a broader perspective: Lewis acids accept electron pairs, while Lewis bases donate them. This definition accommodates a vast array of chemical processes beyond proton transfer, including coordination chemistry, metal–ligand interactions, and organic substitution reactions. It is particularly valuable in explaining reactivity in non-protic solvents, inorganic complexes, and organometallic systems, where traditional acid–base definitions may not apply.

Beyond these classical theories, several specialized models have been developed to address specific chemical environments. The Lux–Flood definition [5], for instance, is used in high-temperature oxide systems and focuses on the transfer of oxide ions (O^{2-}). The Usanovich model [6] generalizes further by encompassing any exchange of positive or negative species, including electrons, but its breadth limits practical use. Importantly, while acidity and basicity are often associated with aqueous systems, they are also quantifiable in the gas phase using thermodynamic parameters such as proton affinity and gas-phase acidity, which provide solvent-independent benchmarks. In applied settings, empirical scales like the Guttmann donor and acceptor numbers [7] and the Drago–Wayland model [8] offer practical tools for ranking Lewis acidity (LA) and Lewis basicity (LB), especially in solvent-dependent or coordination-driven contexts. These diverse definitions set the stage for integrating proton-transfer and electron-pair-transfer acidity and basicity within a unified density-based reactivity framework.

12.1.2 Brief History of Acids and Bases

Early treatments of acidity and basicity were purely empirical and lacked molecular interpretation. It was not until the late nineteenth and early twentieth centuries that formal theories emerged, beginning with Arrhenius [9] and later the Brønsted–Lowry [10–12] and Lewis [13, 14] frameworks.

The first theoretical framework emerged in 1884, when Svante Arrhenius proposed [9] that acids are substances that increase the concentration of hydrogen ions (H^+) in aqueous solution, while bases increase hydroxide ions (OH^-). This model introduced the idea of ionization as central to acid–base behavior and provided a scientific basis for phenomena observed in water. However, its limitation to aqueous systems and to hydroxide-based bases prevented it from explaining broader acid–base chemistry.

In 1923, Johannes Brønsted and Thomas Lowry [10–12], working independently, advanced a more inclusive model by defining acids as proton donors and bases as proton acceptors. This Brønsted–Lowry theory extended acid–base chemistry beyond water and accounted for reactions such as ammonia acting as a base without containing OH^-. That same year, Gilbert N. Lewis [4, 10, 13, 14] introduced an even more general model, in which acids are electron pair acceptors and bases are electron pair donors. The Lewis theory encompasses all Brønsted–Lowry acids and bases while also describing a wide range of non–proton-transfer reactions, such as metal–ligand coordination and organic electrophilic/nucleophilic interactions, making it a cornerstone of chemical reactivity and mechanism analysis.

12.1.3 Integrating Acid–Base Theories: A Density-based Perspective

Although Brønsted–Lowry and Lewis theories were developed to address different chemical contexts, all ultimately describe how electrons reorganize in response to proton or electron-pair transfer. From the standpoint of density-based reactivity theory [15, 16], these models can be viewed as complementary projections of the same underlying phenomenon: changes in the electron density $\rho(\mathbf{r})$ and the external potential $v(\mathbf{r})$.

Brønsted–Lowry acidity reflects how easily electron density reorganizes to release a proton; LA reflects how the density accommodates an incoming electron pair. Thus, the diverse acid–base concepts share a common structure through conceptual density functional theory (CDFT) [17–21] and information-theoretic approach (ITA) [22, 23] descriptors. This unifying viewpoint motivates the second half of this chapter, where molecular electrostatic potential (MEP) [24–26], natural atomic orbital (NAO) energies, ITA quantities, and machine-learning models provide quantitative, transferable descriptors of both Brønsted–Lowry and LA/LB, bridging classical concepts with density-based reactivity theory.

12.2 Brønsted–Lowry Acidity and Basicity

At its heart, the Brønsted–Lowry theory is built around proton transfer: Brønsted–Lowry acid is a species that donates a proton (H^+) and a Brønsted–Lowry base is a species that accepts a proton (H^+). The reaction involves the transfer of a proton from the acid to the base:

$$HA + B \rightleftharpoons A^- + BH^+ \tag{12.1}$$

Here, HA is the acid, B is the base, A^- is the conjugate base, and BH^+ is the conjugate acid. This framework introduces the conjugate acid–base pair concept: every acid has a conjugate base, and every base has a conjugate acid, connected by the reversible loss or gain of a proton.

The Brønsted–Lowry theory is often limited to acid–base reactions in water. In aqueous solution, water itself acts as both an acid and a base:

$$H_2O + H_2O \rightleftharpoons H_3O^+ + OH^- . \tag{12.2}$$

In nonaqueous solvents, such as liquid ammonia (NH_3) or sulfuric acid (H_2SO_4), acid–base strengths change significantly due to differing solvent basicity, polarity, and auto-protolysis constants. Thus, acid–base behavior is relative to the solvent system in question.

12.2.1 Quantification

Acidity, the propensity of a Brønsted–Lowry acid to donate a proton, can be quantified thermodynamically by the acid dissociation reaction:

$$HA \rightleftharpoons H^+ + A^-, \tag{12.3}$$

with the equilibrium constant K_a defined as

$$K_a = \frac{[H^+][A^-]}{[HA]}. \tag{12.4}$$

Acidity is defined as the logarithmic form, the pK_a, provides a more convenient scale:

$$pK_a = -\log_{10} K_a. \tag{12.5}$$

Basicity is the measure of a base's ability to accept a proton (H^+). In aqueous solution, the strength of a base is quantified by its base dissociation constant (K_b), which describes the equilibrium of the reaction between a base (B) and water:

$$B + H_2O \rightleftharpoons BH^+ + OH^-, \tag{12.6}$$

with

$$K_b = \frac{[BH^+][OH^-]}{[B]}, \tag{12.7}$$

and

$$pK_b = -\log_{10} K_b. \tag{12.8}$$

Because every base has a conjugate acid, its basicity is also related to the pK_a of the conjugate acid:

$$pK_a + pK_b = 14 (\text{at } 25\ °\text{C in water}). \tag{12.9}$$

Stronger acids (bases) have smaller $pK_a(pK_b)$ values and larger $K_a(K_b)$ values. Stronger bases are typically conjugate to weaker acids, and vice versa. The relative strength of an acid or base is environment-dependent, particularly influenced by solvent, temperature, and ionic strength.

Relevant to acidity and basicity is the pH value,

$$pH = -\log_{10} a_{H^+} \approx -\log_{10}[H^+], \tag{12.10}$$

where a_{H+} is the activity of hydrogen ions (H^+) in solution. The pH value is directly related to acidity and basicity, but it specifically applies to aqueous solutions and

Brønsted–Lowry acid–base behavior (i.e., involving proton transfer). A solution with pH < 7 is acidic; pH = 7 is neutral (pure water at 25 °C); and pH > 7 is basic.

We can use pH values to categorize acids and bases. A strong acid (e.g., HCl, HNO_3) dissociates H^+ completely, producing more H^+ and lower pH value. A weak acid (e.g., CH_3COOH) dissociates H^+ partially, producing fewer H^+ and a moderate pH value. A strong base (e.g., NaOH) produces more OH^- anions and a high pH value, whereas a weak base (e.g., NH_3) produces fewer OH^- ions and only a slightly basic pH.

Superacids and superbases [27] represent the extremes of Brønsted–Lowry acid–base chemistry, far beyond the strengths of conventional acids like sulfuric acid or bases like sodium hydroxide. A superacid, such as fluoroantimonic acid ($HSbF_6$), is stronger than 100% sulfuric acid and can protonate even hydrocarbons, with acidity quantified by the Hammett acidity function (H_0) rather than pH. For example, sulfuric acid has an H_0 of −12, while $HSbF_6$ has an H_0 value near −28. In contrast, superbases like *n*-butyllithium and lithium diisopropylamide have extremely high proton affinity and can deprotonate very weak acids. These substances challenge the limits of proton donation and acceptance, and they play a crucial role in advanced synthetic chemistry and fundamental studies of reactivity.

12.2.2 Simulations

Computational chemistry provides powerful tools to quantify and predict acidity (via pK_a) and basicity (via pK_b) [24–26, 28–30], which are critical for understanding molecular reactivity, mechanism elucidation, and catalyst design. Simulating these properties typically involves calculating free energy changes $\Delta G°$ associated with proton transfer reactions, Eq. (12.3), in the gas phase or solution.

$$pK_a = \frac{\Delta G°}{2.303RT}, \tag{12.11}$$

where R is the gas constant and T the temperature (usually 298.15 K). Accurate pK_a predictions thus rely on precise calculation of absolute or relative free energies of the acid, base, and their conjugates.

Gas-Phase Approach. In the gas phase [31, 32], the deprotonation free energy is computed directly:

$$\Delta G°_{gas} = G°(A^-) + G°(H^+) - G°(HA), \tag{12.12}$$

where G is the absolute Gibbs free energy including electronic, zero-point, thermal, and entropic contributions, and the free energy of the proton in the gas phase is known, $G(H^+)_{gas} = -6.28$ kcal/mol. This approach yields gas-phase acidity or proton affinity, but not aqueous pK_a, since solvation is ignored.

Thermodynamic Cycle for Aqueous pK_a. To simulate solution-phase acidity (e.g., aqueous pK_a), the following thermodynamic cycle (below) is employed.

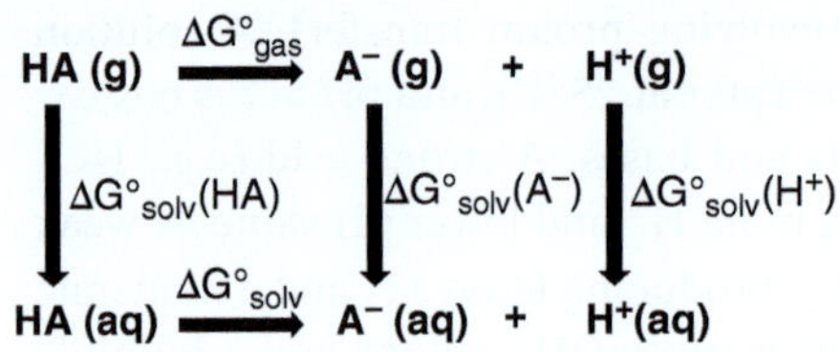

It combines gas-phase free energies with solvation free energies to calculate the pK_a value,

$$pK_a = \frac{1}{2.303RT}\left[\Delta G^{\circ}{}_{\text{gas}} + \Delta G^{\circ}{}_{\text{solv}}\right], \tag{12.13}$$

where

$$\Delta G^{\circ}{}_{\text{solv}} = \Delta G^{\circ}{}_{\text{solv}}(\text{A}^-) + \Delta G^{\circ}{}_{\text{solv}}(\text{H}^+) - \Delta G^{\circ}{}_{\text{solv}}(\text{HA}), \tag{12.14}$$

with $\Delta G^{\circ}{}_{\text{solv}}(\text{H}^+) = -265.9$ kcal/mol at 298K.

Isodesmic and Relative Methods. To reduce systematic errors, relative pK_a calculations use reference compounds with known experimental pK_a:

$$\Delta pK_a = \frac{\Delta\Delta G^{\circ}}{2.303RT}, \tag{12.15}$$

where $\Delta\Delta G^{\circ}$ is the difference in deprotonation free energies between target and reference. This method cancels out some computational errors, especially for systems with a similar structure and solvation behavior.

The accuracy of computational pK_a predictions depends on the choice of method, basis set, and solvation model [33]. Density functional theory (DFT) with medium-sized basis sets often achieves errors of around 1–2 pK_a units for small molecules, but larger errors may occur for complex systems.

Despite significant advancements, computationally simulating Brønsted–Lowry acidity and basicity presents several challenges. The accurate description of solvation effects remains a primary hurdle. Implicit solvent models, while efficient, may not always capture the full complexity of solvent–solute interactions, particularly in highly charged or sterically hindered environments. Explicit solvent models, while more accurate, are computationally demanding, and achieving sufficient sampling of conformational space for reliable free energy calculations can be a significant bottleneck, especially for larger systems. Furthermore, the choice of density functionals (in DFT) or force field parameters (in MM) can significantly impact the accuracy of the calculated pK_a values. There is no single "universal" functional or force field that performs optimally for all chemical systems and environments. Therefore, careful validation against experimental data is crucial. The treatment of zero-point energy and vibrational entropy contributions, particularly for proton transfer, also requires careful consideration.

To address these challenges, advanced computational techniques are continuously developed [34]. Ab initio molecular dynamics (AIMD), where forces are computed on-the-fly from quantum mechanics (QM) calculations, allows for a more dynamic and realistic exploration of the potential energy surface, including the proton transfer event itself. However, AIMD simulations are extremely computationally expensive and typically limited to very short timescales and small systems. Enhanced sampling techniques, such as metadynamics, umbrella sampling, or replica exchange molecular dynamics, are increasingly being combined with both explicit solvent and QM/MM simulations to overcome sampling limitations and efficiently explore the free energy landscape of proton transfer reactions. These methods help to accelerate the exploration of rare events like proton hopping. Machine learning (ML) is also emerging as a powerful tool. ML models, trained on large datasets of experimentally determined or accurately computed pK_a values, can rapidly predict acidity and basicity for new molecules, often with high accuracy. While current ML models are largely data-driven and may lack the atomic-level mechanistic insights provided by ab initio methods, their speed makes them ideal for high-throughput screening.

12.3 CDFT and Brønsted–Lowry Acidity and Basicity

Due to the limitations of directly simulating pK_a values using Eq. (12.11), an alternative approach is to bypass explicit free energy calculations by identifying linear correlations between pK_a values and quantum descriptors. Previously used descriptors include the highest occupied molecular orbital (HOMO) energies, localized reactive orbitals, frontier effective-for-reaction molecular orbitals, electrophilicity, and group-philicity, among others [29, 35]. The underlying rationale is that proton or electron donor–acceptor reactions are largely governed by frontier molecular orbitals. However, such correlations have typically been observed only within structurally related compounds, such as phenols.

Recently, we proposed using two interdependent quantum descriptors to accurately and efficiently estimate molecular pK_a values. These descriptors are the MEP at the nucleus of the acidic atom (e.g., N, O, or S) and the sum of its valence NAOs [24–26, 28]. This approach is grounded in the idea that acidity is a local property centered on the acidic atom, with environmental effects captured through their influence on that atom's electronic structure.

To numerically validate this rationale, we examined a diverse set of 228 molecules, including 154 primary, secondary, and tertiary amines and anilines; 59 carboxylic acids and alcohols; and 15 sulfonic acids and thiols [26]. All calculations were performed using DFT at the B3LYP/6-311+G(2d,2p) level for both geometry optimization and harmonic vibrational analysis. Molecular properties, such as the MEP at each nucleus and NAO energies, were obtained through a full natural bond orbital analysis based on the optimized structures.

Figure 12.1 illustrates strong correlations between experimental pK_a values and each of the two descriptors—MEP and NAO—across the three compound categories. Furthermore, when the MEP of each acidic atom is adjusted by subtracting the corresponding value for the isolated atom, a unique linear relationship emerges between the resulting MEP difference and the experimental pK_a values. This approach enables simultaneous estimation of pK_a values at multiple sites from a single calculation, making it suitable for both small drug-like molecules and amino acids in proteins or other macromolecules. Similar findings have been reported by

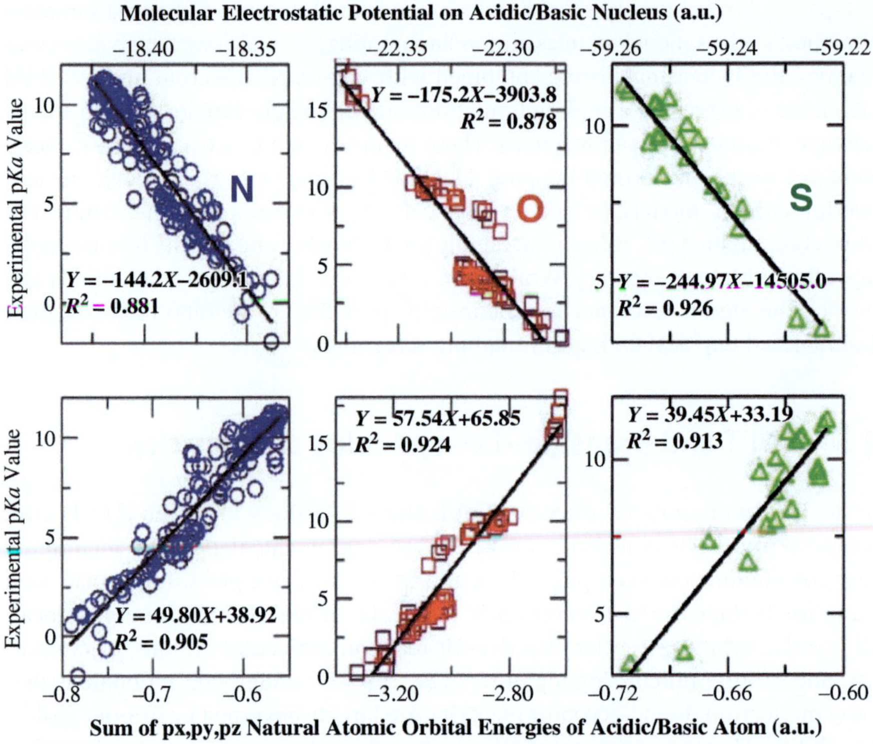

Figure 12.1 Linear relationships between molecular electrostatic potential on acidic nucleus and experimental pK_a values for amines (N), carboxylic acids and alcohols (O), and sulfonic acids and thiols (S) (upper panel); and linear relationships between the sum of three valence NAO 2p/3p orbitals and pK_a values (lower panel). *Source*: Reproduced from [26] with permission of American Chemical Society.

us for various substituted benzoic, sulfinic, seleninic acids, phenols, alkylcarboxylic acids, 20 natural α-amino acids, five DNA/RNA bases, and several other biologically relevant compounds [20]. Together, these results highlight the significant potential of MEP and NAO descriptors in understanding and predicting molecular acidity and basicity.

Why does this approach work so well? What is its theoretical foundation? In CDFT, the total energy change ΔE is related to changes in the number of electrons ΔN and the external potential $\Delta v(\mathbf{r})$, which reads:

$$\begin{aligned}\Delta E &= E[v(\boldsymbol{r}) + \Delta v(\boldsymbol{r}); N + \Delta N] - E[v(\boldsymbol{r}); N] \\ &= \left\{ \left(\frac{\partial E}{\partial N}\right)_{v(\boldsymbol{r})} \Delta N + \int \left(\frac{\delta E}{\delta v(\boldsymbol{r})}\right)_N \Delta v(\boldsymbol{r}) d\boldsymbol{r} \right\} \\ &\quad + \frac{1}{2}\left[\left(\frac{\partial^2 E}{\partial N^2}\right)(\Delta N)^2 + 2\Delta N \int \frac{\delta \partial E}{\delta v(\boldsymbol{r}) \partial N} \Delta v(\boldsymbol{r}) d\boldsymbol{r} \right. \\ &\quad \left. + \int \left(\frac{\delta^2 E}{\delta v(\boldsymbol{r}') \delta v(\boldsymbol{r})}\right)_N \Delta v(\boldsymbol{r}) \Delta v(\boldsymbol{r}') d\boldsymbol{r} d\boldsymbol{r}' \right] \\ &\quad + (\text{third and higher order terms})\end{aligned} \tag{12.16}$$

where the terms in braces {} and brackets [] are the first- and second-order terms, respectively. In Chapter 2, the physical meaning of all derivatives in Eq. (12.16) has been extensively discussed. Focusing on the proton dissociation process of an acid as described in Eq. (12.3), we assume that Eq. (12.16) remains applicable. During this process, the total number of electrons remains fixed with only the external potential $\Delta v(\mathbf{r})$ undergoing change. With $\Delta N = 0$, and retaining only the first-order term in the Taylor expansion, Eq (12.16) is simplified as

$$\Delta E\left[\Delta v(\boldsymbol{r})\right] = \int \rho(\boldsymbol{r})\Delta v(\boldsymbol{r})d\boldsymbol{r}. \tag{12.17}$$

With one proton dissociated from the acid, $\Delta v(\boldsymbol{r})$ can be approximated by

$$\Delta v(\boldsymbol{r}) = \sum_{i \neq H} \frac{Z_i}{|\boldsymbol{R}_i - \boldsymbol{R}_H|} - \frac{1}{|\boldsymbol{r} - \boldsymbol{R}_H|}, \tag{12.18}$$

where $\mathbf{R}_H$ represents the coordinate of the leaving proton and $\mathbf{R}_i$ denotes that of other atoms in the acid. Combining Eqs. (12.17) and (12.18) and approximating pK_a values by this energy change, we have [25]

$$pK_a \propto \Delta E \approx \sum_{i \neq H} \frac{Z_i}{|\boldsymbol{R}_i - \boldsymbol{R}_H|} - \int \frac{\rho(\boldsymbol{r})}{|\boldsymbol{r} - \boldsymbol{R}_H|} d\boldsymbol{r}. \tag{12.19}$$

The physical meaning of the last two quantities is the electrostatic potential experienced by the departing proton in the acid, which has been shown to closely correlate with the potential at the acidic atom itself. This explains why MEP serves as a reliable descriptor for estimating molecular acidity. Since the dominant contribution to the MEP arises from the valence orbitals of both the leaving proton and the acidic atom, it is unsurprising that the sum of NAOs functions as an equivalent descriptor to MEP.

It is important to note that the derivation of Eq. (12.19) relies on several key assumptions: (i) thermodynamic contributions from solvent and temperature effects are considered negligible, and entropic effects are assumed to be minor. As a result, the total electronic energy difference in Eq. (12.17) is used to approximate the Gibbs free energy change in Eq. (12.11). (ii) The first-order term of the Taylor expansion in Eq. (12.16) is deemed sufficient. (iii) Structural relaxation of the conjugate base A^- following proton dissociation in Eq. (12.3) has minimal impact on the external potential, implying that the change in external potential is primarily attributed to the removal of the proton from the acid.

In another study [25], we investigated 27 metal–water complexes of the form $M(H_2O)_n{}^{m+}$, where M represents metal cations, including Al, Ba, Be, Ca, Co, Cr, Fe, Ga, Hf, In, K, Li, Mg, Mn, Na, Ni, Sc, Sr, Ti, Tl, Zn, and Zr, with charges ranging from +1 (e.g., Li^+, K^+, Na^+, Tl^+) to +4 (e.g., Hf^{4+}, Sn^{4+}, Ti^{4+}, Zr^{4+}). Most complexes are hexacoordinated, except $Be(H_2O)_4{}^{2+}$ and $Tl(H_2O)_4{}^{3+}$. The experimental pK_a values for these complexes span nearly 19 orders of magnitude, from −4.0 for $Ti(H_2O)_6{}^{4+}$ to +14.7 for $Na(H_2O)_6{}^{+}$. Strong linear correlations were observed between experimental pK_a values and three quantum descriptors: the MEP at the oxygen nucleus, the sum of valence NAO energies of oxygen, and the chemical potential in CDFT. Similar trends were also found for the departing proton. These linear relationships enable accurate reproduction of experimental pK_a values. Notably, the predictive performance of our method is statistically comparable to that of the conventional thermodynamic cycle approach, while simultaneously offering significant advantages in simplicity and computational efficiency. In later studies [24], we extended to examine basicity and include ITA descriptors to enable more accurate pK_a predictions.

12.4 LA and LB

At its core, Lewis acid–base theory is centered on the fundamental process of electron pair donation and acceptance. A Lewis acid is defined as an electron pair acceptor. This species must have an available empty orbital that can accommodate an electron pair from a donor molecule. The key characteristic of a Lewis acid is its electron deficiency, which creates an electrophilic center capable of forming coordinate bonds. A Lewis base is defined as an electron pair donor. This species must possess at least one lone pair of electrons that can be donated to form a coordinate bond with an acceptor molecule. The Lewis base acts as a nucleophile, seeking electron-deficient centers with which to interact.

For a species to act as a Lewis acid, it must possess an incomplete octet, an energetically accessible empty orbital, and sufficient electrophilic character to attract an electron pair. A Lewis base, in contrast, must contain at least one lone pair, have suitable orbital geometry for overlap with the acceptor, and display adequate nucleophilicity to donate electrons. In molecular-orbital terms, a Lewis acid–base interaction arises from overlap between the base's filled lone-pair orbital and the acid's empty orbital, forming bonding and antibonding combinations; occupation of the bonding orbital stabilizes the system. The interaction strength is governed by the donor–acceptor energy match and the extent of orbital overlap.

Lewis acids commonly include electron-deficient species such as Group 13 halides—BF_3, $AlCl_3$, and related compounds—whose central atoms possess incomplete octets and empty p orbitals capable of accepting electron pairs [36–40]. Metal cations, especially transition-metal ions like Fe^{3+}, Cu^{2+}, Zn^{2+}, and Ag^{+}, are also strong Lewis acids due to their positive charge and accessible d orbitals. Lewis bases, by contrast, are species with lone-pair electron density, exemplified by ammonia, water, alcohols, ethers, and amines, as well as most anions such as halides and hydroxide, whose high charge density promotes electron donation. Unsaturated systems—including alkenes, alkynes, and aromatic rings—can likewise act as Lewis bases through donation of π-electron density, a feature central to metal–π coordination in organometallic chemistry, as seen in benzene complexes like ferrocene.

12.4.1 Quantification

Unlike Brønsted–Lowry acids and bases, which can be quantified using straightforward pH measurements and equilibrium constants, Lewis acids and bases present unique challenges for quantitative assessment. The absence of a universal solvent system and the diversity of Lewis acid–base interactions make it difficult to establish a single, comprehensive scale of LA and LB. Despite these challenges, several approaches to quantify and compare Lewis acid–base strength have been developed [36–38].

Thermodynamic Approaches. The most direct thermodynamic measure of Lewis acid–base strength is the enthalpy change ΔH associated with adduct formation, $A + B \rightarrow A - B$. When a Lewis acid A and a Lewis base B combine to form an adduct $A - B$, the enthalpy of formation (ΔH) provides a direct measure of the interaction strength. More negative ΔH values indicate stronger Lewis acid–base interactions. This approach has been extensively used to rank Lewis acids and bases, particularly in gas-phase studies where complications from solvation are minimized. Calorimetric studies have established extensive databases of formation enthalpies for various acid–base pairs. The equilibrium constant, $K = \left[A - B\right] / \left[A\right]\left[B\right]$, for Lewis acid–base adduct formation provides another quantitative measure. The larger the equilibrium constant, the stronger the Lewis acid–base interaction. The relationship between the equilibrium constant and Gibbs free energy ($\Delta G° = -RT \ln K$) allows for the calculation of interaction strengths under standard conditions.

Spectroscopic Methods. IR spectroscopy provides valuable quantitative information about Lewis acid–base interactions. When a Lewis base coordinates to a Lewis acid, characteristic vibrational frequencies often shift. For example, carbonyl stretching frequencies in metal carbonyls decrease upon coordination, N—H stretching frequencies in amines shift upon coordination to Lewis acids, and the magnitude of these shifts correlates with the strength of the Lewis acid–base interaction. NMR spectroscopy also offers several quantitative measures. For instance, chemical shift changes upon coordination provide information about electron density redistribution, coupling constants can indicate the strength of coordinate bonds, and relaxation times can provide information about the dynamics of Lewis acid–base interactions. Additionally, UV–vis spectroscopy can quantify Lewis acid–base interactions through charge-transfer bands that appear upon complex formation, or changes in d–d transition energies in metal complexes or shifts in $\pi \rightarrow \pi^*$ transitions in organic Lewis bases upon coordination. The Gutmann–Beckett method and the Childs method utilize these spectroscopic changes to gauge Lewis acid strength. The fluorescent Lewis adduct method is another technique that measures Lewis acid strength by observing the degree of a bathochromic shift in a fluorescent probe.

Empirical Scales and Parameters. For example, Gutmann donor and acceptor numbers developed by Viktor Gutmann are one of the most widely used quantitative scales for Lewis acids and bases [7, 39]. A donor number (DN) measures LB by determining the enthalpy of interaction between the Lewis base and antimony pentachloride ($SbCl_5$) in 1,2-dichloroethane solution. The DN scale uses triethylphosphine oxide as a reference (DN = 38.8). Examples of DN values: dimethyl sulfoxide (DMSO): DN = 29.8; acetonitrile: DN = 14.1; and benzene: DN = 0.1. On the other hand, an acceptor number (AN) measures LA by determining the ^{31}P NMR chemical shift of triethylphosphine oxide when coordinated to the Lewis acid. The AN scale uses $SbCl_5$ as a reference (AN = 100). Examples of AN values: boron trifluoride: AN = 89; aluminum trichloride: AN = 87; and water: AN = 54.8.

The ECW Model. The electrostatic-covalent-W (ECW) model, also known as the Drago–Wayland equation [8], is a prominent semi-quantitative model designed to describe and predict the strength of Lewis acid–base interactions. It quantifies the enthalpy of adduct formation ($-\Delta H$) by assigning two parameters to each acid and base: E parameters (E_A and E_B) representing the electrostatic contributions to the bond strength and C parameters (C_A and C_B) denoting the covalent contributions to the bond strength. The model's equation,

$$-\Delta H = E_A E_B + C_A C_B + W, \tag{12.20}$$

where W accounts for a constant energy contribution, such as the cleavage of a dimeric acid or base prior to adduct formation. The ECW model effectively accommodates the limitation of single-parameter descriptions, demonstrating that at least two properties (electrostatic and covalent) are necessary to accurately define the order of Lewis acid or base strength. This model can reproduce and predict the enthalpy of reaction between many acids and bases, serving as a measure of bond strength in the gas phase and weakly solvating media, while largely ignoring entropy effects. The E and C parameters are derived from enthalpies of adduct formation, where the bond between the acid and base is a sigma (σ) interaction and there is no steric repulsion.

12.4.2 Frontiers of Lewis Acidity/Basicity Research

Though simple in definition, Lewis acid–base theory continues to evolve far beyond its classical roots. No longer confined to coordination chemistry or catalysis, it now plays a central role in emerging areas such as metal-free small molecule activation, main group redox systems, and smart materials design. Advances in spectroscopy, theory, and computation have pushed the boundaries of what qualifies as a Lewis acid or base, revealing new reactivity patterns and applications. This discussion highlights some of the most dynamic frontiers in Lewis acid/base research today.

Frustrated Lewis Pairs (FLPs). FLPs [41–43], characterized by Lewis acids and bases in close proximity that are sterically hindered from forming a stable adduct, have emerged as a powerful paradigm for metal-free transformations. This inherent "frustration" keeps the acidic and basic sites reactive, enabling cooperative activation of small molecules such as H_2, CO_2, N_2O, and even CH_4—reactions once thought exclusive to transition metal chemistry. A typical FLP mechanism involves heterolytic H_2 cleavage to generate protic and hydridic species, which then engage in sequential proton and hydride transfer steps. More recently, mechanistic studies have uncovered that some FLPs operate via single electron transfer from base to acid, forming frustrated radical pairs. These radical species open new avenues in organic synthesis, including C—H bond activation and C—C bond formation, and are reshaping our understanding of nonmetal reactivity.

Lewis Superacids and Superbases. Lewis superacids [44] are defined as electron pair acceptors with greater strength than the benchmark Lewis acid, antimony pentafluoride (SbF_5). These highly reactive species are capable of stabilizing elusive intermediates such as carbocations, radical cations, and even free protons in nonaqueous environments. Unlike conventional Lewis acids, superacids operate effectively in weakly coordinating media and facilitate transformations once thought to require transition metal catalysis. Notable examples include perfluorinated boranes, aluminum halides combined with fluoride scavengers, and metal-free systems like $B(C_6F_5)_3$. Their exceptional reactivity has opened new avenues in hydrocarbon activation, superelectrophilic catalysis, and the development of advanced solid acid materials. In contrast, Lewis superbases are extraordinarily strong electron pair donors, capable of deprotonating even very weak acids or engaging in the activation of inert chemical bonds. These include compounds such as proton sponges, guanidines, phosphazenes, and highly basic organolithium or organopotassium reagents.

Single-atom Catalysts (SACs). SACs [45] represent a cutting-edge class of materials in which isolated metal atoms are dispersed on supports, offering maximum atomic efficiency and unique catalytic properties distinct from nanoparticles or bulk metals. These individual atoms often serve as *atomically dispersed Lewis acid or base sites*, enabling highly selective and tunable reactivity at the molecular level. The local coordination environment—such as metal–N_4, metal–O_4, or heteroatom anchoring—plays a crucial role in dictating the acid/base character and catalytic behavior. By precisely engineering these sites, SACs can be designed to mimic enzyme active centers or transition-metal complexes, achieving exceptional performance in applications ranging from CO_2 reduction and hydrogen evolution to selective oxidations and C—H activation. The integration of atomic-level control with acid–base functionality marks a powerful frontier in heterogeneous catalysis and materials chemistry.

Solvent-free and Mechanochemical Lewis Chemistry. It is an emerging frontier that challenges traditional solution-based paradigms by enabling reactions through solid-state grinding, milling, or compression—without the use of solvents [46, 47]. In this context, Lewis acids and bases are employed either as co-reactants or catalysts to facilitate bond formation under mechanical force, offering greener, more sustainable alternatives to conventional methods. Mechanochemistry allows for unique activation modes, enhanced reaction rates, and access to otherwise inaccessible products, all while minimizing waste and avoiding solvation effects that might suppress reactivity. Recent advances have demonstrated successful applications in C—C and C—N bond formation, polymerization, and solid-state organocatalysis, underscoring the potential of this approach to transform synthetic strategies across organic and inorganic chemistry.

Quantifying LA and LB now integrates advanced experimental probes with accurate computational descriptors, enabling more precise and context-dependent assessments [36–38]. Experimentally, methods such as ^{31}P NMR shifts with Et_3PO (Gutmann–Beckett), UV–vis and IR spectral changes, and calorimetry-based donor/acceptor numbers have refined acidity–basicity scales across different environments. Computationally, DFT-derived quantities—including the electrophilicity index (ω), Fukui functions, and electrostatic potential maps—correlate well with observed trends, while binding energies with standardized probes (e.g., NH_3 or F^-) provide direct thermodynamic comparisons. ML models trained on large datasets further extend these predictions across broad chemical spaces. Together, these developments are bringing LA/LB toward more quantitative, transferable, and predictive frameworks that better connect conceptual theory with practical reactivity.

Additionally, the integration of Lewis acid/base principles into advanced materials, such as metal–organic frameworks (MOFs) and novel polymerization techniques, is yielding materials with tailored properties. Crucially, the convergence of computational chemistry and ML, powered by unprecedented datasets like Open Molecules 2025 [48], accelerates the prediction, design, and understanding of Lewis acid/base systems. Finally, the growing recognition of Lewis acid/base roles in biological systems is leading to breakthroughs in enzyme catalysis, drug discovery, and advanced biosensing technologies. These advancements collectively underscore the enduring significance and transformative potential of Lewis acid/base chemistry in addressing complex scientific challenges.

12.5 HSAB Principle

The hard–soft acid–base (HSAB) principle [49–51] offers a qualitative framework to help rationalize the stability, selectivity, and reactivity of chemical interactions between Lewis acids and bases. Proposed by Ralph G. Pearson in 1963 [52–54], the HSAB principle categorizes Lewis acids and bases into hard, soft, and borderline types based on their electronic and structural properties, such as size, charge density, and polarizability. It refines the classical Lewis acid–base theory by introducing the idea that not all acid–base pairs interact with equal affinity, and that matching the "hardness" or "softness" of an acid and a base significantly influences their tendency to form stable complexes. This principle is particularly valuable in inorganic, organic, and coordination chemistry, as it provides insights into why certain acid–base pairs form stable complexes while others do not.

Pearson's HSAB principle grew out of earlier empirical work, particularly the 1958 Ahrland–Chatt–Davies classification of metal ions into "Class (a)" and "Class (b)" types [55]. Class-(a) ions such as alkali metals, alkaline-earths, and high-valent first-row transition metals (e.g., Fe^{3+}, Co^{3+}) prefer hard donor atoms like N, O, and F, whereas Class-(b) ions such as Cu^+, Ag^+, Hg^+, Pt^{2+}, and Pd^{2+} favor softer donors like P, S, and I. Pearson reframed these empirical trends into a general theory by introducing hardness and softness, rooted in fundamental electronic properties such as charge density and polarizability, thereby formalizing the "like binds like" rule and providing a predictive framework for acid–base stability and reactivity.

HSABs differ mainly in size, charge density, electronegativity, and especially polarizability: hard species are small, highly charged or electronegative, and weakly polarizable (e.g., H^+, Li^+, F^-, OH^-), while soft species are larger, less charged or electronegative, and highly polarizable (e.g., Cu^+, Ag^+, I^-, SCN^-). Borderline cases such as Zn^{2+}, Fe^{2+}, Br^-, and pyridine occupy the continuum between these extremes.

These electronic characteristics dictate bonding: hard–hard interactions are predominantly ionic and governed by electrostatics, whereas soft–soft interactions involve stronger covalent and π-bonding contributions. Frontier molecular orbital theory provides support: large HOMO–LUMO gaps correspond to hard species and small gaps to soft ones. Although primarily qualitative, attempts at quantification include Pearson's absolute hardness and the Drago EC model.

Within CDFT, HSAB acquires a formal quantum-mechanical foundation. Pearson's absolute hardness is defined as the second derivative of the electronic energy with respect to electron number, $\eta = (\partial^2 E/\partial N^2)_v$, leading to the practical approximation $\eta \approx (I - A)$, or, under frozen-orbital assumptions, $\eta \approx \varepsilon_{LUMO} - \varepsilon_{HOMO}$. This establishes a direct link between hardness, HOMO–LUMO gap, and polarizability, anchoring the empirical HSAB concept in rigorous electronic structure theory.

HSAB theory has wide-ranging applications: it predicts metal–ligand preferences, complex stability, linkage isomerism (e.g., SCN^- binding modes), and symbiotic effects in coordination chemistry; rationalizes nucleophilic site selectivity in organic reactions, such as 1,2- versus 1,4-addition in α,β-unsaturated carbonyls; and explains biological phenomena like heavy-metal toxicity and the design of effective chelators. It also guides catalyst and ligand design, accounts for solubility trends (e.g., insoluble AgI from soft–soft interactions), and helps rationalize the geochemical distribution of metal ores. Despite this utility, HSAB remains qualitative: borderline species behave inconsistently, steric and solvent effects can override hardness/softness, classifications do not reflect intrinsic acid/base strength, and well-known exceptions (e.g., H_2 stability or cases conflicting with Fajans's rules) highlight its role as a flexible heuristic rather than a strict predictive law.

12.6 CDFT, ITA, and LA and LB

The success of MEP and NAO descriptors [24–26] in predicting Brønsted–Lowry pK_a values naturally raises the question of whether these local density-based quantities also encode LA and LB. Conceptually, proton transfer and electron-pair donation both probe the electron density distribution near the reactive center. Hence, descriptors rooted in $\rho(\mathbf{r})$, its gradients, and local external potential should be transferable across these acid–base frameworks. In Section 12.3, we used CDFT to analyze Brønsted–Lowry acidity and basicity, while Chapter 10 employed ITA to study electrophilicity and nucleophilicity [56, 57]. These two sets of concepts are closely connected to LA and LB. In this section, we integrate both CDFT and ITA frameworks to provide a quantitative description of LA and LB.

In this work [58], LA and LB are quantified through the equilibrium constant K of the association reaction between a Lewis acid A and a Lewis base B,

$$A + B \overset{K}{\Longleftrightarrow} AB, \tag{12.21}$$

with $K = [AB]/[A][B]$ using equilibrium concentrations. This reaction represents the formation of a Lewis acid–base adduct via the interaction of an electron pair acceptor with an electron pair donor. A larger equilibrium constant corresponds to a stronger acid–base interaction, and thus greater LA or LB. Following the formulation by Mayr and coworkers, this relationship can be expressed as

$$\lg K = \text{LA} + \text{LB}. \tag{12.22}$$

Because the values of LA and LB are defined in terms of specific acid–base pairings, there is no absolute or universal scale for LA or LB; instead, these quantities are inherently relative and context-dependent.

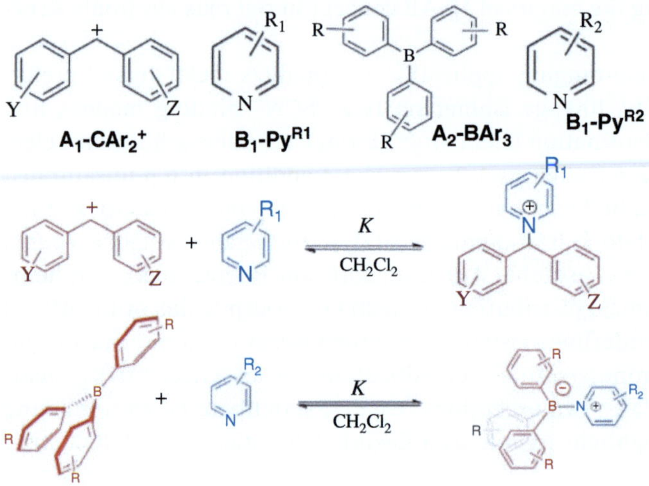

Shown above are the systems examined in this work, whose experimental scales of LA and LB are available in the literature. It consists of two sets of Lewis acids, including 14 benzhydryl cations (A_1-CAr_2^+) and 7 triarylboranes (A_2-BAr_3), respectively, interacting with two sets of Lewis bases, including 18 pyridines $\left(B_1\text{-}Py^{R1}\right)$ and 13 pyridines $\left(B_2\text{-}Py^{R2}\right)$.

We begin by evaluating the effectiveness of MEP and NAO descriptors in capturing LA and LB. As shown in Figure 12.2a and b, MEP exhibits strong linear correlations with both LA and LB, with correlation coefficients $\left(R^2\right)$ exceeding 0.91. Specifically, Figure 12.2a correlates the MEP at the acidic carbon center in the A1 – CAr_2^+ series with LA, while Figure 12.2b relates the MEP at the basic nitrogen center in the B1–pyR_1 series to LB. These trends mirror our earlier findings for Brønsted–Lowry acidity and basicity. Notably, the slopes of these correlations are opposite in sign—positive for acids and negative for bases—highlighting the directional nature of electrostatic potential in acid–base interactions. Furthermore, the strong linear relationships observed in Figure 12.2c and d demonstrate that NAO descriptors correlate similarly with LA and LB, reinforcing their close parallelism with MEP. Together, these results confirm that both MEP and NAO values at the reactive centers are reliable descriptors for assessing not only Brønsted–Lowry acidity and basicity but also their Lewis counterparts.

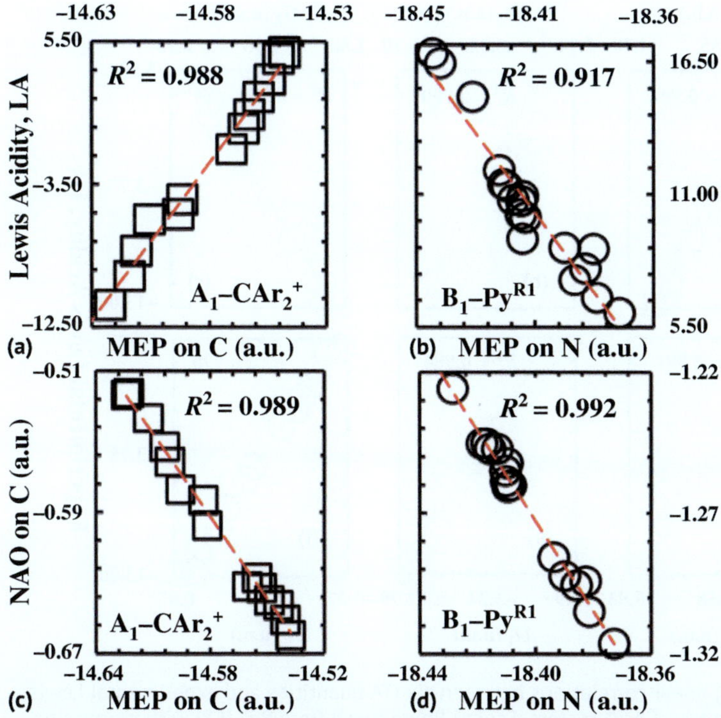

Figure 12.2 Strong linear correlations between MEP and NAO values at reactive centers and experimental Lewis acidity/basicity values. *Source*: Reproduced from Ref. [58] with permission of John Wiley & Sons.

Turning to ITA descriptors, Figure 12.3 presents a series of strong linear correlations between LA for the A1 – CAr_2^+ series and six ITA quantities evaluated at the acidic carbon center: Shannon entropy (Figure 12.3a), relative Rényi entropy of order 2 (Figure 12.3b), G_2 function (Figure 12.3c), Ghosh–Berkowitz–Parr entropy (Figure 12.3d), alternative relative Fisher information (Figure 12.3e), and Fisher information (Figure 12.3f). Each correlation exhibits an R^2 value greater than 0.98, indicating excellent agreement between the descriptors and experimentally derived LA values. Similar results were obtained for other Lewis acids and bases.

With only a few exceptions, the majority of ITA descriptors show consistently strong linear correlations with experimental measures across all systems studied. Notably, these correlations are generally stronger and more uniform than those observed for electrophilicity and nucleophilicity, reflecting the more intrinsic thermodynamic nature of LA and LB. In contrast, electrophilicity and nucleophilicity appear more sensitive to kinetic influences and specific reaction environments. These results underscore the effectiveness of ITA quantities—along with MEP and NAO descriptors—as powerful and reliable tools for quantifying LA and LB.

Although LA and LB are inherently relational—lacking a universal scale due to their dependence on specific acid–base pairings—it is still worth asking whether a unified predictive model can be constructed by combining LA and LB values. To investigate this possibility, we utilized the two LA/LB datasets from this study to generate 343

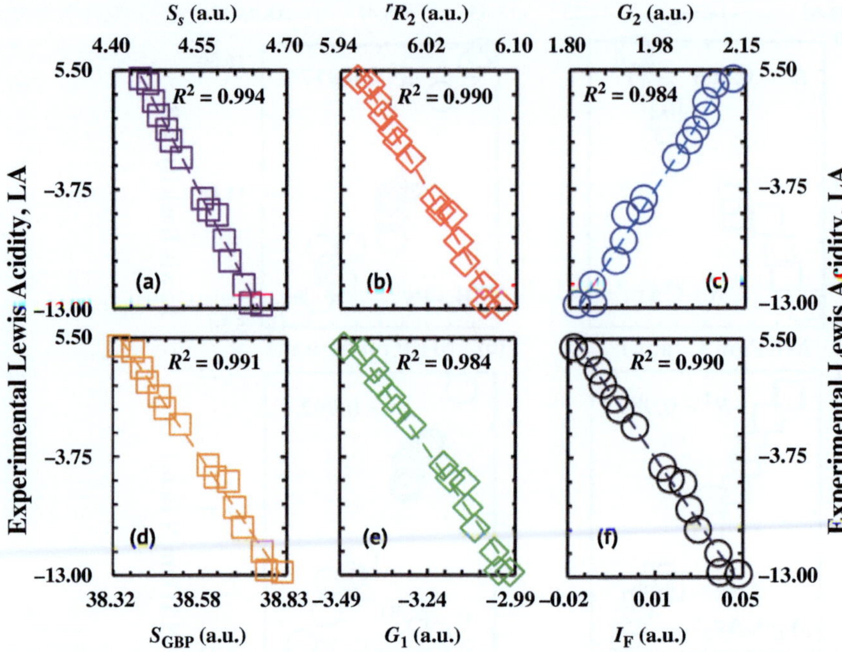

Figure 12.3 Strong linear correlations between six ITA quantities and experimental Lewis acidity scales for the $A_1 - CAr_2^+$ system. *Source*: Reproduced from Ref. [58] with permission of John Wiley & Sons.

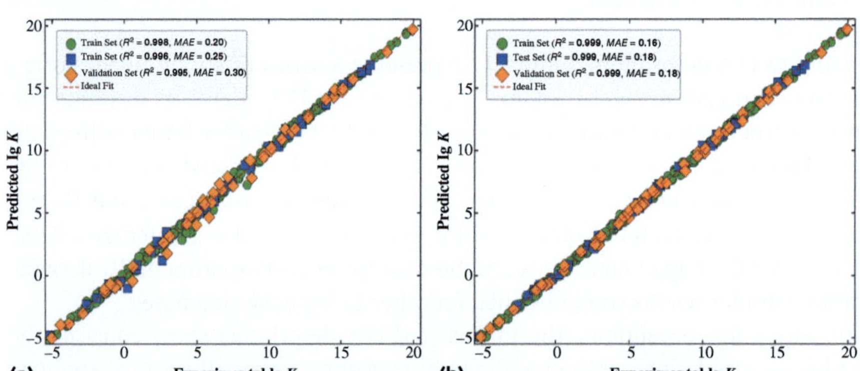

Figure 12.4 Comparison of predicted and experimental lg K values using the XGBoost model from machine learning, with two descriptors (a) MEP_a+MEP_b; (b) S_{GBPa}+S_{GBPb} as the features. *Source*: Reproduced from Ref. [58] with permission of John Wiley & Sons.

equilibrium constant data points ($14\times18 + 7\times13$ for lg K) and applied an XGBoost ML model to assess the feasibility of such an approach. Remarkably, the model was able to accurately predict lg K using any combination of one descriptor from the Lewis acid and one from the Lewis base as input features. Two representative cases are shown in Figure 12.4: one using MEP values (Figure 12.4a) and the other using S_{GBP}

values (Figure 12.4b), both achieving excellent predictive performance with R^2 values exceeding 0.99 and mean absolute errors below 0.30. These results demonstrate that, despite the absence of absolute LA or LB scales, any well-chosen pair of descriptors capturing the essential electronic characteristics of the acid and base can be effectively employed in ML models to predict reaction equilibria with high accuracy.

12.7 Outlook: A Unified View of Acid–Base Chemistry

The developments reviewed in this chapter reveal that classical and modern views of acidity and basicity share a common density-based foundation. Acidity and basicity have traditionally been introduced through multiple conceptual frameworks—Brønsted–Lowry proton transfer, Lewis electron-pair donation, and the HSAB principle. Each highlights a particular projection of chemical reactivity, but none alone captures the full complexity of how molecules reorganize their electron density when forming or breaking acid–base interactions. The results presented in this chapter reveal that these classical viewpoints, rather than being independent or competing theories, can be understood as different manifestations of a single, deeper principle: acid–base chemistry is governed by how electron density $\rho(\mathbf{r})$ responds to changes in the external potential $v(\mathbf{r})$ at reactive centers.

Within this density-based perspective, the unifying quantities are the MEP, the valence NAOs, and the information-theoretic descriptors that collectively encode the local electronic environment. These descriptors naturally bridge Brønsted–Lowry and Lewis frameworks, capturing both proton transfer and electron-pair donation within the same mathematical and conceptual structure. Their strong linear correlations with experimental acidity/basicity scales reflect a universal dependence on local electronic structure rather than mechanistic details of a particular reaction class. Meanwhile, CDFT provides a rigorous foundation for the HSAB principle, quantifying hardness, softness, and charge-transfer propensity from energy derivatives. Together, these developments demonstrate that acid–base reactivity is fundamentally a density-driven phenomenon.

The success of this unified treatment also clarifies methodological implications. Explicit free-energy simulations, while valuable for benchmarking and mechanistic analysis, are no longer necessary as the default strategy for acidity/basicity prediction. Local descriptors—MEP, NAOs, ITA indices, and CDFT quantities—offer a dramatically simpler, transferable, and computationally efficient alternative. Moreover, machine-learning models built on these descriptors can predict equilibrium constants across acid–base pairs with high accuracy, even in the absence of universal LA/LB scales. This synergy between physical insight and data-driven approaches provides a powerful route for rapid screening, catalyst design, and reactivity prediction in complex chemical spaces.

Looking ahead, extending these density-based descriptors and ML models to excited states, complex environments (enzymes, interfaces, porous materials), and strongly cooperative systems may reveal new organizing principles for proton and electron-pair transfer. The interplay between local descriptors and emergent phenomena such as cooperativity, frustration, and hierarchical organization discussed in other chapters of this book offers a promising route toward a genuinely predictive and conceptually rich acid–base chemistry.

References

1 Himmel D, Radtke V, Butschke B, Krossing I. Basic remarks on acidity. Angew Chem Int Ed. 2018 Apr 9;57(16):4386–411.

2 Morris RH. Brønsted-Lowry acid strength of metal hydride and dihydrogen complexes. Chem Rev. 2016 Aug 10;116(15):8588–654.

3 Vallejo Narváez WE, Jiménez EI, Romero-Montalvo E, Sauza-de la Vega A, Quiroz-García B, Hernández-Rodríguez M, et al. Acidity and basicity interplay in amide and imide self-association. Chem Sci. 2018 May 21;9(19):4402–13.

4 Jensen WB. The Lewis acid–base definitions: a status report. Chem Rev. 1978 Feb;78(1):1–22.

5 Ottonello G, Moretti R. Lux-Flood basicity of binary silicate melts. J Phys Chem Solids. 2004 Aug;65 (8–9):1609–14.

6 Reiners CS, Marniok K, Müller S. The Usanovich definition of acids and bases – a forgotten pioneer of the donor—acceptor principle. CHEMKON. 2022 Mar;29(2):57–63.

7 Schmeisser M, Illner P, Puchta R, Zahl A, van Eldik R. Gutmann donor and acceptor numbers for ionic liquids. Chem Eur J. 2012 Aug 27;18(35):10969–82.

8 Sarkar M. The Drago–Wayland equation. Reson. 2024 May 7;29(4):503–15.

9 Kolb D. Acids and bases. J Chem Educ. 1978 Jul;55(7):459.

10 Paik S-H. Understanding the relationship among Arrhenius, Brønsted–Lowry, and Lewis theories. J Chem Educ. 2015 Sep 8;92(9):1484–9.

11 Kauffman GB. The Bronsted-Lowry acid base concept. J Chem Educ. 1988 Jan;65(1):28.

12 Puchtler H, Meloan SN, Spencer M. Current chemical concepts of acids and bases and their application to anionic ("acid") and cationic ("basic") dyes. Histochemistry. 1985;82(4):301–6.

13 Laurence C, Graton J, Gal J-F. An overview of Lewis basicity and affinity scales. J Chem Educ. 2011 Dec 1;88(12):1651–7.

14 Shaffer AA. Let us give Lewis acid–base theory the priority it deserves. J Chem Educ. 2006 Dec;83(12):1746.

15 Rong C, Zhao D, He X, Liu S. Development and applications of the density-based theory of chemical reactivity. J Phys Chem Lett. 2022 Dec 8;13(48):11191–200.

16 He X, Li M, Rong C, Zhao D, Liu W, Ayers PW, et al. Some recent advances in density-based reactivity theory. J Phys Chem A. 2024 Feb 22;128(7):1183–96.

17 Parr RG, Yang W. Density-functional theory of atoms and molecules. Oxford University Press; 1989.

18 Geerlings P, De Proft F. Conceptual DFT: the chemical relevance of higher response functions. Phys Chem Chem Phys. 2008 Jun 7;10(21):3028–42.

19 Liu S. Conceptual density functional theory and some recent developments. Acta Phys-Chim Sin. 2009;25(3):590–600.

20 Liu S, editor. Conceptual Density Functional Theory: Towards a New Chemical Reactivity Theory. 1st ed. Wiley-VCH GmbH: Wiley; 2022.

21 Geerlings P, Chamorro E, Chattaraj PK, De Proft F, Gázquez JL, Liu S, et al. Conceptual density functional theory: status, prospects, issues. Theor Chem Acc. 2020 Feb;139(2):36.

22 Rong C, Wang B, Zhao D, Liu S. Information-theoretic approach in density functional theory and its recent applications to chemical problems. WIREs Comput Mol Sci. 2020 Jul;10(4):e1461.

23 Rong C, Yu D, Liu S. Information-theoretic approach. In: Liu S, editor. Conceptual Density Functional Theory: Towards a New Chemical Reactivity Theory. Wiley; 2022. p. 281–99.

24 Cao X, Rong C, Zhong A, Lu T, Liu S. Molecular acidity: an accurate description with information-theoretic approach in density functional reactivity theory. J Comput Chem. 2018 Jan 15;39(2):117–29.

25 Liu S, Schauer CK, Pedersen LG. Molecular acidity: a quantitative conceptual density functional theory description. J Chem Phys. 2009 Oct 28;131(16):164107.

26 Liu S, Pedersen LG. Estimation of molecular acidity via electrostatic potential at the nucleus and valence natural atomic orbitals. J Phys Chem A. 2009 Apr 16;113(15):3648–55.

27 Srivastava AK, Misra N. Superbases and superacids form supersalts. Chem Phys Lett. 2016 Jan;644:1–4.

28 Huang Y, Liu L, Liu W, Liu S, Liu S. Modeling molecular acidity with electronic properties and Hammett constants for substituted benzoic acids. J Phys Chem A. 2011 Dec 29;115(51):14697–707.

29 Burger SK, Liu S, Ayers PW. Practical calculation of molecular acidity with the aid of a reference molecule. J Phys Chem A. 2011 Feb 24;115(7):1293–304.

30 Yu D, Du R, Xiao J-C, Xu S, Rong C, Liu S. Theoretical study of pK_a values for trivalent rare-earth metal cations in aqueous solution. J Phys Chem A. 2018 Jan 18;122(2):700–7.

31 Huang Y, Liu L, Liu S. Towards understanding proton affinity and gas-phase basicity with density functional reactivity theory. Chem Phys Lett. 2012 Feb;527:73–8.

32 Yang W, Mortier WJ. The use of global and local molecular parameters for the analysis of the gas-phase basicity of amines. J Am Chem Soc. 1986 Sep 1; 108(19):5708–11.

33 Seybold PG, Shields GC. Computational estimation of pK_a values. WIREs Comput Mol Sci. 2015 May;5(3):290–7.

34 Simon C, Ciccotti G, Klein ML. Computing the acidity of liquids via ab initio molecular dynamics. ChemPhysChem. 2007 Oct 8;8(14):2072–6.

35 Ramalho TC, Pereira DH. Understanding the substituent effect on the acidity of alcohols and *para*-substituted phenols. Mol Simul. 2009 Dec;35(15):1269–78.

36 Mayer RJ, Hampel N, Ofial AR. Lewis acidic boranes, Lewis bases, and equilibrium constants: a reliable scaffold for a quantitative Lewis acidity/basicity scale. Chem Eur J. 2021 Feb 24;27(12):4070–80.

37 Mayr H, Ammer J, Baidya M, Maji B, Nigst TA, Ofial AR, et al. Scales of Lewis basicities toward C-centered Lewis acids (carbocations). J Am Chem Soc. 2015 Feb 25;137(7):2580–99.

38 Mayer RJ, Ofial AR, Mayr H, Legault CY. Lewis acidity scale of diaryliodonium ions toward oxygen, nitrogen, and halogen Lewis bases. J Am Chem Soc. 2020 Mar 18;142(11):5221–33.

39 Sivaev IB, Bregadze VI. Lewis acidity of boron compounds. Coord Chem Rev. 2014 Jul; 270–271:75–88.

40 Li HL, Kuninobu Y, Kanai M. Lewis acid–base interaction-controlled ortho-selective C-H borylation of aryl sulfides. Angew Chem Int Ed. 2017 Feb 1;56(6):1495–9.

41 Stephan DW, Erker G. Frustrated Lewis pair chemistry: development and perspectives. Angew Chem Int Ed. 2015 May 26;54(22):6400–41.

42 Stephan DW. Frustrated Lewis pairs: from concept to catalysis. Acc Chem Res. 2015 Feb 17;48(2):306–16.

43 Stephan DW. The broadening reach of frustrated Lewis pair chemistry. Science. 2016 Dec 9;354(6317):1248.

44 Lv W, Dai Y, Guo R, Su Y, Ruiz DA, Liu LL, et al. Geometrically constrained organoboron species as Lewis superacids and organic superbases. Angew Chem Int Ed. 2023 Sep 4;62(36):e202308467.

45 Yang X-F, Wang A, Qiao B, Li J, Liu J, Zhang T. Single-atom catalysts: a new frontier in heterogeneous catalysis. Acc Chem Res. 2013 Aug 20;46(8):1740–8.

46 Bui M, Heinekamp C, Fuhry E, Weidner S, Radnik J, Ahrens M, et al. Lewis-acid induced mechanochemical degradation of polyvinylidene fluoride: transformation into valuable products. Chem Sci. 2025 Oct 15;16(40):18903–10.

47 Do J-L, Friščić T. Mechanochemistry: a force of synthesis. ACS Cent Sci. 2017 Jan 25;3(1):13–9.

48 Levine DS, Shuaibi M, Spotte-Smith EWC, Taylor MG, Hasyim MR, Michel K, et al. The open molecules 2025 (OMol25) dataset, evaluations, and models. arXiv. 2025; arXiv:2505.08762.

49 Ho T-L. Hard soft acids bases (HSAB) principle and organic chemistry. Chem Rev. 1975 Feb 1;75(1):1–20.

50 Chattaraj PK, Lee H, Parr RG. HSAB principle. J Am Chem Soc. 1991 Feb;113(5):1855–6.

51 Ayers PW, Parr RG, Pearson RG. Elucidating the hard/soft acid/base principle: a perspective based on half-reactions. J Chem Phys. 2006 May 21;124(19):194107.

52 Datta D. On Pearson's HSAB principle. Inorg Chem. 1992 Jun;31(13):2797–800.

53 Pearson RG. The principle of maximum hardness. Acc Chem Res. 1993 May;26(5):250–5.

54 Parr RG, Pearson RG. Absolute hardness: companion parameter to absolute electronegativity. J Am Chem Soc. 1983 Dec;105(26):7512–6.

55 Ahrland S, Chatt J, Davies NR. The relative affinities of ligand atoms for acceptor molecules and ions. Q Rev, Chem Soc. 1958;12(3):265.

56 Liu S, Rong C, Lu T. Information conservation principle determines electrophilicity, nucleophilicity, and regioselectivity. J Phys Chem A. 2014 May 22;118(20):3698–704.

57 Liu S, Rong C, Lu T. Electronic forces as descriptors of nucleophilic and electrophilic regioselectivity and stereoselectivity. Phys Chem Chem Phys. 2017 Jan 4;19(2):1496–503.

58 Zhuo L, Zheng Y, Zeng L, Zhao Y, Li M, Rong C, et al. Quantification of Lewis acidity and Lewis basicity: a density-based reactivity theory study. J Comput Chem. 2025 Aug 15;46(22):e70212.

13

Aromaticity and Antiaromaticity

TABLE OF CONTENTS

13.1 Introduction

Following the previous chapters on stereoelectronic effects and chiroptical behavior, we now turn to aromaticity—another foundational organizing concept in chemical reactivity [1–15]. Although historically associated with benzene's unusual properties, aromaticity today encompasses a wide variety of delocalized cyclic systems across organic, inorganic, main-group, and organometallic chemistry. Hückel's

$4n + 2$ rule [16, 17] and its $4n$ antiaromatic counterpart provide the classical starting point: planar, cyclic, conjugated systems with $(4n + 2)$ π-electrons exhibit exceptional stabilization, whereas those with $4n$ electrons often distort or react readily [4, 11].

Over the past few decades, however, the concept has expanded far beyond simple π-rings. Excited-state aromaticity [18–23], Möbius and topological aromaticity [24–27], three-dimensional aromaticity [28–33], σ-aromaticity [15, 34–37], and metalloaromaticity [2, 13, 38, 39] show that cyclic delocalization can arise from different dimensionalities, orbital types, spin states, and connectivities. These diverse manifestations reveal aromaticity as a broad electronic phenomenon rather than a single structural motif.

Because each manifestation reflects a different physical aspect—geometric uniformity, energetic stabilization, magnetic response, or electron delocalization—no single quantitative criterion fully captures aromatic behavior [7, 11, 31]. Multiple indicators are therefore needed, each probing a complementary facet of cyclic delocalization.

Taken together, these developments show how aromaticity extends far beyond classical π-rings. This chapter reviews these developments and then integrates them into the density-based reactivity framework developed in this book [40–52]. Aromaticity provides an ideal testbed: it is conceptually rich, experimentally accessible, theoretically multifaceted, and highly sensitive to how electron density is organized. By examining aromaticity through structural, energetic, magnetic, real-space, conceptual density functional theory (CDFT) [42–47], and information-theoretic approach (ITA) [48–52] descriptors, we show that a density-based perspective naturally unifies classical aromaticity with its modern generalizations.

13.2 Aromaticity and Antiaromaticity in Ground State

13.2.1 Criteria of Aromaticity

The concepts of aromaticity and antiaromaticity originated from empirical observations of the unusual behavior of certain organic compounds in their ground state. Ground-state aromaticity and ground-state antiaromaticity describe the aromatic or antiaromatic character of a molecule in its lowest-energy electronic state—typically the singlet ground state (S_0) for neutral organic molecules.

A molecule is aromatic in the ground state if it meets the following four criteria: (i) planar (or nearly planar) structure; (ii) cyclic arrangement of atoms; (iii) fully conjugated π-electron system (continuous overlap of p-orbitals); and (iv) contains $(4n + 2)$ π-electrons (Hückel's rule), where n is a nonnegative integer. Such systems are thermodynamically stabilized by electron delocalization.

The criteria for aromaticity are interconnected and mutually reinforcing. Planarity is essential for p-orbital overlap, which enables full conjugation, leading to electron delocalization and the resulting exceptional stability. This precise alignment and continuous overlap of p-orbitals allow the π-electrons to delocalize across the entire ring, leading to the highly stable, filled bonding molecular orbitals described by Hückel's molecular orbital theory. Any deviation from these conditions, such as a molecule becoming nonplanar like cyclooctatetraene, breaks the delocalized system and eliminates aromaticity, causing it to behave as a nonaromatic compound. This highlights the delicate and interdependent nature of the aromatic criteria.

Classic examples include benzene (6 π-electrons), naphthalene (10 π-electrons), and pyridine (6 π-electrons, with a nitrogen lone pair not in the π system). These molecules show characteristic behavior like bond length equalization, diatropic ring currents in NMR, and electrophilic aromatic substitution reactions. Detailed energetic, magnetic, and electronic consequences of these criteria are discussed in Section 13.4, which summarizes the major aromaticity descriptors.

13.2.2 Manifestations of Aromaticity

Structural Features. Aromatic compounds exhibit distinctive geometrical features that reflect their high degree of electron delocalization. One of the most striking characteristics is *planarity*, which allows for effective overlap of adjacent p-orbitals and continuous π-electron delocalization around the ring. This delocalization leads to *bond length equalization*: rather than displaying alternating single and double bonds, as suggested by localized structures, aromatic systems such as benzene possess intermediate and nearly identical bond lengths throughout the ring—1.39 Å in benzene, for example. This uniformity contrasts sharply with the expected values for isolated single (~1.54 Å) and double (~1.34 Å) carbon–carbon bonds. The rigid, symmetrical ring geometry of aromatic compounds not only contributes to their thermodynamic stability but also influences their electronic, magnetic, and spectroscopic properties, making geometry a critical component of aromatic character.

Exceptional Stability. The most defining characteristic of aromatic compounds is their unusual thermodynamic stability, often referred to as resonance energy. This stability arises from the extensive delocalization of π-electrons across the entire cyclic conjugated system. According to Hückel's molecular orbital theory, this $4n + 2$ electron count ensures that all bonding molecular orbitals are filled with paired electrons, leading to a lower energy configuration compared to what would be predicted for a hypothetical nonaromatic analogue. This additional stability is evident when comparing the hydrogenation energies of aromatic compounds to those of simple alkenes. The "special stability" of aromatic compounds is not merely an abstract concept; it is quantifiable as "resonance energy," which manifests in their resistance to addition reactions and preference for substitution, fundamentally altering their chemical behavior compared to simple alkenes. This means that electron delocalization leads to resonance energy, which confers high stability and drives the characteristic preference for substitution over addition.

Unique Reactivity. The high stability of the aromatic system profoundly influences its reactivity. Aromatic compounds are generally less reactive than nonaromatic compounds. They tend to undergo reactions that preserve the aromatic ring system rather than reactions that would disrupt it. The most characteristic reaction is electrophilic aromatic substitution (EAS), where an electrophile replaces a hydrogen atom on the ring. The delocalized π-system plays a crucial role in stabilizing the intermediate σ-complex in EAS reactions, facilitating the reaction and influencing its regioselectivity. The influence of aromaticity extends beyond ground-state stability to actively participating in reaction mechanisms, where it can kinetically and thermodynamically promote reactions by stabilizing transition states or products. This mechanistic understanding reveals how aromaticity governs chemical transformation processes.

Physical Properties. Aromatic compounds exhibit distinct physical properties that reflect their delocalized electron systems. They display large anisotropic diamagnetic susceptibilities and can sustain ring currents under applied magnetic fields, which are observable through techniques like NMR spectroscopy. The delocalized π-system also leads to characteristic UV–vis absorption bands, which are useful for spectroscopic identification. For instance, benzene shows a characteristic absorption band around 254 nm due to the transition from its highest-occupied molecular orbital (HOMO) to lowest-unoccupied molecular orbital (LUMO).

13.2.3 Antiaromaticity

The concept of antiaromaticity was formally introduced by Ronald Breslow in the 1960s [53, 54]. His work sought to explain the unusual reactivity and inherent instability observed in certain cyclic compounds that, despite being cyclic and conjugated, did not exhibit aromatic stability. Breslow's contributions established antiaromaticity as a fundamental concept in physical organic chemistry, providing a framework to understand the limits of aromaticity itself.

Similar to aromatic compounds, antiaromatic compounds must meet specific structural and electronic criteria: (i) cyclic: The molecule must form a closed ring structure; (ii) planar: The ring must be planar, allowing for continuous overlap of p-orbitals; (iii) fully conjugated: There must be a continuous, uninterrupted system of p-orbitals throughout the ring; and (iv) $4n$ π-electrons: The cyclic conjugated system must contain a number of π-electrons that is a multiple of four (4, 8, 12, 16, etc.), where "n" is a positive integer.

Antiaromatic compounds are defined by their pronounced instability and high reactivity, which stem from their unfavorable electronic configuration. According to molecular orbital theory, cyclic, conjugated systems with $4n$ π-electrons possess a high-energy HOMO and a small HOMO–LUMO gap, resulting in elevated ground-state energies relative to nonaromatic or open-chain analogs. This electronic destabilization creates a strong thermodynamic driving force for structural distortion or chemical transformation to relieve antiaromatic strain. As a result, antiaromatic compounds frequently undergo dimerization, addition, or rearrangement reactions—kinetic consequences of their inherent thermodynamic instability. A classic example is cyclobutadiene, which is so unstable that it spontaneously dimerizes via a 2 + 2 cycloaddition with virtually no activation barrier. In addition to their chemical reactivity, antiaromatic systems exhibit paratropic ring currents detectable by NMR spectroscopy, causing unusual shielding patterns. Their instability can also influence other molecular properties, such as acidity; for instance, the cyclopropenyl anion is significantly less acidic than predicted due to its antiaromatic character.

13.2.4 Theoretical Frameworks

Hückel Molecular Orbital (HMO) Theory. The earliest and most fundamental theoretical basis for aromaticity comes from HMO theory, developed by Erich Hückel in 1931 [17]. Hückel's theory applies to monocyclic, planar, conjugated π-systems and predicts that molecules with $(4n + 2)$ π-electrons are aromatic, while those with $4n$ π-electrons are antiaromatic. This rule arises from solving the π-electron energy levels for cyclic systems, where $(4n + 2)$ configurations fill bonding molecular orbitals completely, resulting in a closed-shell, delocalized, and stable electronic structure. In contrast, $4n$ systems often have degenerate or partially filled nonbonding or antibonding orbitals, leading to electronic destabilization. Despite its simplicity and limitations, Hückel's rule remains a cornerstone for evaluating aromaticity and antiaromaticity, particularly in small rings.

Molecular Orbital Theory (MOT). Beyond Hückel's simplified treatment, modern MOT provides a more generalized and detailed understanding of electron delocalization in conjugated systems [55]. MOT allows us to visualize how π-electrons occupy molecular orbitals that extend over an entire ring. Aromatic systems exhibit fully filled bonding orbitals with large HOMO–LUMO gaps, while antiaromatic systems have electrons populating destabilized antibonding orbitals or small HOMO–LUMO gaps. Computational methods based on MOT are commonly used to predict and quantify aromaticity and antiaromaticity in increasingly complex molecules, including those with heteroatoms, fused rings, and nonplanar structures.

Valence Bond Theory (VBT) and Resonance. VBT, particularly the concept of resonance, provides another lens for interpreting aromaticity [56]. In VBT terms, aromatic compounds are understood as hybrids of multiple contributing structures, each representing different arrangements of π-bonds. The more significant and symmetric the resonance contributors, the greater the delocalization and the resulting stabilization. For example, benzene is often depicted as an equal hybrid of two Kekulé structures. While VBT is more intuitive and commonly used for pedagogical purposes, it lacks the quantitative power of MOT and is less effective for nonclassical or metal-containing systems.

Main Challenges. Aromaticity and antiaromaticity are intrinsically multifaceted. Since they are not directly measurable, they must be inferred from multiple criteria—electronic, magnetic, structural, and energetic—that often yield conflicting results. This makes it difficult to define a universal standard [57], especially for molecules with partial, local, or ambiguous aromatic character [7]. The classical $4n + 2$ and $4n$ rules also break down in nonplanar, heteroatomic, polycyclic, or excited-state systems, where concepts like Baird's rule or adaptive aromaticity apply. Another challenge is quantifying antiaromaticity. While aromaticity is associated with stability, antiaromatic systems are often transient and highly reactive, making experimental characterization difficult. Cyclobutadiene, for instance, requires stabilization by bulky substituents or low temperatures to be studied, complicating direct measurements. Theoretical models must account for dynamic effects, such as distortions from planarity, which reduce antiaromatic character but obscure clear classification. Additionally, the extension of aromaticity to new domains—such as Möbius, 3D, σ-, hyperconjugative, and metalloaromatic systems—requires specialized theoretical tools, many of which are method-sensitive and computationally demanding. Together, these issues highlight the need for more unified and robust frameworks to describe aromatic behavior across diverse chemical contexts.

13.3 Other Types of Aromaticity and Antiaromaticity

Beyond classical Hückel aromaticity, numerous frameworks have been developed to describe situations in which cyclic electron delocalization arises from alternative orbital topologies, dimensionalities, or electronic states. These variants broaden the scope of aromaticity and deepen its conceptual foundations, illustrating that cyclic delocalization is a general electronic phenomenon rather than a feature limited to planar π-systems. The most prominent extensions are summarized below.

13.3.1 Excited State Aromaticity and Antiaromaticity

Excited-state aromaticity [18–21] extends classical concepts to electronically promoted molecules. Baird's seminal 1972 proposal [22, 23] predicts that in the lowest $\pi \rightarrow \pi^*$ triplet state (T_1), aromaticity inverts: $4n$ systems become aromatic while $(4n + 2)$ systems become antiaromatic. Electron promotion from a bonding π-MO to a π^*-MO alters orbital occupancy, producing geometric and energetic behavior opposite to the ground state. Thus, benzene becomes antiaromatic in T_1, whereas cyclobutadiene becomes aromatic. This inversion affects planarity, photostability, and reactivity and provides mechanistic insight into electrocyclizations, photorearrangements, and singlet-fission materials [18–21]. Extensions to S_1 states have similar implications but remain more system-dependent. Because excited-state aromaticity cannot be probed with ground-state indicators alone, analysis often requires TD-DFT, CASSCF, or time-resolved spectroscopy.

13.3.2 Möbius Aromaticity

Möbius aromaticity [24–27] introduces topology into aromaticity by considering cyclic conjugated systems with a half-twist. Heilbronner proposed that such twisted systems exhibit aromaticity with $4n$ π-electrons and antiaromaticity with $(4n + 2)$ π-electrons due to orbital phase inversion along the loop. Experimental realizations—mostly in expanded annulenes, metallacycles, and porphyrinoids—confirm that Möbius topology can sustain delocalization, though conformational flexibility complicates unambiguous assignment. Demonstrating true Möbius aromaticity requires verifying orbital-phase inversion or ring currents, not geometry alone.

13.3.3 Three-dimensional Aromaticity

Three-dimensional aromaticity [28–33] generalizes delocalization to polyhedral and spherical frameworks. Closo-boranes such as $[B_{12}H_{12}]^{2-}$ illustrate delocalization over 3D cages, rationalized via Wade–Mingos electron-counting rules. Fullerenes provide additional cases: certain charged or substituted derivatives approximate spherical aromaticity following Hirsch's $2(n + 1)^2$ rule. Metal clusters (e.g., Au_n, Sn_n) also show 3D σ- or d-orbital delocalization. These systems demonstrate that aromaticity need not be planar or π-based; rather, it reflects symmetry-supported delocalization within a closed 3D shell.

13.3.4 σ-Aromaticity

σ-Aromaticity [15, 34–37] refers to systems in which cyclic stabilization arises from σ-electron delocalization, typically through multicenter bonds. Cyclopropane and electron-deficient boranes (e.g., diborane) exhibit stabilizing 3c − 2e σ-frameworks analogous to π-aromatic delocalization. Characterizing σ-aromaticity is more challenging because classical π-based indicators (NICS, HOMA) may not directly apply, requiring complementary tools such as natural bond orbital (NBO), AdNDP, electron localization function (ELF), and current-density analyses.

13.3.5 Metalloaromaticity

Metalloaromaticity [2, 13, 38, 39] arises when cyclic delocalization involves metal d-, s-, or p-orbitals. Metallabenzenes, in which a metal replaces a carbon atom of a benzene ring, preserve many aromatic signatures—planarity, bond uniformity, and diatropic currents. Small metal clusters (Cu_3^+, Ag_3^+, Au_n) form σ- or π-aromatic rings depending on orbital symmetry. These species underscore the versatility of aromatic delocalization in organometallic and cluster chemistry.

13.3.6 Adaptive Aromaticity

Adaptive aromaticity [58–60] describes systems aromatic in both ground and excited states, defying the usual Hückel/Baird complementarity. Observed primarily in organometallic rings (e.g., osmapentalenes), such systems maintain delocalization because orbital reorganizations preserve cyclic electron flow across states. Although still rare, adaptive aromaticity reveals the dynamic, state-dependent nature of delocalization.

13.3.7 Miscellaneous Types of Aromaticity and Antiaromaticity

Other forms include homoaromaticity, δ-aromaticity, hyperconjugative aromaticity [61–64], charge-transfer aromaticity, excited-state multiring aromaticity, and magnetic aromaticity. These special cases highlight how diverse orbital interactions—through-space, d-orbital, $\sigma \rightarrow \pi^*$ donation, or long-range charge transfer—can generate cyclic delocalization patterns.

13.4 Descriptors of Aromaticity and Antiaromaticity

A wide variety of descriptors has been developed to quantify aromaticity, each probing a different physical manifestation of cyclic electron delocalization. These classes of descriptors reflect different physical principles and often yield complementary insights. No single index is universally reliable, because aromaticity involves structural, energetic, magnetic, and electronic features that need not correlate perfectly. Accordingly, the following subsections group the most commonly used descriptors according to the physical property they measure.

13.4.1 Structural Descriptors

The Harmonic Oscillator Model of Aromaticity (HOMA). The HOMA index [65], introduced by Krygowski and coworkers, is a quantitative structural descriptor of aromaticity based on bond length deviation from an optimal aromatic reference value. It is defined as:

$$\mathrm{HOMA} = 1 - \frac{\alpha}{n}\sum_{i=1}^{n}(R_i - R_{\mathrm{opt}})^2, \tag{13.1}$$

where R_i is the bond length of the i-th bond in the ring, R_{opt} is the optimal bond length corresponding to maximum aromaticity (e.g., 1.388 Å for C—C in benzene), n is the number of bonds considered, and α is an empirical constant that normalizes the index. HOMA values range from 1 (perfect aromaticity with no bond alternation) to 0 (nonaromatic), and negative values indicate severe bond length alternation, which often accompanies antiaromatic or strongly distorted systems. Importantly, HOMA is sensitive to both experimental and computationally derived structures, making it widely applicable.

Bond Length Alternation (BLA). An alternative structural descriptor particularly useful in linear and cyclic π-conjugated systems is BLA. Defined as the average difference between adjacent single and double bonds, BLA [66] provides insight into the degree of delocalization:

$$\mathrm{BLA} = \frac{1}{n}\sum_{i=1}^{n}\left|R_{i,\mathrm{single}} - R_{i,\mathrm{double}}\right|, \tag{13.2}$$

where $R_{i,\mathrm{single}}$ and $R_{i,\mathrm{double}}$ are single and double bond lengths in the ring structure, respectively. For fully delocalized systems like benzene or annulenes with $4n + 2$ π electrons, BLA is close to zero. In contrast, in polyenes or antiaromatic systems, significant BLA indicates localization and loss of aromatic character. Although simple, BLA is effective in systems where HOMA may be less applicable, such as non–six-membered rings or heterocycles.

Ring Planarity and Geometrical Distortion. A critical prerequisite for π-electron delocalization is planarity [5, 67]. Aromatic systems typically enforce strict coplanarity to allow full p-orbital overlap. Deviations from planarity can therefore be interpreted as disruptions in aromatic conjugation. In contrast, antiaromatic systems often distort out of planarity to relieve destabilization from π-electron delocalization. For instance, cyclooctatetraene (COT), which formally has 8 π-electrons (a 4n system), adopts a tub-shaped nonplanar conformation to avoid antiaromaticity. This leads to the definition of out-of-plane distortion angles, puckering parameters, and dihedral angles as structural descriptors. These geometric indicators can be measured via crystallography or optimized computational geometries to assess aromatic character.

Limitations. Although structural descriptors provide intuitive and often accurate assessments of aromaticity, they have limitations. For example, HOMA depends heavily on reference bond lengths and may not account for heteroatoms or substitution patterns; BLA loses significance in non-alternant systems or those with resonance contributors of differing weights; and planarity-based measures may fail to distinguish between antiaromaticity and mere steric distortion. Despite these limitations, structural descriptors remain vital for classifying and understanding aromatic and antiaromatic systems. In computational studies, they complement other indices to yield a comprehensive picture of (anti)aromatic behavior.

13.4.2 Energetic Descriptors

Aromatic Resonance Energy (ARE). The ARE index [68] also known as delocalization energy, is the extra stabilization a compound experiences due to delocalized π-electron conjugation compared to a hypothetical localized structure. One of the earliest and most influential estimates of ARE came from the heats of hydrogenation of benzene and related compounds. For example, the hypothetical hydrogenation of 1,3,5-cyclohexatriene (with localized double bonds) would release three times the heat of hydrogenation of cyclohexene (~−28.6 kcal/mol), that is, about −85.8 kcal/mol. In reality, the hydrogenation of benzene releases only ~−49.8 kcal/mol, indicating a resonance energy of −36.0 kcal/mol. This ~36 kcal/mol difference reflects the energetic stabilization due to aromatic delocalization. Conversely, if the real heat of hydrogenation were more exothermic than expected, it would suggest antiaromatic destabilization, as is sometimes seen in cyclobutadiene or planar cyclooctatetraene. However, resonance energy is context-dependent and cannot be measured directly without carefully chosen reference systems. This motivates the use of isodesmic, homodesmotic, and hypohomodesmotic reactions.

Isodesmic and Homodesmotic Reaction Energies. To quantify stabilization energies without relying on hypothetical species, isodesmic and homodesmotic reaction energies can be used [69, 70]. An isodesmic reaction preserves the number and type of bonds, allowing the comparison of a molecule with its substituted analogs. For example,

$$C_6H_6 + 3\,CH_3CH = CH_2 \rightarrow 3\,CH_3CH_2CH_3 + C_6H_5CH = CH_2. \qquad (13.3)$$

A homodesmotic reaction goes further, matching not only bond types but also hybridization and bonding environments. These reactions reduce strain and substituent effects, isolating the influence of delocalization. The reaction energy ΔE from such schemes provides a more chemically meaningful estimate of aromatic stabilization or antiaromatic destabilization.

Isomerization Stabilization Energy (ISE). The ISE index [71–73] is a particularly clean method for estimating aromatic stabilization in conjugated ring systems. It involves calculating the energy difference between an aromatic compound and a nonaromatic isomer with the same number of π-electrons but disrupted conjugation:

$$ISE = E_{\text{non-aromatic isomer}} - E_{\text{aromatic isomer}}. \quad (13.4)$$

For instance, ISE can be estimated by comparing benzene with its Dewar benzene or bicyclo[2.2.0]hexatriene isomers. A large, positive ISE indicates strong aromatic stabilization. Similarly, for antiaromatic systems, a negative ISE indicates destabilization relative to a nonconjugated isomer. ISE is widely used in computational studies and has the advantage of not requiring experimental reference data.

Aromatic Stabilization Energy (ASE). ASE [69] is a more general term referring to the energetic advantage of a compound due to aromatic delocalization, often estimated via thermochemical cycles. ASE can be derived using heats of formation, combustion, or hydrogenation and often overlaps conceptually with ARE. For example, for polycyclic aromatic hydrocarbons (PAHs), ASE quantifies the degree of extra stabilization from aromatic conjugation across multiple rings, as in:

$$ASE_{PAH} = n \cdot E_{\text{monocycle}} - E_{PAH}, \quad (13.5)$$

where n is the number of monocyclic aromatic rings. ASE values are also additive in certain fused systems, but this additivity breaks down in systems exhibiting strong localization or antiaromatic character.

Limitations. Energetic descriptors are quantitative and conceptually grounded but require careful choice of reference states. The ambiguity in defining a "nonaromatic" baseline often leads to different stabilization energies depending on the method employed. In addition, thermochemical methods can be influenced by strain, substituent effects, or solvation; MO-based estimations rely on idealized approximations that may not capture real electronic effects in complex systems; and for nonplanar or fluxional systems, energy differences may reflect geometry rather than aromaticity.

13.4.3 Property Descriptors

Magnetic Descriptors: Nucleus-Independent Chemical Shift (NICS). One of the most powerful and widely used property-based indicators of aromaticity is the NICS, introduced by Schleyer et al. [74, 75]. It is based on the magnetic shielding calculated at or near the center of a ring system, and it reflects the ring current induced by an external magnetic field. Aromatic systems generate a diatropic ring current that shields the ring center, leading to negative NICS values, whereas antiaromatic systems generate a paratropic ring current, resulting in positive NICS values.

The NICS value at the ring center is given by:

$$NICS(0) = -\sigma_{iso} = -\frac{1}{3}\left(\sigma_{xx} + \sigma_{yy} + \sigma_{zz}\right), \quad (13.6)$$

where σ_{iso} is the isotropic magnetic shielding at the ring center, and σ_{xx}, σ_{yy}, and σ_{zz} are diagonal components of the magnetic shielding tensor. To reduce σ-electron contributions, NICS values are often evaluated above the ring plane (e.g., at 1 Å), denoted as NICS (1). Further refinements involve decomposing shielding into tensor components, particularly the out-of-plane component σ_{zz}, yielding NICS$(1)_{zz}$, which is especially sensitive to π-electron delocalization. Typical guidelines are aromatic: NICS$(1)_{zz} < -10$ ppm; nonaromatic: NICS$(1)_{zz} \approx 0$ ppm; and antiaromatic: NICS$(1)_{zz} > +10$ ppm.

ACID and GIMIC. Anisotropy of the induced current density (ACID) and gauge-including magnetically induced currents (GIMIC) [76] are two powerful computational tools used to visualize aromaticity and antiaromaticity by directly modeling the magnetically induced current densities in molecular rings. When a molecule is exposed to an external magnetic field, aromatic systems generate diatropic (clockwise) ring currents, while antiaromatic systems exhibit paratropic (counterclockwise) currents. ACID plots display current density vectors superimposed on molecular structures, providing intuitive visual evidence of electron delocalization pathways, especially π-currents. GIMIC, on the other hand, calculates and quantifies current strengths along defined paths in the molecule, enabling numerical comparison between different systems or regions within the same molecule. Together, ACID and GIMIC offer complementary insights into the magnetic response and electron delocalization, making them invaluable for assessing aromaticity in complex, nonplanar, or heterocyclic systems.

Magnetic Susceptibility Exaltation ($\Delta\chi$). Magnetic susceptibility exaltation [4, 76] is another classic aromaticity descriptor, quantifying the difference between the experimental magnetic susceptibility of a conjugated ring and the sum of its nonconjugated components:

$$\Delta\chi = \chi_{\mathrm{exp}} - \sum\chi_{\mathrm{ref}}. \tag{13.7}$$

A negative exaltation (i.e., enhanced diamagnetism) signals aromaticity, and a positive exaltation (reduced diamagnetism or paramagnetism) suggests antiaromaticity. This descriptor was historically important for confirming benzene's aromatic nature and continues to be useful for comparing homologous series (e.g., $[4n+2]$ versus $[4n]$ annulenes).

Electronic Descriptors. Electronic descriptors focus on the distribution and delocalization of electrons within the cyclic system. Aromaticity is intrinsically linked to the cyclic delocalization of π-electrons, while antiaromaticity implies a tendency toward electron localization or a disruptive delocalization pattern. *Delocalization indices (DIs)* [77], often derived from quantum theory of atoms in molecules (QTAIM) or NBO analysis, quantify the number of electrons shared or delocalized between two atomic centers. In aromatic systems, the DIs between adjacent atoms within the ring are typically more uniform and higher than in nonaromatic systems, reflecting extensive electron sharing. For antiaromatic systems, DIs may show more significant alternation, indicating electron localization. The *aromatic fluctuation index (FLU)* [14] is an electronic descriptor that quantifies the degree of electron delocalization.

$$\mathrm{FLU} = \frac{1}{n}\sum_{i=1}^{n}\left(\frac{\delta_{i,i+1} - \bar{\delta}}{\bar{\delta}}\right)^2, \tag{13.8}$$

where $\delta_{i,i+1}$ is the delocalization index between adjacent atoms and $\bar{\delta}$ is the average. It is based on the fluctuation (variance) in the electron population of atomic basins. A smaller fluctuation (more uniform electron distribution) indicates higher aromaticity.

Spectroscopic Descriptors. Spectroscopic properties [78] provide indirect yet useful information about aromaticity. UV–vis absorption spectra of aromatic systems often show highly allowed $\pi - \pi^*$ transitions, while antiaromatic systems display weak and broadened bands due to excited-state instability. Infrared (IR) spectroscopy can reveal vibrational delocalization; for example, C=C stretch bands in aromatic rings are narrower and shifted compared to localized alkenes. These spectral features can be quantitatively interpreted using TD-DFT simulations, aiding in the assignment of aromatic or antiaromatic character, especially in polycyclic or heterocyclic compounds.

13.4.4 Other Descriptors

Many other descriptors and indices have been developed in the literature to capture specific aspects of aromaticity and antiaromaticity, particularly in complex or unconventional systems. Among them, the *multicenter index (MCI)* [79, 80] and the *para-delocalization index (PDI)* [77, 81] are rooted in electron delocalization theory. MCI measures electron sharing among three or more atoms in a ring and is defined as a higher-order generalization of the delocalization index. It quantitatively reflects cyclic delocalization and correlates well with aromatic stabilization. Similarly, PDI focuses on delocalization between para-related atoms in six-membered rings (e.g., benzene), providing a simplified but effective measure of aromatic delocalization:

$$\mathrm{PDI} = \frac{1}{n}\sum_{i}\delta_{\mathrm{para}(i)}, \tag{13.9}$$

where $\delta_{\mathrm{para}(i)}$ is the delocalization index between para-positioned atoms in the ring. Both MCI and PDI are computed from electron density matrices and are particularly useful when dealing with heteroaromatic or polynuclear systems where traditional methods may fail.

Another important class includes *topological* and *graph-theoretical descriptors*, such as the Irving–Williams Index and graph energy-based aromaticity indices (e.g., topological resonance energy, TRE) [82]. These descriptors analyze aromaticity based on the structure and connectivity of π-networks. Lastly, *electron density of delocalized bonds (EDDB)* [83] is a newer real-space method that extracts only the delocalized electron density associated with aromatic circuits, offering chemically intuitive visualizations and robust numerical indices that correlate with other aromaticity measures. These complementary tools enrich the aromaticity toolbox, especially for ambiguous or borderline systems.

Despite the wide variety of descriptors and indices for aromaticity and antiaromaticity, each carries inherent limitations that can affect reliability and comparability across different systems. Many descriptors, such as HOMA or NICS, are sensitive to molecular geometry, computational method, and choice of reference values, which may lead to inconsistent results in nonstandard systems like heterocycles or nonplanar rings. Magnetic descriptors (e.g., NICS, ACID, GIMIC) can be gauge- or position-dependent, and may misrepresent aromaticity when σ- and π-contributions are not properly separated. Energetic indices rely on arbitrary reference systems or hypothetical isodesmic reactions, making cross-comparison problematic. Electron delocalization-based indices like MCI or FLU depend on electron density partitioning schemes that may not be universally applicable. Moreover, many indices are not easily interpretable in isolation or for systems with competing local and global aromatic characters (e.g., polycyclic or Möbius aromatic systems). Consequently, no single descriptor is universally sufficient, and a multidimensional approach combining multiple descriptors is often necessary.

13.5 QTAIM, DFT, CDFT, and Aromaticity

While traditional descriptors capture specific manifestations of aromaticity, density-based approaches probe directly how electron density is organized in space. Density-based approaches for evaluating aromaticity and antiaromaticity have a long and evolving history, transitioning from early qualitative concepts of electron delocalization to advanced, quantitatively rigorous frameworks. These methods focus on analyzing the electron density distribution within a molecule to assess the extent, continuity, and nature of π-electron delocalization—central to understanding aromatic character. Descriptors such as the DI and the FLU (aromatic fluctuation index) quantify how electrons are shared between bonded atoms or fluctuate around a mean delocalization pattern, thus distinguishing aromatic systems, which exhibit uniform delocalization, from antiaromatic ones, which tend to localize electron density. In the following section, we highlight using Bader's QTAIM, ELF, and CDFT descriptors such as hardness and softness, to provide complementary and insightful measures of electron pairing and reactivity relevant to aromaticity and antiaromaticity.

Bader's QTAIM [84] provides valuable insight into electron delocalization and bonding continuity within cyclic systems. Aromatic rings typically exhibit a continuous network of bond paths and ring critical points, indicating a closed loop of electron density consistent with π-delocalization. Furthermore, properties at bond critical points can distinguish between localized and delocalized bonding: aromatic systems generally show uniform values around the ring, whereas antiaromatic systems exhibit alternation or disruption in these values.

ELF [85, 86] is a powerful real-space tool originally developed by Becke and Edgecombe to describe the probability of finding an electron in the neighborhood of another electron with the same spin. It has since become widely used to assess chemical bonding, particularly electron pairing and delocalization, and has proven valuable for evaluating aromaticity and antiaromaticity. ELF is defined as:

$$ELF = \frac{1}{1 + \chi^2(\mathbf{r})}, \tag{13.10}$$

where $\chi(\mathbf{r})$ is a dimensionless localization index derived from the curvature of the conditional pair probability. High ELF values (close to 1) indicate regions of localized electron pairs (as in lone pairs or covalent bonds), while low ELF values (approaching 0) indicate delocalized regions, such as those found in π-conjugated systems. In aromatic rings, ELF maps typically reveal a continuous torus-like region of moderate localization above and below the ring plane, reflecting π-electron delocalization. Conversely, in antiaromatic systems, ELF plots show fragmented or more localized domains, signaling disrupted or uneven electron delocalization.

To quantitatively compare systems, topological analysis of ELF can be performed by integrating over basins associated with π-electrons or by tracking ELF-π bifurcation values, which signal the emergence of delocalized π-domains. These features correlate strongly with aromaticity: lower bifurcation values and more extended π-basins are associated with greater aromatic character. Importantly, ELF offers a model-independent, density-based visualization of aromaticity, free from reliance on resonance structures or magnetic perturbations. It also works effectively in heterocycles, nonplanar rings, and even transition states or excited states, making it a versatile complement to other aromaticity descriptors such as NICS or HOMA.

CDFT [42] provides another density-based framework for interpreting molecular stability and chemical reactivity. Within this framework, global descriptors such as chemical hardness (η) and softness (S) have been applied to evaluate aromaticity and antiaromaticity, offering an energetic and reactivity-based perspective. Chemical hardness reflects a system's resistance to charge transfer and is defined as $\eta = (I - A)$, where I is the ionization potential and A is the electron affinity. Aromatic systems, due to their enhanced electronic stability and delocalization, typically exhibit higher hardness values, indicating resistance to perturbations. In contrast, antiaromatic systems tend to be softer (i.e., have smaller η), reflecting their greater polarizability and electronic instability. The inverse of hardness gives the global softness: $S = 1/\eta$, which quantifies a system's ability to deform or redistribute electron density in response to external perturbations. These CDFT descriptors are particularly useful for comparing series of isoelectronic or conjugated systems, where increasing softness can signal a breakdown in delocalization—often associated with antiaromaticity. Moreover, electrophilicity index ($\omega = \mu^2/2\eta$, where μ is the chemical potential) can further distinguish aromatic from antiaromatic species, as antiaromatic systems often exhibit higher electrophilicity due to their low stability and high reactivity. Thus, CDFT descriptors provide a quantitative energetic lens to complement structural and magnetic indicators of aromaticity.

13.6 ITA, Aromaticity, and Antiaromaticity

Because aromaticity fundamentally reflects electron localization versus delocalization, information-theoretic approach (ITA)—grounded in probability distributions and information content—is naturally positioned to reveal deeper patterns and organizing principles. In a few recent studies [23, 27, 87–90], we applied ITA to tackle the issue of aromaticity and antiaromaticity. In the first study, we examined fulvene derivatives of different ring sizes with different substituents (see structures below). We chose fulvene derivatives because their aromatic or antiaromatic nature can be tuned systematically by substituent and ring-size variations. We also considered four representative aromaticity indices (ASE, FLU, HOMA, and NICS) and eight ITA quantities.

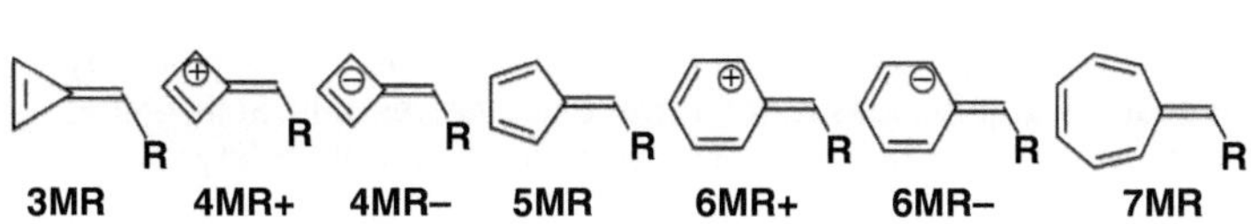

We found that aromaticity indices tend to correlate well with one another, and a similar internal consistency is observed among ITA quantities. These correlations hold consistently across various series of fulvene derivatives featuring different ring structures. However, when examining cross-correlations between ITA descriptors and traditional aromaticity indices, we discovered two distinct and opposite correlation patterns (Figure 13.1), indicating that a universal correlation does not exist across all fulvene systems. Instead, the direction and strength of these correlations depend critically on the electronic nature of the ring structure. Interestingly, the two sets of systems—each exhibiting internally consistent but opposite cross-correlation trends—correspond to molecules with a total of either $4n + 2$ or $4n$ π electrons, aligning with Hückel's rule for aromaticity and antiaromaticity, respectively. This finding underscores the fundamental influence of electron count and ring topology on the interplay between ITA descriptors and conventional aromaticity measures.

This inversion of correlation signs arises because ITA quantities fundamentally track electron localization versus delocalization. Aromatic ($4n + 2$) systems stabilize delocalization, lowering information-theoretic measures, whereas antiaromatic ($4n$) systems destabilize delocalization, increasing information measures. Consequently, the same ITA descriptor increases aromaticity in one regime and decreases it in the other, producing symmetric but opposite-signed linear correlations.

This opposite cross-correlation pattern was further validated by examining the excited-state aromaticity of a series of fulvene derivatives [23, 89]. In the triplet state, aromatic and antiaromatic behavior follow Baird's $4n$ rule, contrasting with Hückel's $4n + 2$ rule that governs the singlet ground state. As illustrated in Figure 13.2, the cross-correlation between HOMA and relative Rényi entropy in the triplet state exhibits a slope that is opposite in sign to that observed in the ground state. This reversal clearly indicates that the governing principles of aromaticity switch between electronic states,

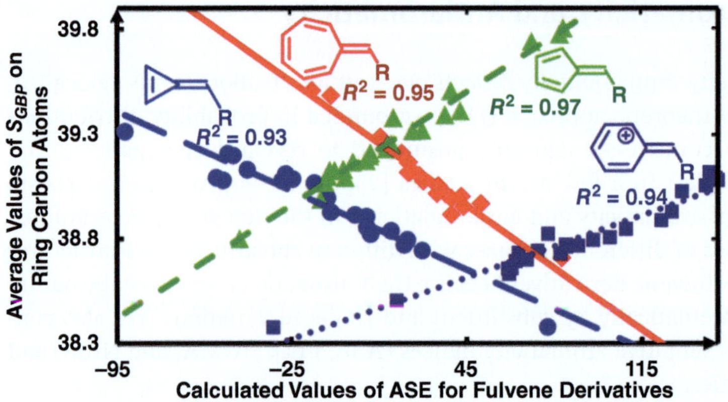

Figure 13.1 Illustrative examples of strong linear correlations between the aromaticity index ASE and the ITA quantity S_{GBP} for substituted fulvene derivatives 3MR, 5MR, 6MR+, and 7MR. *Source*: Reproduced from Ref. [87] with permission of The Royal Society of Chemistry.

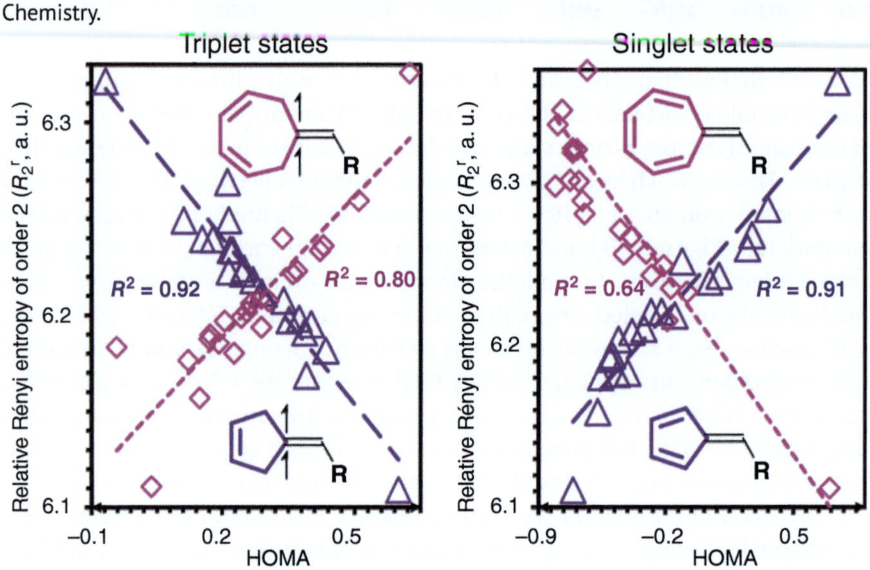

Figure 13.2 Cross correlations of ITA quantities with HOMA for singlet and triplet states. *Source*: Reproduced from Ref. [23] with permission of American Chemical Society.

with Hückel's rule applying to singlet ground states and Baird's rule to triplet excited states. The strong, linear yet oppositely signed correlations in Figure 13.2 provide compelling evidence for this fundamental shift in aromatic character between different spin states.

Extending our analysis to Möbius aromaticity, we examined four porphyrinoid derivatives—each differing in spin state and total charge—as illustrative models (Figure 13.3) [21, 22, 91]. By systematically rotating two benzene rings within each molecule, we explored how conformational changes influence both stability and aromaticity (Figure 13.4). Our findings reveal that the aromatic and antiaromatic tendencies observed throughout these conformational transitions align with the

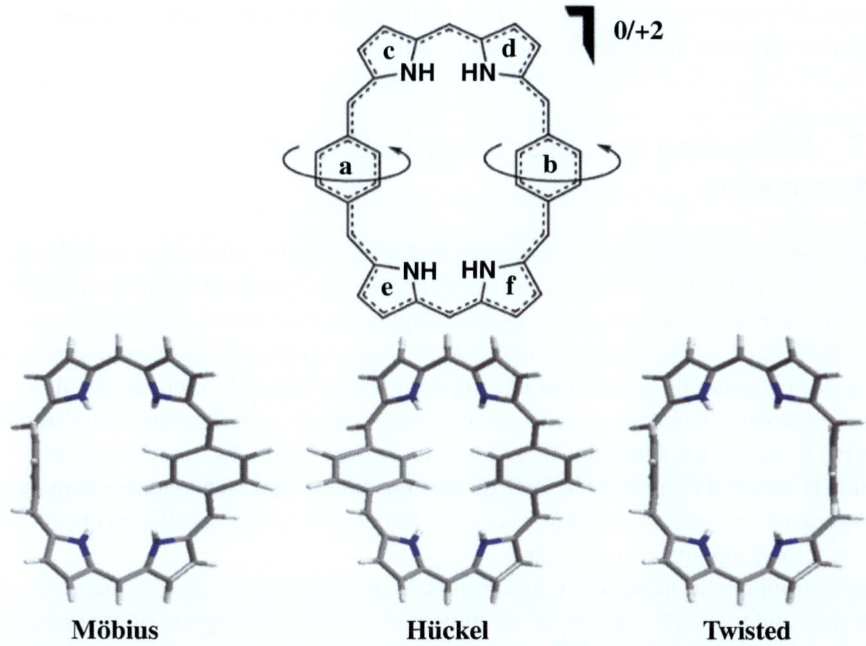

Figure 13.3 Four porphyrinoid derivatives studied. By rotating the two benzene rings (a and b), three distinct conformations, Möbius, Hückel, and twisted, can be generated. *Source*: Reproduced from Ref. [91] with permission of Springer Nature.

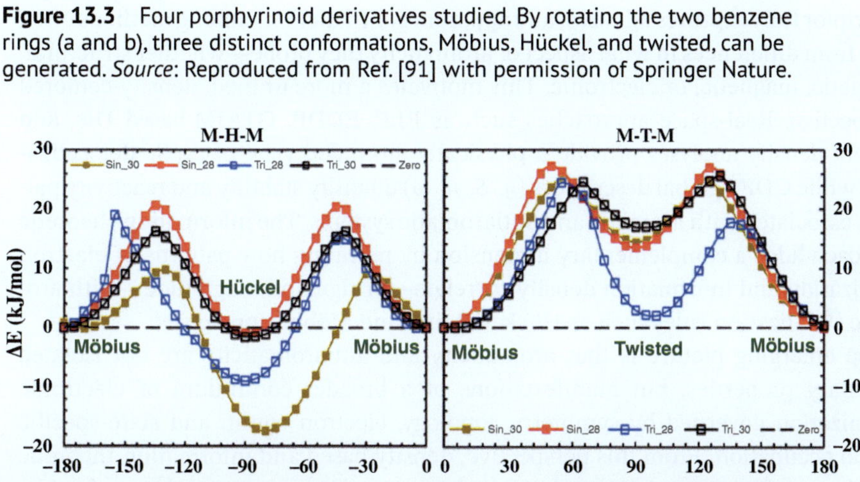

Figure 13.4 The energy profile of the four model systems as a function of the two benzene ring rotations with the Möbius conformation as the reference. (a) M-H-M represents the rotation of ring a, and (b) M-T-M denotes that of ring b. *Source*: Reproduced from Ref. [91] with permission of Springer Nature.

predictions of Hückel's rule in the ground state, Baird's rule in the excited state, and the Möbius rule for twisted topologies. Notably, we uncovered strong linear correlations between aromaticity indicators and ITA quantities, affirming ITA's utility in tracking aromatic character. Most significantly, this study revealed a reversal in the correlation between aromaticity propensity and ITA measures depending on the direction of conformational change, highlighting the intricate interplay between molecular structure, aromaticity, and information.

13.7 Concluding Remarks: Toward a Unified Understanding

Aromaticity and antiaromaticity have evolved from simple rules about π-electron counts into a rich and multifaceted set of concepts that govern stability, structure, reactivity, and electronic behaviors. In this chapter, we have traced this evolution—from Hückel's pioneering insights and the classical manifestations of aromaticity, to the diverse generalizations that encompass excited states, different dimensionalities, topologically twisted systems, σ- and δ-aromaticity, metalloaromaticity, and adaptive frameworks. Although these variants appear highly heterogeneous, they ultimately reflect a common physical theme: the extent to which electron density is delocalized or destabilized within a cyclic framework under specific electronic, geometric, and topological constraints.

The multiplicity of aromaticity descriptors from structural, energetic, magnetic, electronic, and topological perspectives illustrates both the conceptual richness and methodological challenges of defining a universal measure of aromaticity. No single descriptor fully captures all cases, and apparent contradictions among methods often arise from differences in what aspect of aromaticity they probe—whether structural, energetic, magnetic, or electronic. This motivates a more unified, density-centered perspective. Real-space approaches such as ELF, EDDB, QTAIM-based DIs, and current-density analyses provide a physically grounded view of cyclic delocalization, while CDFT global descriptors (η, S, μ, ω) quantify stability and reactivity patterns associated with aromatic and antiaromatic systems. The information-theoretic approach adds a complementary dimension by revealing how patterns of electron localization and information density correlate—in sign and magnitude—with aromatic stabilization rules such as Hückel, Baird, and Möbius topologies.

The emerging picture is that aromaticity and antiaromaticity are not isolated or binary properties, but manifestations of a broader continuum of electronic organization governed by symmetry, topology, electron count, and state-specific orbital occupation. From this perspective, density-based and information-theoretic descriptors offer a promising route toward constructing more general, transferable, and quantifiable criteria that apply seamlessly across classical, excited-state, and topologically nontrivial systems. Continued integration of these frameworks with machine learning, multiscale simulations, and experimental data will likely yield even deeper insights, including predictive models that unify classical aromaticity with its modern extensions.

As aromaticity continues to expand into new chemical spaces—metalloaromatic clusters, π-conjugated materials, magnetic and optoelectronic systems, and responsive molecular architectures—the need for coherent, quantitative, and conceptually grounded descriptors will only grow. Density-based reactivity theory provides a natural foundation for this unification, offering a path toward a more comprehensive understanding of aromaticity and antiaromaticity across states, structures, and chemical environments. In this sense, aromaticity becomes a natural testbed illustrating the power of density-based reactivity theory to unify classical concepts with modern generalizations.

References

1. De Proft F, Geerlings P. Conceptual and computational DFT in the study of aromaticity. Chem Rev. 2001 May;101(5):1451–64.
2. Boldyrev AI, Wang L-S. All-metal aromaticity and antiaromaticity. Chem Rev. 2005 Oct;105(10):3716–57.
3. Krygowski TM, Cyrañski MK, Czarnocki Z, Häfelinger G, Katritzky AR. Aromaticity: a theoretical concept of immense practical importance. Tetrahedron. 2000 Mar;56(13):1783–96.
4. von Ragué Schleyer P. Introduction: aromaticity. Chem Rev. 2001 May 9;101(5):1115–8.
5. Krygowski TM, Cyrański MK. Structural aspects of aromaticity. Chem Rev. 2001 May;101(5):1385–420.
6. Zhao L, Grande-Aztatzi R, Foroutan-Nejad C, Ugalde JM, Frenking G. Aromaticity, the Hückel 4 n+2 rule and magnetic current. ChemistrySelect. 2017 Jan 23;2(3):863–70.
7. Solà M. Aromaticity rules. Nat Chem. 2022 Jun;14(6):585–90.
8. Sablon N, De Proft F, Solà M, Geerlings P. The linear response kernel of conceptual DFT as a measure of aromaticity. Phys Chem Chem Phys. 2012 Mar 21;14(11):3960–7.
9. von Ragué Schleyer P, Wu JI, Cossío FP, Fernández I. Aromaticity in transition structures. Chem Soc Rev. 2014 Jul 21;43(14):4909–21.
10. Grèbol-Tomàs J, Matito E, Salvador P. Can aromaticity be evaluated using atomic partitions based on the Hilbert-space? Chem Eur J. 2024 May 21;30(44):e202401282.
11. Feixas F, Matito E, Poater J, Solà M. Quantifying aromaticity with electron delocalisation measures. Chem Soc Rev. 2015 Sep 21;44(18):6434–51.
12. Agranat I. Aromaticity – a theoretical notion. Struct Chem. 2024 Jun;35(3):715–20.
13. Zubarev DY, Averkiev BB, Zhai H-J, Wang L-S, Boldyrev AI. Aromaticity and antiaromaticity in transition-metal systems. Phys Chem Chem Phys. 2008 Jan 14;10(2):257–67.
14. Matito E, Duran M, Solà M. The aromatic fluctuation index (FLU): a new aromaticity index based on electron delocalization. J Chem Phys. 2005 Jan 1;122(1):14109.
15. Fias S, Geerlings P, Ayers P, De Proft F. σ, π aromaticity and anti-aromaticity as retrieved by the linear response kernel. Phys Chem Chem Phys. 2013 Feb 28;15(8):2882–9.
16. Martín N, Scott LT. Challenges in aromaticity: 150 years after Kekulé's benzene. Chem Soc Rev. 2015;44(18):6397–400.

17 Kutzelnigg W. What I like about Hückel theory. J Comput Chem. 2007 Jan 15;28(1):25–34.
18 Ottosson H. Exciting excited-state aromaticity. Nat Chem. 2012 Dec;4(12):969–71.
19 Rosenberg M, Dahlstrand C, Kilså K, Ottosson H. Excited state aromaticity and antiaromaticity: opportunities for photophysical and photochemical rationalizations. Chem Rev. 2014 May 28;114(10):5379–425.
20 Fallon KJ, Budden P, Salvadori E, Ganose AM, Savory CN, Eyre L, et al. Exploiting excited-state aromaticity to design highly stable singlet fission materials. J Am Chem Soc. 2019 Sep 4;141(35):13867–76.
21 Kim J, Oh J, Osuka A, Kim D. Porphyrinoids, a unique platform for exploring excited-state aromaticity. Chem Soc Rev. 2022 Jan 4;51(1):268–92.
22 Liu C, Ni Y, Lu X, Li G, Wu J. Global aromaticity in macrocyclic polyradicaloids: Hückel's rule or Baird's rule? Acc Chem Res. 2019 Aug 20;52(8):2309–21.
23 Yu D, Rong C, Lu T, De Proft F, Liu S. Baird's rule in substituted fulvene derivatives: an information-theoretic study on triplet-state aromaticity and antiaromaticity. ACS Omega. 2018 Dec 31;3(12):18370–9.
24 Rzepa HS. Möbius aromaticity and delocalization. Chem Rev. 2005 Oct;105(10):3697–715.
25 Yoon ZS, Osuka A, Kim D. Möbius aromaticity and antiaromaticity in expanded porphyrins. Nat Chem. 2009 May;1(2):113–22.
26 Mauksch M, Tsogoeva SB. Demonstration of "Möbius" aromaticity in planar metallacycles. Chem Eur J. 2010 Jul 12;16(26):7843–51.
27 Yu D, Rong C, Lu T, Geerlings P, De Proft F, Alonso M, et al. Switching between Hückel and Möbius aromaticity: a density functional theory and information-theoretic approach study. Phys Chem Chem Phys. 2020 Feb 28;22(8):4715–30.
28 Aihara J. Three-dimensional aromaticity of polyhedral boranes. J Am Chem Soc. 1978 May;100(11):3339–42.
29 Fokin AA, Kiran B, Bremer M, Yang X, Jiao H, von Ragué Schleyer P, et al. Which electron count rules are needed for four-center three-dimensional aromaticity? Chem Eur J. 2000 May 2;6(9):1615–1628.
30 King RB. Three-dimensional aromaticity in polyhedral boranes and related molecules. Chem Rev. 2001 May;101(5):1119–52.
31 Poater J, Solà M, Viñas C, Teixidor F. π aromaticity and three-dimensional aromaticity: two sides of the same coin? Angew Chem. 2014 Nov 3;126(45):12387–91.
32 Nozawa R, Kim J, Oh J, Lamping A, Wang Y, Shimizu S, et al. Three-dimensional aromaticity in an antiaromatic cyclophane. Nat Commun. 2019 Aug 8;10(1):3576.
33 Kawashima H, Ukai S, Nozawa R, Fukui N, Fitzsimmons G, Kowalczyk T, et al. Determinant factors of three-dimensional aromaticity in antiaromatic cyclophanes. J Am Chem Soc. 2021 Jul 21;143(28):10676–85.
34 Dewar MJS. σ-conjugation and σ-aromaticity. Bull Soc Chim Belges. 1979 Jan;88(12):957–67.
35 Cremer D. Pros and cons of σ-aromaticity. Tetrahedron. 1988 Jan;44(24):7427–54.
36 Popov IA, Starikova AA, Steglenko DV, Boldyrev AI. Usefulness of the σ-aromaticity and σ-antiaromaticity concepts for clusters and solid-state compounds. Chem Eur J. 2018 Jan 9;24(2):292–305.

37 Wu J, Liu X, Hao Y, Chen H, Su P, Wu W, et al. σ-aromaticity in a fully unsaturated ring. Chem Asian J. 2018 Dec 4;13(23):3691–6.

38 Masui H. Metalloaromaticity. Coord Chem Rev. 2001 Oct; 219–221:957–92.

39 Feixas F, Matito E, Poater J, Solà M. Metalloaromaticity. WIREs Comput Mol Sci. 2013 Mar;3(2):105–22.

40 Rong C, Zhao D, He X, Liu S. Development and applications of the density-based theory of chemical reactivity. J Phys Chem Lett. 2022 Dec 8;13(48):11191–200.

41 He X, Li M, Rong C, Zhao D, Liu W, Ayers PW, et al. Some recent advances in density-based reactivity theory. J Phys Chem A. 2024 Feb 22;128(7):1183–96.

42 Parr RG, Yang W. Density-functional theory of atoms and molecules. Oxford University Press; 1989.

43 Liu S, editor. Conceptual Density Functional Theory: Towards a New Chemical Reactivity Theory. 1st ed. Wiley-VCH GmbH: Wiley; 2022.

44 Liu S. Conceptual density functional theory and some recent developments. Acta Phys-Chim Sin. 2009;25(3):590–600.

45 Geerlings P, De Proft F, Langenaeker W. Conceptual density functional theory. Chem Rev. 2003 May;103(5):1793–873.

46 Geerlings P, Chamorro E, Chattaraj PK, De Proft F, Gázquez JL, Liu S, et al. Conceptual density functional theory: status, prospects, issues. Theor Chem Acc. 2020 Feb;139(2):36.

47 Geerlings P, Fias S, Boisdenghien Z, De Proft F. Conceptual DFT: chemistry from the linear response function. Chem Soc Rev. 2014 Dec 11;43(14):4989–5008.

48 Heidar-Zadeh F, Ayers PW, Verstraelen T, Vinogradov I, Vöhringer-Martinez E, Bultinck P. Information-theoretic approaches to atoms-in-molecules: Hirshfeld family of partitioning schemes. J Phys Chem A. 2018 May 3;122(17):4219–45.

49 Zhao Y, Zhao D, Liu S, Rong C, Ayers PW. Why are information-theoretic descriptors powerful predictors of atomic and molecular polarizabilities. J Mol Model. 2024 Oct 3;30(11):361.

50 Rong C, Wang B, Zhao D, Liu S. Information-theoretic approach in density functional theory and its recent applications to chemical problems. WIREs Comput Mol Sci. 2020 Jul;10(4):e1461.

51 Liu S. Information-theoretic approach in density functional reactivity theory. Acta Phys-Chim Sin. 2016;32(1):98–118.

52 Zhao Y, Zhao D, Rong C, Liu S, Ayers PW. Information theory meets quantum chemistry: a review and perspective. Entropy (Basel). 2025 Jun 16;27(6):644.

53 Breslow R. Antiaromaticity. Acc Chem Res. 1973 Dec;6(12):393–8.

54 Breslow R, Brown J, Gajewski JJ. Antiaromaticity of cyclopropenyl anions. J Am Chem Soc. 1967 Aug;89(17):4383–90.

55 Pierrefixe SCAH, Bickelhaupt FM. Aromaticity: molecular-orbital picture of an intuitive concept. Chem Eur J. 2007;13(22):6321–8.

56 Kuwajima S. Valence bond theory of aromaticity. J Am Chem Soc. 1984 Oct;106(22):6496–502.

57 Solà M. Connecting and combining rules of aromaticity. Towards a unified theory of aromaticity. WIREs Comput Mol Sci. 2019 Jul;9(4):e1404.

58 Chen D, Qiu R, Dong S, Zhu J. Adaptive aromaticity in ruthenacycles. Theor Chem Acc. 2020 Feb;139(2):21.

59 Huang Y, Chen D, Zhu J. Adaptive aromaticity in metallasilapentalynes. Organometallics. 2021 Apr 12;40(7):899–906.

60 Deng Q, Zhu J. Adaptive σ aromaticity in the rhenacyclopropene rings. J Comput Chem. 2023 Nov 5;44(29):2294–301.

61 Levandowski BJ, Zou L, Houk KN. Schleyer hyperconjugative aromaticity and Diels-Alder reactivity of 5-substituted cyclopentadienes. J Comput Chem. 2016 Jan 5;37(1):117–23.

62 Xiao K, Zhao Y, Zhu J, Zhao L. Hyperconjugative aromaticity and protodeauration reactivity of polyaurated indoliums. Nat Commun. 2019 Dec 10;10(1):5639.

63 Chen D, Xie Q, Zhu J. Unconventional aromaticity in organometallics: the power of transition metals. Acc Chem Res. 2019 May 21;52(5):1449–60.

64 Chen S, Imran S, Zhao Y, Zhu J. Probing the limit of the number of saturated atoms for achieving hyperconjugative aromaticity. Inorg Chem. 2024 Jul 29;63(30):14162–70.

65 Dobrowolski JC. Three queries about the HOMA index. ACS Omega. 2019 Nov 12;4(20):18699–710.

66 Choi CH, Kertesz M. Bond length alternation and aromaticity in large annulenes. J Chem Phys. 1998 Apr 22;108(16):6681–8.

67 Krygowski TM, Szatylowicz H, Stasyuk OA, Dominikowska J, Palusiak M. Aromaticity from the viewpoint of molecular geometry: application to planar systems. Chem Rev. 2014 Jun 25;114(12):6383–422.

68 Herndon WC. Resonance energies of aromatic hydrocarbons. Quantitative test of resonance theory. J Am Chem Soc. 1973 Apr;95(7):2404–6.

69 Glukhovtsev M. Aromaticity today: energetic and structural criteria. J Chem Educ. 1997 Jan;74(1):132.

70 Suresh CH, Koga N. An isodesmic reaction based approach to aromaticity of a large spectrum of molecules. Chem Phys Lett. 2006 Feb;419 (4–6):550–6.

71 Zhu J, An K, von Ragué Schleyer P. Evaluation of triplet aromaticity by the isomerization stabilization energy. Org Lett. 2013 May 17;15(10):2442–5.

72 Huang D, Liu W, Zheng Y, Feng R, Chai Z, Wei J, et al. Nonplanar aromaticity of dinuclear rare-earth metallacycles. J Am Chem Soc. 2024 May 22;146(22):15609–15618.

73 Zhao D, Liu S, Rong C, Zhong A, Liu S. Toward understanding the isomeric stability of fullerenes with density functional theory and the information-theoretic approach. ACS Omega. 2018 Dec 31;3(12):17986–90.

74 Chen Z, Wannere, CS, Corminboeuf, C, Puchta R, von Ragué Schleyer P. Nucleus-Independent chemical shifts (NICS) as an aromaticity criterion. Chem Rev. 2005 Sept 29; 105(10):3842–88.

75 Inostroza D, García V, Yañez O, Torres-Vega JJ, Vásquez-Espinal A, Pino-Rios R, et al. On the NICS limitations to predict local and global current pathways in polycyclic systems. New J Chem. 2021;45(18):8345–51.

76 Gershoni-Poranne R, Stanger A. Magnetic criteria of aromaticity. Chem Soc Rev. 2015 Sep 21;44(18):6597–615.

77 Poater J, Fradera X, Duran M, Solà M. The delocalization index as an electronic aromaticity criterion: application to a series of planar polycyclic aromatic hydrocarbons. Chem Eur J. 2003 Jan 20;9(2):400–6.

78 Kalescky R, Kraka E, Cremer D. Description of aromaticity with the help of vibrational spectroscopy: anthracene and phenanthrene. J Phys Chem A. 2014 Jan 9;118(1):223–37.

79 Giambiagi M, Segre de Giambiagi M, dos Santos Silva CD, Paiva de Figueiredo A. Multicenter bond indices as a measure of aromaticity. Phys Chem Chem Phys. 2000;2(15):3381–92.

80 Bultinck P, Ponec R, Van Damme S. Multicenter bond indices as a new measure of aromaticity in polycyclic aromatic hydrocarbons. J Phys Org Chem. 2005 Aug;18(8):706–18.

81 Güell M, Matito E, Luis JM, Poater J, Solà M. Analysis of electron delocalization in aromatic systems: individual molecular orbital contributions to para-delocalization indexes (PDI). J Phys Chem A. 2006 Oct 12;110(40):11569–74.

82 Aihara J-I. Magnetic resonance energy and topological resonance energy. Phys Chem Chem Phys. 2016 Apr 28;18(17):11847–57.

83 Szczepanik DW, Andrzejak M, Dominikowska J, Pawełek B, Krygowski TM, Szatylowicz H, et al. The electron density of delocalized bonds (EDDB) applied for quantifying aromaticity. Phys Chem Chem Phys. 2017 Nov 1;19(42):28970–81.

84 Bader RFW. Atoms in Molecules. Oxford University Press; 1989.

85 Becke AD, Edgecombe KE. A simple measure of electron localization in atomic and molecular systems. J Chem Phys. 1990 May;92(9):5397–403.

86 Savin A, Nesper R, Wengert S, Fässler TF. ELF: the electron localization function. Angew Chem Int Ed. 1997 Sep 17;36(17):1808–32.

87 Yu D, Rong C, Lu T, Chattaraj PK, De Proft F, Liu S. Aromaticity and antiaromaticity of substituted fulvene derivatives: perspectives from the information-theoretic approach in density functional reactivity theory. Phys Chem Chem Phys. 2017 Jul 19;19(28):18635–45.

88 Deng Y, Yu D, Cao X, Liu L, Rong C, Lu T, et al. Structure, aromaticity and reactivity of corannulene and its analogues: a conceptual density functional theory and density functional reactivity theory study. Mol Phys. 2017 Nov 30;116:1–13.

89 He X, Yu D, Wu J, Wang B, Rong C, Chattaraj PK, et al. Towards understanding metal aromaticity in different spin states: a density functional theory and information-theoretic approach analysis. Chem Phys Lett. 2020 Dec;761:138065.

90 Yu D, Stuyver T, Rong C, Alonso M, Lu T, De Proft F, et al. Global and local aromaticity of acenes from the information-theoretic approach in density functional reactivity theory. Phys Chem Chem Phys. 2019 Aug 21;21(33):18195–210.

91 He X, Li M, Yu D, Wang B, Zhao D, Rong C, et al. Conformational changes for porphyrinoid derivatives: an information-theoretic approach study. Theor Chem Acc. 2021 Sep;140(9):123.

14

Catalysis

TABLE OF CONTENTS

14.1 Introduction

Following the preceding chapters that examined how electronic structure and density-based principles govern chemical reactivity, we now turn to catalysis—the most powerful and far-reaching manifestation of controlled chemical change.

A brief remark on scope is warranted. Although catalysis is a vast and mature discipline, the author's own research contributions to this field have been modest and primarily conceptual, emerging from density-based reactivity theory rather than traditional catalyst design or mechanistic specialization. Accordingly, this chapter does not attempt to survey catalysis in its full breadth. Instead, its aim is to introduce key principles, clarify the conceptual foundations, and show how density-based approaches can provide a unified and powerful hierarchical framework for understanding catalytic reactivity.

14.1.1 What Is Catalysis?

Catalysis lies at the heart of chemistry [1–4]. It is the science—and the art—of accelerating chemical transformations by means of a substance, the catalyst, which enables the reaction without being consumed in the overall process. A catalyst provides a more kinetically accessible reaction pathway, enabling the system to reach equilibrium more rapidly without altering its thermodynamic endpoint. This rate enhancement can be dramatic, transforming processes that might take years under normal conditions into reactions that complete in seconds or even shorter.

The defining feature of a catalyst is *regenerability*: it participates in the reaction mechanism but is restored at the end of each catalytic cycle, ready to act again. This distinguishes catalysis from stoichiometric reagents and makes it fundamentally more efficient, particularly in systems where sustainability, selectivity, and energy economy are paramount.

Catalysis is not merely a tool of chemistry—it is a unifying principle across the natural and technological worlds. In biological systems, enzymes, nature's catalysts, govern every facet of life, from metabolism to DNA replication, operating with astounding precision and speed under mild conditions. In industry, catalysts are essential to over 90% of all chemical manufacturing processes, playing crucial roles in the synthesis of fertilizers (e.g., the Haber–Bosch process), fuels, polymers, and pharmaceuticals. Heterogeneous catalysts [2], typically solids acting on gaseous or liquid reactants, are the workhorses of petrochemical and environmental chemistry, while homogeneous catalysts, often soluble metal complexes or organic molecules, enable finely controlled molecular transformations with exceptional specificity.

14.1.2 A Brief History of Catalysis

The concept of catalysis, though rooted in ancient practices such as fermentation and metallurgy, was not formally articulated until the nineteenth century. It was the Swedish chemist Jöns Jakob Berzelius [5], who, in 1835, first introduced the term *catalysis* (from the Greek *kata-*, meaning "down," and *lysis*, meaning "loosening" or "breaking"), to describe the action of certain substances that accelerate chemical reactions without themselves being consumed. Berzelius noted phenomena such as the decomposition of hydrogen peroxide in the presence of manganese dioxide, recognizing that some reactions proceeded only when a "foreign substance" was present.

However, the precise nature of this mysterious facilitation remained unclear until later in the century. It was Wilhelm Ostwald [6], a German physical chemist and one of the founders of modern chemical kinetics, who gave catalysis a rigorous scientific definition. In 1894, Ostwald proposed that a catalyst is a substance that changes the rate of a chemical reaction without altering its final equilibrium—a definition that emphasized the kinetic rather than thermodynamic nature of catalytic action. This conceptual breakthrough earned Ostwald the 1909 Nobel Prize in Chemistry, and his formulation remains foundational in chemical theory.

Another milestone was achieved by Paul Sabatier [7], who was awarded the 1912 Nobel Prize for his work on the hydrogenation of organic compounds using finely divided nickel. Sabatier's investigations into surface-catalyzed reactions laid the groundwork for heterogeneous catalysis, which would later dominate industrial applications. He demonstrated not only that metals could facilitate hydrogen addition but that catalytic activity could be systematically studied and optimized.

The twentieth century saw catalysis emerge as an industrial powerhouse. Among the most transformative developments was the Haber–Bosch process [8], developed by Fritz Haber and Carl Bosch in the early 1900s. This high-pressure synthesis of ammonia from nitrogen and hydrogen, using an iron-based catalyst, revolutionized global agriculture by making nitrogen fertilizers widely available, effectively feeding a growing world population. It remains one of the most energy-intensive and essential catalytic processes in human history.

In the mid-twentieth century, Karl Ziegler [9] and Giulio Natta [10] introduced a new era with their discovery of catalysts for the polymerization of olefins. Their Ziegler–Natta catalysts [9], based on transition metals like titanium and aluminum compounds, enabled the controlled synthesis of polyethylene and polypropylene—materials that now form the backbone of the plastics industry. Their achievement, recognized by the 1963 Nobel Prize in Chemistry, showcased the power of homogeneous and supported catalysts in controlling molecular architecture with precision.

The late twentieth and early twenty-first centuries have witnessed the expansion and diversification of catalysis. The rise of organocatalysis, popularized by scientists like Benjamin List and David MacMillan (who shared the 2021 Nobel Prize) [11], introduced small organic molecules as efficient catalysts for asymmetric reactions. Meanwhile, biocatalysis has flourished through advances in enzyme engineering and synthetic biology, enabling environmentally benign and highly selective transformations.

Photocatalysis and electrocatalysis have also surged to the forefront, especially in the context of renewable energy and climate change mitigation. These fields aim to drive key reactions—such as water splitting, CO_2 reduction, and nitrogen fixation—using solar or electrical energy. Modern catalyst design increasingly integrates principles from materials science, computational chemistry, and machine learning (ML), marking a shift from empirical discovery to rational, theory-guided development.

Today, catalysis is inseparable from the vision of green chemistry and sustainable development. It underpins technologies for cleaner energy, closed-loop recycling, pollution control, and the valorization of waste streams. As we move toward a carbon-neutral future, catalytic innovation remains central to solving global challenges.

What began as a curious acceleration in chemical reactions observed by nineteenth-century pioneers has evolved into a cornerstone of modern science and technology. The history of catalysis is not just a chronicle of scientific achievements—it is a testament to chemistry's power to reshape the world through controlled transformation.

14.1.3 Types of Catalysis

Catalysis manifests in several distinct forms [1], each defined by the physical phase of the catalyst relative to the reactants and the nature of catalytic interaction. The two classical categories are homogeneous and heterogeneous catalysis. In *homogeneous catalysis*, the catalyst and reactants exist in the same phase—typically as molecules dissolved in a solution. This type of catalysis allows for precise control of reaction conditions and is common in organometallic and coordination chemistry. *Heterogeneous catalysis*, by contrast, involves a catalyst in a different phase, most often a solid interacting with gaseous or liquid reactants. These systems dominate industrial applications like petrochemical refining, catalytic converters, and ammonia synthesis, where surface interactions between reactants and catalysts play a central role.

Beyond these classical divisions, enzymatic catalysis, or biocatalysis, leverages nature's most refined molecular machines: enzymes. These protein-based catalysts operate under mild conditions with remarkable selectivity, often achieving reaction rates orders of magnitude faster than synthetic counterparts. The specificity of enzyme–substrate recognition is governed by complex three-dimensional folding, dynamic conformational changes, and intricate active site environments. Advances in protein engineering and synthetic biology have expanded the use of enzymes in pharmaceuticals, green chemistry, and even nonnatural transformations, making biocatalysis a vibrant frontier of research and application.

More recently, the field has expanded to prominently include organocatalysis, photocatalysis, and electrocatalysis. *Organocatalysis* utilizes small organic molecules—such as proline or imidazoles—as catalysts for enantioselective and carbon–carbon bond-forming reactions, often without the need for metals. *Photocatalysis* employs light energy (typically from visible or UV sources) to excite catalysts into reactive states, enabling transformations such as water splitting and CO_2 reduction. *Electrocatalysis* uses applied electric potentials to drive redox reactions, crucial for sustainable energy technologies like fuel cells and electrolyzers. These emerging types of catalysis increasingly blur the boundaries between disciplines, requiring an integrated understanding of chemistry, materials science, and physics to unlock their full potential.

14.1.4 Key Parameters and Concepts in Catalysis

At the core of catalytic science are several fundamental parameters that define how efficiently and selectively a catalyst promotes a chemical reaction. One of the most crucial concepts is the activation energy (E_a)—the energy barrier that must be overcome for reactants to transform into products. A catalyst lowers this barrier by providing an alternative reaction pathway, thereby increasing the rate at which equilibrium is reached without altering the thermodynamics of the system. This principle underscores the kinetic nature of catalysis: catalysts do not change the position of equilibrium; they simply allow the system to get there faster.

Two quantitative metrics widely used to assess catalyst performance are the turnover number (TON) and turnover frequency (TOF). TON refers to the total number of catalytic cycles a catalyst can perform before deactivation, effectively measuring its lifetime or robustness. A catalytic cycle is a sequence of elementary steps through which a catalyst facilitates a chemical reaction, returning to its original state at the end of each cycle. TOF, on the other hand, represents the number of catalytic cycles per unit time—typically expressed in s^{-1}—and thus serves as a direct indicator of catalytic activity. High TON and TOF values are desirable in both academic and industrial contexts, as they imply efficient utilization of catalyst materials and reduced operational costs.

Equally important are the concepts of selectivity and efficiency. Selectivity describes the catalyst's ability to favor one reaction pathway or product over others, which is particularly critical in complex organic or biochemical transformations where multiple competing reactions may occur. In practice, efficiency encompasses both selectivity and activity and often incorporates environmental and economic considerations such as atom economy, energy input, and byproduct minimization. Together, these parameters—activation energy, TON, TOF, selectivity, and efficiency—form the essential vocabulary of catalytic science and provide the basis for comparing and designing catalysts across diverse chemical systems.

Beyond global rate and selectivity metrics such as TON and TOF, catalytic performance also depends critically on catalyst speciation. In catalytic systems, the resting state refers to the most abundant form of the catalyst under steady-state conditions, often serving as a stable reservoir rather than the active species itself. Identifying it is critical for accurate TOF calculations, understanding catalyst efficiency, and guiding rational design. In contrast, off-cycle species are catalyst-derived forms that do not participate in the productive catalytic cycle—such as decomposition products or kinetically trapped intermediates—which can cause inefficiencies, promote catalyst deactivation, or complicate purification and recovery. Detecting and minimizing off-cycle pathways is therefore essential for improving catalyst performance and stability, often using tools like in situ spectroscopy and computational modeling.

14.1.5 Catalysis: The Holy Grail of Chemistry

Catalysis is often described as the *holy grail of chemistry*—a term that reflects both its immense practical importance and its profound scientific challenge. At its heart, catalysis is the art of accelerating chemical reactions by providing alternative pathways with lower barriers, all while remaining unchanged in the process. This seemingly simple concept has far-reaching implications, touching every corner of chemistry, industry, and sustainable development [12].

From the synthesis of fertilizers via the Haber–Bosch process to the production of fuels, pharmaceuticals, and plastics, catalysis enables reactions that would otherwise be too slow, inefficient, or economically unfeasible. Its ability to increase reaction rates, reduce energy demands, and steer selectivity toward desired products makes it the engine of chemical innovation and manufacturing.

But the importance of catalysis extends far beyond efficiency. It is central to building a sustainable, circular chemical economy. Catalysts are indispensable in carbon capture and utilization, hydrogen production from water splitting, biomass conversion, and the breakdown of persistent pollutants. Without advances in catalysis, addressing climate change, resource scarcity, and environmental degradation would be virtually impossible.

Scientifically, catalysis represents one of chemistry's grandest challenges. Designing a catalyst that is active, selective, stable, and made from abundant materials requires control over atomic and electronic structure, surface properties, and dynamic reaction environments. This demands deep integration of quantum chemistry, materials science, kinetics, spectroscopy, and increasingly, data science and ML. Catalysis is where theory meets application—where fundamental understanding translates directly into transformative technologies.

Perhaps most compelling is the elegance of catalysis itself. A good catalyst doesn't just make a reaction faster; it makes it *better*—cleaner, more controlled, more purposeful. It embodies the ideals of atom economy, precision synthesis, and minimal waste. In asymmetric catalysis, for example, chiral products can be made with exquisite selectivity, enabling the efficient production of life-saving drugs.

Catalysis is the holy grail not only for what it accomplishes, but also for what it aspires to: the ideal of perfect control over chemical change. It is both a tool and a dream—powerful, essential, and still, in many ways, elusive. As such, it remains the central quest in chemistry's pursuit to understand and engineer the molecular world.

14.2 Theoretical Aspects of Catalysis

Having outlined the conceptual foundations of catalysis, we now turn to the theoretical principles that govern catalytic reactivity and mechanism. Theoretical catalysis [13–16] provides the foundational framework that links catalytic reactivity to atomic structure, energy landscapes, and mechanistic pathways. By integrating physical principles with chemical intuition, theory enables the interpretation of experimental data, the design of new catalysts, and the prediction of reactivity trends, connecting thermodynamic and kinetic quantities to atomic and electronic structures in models that reveal the inner workings of catalytic systems. This section outlines the principal theories, models, and approaches that underpin our understanding of catalysis.

Potential Energy Surfaces and Reaction Coordinates. Catalyzed reactions proceed through a complex landscape of energetic states. The potential energy surface (PES) represents the multidimensional energy landscape that maps nuclear configurations to potential energy. A reaction coordinate traces the minimum energy path between reactants and products via a transition state. Catalysts function by reshaping this PES, typically lowering the energy of the transition state and thereby reducing the activation energy (E_a). Understanding the topology of the PES is crucial for identifying intermediates, transition states, and the overall reaction mechanism.

Transition State Theory (TST). One of the foundational pillars of theoretical catalysis is TST [17, 18], which describes how molecules surmount energy barriers to form products. According to TST, the rate of a reaction is governed by the concentration of species at the top of the energy barrier (the transition state) and their frequency of crossing to form products. The Eyring equation quantifies this rate, accounting for entropic and enthalpic contributions. Catalysts enhance reaction rates by stabilizing the transition state, effectively reducing the free energy difference between it and the reactants. Although TST provides the dominant classical description of barrier crossing, it does not account for quantum mechanical effects such as tunneling or nonadiabatic transitions, which can become significant in enzymatic, low-temperature, or light-driven catalytic systems. These nonclassical pathways motivate extensions of TST and the use of quantum dynamics in certain catalytic contexts.

Sabatier Principle and Volcano Plots. A guiding principle in heterogeneous catalysis is the Sabatier Principle [1, 19, 20], which states that optimal catalytic activity occurs when the interaction between catalyst and reactants is neither too strong nor too weak. This concept gives rise to volcano plots, which graph catalytic activity as a function of binding strength for a given intermediate. The apex of the volcano represents the best compromise between activation and desorption energies. This principle has been instrumental in understanding catalytic performance in processes like hydrogen evolution and ammonia synthesis.

Brønsted–Evans–Polanyi (BEP) Relationships. The BEP relationship [21, 22] posits a linear correlation between the activation energy of a reaction and the enthalpy change (or adsorption energy) associated with that reaction. This empirical rule allows researchers to estimate kinetic parameters from thermodynamic data, streamlining the search for active catalysts. BEP relationships form the basis for many scaling laws and are integral to high-throughput computational screening of catalyst materials.

Scaling Relationships and Descriptor-based Design. Linear scaling laws, such as those relating adsorption energies of chemically similar intermediates, are widely used in heterogeneous catalysis. These relationships simplify complex reaction networks by reducing the number of independent variables. Combined with BEP relationships, they enable the identification of key descriptors (e.g., adsorption energy of a single species) that correlate strongly with catalytic activity. This has led to rational catalyst design strategies and activity maps for various catalytic reactions.

Microkinetic Modeling. Microkinetic models [15, 20, 23] incorporate detailed mechanisms consisting of elementary steps, each with associated rate constants and thermodynamic parameters. These models solve the coupled differential equations governing surface coverages and reaction rates under steady-state or dynamic conditions. Microkinetic modeling enables the identification of rate-determining steps (RDSs), reaction bottlenecks, and optimal operating conditions. It provides a quantitative link between theory and experiment, especially when paired with density functional theory (DFT)–derived energetics.

Computational Chemistry and First-Principles Simulations. DFT has revolutionized theoretical catalysis by allowing the calculation of electronic structure, adsorption energies, and activation barriers with reasonable computational cost. DFT provides atomistic insight into surface-adsorbate interactions, electronic properties, and charge transfer processes. It plays a central role in generating input for microkinetic models and testing hypotheses about reaction mechanisms. Despite its limitations (e.g., treatment of dispersion forces, solvent effects), DFT remains the workhorse of computational catalysis.

Surface Reaction Models. In heterogeneous catalysis, classical surface reaction models such as Langmuir–Hinshelwood (both reactants adsorb), Eley–Rideal (one reactant adsorbed), and Mars–van Krevelen (involving lattice oxygen) provide mechanistic frameworks for interpreting reaction pathways. These models describe how species adsorb, migrate, react, and desorb from catalyst surfaces. They are often combined with kinetic expressions to derive rate laws used in reactor design and catalyst screening.

Enzymatic Catalysis and Biocatalysis Models. Enzymes are nature's catalysts, and their theoretical treatment involves models such as the lock-and-key and induced-fit hypotheses. Michaelis–Menten kinetics provides a simplified but widely applicable framework for enzymatic rate analysis. Advanced models also incorporate transition state stabilization, quantum tunneling, and protein dynamics, offering insights into how biological catalysts achieve extraordinary rates and specificity. In certain catalytic systems, especially at low temperatures or in enzymatic contexts, quantum mechanical effects like tunneling can become significant. Nonclassical isotope effects and zero-point energy contributions must then be considered. Molecular dynamics and path-integral approaches allow the exploration of such phenomena beyond static PES models.

Emerging Theories and Data-Driven Approaches. Recent years have seen the integration of ML and data science into theoretical catalysis [24, 25]. ML models trained on DFT datasets can rapidly predict key descriptors, explore chemical space, and even suggest new catalyst compositions. Multiscale modeling approaches combine quantum mechanics, molecular mechanics, and continuum models to bridge length and time scales. These innovations promise to enhance predictive accuracy and design efficiency in catalysis research.

Theoretical catalysis is thus a rich and rapidly evolving field, encompassing both fundamental principles and cutting-edge methodologies. Together, these theories and models offer a roadmap for understanding how catalysts work and how they can be improved to meet future challenges in energy, sustainability, and chemical manufacturing.

14.3 Understanding Catalysis with Physicochemical Effects

To complement the theoretical foundations outlined in Section 14.2, we now turn to a density-based perspective [26, 27] on catalytic reactivity. While conventional models such as TST, BEP relationships, and scaling laws describe how energies and rates correlate, they provide limited insight into the underlying physicochemical factors that govern reactivity along the reaction coordinate. To overcome this limitation, we recently introduced a density-based reactivity model [28, 29] that decomposes the intrinsic reaction coordinate (IRC) into physically meaningful segments and then analyzes each using conceptual density functional theory (CDFT) [30, 31] and information-theoretic approach (ITA) [32, 33] descriptors. This approach reveals the distinct steric and electronic factors that control reactivity at each stage of the catalytic process.

To that end, we recently developed an analytical approach that partitions the PES along the IRC into three distinct segments [28, 29]: the pre-TS (before the transition state), the TS (at the transition state), and the post-TS (after the transition state) regions (Figure 14.1). For each segment, we applied density-based descriptors from

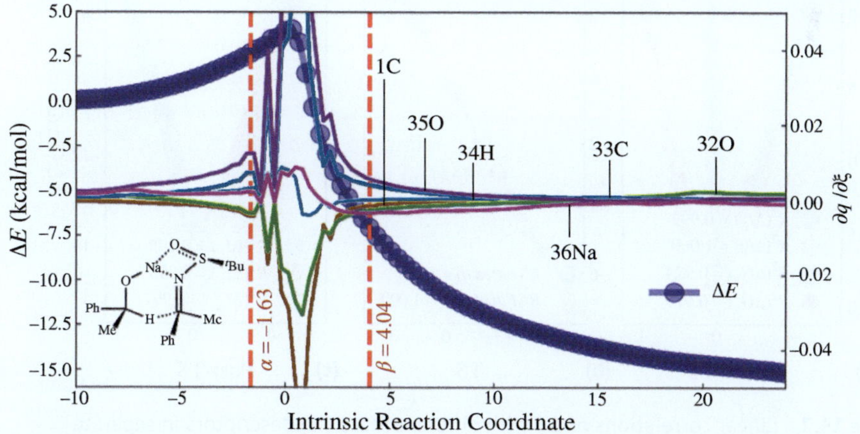

Figure 14.1 The PES and charge fluctuation of six atoms of transition-metal-free *N*-alkylation of amines reaction C01 along the intrinsic reaction path.
Source: Reproduced from Ref. [29] with permission of John Wiley & Sons.

CDFT and ITA to identify the dominant reactivity-driving factors. Our analysis reveals a clear trend: steric effects play a primary role in both the pre-TS and post-TS regions, while the intrinsic electrophilicity and nucleophilicity of the transition state itself are the key determinants of reactivity in the TS region. This segmentation offers a more nuanced understanding of how structure and electronic properties govern catalysis at different stages of a reaction path.

To illustrate the strong correlation between reaction energy (ΔE) and the dominant physicochemical effects in different segments of the reaction pathway, Figure 14.2 presents linear regression plots for each region: Figure 14.2a shows ΔE versus three steric effect descriptors in the pre-TS region; Figure 14.2b displays ΔE versus two intrinsic electrophilic and nucleophilic propensity descriptors for the TS region; and Figure 14.2c presents ΔE versus two steric descriptors for the post-TS region. All correlations exhibit linear correlation coefficients exceeding 0.9, providing strong quantitative support for two key conclusions: (i) steric effects and intrinsic charge-transfer propensities serve as reliable and robust reactivity descriptors within their respective regions, and (ii) multiple descriptors may capture similar underlying effects—whether steric or electronic—highlighting the richness and complementarity of conceptual reactivity tools. The validity and broad applicability of this result have been demonstrated through its successful application to 15 simple organic and organometallic reactions and to 22 more complex transformations from homogeneous catalysis (Figure 14.3).

These findings imply that catalyst optimization should target steric control in pre-TS and post-TS regions—for example, ligand bulk, coordination geometry, confinement effects—while tuning electronic properties primarily influences the TS region. This segmentation therefore provides a design principle linking mechanistic

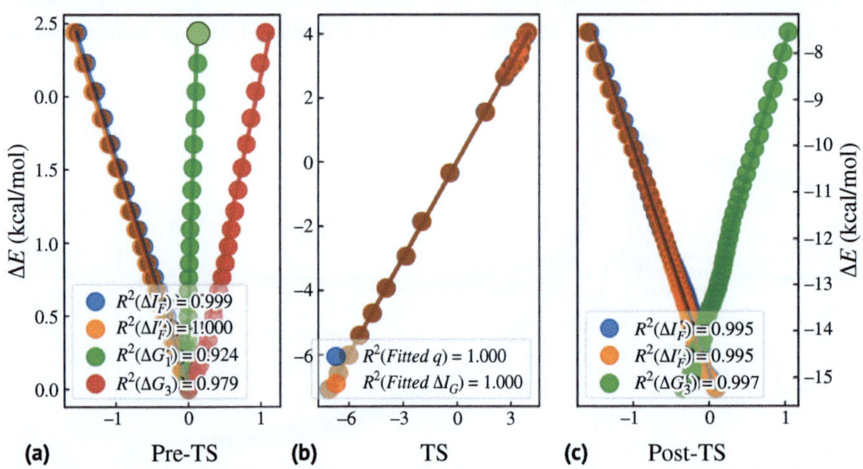

Figure 14.2 Linear correlations of ΔE with different reactivity descriptors in separate segments. (a) For pre-TS region, ΔE versus ΔI_F, $\Delta I'_F$, ΔG_1 and ΔG_3; (b) for the TS region, ΔE versus multivariate fitting with the change of both Hirshfeld charge and information gain; and (c) for the post-TS region, ΔE versus ΔI_F, $\Delta I'_F$ and ΔG_3. *Source*: Reproduced from Ref. [29] with permission of John Wiley & Sons.

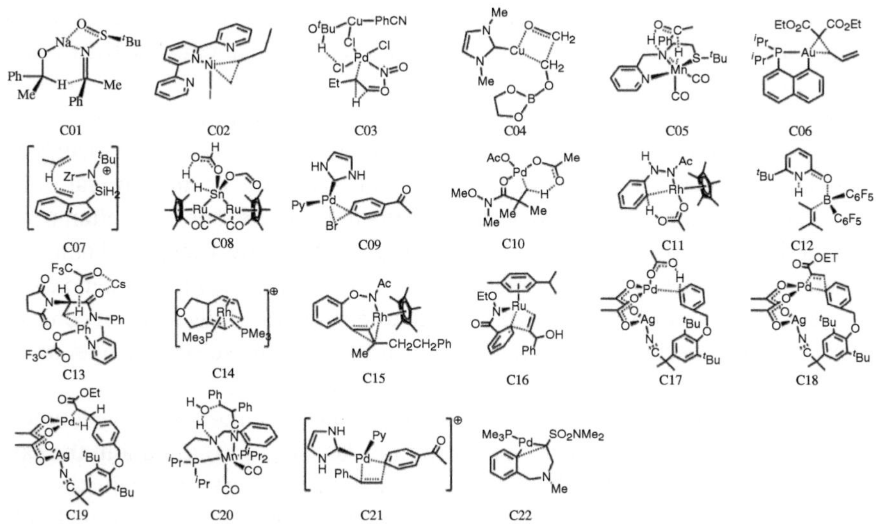

Figure 14.3 Selected catalysis systems studied for the reactivity partition scheme. *Source*: Reproduced from Ref. [29] with permission of John Wiley & Sons.

stages with catalyst structure. These density-based insights into reactivity now can set the stage for examining how catalysts operate through organized sequences of elementary steps—namely, catalytic cycles.

14.4 Catalytic Cycle and Mechanistic Aspects of Catalysis

14.4.1 Anatomy of a Catalytic Cycle

A catalytic cycle [34, 35] is a conceptual and mechanistic representation of the sequence of elementary steps by which a catalyst facilitates the transformation of reactants into products. In this cycle, the catalyst participates in one or more intermediate forms but is regenerated at the end of each cycle, enabling it to be reused continuously. Each step typically involves bond-making and bond-breaking events, such as substrate coordination, activation, transformation, and product release. The cycle highlights not only the roles played by the catalyst but also provides insight into how specific reaction intermediates are stabilized or destabilized during the reaction.

The observed rate of a catalytic cycle depends on the kinetics of the individual elementary steps that compose the cycle. Under steady-state conditions—where concentrations of reactive intermediates remain relatively constant—the overall rate is often governed by the RDS, which is typically the slowest and highest-energy transition in the cycle. In simple cases, the rate law can be derived based on the concentration of the reactants and the kinetics of the RDS, sometimes modified by equilibrium or pre-equilibrium steps that precede it. In more complex catalytic networks, microkinetic modeling is employed to account for the interplay between multiple pathways and steps, allowing a more accurate prediction of the overall reaction rate.

Importantly, experimental measurement of catalytic rates often involves macroscopic observables such as TOF or TON, which integrate over many catalytic cycles. These values reflect the efficiency and activity of the catalyst in real time, but to relate them to individual molecular events in the cycle, one must often combine kinetic data with mechanistic studies (e.g., spectroscopy, isotopic labeling, or computational modeling). Thus, determining the observed rate of a catalytic cycle involves not only identifying the RDS but also understanding how reactant concentrations, catalyst speciation, and environmental factors (e.g., solvent, temperature, and pressure) influence the full catalytic landscape.

Although catalytic cycles vary in their molecular specifics, they generally follow a common mechanistic framework comprising five key stages: (i) initiation: the catalyst is activated, often through processes such as ligand exchange, reduction, or solvation, to generate its reactive form; (ii) substrate activation: the substrate binds to the catalyst, positioning itself for subsequent transformation; (iii) transformation of intermediates: critical bond-forming or bond-breaking events take place, driving the chemical conversion forward; (iv) product release: the newly formed product dissociates from the catalyst, freeing up the coordination site; and (v) catalyst regeneration: the catalyst returns to its original active state, ready to enter a new catalytic cycle.

A well-known illustration of this sequence is found in the Wilkinson hydrogenation cycle. In this homogeneous catalytic system, (i) the oxidative addition of dihydrogen to the Rh(I) complex occurs first, (ii) followed by coordination of the alkene substrate, (iii) migratory insertion of the alkene into the metal–hydride bond, and, finally, (iv) reductive elimination releases the alkane product while regenerating the Rh(I) catalyst. This archetype exemplifies the elegance and efficiency of organometallic catalysis (Figure 14.4).

14.4.2 Common Elementary Steps in Homogeneous Catalysis

Building on the general architecture of catalytic cycles introduced in Section 14.4.1, we now focus on the elementary organometallic transformations that constitute the mechanistic vocabulary of homogeneous catalysis. These steps govern how the catalyst activates substrates, forms key intermediates, and releases products. Homogeneous catalysis, where the catalyst and reactants are in the same phase (typically liquid), often involves molecular complexes. The elementary steps here focus on bond making and breaking within coordination spheres of metal centers. Common steps include oxidative addition, reductive elimination, migratory insertion, β-hydride elimination, ligand exchange, acid/base activation, and π-complexation.

Oxidative Addition and Reductive Elimination. Among the most defining steps in organometallic catalysis are oxidative addition and reductive elimination. These two

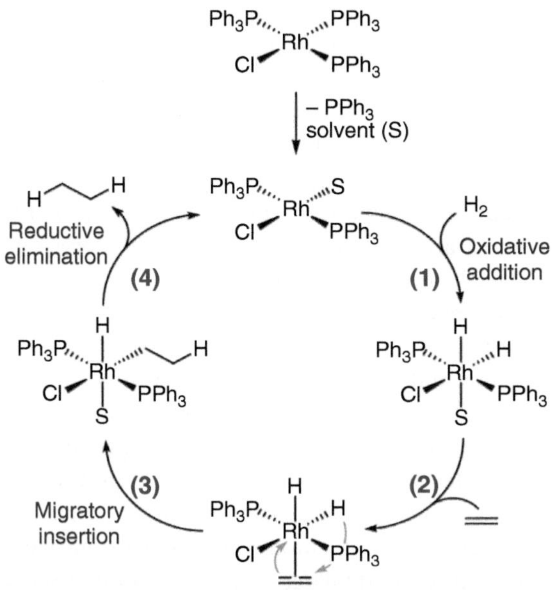

Figure 14.4 The catalytic cycle of Wilkinson hydrogenation.

reactions are mechanistically complementary and often bookend catalytic cycles such as those involved in cross-coupling reactions (e.g., Suzuki, Heck, and Negishi reactions). In oxidative addition, a metal complex inserts into a σ-bond—typically H—H, C—H, or C—X (X = halogen)—resulting in an increase in both oxidation state (typically by two units) and coordination number of the metal center. This step activates otherwise inert bonds and transforms the metal into a more reactive species capable of subsequent transformations. Conversely, reductive elimination forms a new σ-bond between two ligands bound to the metal center, thereby reducing its oxidation state and often restoring its original coordination environment. This step is typically responsible for product release in catalytic cycles. The efficiency of reductive elimination is influenced by the electronic and steric environment around the metal, as well as the relative positioning of the ligands involved.

Migratory Insertion and β-hydride Elimination. Migratory insertion involves the insertion of a π-bound ligand—such as CO, alkenes, or alkynes—into a metal–ligand σ-bond. For instance, in alkene insertions, the alkyl group migrates to the alkene, forming a new carbon–carbon bond and a metal–alkyl intermediate. This step is central to polymerization, hydroformylation, and carbonylation reactions, among others. β-Hydride elimination, on the other hand, is often a decomposition or rearrangement pathway, but it can also serve as a strategic bond-breaking step in synthetic catalysis. In this step, a β-hydrogen (on the carbon adjacent to the metal-bound carbon) is transferred to the metal, forming a metal–hydride species and an alkene product. This reaction plays a critical role in alkene formation, chain termination, and isomerization processes.

Ligand Exchange Mechanisms. Catalytic efficiency is highly dependent on the ability of the metal center to exchange ligands—enabling substrate coordination and product release. Ligand exchange can proceed through associative, dissociative, or interchange mechanisms. Associative pathways involve the addition of a new ligand before the displacement of the original, typically increasing the coordination number temporarily. In contrast, dissociative pathways involve the loss of a ligand prior to binding a new one, reducing the metal's coordination number in the intermediate state. These mechanisms are heavily influenced by the steric and electronic properties of the metal and ligands, as well as by solvent polarity and donor capability. Understanding the operative exchange pathway is critical for tuning catalyst lifetime and selectivity.

Acid/Base Activation. In both organocatalysis and biocatalysis, proton transfer events—central to acid/base catalysis—activate substrates toward nucleophilic or electrophilic attack. This activation can involve direct protonation or deprotonation or the formation of reactive intermediates such as enamines or iminium ions. In enzymatic catalysis, amino acid residues such as histidine, glutamate, or aspartate often function as general acids or bases, orchestrating proton shuttling during catalysis. Organocatalytic systems, such as those employing proline or chiral amines, similarly rely on acid/base interactions to modulate substrate reactivity and stereoselectivity. These transformations underscore the importance of pK_a matching, solvent environment, and intramolecular hydrogen bonding.

π-Complexation. A hallmark of many transition metal catalysts is their ability to engage in π-complexation with unsaturated molecules like alkenes, alkynes, and arenes. Through π-bond donation from the substrate to the metal and subsequent π-back-donation from the metal to the substrate's antibonding orbitals, the electron density on the substrate is perturbed, rendering it more susceptible to nucleophilic or electrophilic attack. This dynamic interaction modulates the reactivity of the π-system and facilitates transformations such as olefin metathesis, cycloaddition, and electrophilic activation. The strength and nature of π-complexation are sensitive to both the electronic properties of the metal and the ligand field, providing a tunable handle for catalytic design.

Together, these elementary steps compose the mechanistic vocabulary of catalysis. Mastery of these transformations is key not only for interpreting mechanistic pathways but also for designing new catalytic systems with enhanced activity, selectivity, and sustainability.

14.4.3 Common Elementary Steps in Heterogeneous Catalysis

Heterogeneous catalysis, where the catalyst and reactants are in different phases (typically a solid catalyst and gaseous or liquid reactants), is particularly prevalent in industrial applications. The elementary steps in heterogeneous catalysis often involve interactions at the catalyst surface. The most common elementary steps include:

Adsorption (Chemisorption and Physisorption). The very first step in many heterogeneous catalytic reactions is the adsorption of reactant molecules onto the catalyst surface. This can occur via two primary mechanisms. (i) *Physisorption*: A weak, nonspecific interaction driven by van der Waals forces. It's reversible and typically occurs at low temperatures, forming multiple layers of adsorbed molecules. Physisorption is generally not directly involved in bond breaking or forming for the catalytic reaction itself, but it can be a precursor to chemisorption. (ii) *Chemisorption*. A stronger, more specific interaction involving the formation of chemical bonds (covalent or ionic) between the adsorbate and the catalyst surface. This process is often irreversible and typically forms a monolayer. Chemisorption is crucial because it activates the reactant molecules, weakening existing bonds or creating new reactive sites, making them more susceptible to reaction. For example, hydrogen molecules (H_2) might dissociatively chemisorb onto a metal surface, breaking into individual hydrogen atoms (H) that are highly reactive.

Surface Diffusion. Once adsorbed, molecules or atoms are not necessarily static. They can move across the catalyst surface, searching for active sites or other adsorbed species with which to react. This surface diffusion is essential for bringing reactants into close proximity and facilitating subsequent reaction steps. The mobility of adsorbed species depends on factors like temperature, the strength of the adsorption bond, and the surface topography.

Surface Reaction (Reaction between Adsorbed Species). Surface reaction (reaction between adsorbed species). This is where the actual chemical transformation takes place. As introduced in Section 14.2, common mechanisms include the Langmuir–Hinshelwood pathway (two adsorbed species react on the surface), the Eley–Rideal pathway (a gas-phase molecule reacts directly with an adsorbed species), and the Mars–van Krevelen mechanism (in which lattice oxygen participates and is subsequently replenished).

Desorption. After the surface reaction, the product molecules must desorb from the catalyst surface to free up active sites for new reactant molecules. Desorption is essentially the reverse of adsorption. If the product remains strongly adsorbed, it can poison the catalyst by blocking active sites, hindering further catalytic activity. The rate of desorption is influenced by the strength of the product–surface bond and temperature.

14.4.4 Theoretical Models for Catalytic Cycles

Several theoretical models have been developed to understand and predict the behavior of catalytic cycles. Among the most prominent is the *energetic span model* (ESM) [34], which links the kinetics of a catalytic cycle to the energy differences between key intermediates and TSs. By identifying the turnover-determining transition state and turnover-determining intermediate (TDI), ESM allows for the calculation of TOF from energy profiles. This model is especially valuable when combined with computational chemistry methods like DFT, as it translates energy landscapes into mechanistic insight and catalytic performance.

Microkinetic modeling provides a more comprehensive, quantitative approach [15, 20, 23]. It treats each elementary step in the cycle as a rate equation, building a system of differential equations that describe the full reaction network. This approach is widely used in heterogeneous catalysis, where surface coverages and adsorption equilibria play a major role. Microkinetic models can predict overall rates, product selectivity, and the effect of changing conditions, making them indispensable in catalyst design and reactor optimization. Related approaches, such as steady-state approximations and rate-determining-step analyses, simplify these complex systems when full kinetic data are unavailable.

Other tools such as linear free energy relationships (LFERs), kinetic isotope effects (KIEs), and TST offer ways to probe specific steps in a catalytic cycle. LFERs help rationalize how electronic effects influence barriers and reactivity, while KIEs and TST provide mechanistic insights into bond-breaking/forming steps. Additionally, PES mapping using quantum chemical methods underpins many of these models. Collectively, these frameworks offer a rich theoretical foundation to interpret, predict, and refine catalytic processes across diverse systems.

14.5 Emerging Areas and Main Challenges of Catalysis

14.5.1 Emerging Areas in Catalysis

A growing suite of emerging catalytic strategies is reshaping how chemists think about reaction design, sustainability, and functional materials. These new approaches are often interdisciplinary, integrating insights from physics, biology, materials science, and computer science to create innovative solutions for energy conversion, environmental remediation, and precision synthesis [36].

Photocatalysis and Electrocatalysis. Photocatalysis and electrocatalysis [37–39] are two of the most intensively studied fields in sustainable catalysis. In photocatalysis, light energy—usually from sunlight or LEDs—is used to activate a catalyst that can drive redox reactions under mild conditions. Semiconductors like TiO_2 and organic dyes have been used to split water, reduce CO_2, or degrade pollutants. Photocatalysis is particularly promising for artificial photosynthesis, where the long-term goal is to mimic natural photosynthetic systems to convert solar energy into fuels or valuable chemicals. Electrocatalysis, on the other hand, couples chemical transformations to an electric current. It is central to technologies such as fuel cells, electrolyzers, and batteries. In electrocatalysis, reactions such as the hydrogen evolution reaction, oxygen evolution reaction, nitrogen reduction, and CO_2 reduction are driven by precisely controlled potentials. The field has benefited significantly from high-throughput computational screening, enabling the design of electrocatalysts with tailored binding energies and reaction profiles. Emerging materials such as 2D transition metal dichalcogenides, doped carbon frameworks, and single-atom catalysts (SACs) are unlocking unprecedented control over surface chemistry and electron transfer.

SAC and Atomically Precise Materials. SAC [40, 41] represents a convergence of heterogeneous and homogeneous catalysis, where isolated metal atoms are anchored on solid supports like graphene, metal–organic frameworks (MOFs), or oxides. These catalysts combine the high selectivity of molecular catalysts with the durability of solid materials. Because each atom is a catalytically active site, SACs offer maximized atom efficiency, a critical metric in both economic and environmental terms. Theoretical models such as scaling relations and microkinetic simulations are crucial in understanding how coordination environment, charge transfer, and support interactions dictate activity. A related frontier involves atomically precise nanoclusters and quantum dots, whose catalytic behavior arises not only from their composition but also from their discrete electronic structure. These systems challenge traditional assumptions of structure–activity relationships and demand a quantum mechanical description of reactivity. Such precision enables fine-tuning of catalytic pathways and opens up new applications in photocatalysis, electrocatalysis, and bioinorganic chemistry.

Mechanocatalysis and Catalyst Activation by Force. Mechanochemistry [42] has reemerged as a green alternative to solution-based chemistry, and mechanocatalysis extends this idea by using mechanical energy as a direct driver of catalytic transformation. In mechanocatalytic systems, external forces—such as shear, grinding, or pressure—alter the electronic structure or geometry of the catalyst to activate substrates or induce otherwise inaccessible reactivity. For example, in tribocatalysis, surface interactions during mechanical contact generate highly reactive sites for redox reactions, including degradation of pollutants or activation of small molecules. Mechanocatalysis is also being explored at the molecular scale, where force-modified ligands or metal centers undergo strain-induced activation. This approach aligns with the broader interest in out-of-equilibrium catalysis, where energy inputs other than heat—light, electricity, force, or concentration gradients—are harnessed to regulate catalytic function. Mechanocatalysis is especially appealing for its solvent-free operation, minimal waste, and potential for in situ activation and deactivation of catalysts, making it attractive for both green chemistry and materials science.

Biocatalysis Beyond Enzymes and Artificial Catalysts. While enzymes remain nature's supreme catalysts, recent developments in biocatalysis have moved beyond native enzymes to include engineered proteins, ribozymes, and artificial metalloenzymes. Directed evolution, computational protein design, and metallocofactor insertion have created enzymes that catalyze abiological reactions such as carbene and nitrene insertion, olefin metathesis, and cyclopropanation. These systems bridge the gap between biology and organometallic chemistry and offer unparalleled selectivity and sustainability. Beyond biomacromolecules, DNA-based catalysts, aptamer-catalyst hybrids, and supramolecular biocatalysts are expanding the biocatalytic toolbox. These constructs exploit the chiral and recognition properties of biological scaffolds but integrate synthetic catalytic centers to achieve new-to-nature transformations.

ML and AI in Catalyst Discovery. A transformative trend across all catalytic fields is the integration of ML and artificial intelligence (AI). These technologies are accelerating catalyst discovery, mechanism elucidation, and reaction optimization. ML models can predict key properties such as adsorption energies, activation barriers, or turnover frequencies based on molecular descriptors, DFT data, or experimental databases. Reinforcement learning algorithms are being deployed to navigate vast chemical spaces, suggesting optimal catalyst compositions or reaction conditions with minimal experimental input. Notably, AI-driven approaches are making headway in inverse design: instead of testing candidates one by one, the target reactivity is specified, and the algorithm proposes a catalyst that should achieve it. Combined with automated synthesis and high-throughput screening, this leads to self-driving laboratories for catalyst development. Still, theoretical and interpretative frameworks remain essential to ensure that AI predictions are chemically meaningful and generalizable.

Miscellaneous Catalysis. Autocatalysis, where a reaction product acts as its own catalyst, plays critical roles in prebiotic chemistry and nonlinear reaction dynamics. A classic example is the hydrolysis of esters catalyzed by the generated carboxylic acid. *Cooperative catalysis* leverages the synergy between multiple catalytic sites—often combining Lewis acids and bases, metals and organocatalysts, or enzymes and synthetic catalysts—to enable transformations that neither catalyst could achieve alone. A related concept is *frustrated Lewis pairs (FLPs)*, which involve spatially separated Lewis acid–base pairs that cannot quench each other and instead activate small molecules like H_2 or CO_2 under mild conditions. *Cascade (or tandem) catalysis* integrates multiple catalytic steps into a single operation without isolating intermediates, improving efficiency and selectivity in complex molecule synthesis. These emerging paradigms highlight a shift from isolated catalytic steps to systems-level design, where dynamic interactions, multiple activation modes, and reaction network integration offer new opportunities for control, sustainability, and reactivity.

14.5.2 Main Challenges of Catalysis

Catalysis remains one of the most powerful tools in chemical science, yet it faces a number of pressing challenges that limit its broader impact, efficiency, and predictability. These challenges span practical, mechanistic, and theoretical domains, and addressing them is critical for advancing chemical synthesis, energy conversion, environmental remediation, and sustainable industry. This section highlights several of the most significant issues currently confronting the field.

Catalyst Stability and Longevity. One of the foremost challenges in catalysis is ensuring that catalysts remain active and selective over extended periods. Many catalysts degrade under operating conditions due to poisoning, aggregation, leaching, or structural rearrangement. This is particularly problematic in industrial processes where catalyst replacement is costly. Understanding and mitigating deactivation pathways, especially in harsh or dynamic environments, remains an ongoing area of research. Advanced spectroscopic and computational tools are increasingly employed to identify degradation mechanisms and inform strategies for catalyst stabilization.

Selectivity and Control of Reaction Pathways. Highly selective catalysis is essential for minimizing byproducts, improving atom economy, and reducing downstream purification steps. However, many catalytic systems operate under kinetic or thermodynamic conditions that permit multiple competing pathways. Achieving exquisite control over chemo-, regio-, and stereoselectivity is particularly difficult in complex molecular environments or multifunctional substrates. Strategies such as ligand design, confinement effects (e.g., in porous materials), and cooperative or dual catalysis are being developed to tackle this issue.

Earth-abundant and Sustainable Catalysts. A critical challenge for green chemistry and sustainable development is the replacement of rare or toxic elements (such as platinum, palladium, or rhodium) with Earth-abundant metals like iron, nickel, cobalt, or copper. While promising, these first-row transition metals often suffer from lower activity, poorer selectivity, or instability under reaction conditions. Rational ligand design, redox tuning, and the discovery of new coordination environments are key to unlocking the full potential of sustainable catalysts.

Understanding Complex Mechanisms. Mechanistic understanding is foundational to rational catalyst design, yet many catalytic cycles involve transient intermediates, multiple competing pathways, and dynamic speciation that elude direct observation. Isolating or identifying on-cycle versus off-cycle species, detecting short-lived transition states, and mapping out full reaction networks remain formidable tasks. Emerging techniques such as operando spectroscopy, kinetic modeling, isotopic labeling, and computational simulations have improved mechanistic insight, but challenges persist in correlating this data with catalytic performance.

Scalability and Industrial Translation. Even when catalytic processes are well-characterized in the laboratory, scaling them for industrial use introduces new difficulties. Reaction conditions must be robust and reproducible, catalysts must be recyclable or easily separated, and costs must be competitive. Issues such as heat and mass transfer limitations, reactor design, and catalyst support integration often require bespoke engineering solutions. Bridging the gap between academic discovery and industrial application remains a challenge that calls for closer collaboration between chemists and chemical engineers.

Theoretical Limitations and the Absence of a General Theory. Perhaps the most fundamental challenge is the lack of a universal, predictive theory of catalysis. Despite the success of frameworks like TST, the Sabatier principle, linear scaling relations, and volcano plots, these approaches are limited in scope and often system-specific. Catalytic behavior is shaped by multiple variables—electronic structure, entropy, solvation, and dynamic evolution—that are difficult to generalize. There is no single theory that can predict catalytic activity and selectivity across all forms of catalysis (homogeneous, heterogeneous, enzymatic, etc.) from first principles. The development of such a theory would revolutionize catalyst design, enabling it to move from empirical exploration to predictive science. Progress in this direction includes the energy span model, microkinetic modeling, catalysis informatics, and the use of AI and ML to identify patterns across large datasets.

In summary, the field of catalysis must grapple with technical, conceptual, and practical challenges. Solving them will not only improve existing processes but also enable entirely new reactivity, opening doors to sustainable chemistry, efficient energy use, and advanced materials. Tackling these issues requires both deep fundamental research and innovative interdisciplinary approaches.

14.6 Toward the Ultimate Theory of Catalysis

Despite its ubiquity and transformative power, catalysis still lacks a comprehensive, unified theoretical framework—an "ultimate theory of catalysis"—that can both explain and predict the behavior of catalytic systems across all domains. To envision such a theory, one must draw from diverse intellectual traditions: not only from physical chemistry and mechanistic modeling but also from the philosophical dimensions of *ontology* (what exists), *epistemology* (how it is known), *understanding* (why catalytic behavior occurs) [43, 44], as well as from systems theory, mathematics, and ML. This section explores the foundational and functional requirements such a theory must meet and the pathways through which it might be approached.

14.6.1 Ontological and Epistemological Requirements

Catalysis challenges fundamental notions of causality and identity in chemistry: a catalyst alters reaction outcomes without being consumed, yet exists transiently in multiple forms throughout the catalytic cycle. This raises ontological questions about what entities genuinely participate in catalysis—catalysts, substrates, intermediates, transition states, environments—and how to treat species that are fleeting, virtual, or probabilistic. A complete theory must therefore provide a layered account of these entities, situating intermediates, off-cycle species, and emergent properties such as cooperativity within a coherent ontological framework.

Epistemologically, catalysis exemplifies how scientific understanding is constructed from indirect evidence. Mechanistic inference draws on kinetics, spectroscopy, and theory, yet rarely grants full access to the underlying microstates. An ultimate theory must clarify what constitutes valid knowledge—distinguishing data, models, and mechanistic understanding—and integrate deductive reasoning from physical principles with inductive patterns extracted from experiments and ML.

Finally, such a theory must accommodate multiple levels of abstraction. The same catalytic event must be intelligible in terms of electronic structure, potential-energy landscapes, rate laws, and macroscopic observables such as TOF. It must also explain how catalytic behavior emerges from noncatalytic systems, requiring attention to system dynamics, feedback, and self-organization. In short, the ontological and epistemological foundations of catalysis must support a coherent, multiscale account of what exists in a catalytic system and how we come to know it.

14.6.2 Functional Requirements: What Should the Theory Do?

Beyond philosophical grounding, an ultimate theory of catalysis must satisfy key functional criteria. It must be predictive—able to forecast reactivity, selectivity, optimal conditions, and catalyst structures—and retrodictive, explaining why known catalysts succeed or fail. It must also remain mechanistically flexible: applicable to pathways involving classical transition states, surface hopping, tunneling, photochemical excitation, or dynamically evolving environments, without committing to a single mechanistic template.

The theory must unify thermodynamics and kinetics, clarifying when relationships such as LFERs, BEP correlations, and volcano plots emerge and why they break down. It should connect energy landscapes with rate expressions in a principled manner that generalizes across catalytic domains.

Equally important, the theory must be computable. It should guide algorithms, simulations, and data-driven models, enabling feedback between quantum calculations, microkinetic modeling, ML, and experiment. A functional theory therefore cannot remain purely conceptual—it must be implementable, operationalizable, and capable of informing catalyst design.

14.6.3 Conceptual Requirements: From Chemical Concepts to Understanding

Chemists reason through concepts—ligand effects, substrate activation, transition-state stabilization—not merely equations. An ultimate theory of catalysis must therefore preserve the intelligibility and transferability of such concepts while grounding them in rigorous principles. Concepts serve as cognitive tools that connect empirical patterns to mechanistic models; a robust theory should clarify their scope, limits, and interrelations rather than replace them with purely formal descriptors.

Understanding in catalysis requires more than computing activation energies; it involves seeing *why* a catalyst works, how it can fail, and how its behavior generalizes across systems. Thus, the theory must yield mechanistically transparent models that support heuristic reasoning. Frameworks such as microscopic reversibility, Curtin–Hammett scenarios, and Sabatier-type arguments must remain usable as organizing principles within the broader theoretical structure.

Importantly, the theory must bridge operational and conceptual levels. It must relate experimental observables and data-analysis tools to higher-level mechanistic insight, and support the emergence of new concepts—such as FLPs or cooperative activation—by embedding them within a coherent explanatory architecture. In this way, the theory promotes chemical understanding as much as prediction.

14.6.4 Additional Theoretical and Methodological Requirements

Several further criteria follow from the broader methodological landscape of catalysis. First, the theory must rest on rigorous mathematical foundations capable of handling nonlinear kinetics, coupled equilibria, reaction networks, and stochastic or dynamic behavior. Tools from graph theory, topology, and differential equations will likely play central roles.

Second, the theory must be modular and scalable, allowing molecular-level catalytic units to be embedded within larger systems—enzymes, materials, reactors—and enabling seamless transitions across length and time scales. Such modularity is essential for unifying homogeneous, heterogeneous, and biological catalysis.

Third, adaptability is crucial. With AI and ML increasingly central to catalyst discovery, the theory must incorporate mechanisms for learning from new data, adjusting internal parameters, and refining its abstractions. A static theory cannot accommodate the evolving frontier of catalysis.

Fourth, the theory must balance universality with domain-specific insight. It should extend across all forms of catalysis while still explaining their distinctive features—for example, the exceptional selectivity of enzymes or the unique behavior of SACs.

14.6.5 Catalyst as Waterwheel: A Hierarchical Model

A refined way to conceptualize an ultimate theory of catalysis is to reexamine the energetic structure of a chemical transformation. Any reaction from reactants to products is not purely downhill: it must first overcome an uphill climb to reach the TS, after which the system descends toward the products. In the uncatalyzed pathway, this uphill segment occurs in one steep ascent, corresponding to a single large activation barrier. This situation is analogous to attempting to scale a tall cliff in a single leap—possible only when substantial thermal energy is available.

A catalyst reshapes this landscape not simply by lowering the highest barrier, but more fundamentally by converting a single, prohibitive uphill climb into a series of smaller, manageable ascents. This motivates the analogy of a catalyst as a *waterwheel.* A waterwheel is a mechanical device consisting of a large wheel fitted with paddles or buckets that is turned by the flow of water. Commonly installed along riverbanks, it converts the kinetic energy of moving water into rotational motion. Traditionally, this rotation is used to lift water to higher ground for irrigation or to drive simple machinery, allowing for the transport of water uphill without additional energy input. Whereas a waterfall releases its potential energy in a single drop, a waterwheel partitions this energy across many paddles. Each paddle captures only a fraction of the total driving force, allowing controlled work to be performed in a stepwise manner. In a catalytic cycle, the intermediate states serve as these paddles: each represents a partial reorganization of bonding or charge distribution, requiring a correspondingly smaller energetic investment to advance the system upward toward the TS. The rate-limiting step is the "largest paddle," the most energetically demanding subproblem whose difficulty is still real but no longer prohibitive because it is embedded within a structured multistep cycle.

In the hierarchical picture, the catalyst enables the system to climb the necessary uphill energetic pathway incrementally rather than in a prohibitively large single step. The barrier is still present—catalysts do not eliminate the need to ascend—but its height is partitioned into a sequence of accessible sub-steps. This stratification is precisely what makes a catalytic cycle an instance of hierarchical modeling: each intermediate constitutes a lower-level energetic subproblem, while their ordered arrangement forms a higher-level organization whose emergent behavior—reduced effective activation and controlled energy flow—cannot be inferred from any single step alone. The waterwheel metaphor accurately captures this idea of distributed energy management: the catalyst orchestrates a multistage progression toward the TS by breaking the activation requirement into digestible portions.

This analogy becomes even more powerful when considering uphill net reactions. A waterwheel can be driven in reverse when supplied with external energy—lifting water for irrigation or mechanical work. Likewise, catalytic systems can promote globally uphill transformations, provided energy is supplied from heat, photons, electrical potential, or chemical fuels. Here, the catalyst acts as an energy transducer, converting external energy into chemical potential and distributing this input across the cycle so that each intermediate step becomes feasible. Photocatalysis, electrocatalysis, and enzymatic ATP-driven reactions exemplify this principle. The catalyst, functioning like a powered waterwheel, makes thermodynamically unfavorable transformations possible by enabling the system to climb an otherwise inaccessible energetic slope.

The catalyst-as-waterwheel perspective also implies that cooperativity and frustration are inherent features of catalytic cycles. Because each elementary step participates in a coupled energetic flow, improvements or perturbations at one stage inevitably influence the performance of others, giving rise to positive or negative cooperativity across the cycle. At the same time, the catalyst must satisfy competing energetic, structural, and kinetic demands—binding versus release, activation versus stabilization—leading to unavoidable forms of frustration that shape the overall free-energy landscape. In this hierarchical view, cooperativity provides the cycle's dynamical coherence, while frustration imposes its structural constraints, together defining the emergent efficiency of catalysis.

The hierarchical structure of the waterwheel model also allows classical catalytic principles to be rediscovered from a unified energetic perspective. In a Curtin–Hammett scenario, rapidly interconverting intermediates function like neighboring paddles whose energetic depths are effectively irrelevant; the catalytic flow depends instead on the height of the next uphill micro-step, so product distribution is governed entirely by the relative transition-state energies. The Sabatier principle likewise emerges as a condition on optimal energy partitioning: intermediates that bind too weakly correspond to paddles too shallow to capture and transmit energetic work, whereas intermediates that bind too strongly become deep wells that impede turnover. Maximum activity occurs when the energetic "depths" of intermediates are balanced to permit both stabilization and facile release. Thus, both Curtin–Hammett behavior and the Sabatier optimum arise naturally as manifestations of how the catalytic cycle organizes and distributes activation work within its hierarchical architecture.

Viewed through this lens, an ultimate theory of catalysis frames the catalyst as a hierarchical energy-management device—one that partitions the uphill activation work, channels internal and external energy through a structured sequence of intermediates, and directs the overall flow of energy toward productive chemical change. This unified perspective integrates classical barrier-lowering, reaction pathways, kinetic control, and energy transduction into a single conceptual architecture of a hierarchical model, offering a more comprehensive foundation for understanding catalysis across molecular, materials, and biological systems.

It is useful to situate the waterwheel model within the broader requirements outlined in Sections 14.6.1–14.6.4. Ontologically and epistemologically, the model provides a layered account of the entities and forms of knowledge involved in catalysis: intermediates serve as lower-level energetic subproblems, while their ordered coordination constitutes a higher-level mechanism whose emergent behavior—barrier reduction, controlled energy flow, and feasibility of uphill transformations—cannot be inferred from any single step alone. Conceptually, the model offers a chemically intelligible and mechanistically transparent framework that links stepwise energy management with catalytic function, thereby satisfying the requirement that an ultimate theory support genuine chemical understanding. Functionally and methodologically, the model is qualitative rather than predictive: it unifies thermodynamic and kinetic perspectives and applies across catalytic domains, but it does not yet provide quantitative formalisms or computable algorithms. In this sense, the waterwheel model meets the conceptual and representational aspects of the requirements for an ultimate theory of catalysis while pointing toward the mathematical and data-driven developments needed for a fully predictive framework.

14.7 Conclusions

Catalysis stands as one of the most powerful and unifying ideas in chemistry, connecting molecular structure, energetics, reaction kinetics, materials behavior, and even biological function. Throughout this chapter, we have traced its development from empirical origins to a conceptually rich discipline grounded in theory, computation, and interdisciplinary innovation. Classical models, such as transition-state theory, the Sabatier principle, BEP relationships, scaling laws, and microkinetic modeling, provide indispensable tools for interpreting activity and selectivity, while contemporary advances in electrocatalysis, photocatalysis, SAC, biocatalysis, and machine-learning-driven discovery continue to expand the boundaries of what catalysis can achieve.

The density-based reactivity theory adds a complementary layer of insight, revealing how steric, electrophilic, and nucleophilic factors influence different segments of a reaction path. By decomposing the IRC into pre-TS, TS, and post-TS regions, this approach exposes the physicochemical origins of catalytic reactivity and bridges classical energy landscapes with conceptual descriptors from CDFT and ITA. Its success across diverse catalytic systems suggests a promising direction for unifying mechanistic interpretation with quantitative prediction.

Yet the broader challenge remains: catalysis still lacks a single, coherent theoretical framework capable of integrating microscopic electronic structure, macroscopic kinetics, dynamic environments, and conceptual understanding. The ontological, epistemological, and functional requirements outlined above suggest that an "ultimate theory of catalysis" will likely emerge not from any single model, but from the synthesis of hierarchical representations, appropriate abstraction, and data-driven learning with chemically meaningful concepts. Such a theory must be predictive, interpretable, mechanistically flexible, and capable of accommodating the diverse phenomena that fall under the umbrella of catalysis.

The *waterwheel* model discussed earlier provides a concrete example of how such a hierarchical theory may operate in catalysis. By partitioning the uphill activation work into smaller, manageable steps and enabling energy transduction from external sources when necessary, the catalyst functions as a structured energy-management device. This framework unifies classical barrier lowering with stepwise mechanistic organization and externally powered uphill reactivity, offering a conceptually intuitive bridge between traditional potential-energy landscapes and a future hierarchical theory of catalysis. In doing so, the model reframes catalysis not merely as barrier lowering but as an emergent organizational strategy for energy flow, a perspective that aligns catalysis with broader hierarchical principles found across chemical systems.

Catalysis, therefore, continues to serve not only as a technological engine but also as a conceptual frontier—an area where fundamental questions about reactivity, mechanism, and understanding intersect. As chemistry advances toward more sustainable, selective, and intelligent systems, catalysis will remain central to the discipline's deepest aspirations and most transformative achievements.

References

1 Roduner E. Understanding catalysis. Chem Soc Rev. 2014 Dec 21;43(24):8226–39.
2 Schlögl R. Heterogeneous catalysis. Angew Chem Int Ed. 2015 Mar 9;54(11):3465–520.
3 Beletskaya IP, Nájera C, Yus M. Stereodivergent catalysis. Chem Rev. 2018 May 23;118(10):5080–200.
4 Sak MH, Jacobsen EN. Selective noncovalent catalysis with small molecules. Chem Rev. 2025 May 15;125(17):8319–69.
5 Winderlich R. Jons Jakob Berzelius. J Chem Educ. 1948 Sep;25(9):500.
6 Janakiraman D. Wilhelm Ostwald, the father of physical chemistry. Reson. 2012 May;17(5):454–66.
7 Taylor H. Paul Sabatier 1854–1941. J Am Chem Soc. 1944 Oct;66(10):1615–7.
8 Humphreys J, Lan R, Tao S. Development and recent progress on ammonia synthesis catalysts for Haber–Bosch process. Adv Energy Sustain Res. 2021 Jan;2(1):2000043.

9 Kießling C. Karl Ziegler–early career of a scientist. Angew Chem Int Ed. 2025 Jan 2;64(1):e202403180.

10 Allegra G. Giulio Natta: The man and the scientist. Macromol Symp. 1995 Jul;98(1):1213–8.

11 Hargittai I. The 2021 chemistry Nobel laureates and asymmetric organocatalysis. Struct Chem. 2022 Feb;33(1):303–5.

12 Thomas JM. Turning points in catalysis. Angew Chem Int Ed Engl. 1994 May 18;33(9):913–37.

13 Martí S, Roca M, Andrés J, Moliner V, Silla E, Tuñón I, et al. Theoretical insights in enzyme catalysis. Chem Soc Rev. 2004 Feb 20;33(2):98–107.

14 Greeley J. Theoretical heterogeneous catalysis: scaling relationships and computational catalyst design. Annu Rev Chem Biomol Eng. 2016 Jun 7;7:605–35.

15 Chen BWJ, Xu L, Mavrikakis M. Computational methods in heterogeneous catalysis. Chem Rev. 2021 Jan 27;121(2):1007–48.

16 Ramos MJ, Fernandes PA. Computational enzymatic catalysis. Acc Chem Res. 2008 Jun;41(6):689–98.

17 Truhlar DG, Hase WL, Hynes JT. Current status of transition-state theory. J Phys Chem. 1983 Jul;87(15):2664–82.

18 Bao JL, Truhlar DG. Variational transition state theory: theoretical framework and recent developments. Chem Soc Rev. 2017 Dec 11;46(24):7548–96.

19 Ooka H, Huang J, Exner KS. The Sabatier principle in electrocatalysis: basics, limitations, and extensions. Front Energy Res. 2021 May 4;9:654460.

20 Zhu X, Huang J, Eikerling M. Hierarchical modeling of the local reaction environment in electrocatalysis. Acc Chem Res. 2024 Jul 20;57(15):2080–92.

21 Bligaard T, Nørskov JK, Dahl S, Matthiesen J, Christensen CH, Sehested J. The Brønsted–Evans–Polanyi relation and the volcano curve in heterogeneous catalysis. J Catal. 2004 May 15;224(1):206–17.

22 Vojvodic A, Calle-Vallejo F, Guo W, Wang S, Toftelund A, Studt F, et al. On the behavior of Brønsted-Evans-Polanyi relations for transition metal oxides. J Chem Phys. 2011 Jun 28;134(24):244509.

23 Besora M, Maseras F. Microkinetic modeling in homogeneous catalysis. WIREs Comput Mol Sci. 2018 Nov;8(6):e1372.

24 Mou L-H, Han T, Smith PES, Sharman E, Jiang J. Machine learning descriptors for data-driven catalysis study. Adv Sci (Weinh). 2023 Aug;10(22):e2301020.

25 Suvarna M, Pérez-Ramírez J. Embracing data science in catalysis research. Nat Catal. 2024 Apr 23;7(6):624–35.

26 Rong C, Zhao D, He X, Liu S. Development and applications of the density-based theory of chemical reactivity. J Phys Chem Lett. 2022 Dec 8;13(48):11191–200.

27 He X, Li M, Rong C, Zhao D, Liu W, Ayers PW, et al. Some recent advances in density-based reactivity theory. J Phys Chem A. 2024 Feb 22;128(7):1183–96.

28 Wang B, Rong C, Lei M, Liu S, De Proft F. Mechanistic study and conceptual chemical reactivity analysis of hydroboration of carbon dioxide catalyzed by a manganese(I)-PNP-pincer complex. Inorg Chem. 2023 May 15;62(19):7366–75.

29 Wang B, Liu S, Lei M, De Proft F. Steric effect and intrinsic electrophilicity and nucleophilicity from a conceptual density functional theory and

information-theoretic approach as quantitative probes of chemical reactions. Chem Eur J. 2024 Jul 3; e202401295.

30 Parr RG, Yang W. Density-Functional Theory of Atoms and Molecules. Oxford University Press; 1989.

31 Geerlings P, Chamorro E, Chattaraj PK, De Proft F, Gázquez JL, Liu S, et al. Conceptual density functional theory: status, prospects, issues. Theor Chem Acc. 2020 Feb;139(2):36.

32 Rong C, Wang B, Zhao D, Liu S. Information-theoretic approach in density functional theory and its recent applications to chemical problems. WIREs Comput Mol Sci. 2020 Jul;10(4):e1461.

33 Liu S. Information-theoretic approach in density functional reactivity theory. Acta Phys-Chim Sin. 2016;32(1):98–118.

34 Kozuch S, Shaik S. How to conceptualize catalytic cycles? The energetic span model. Acc Chem Res. 2011 Feb 15;44(2):101–10.

35 Kozuch S. A refinement of everyday thinking: the energetic span model for kinetic assessment of catalytic cycles. WIREs Comput Mol Sci. 2012 Sep;2(5):795–815.

36 Bogaerts A, Centi G, Hessel V, Rebrov E. Challenges in unconventional catalysis. Catal Today. 2023 Aug;420:114180.

37 Dai L, Fu Y, Wei M, Wang F, Tian B, Wang G, et al. Harnessing electro-descriptors for mechanistic and machine learning analysis of photocatalytic organic reactions. J Am Chem Soc. 2024 Jul 4;146(28):19019–29.

38 Windle CD, Perutz RN. Advances in molecular photocatalytic and electrocatalytic CO2 reduction. Coord Chem Rev. 2012 Nov;256 (21–22):2562–70.

39 Ochedi FO, Liu D, Yu J, Hussain A, Liu Y. Photocatalytic, electrocatalytic and photoelectrocatalytic conversion of carbon dioxide: a review. Environ Chem Lett. 2021 Apr;19(2):941–67.

40 Yang X-F, Wang A, Qiao B, Li J, Liu J, Zhang T. Single-atom catalysts: a new frontier in heterogeneous catalysis. Acc Chem Res. 2013 Aug 20;46(8):1740–8.

41 Fan Y, Liu S, Yi Y, Rong H, Zhang J. Catalytic nanomaterials toward atomic levels for biomedical applications: from metal clusters to single-atom catalysts. ACS Nano. 2021 Feb 23;15(2):2005–37.

42 Hwang S, Grätz S, Borchardt L. A guide to direct mechanocatalysis. Chem Commun. 2022 Feb 3;58(11):1661–71.

43 Liu S. The evolving quest for chemical understanding in the quantum age. J Chem Theory Comput. 2025 Oct 9;21(20):10068–79.

44 Liu S. Harvesting chemical understanding with machine learning and quantum computers. ACS Phys Chem Au. 2024 Mar 27;4(2):135–42.

15

Excited States

TABLE OF CONTENTS

Having developed density-based concepts for ground-state structure and reactivity in previous chapters, we now extend these ideas to excited states, where electronic rearrangements produce new patterns of bonding, stability, and reactivity behaviors. While excited-state theory is a broad and specialized field, the treatment in this chapter reflects a conceptual and density-based perspective consistent with the author's research background.

15.1 Fundamentals of Excited States

15.1.1 The Role and Importance of Excited States

While the electronic ground state of a molecule or material forms the foundation for most chemical descriptions, it represents only part of the full picture. Many of the most compelling chemical and physical phenomena emerge when systems are promoted to excited electronic states—configurations accessed by absorption of light, electron impact, or other perturbations. These states drive light-induced chemical and physical processes across diverse molecular and materials systems that range from the design of solar energy materials and organic light-emitting diodes to photocatalysis, biological light sensing, and artificial photosynthesis [1–19].

A rigorous understanding of excited states is essential for interpreting and predicting spectroscopic signatures such as UV–vis absorption, fluorescence, phosphorescence, and X-ray absorption; modeling biological light-driven reactions; and describing nonadiabatic events where electronic and nuclear motions are strongly coupled. Unlike the ground state, excited states present distinctive theoretical challenges because the variational principle that guarantees the accuracy of many ground-state methods does not generally apply. As a result, multiple computational strategies have been developed, including linear-response theories, equation-of-motion methods, state-specific approaches such as ΔSCF, and multireference formalisms, each offering different balances between accuracy, computational cost, and range of applicability.

15.1.2 Electronic Structure, Excitation Types, and Geometry Effects

An electronic state corresponds to a particular distribution of electrons in molecular orbitals, characterized by total spin and spatial symmetry. The spin multiplicity, given by $2S + 1$, classifies states by their total spin S. Singlet states ($S = 0$) have paired electrons and zero net spin, whereas triplet states ($S = 1$) possess two unpaired electrons with parallel spins, giving rise to a multiplicity of three. Transitions between states of different multiplicity are often spin-forbidden, though spin–orbit coupling can mediate weak intersystem crossing, a key mechanism in phosphorescence and photochemistry.

Excited states are also distinguished by the nature of the underlying electron promotion. Valence excitations involve transitions between bonding or nonbonding valence orbitals, such as $\pi \rightarrow \pi^*$ or $n \rightarrow \pi^*$. Rydberg excitations promote electrons to diffuse, high-energy orbitals resembling atomic Rydberg states. Charge-transfer excitations relocate electrons between spatially distinct regions, a process especially important in donor–acceptor complexes. Core excitations involve inner-shell electrons and are central to X-ray spectroscopies. Each excitation type imposes specific requirements on basis sets, electron correlation treatment, and computational protocols.

Upon absorption of a photon, electronic transitions occur much faster than nuclear motions, in accordance with the Born–Oppenheimer approximation.

The vertical excitation energy is the energy gap between the ground and excited states at the ground-state geometry, whereas the adiabatic excitation energy incorporates nuclear relaxation to the equilibrium geometry of the excited state. The Franck–Condon principle explains why observed spectra often display vibrational structure, reflecting the nuclear wavefunction overlap between initial and final states. In certain regions of the potential energy landscape—most notably at conical intersections—the Born–Oppenheimer approximation breaks down, allowing ultrafast nonradiative decay pathways that mix electronic states and redistribute energy efficiently.

15.1.3 Linking Theory to Experiment

For theories to be relevant to experiments, they must predict measurable quantities with accuracy and reliability. The most fundamental of these is the excitation energy, the difference in total energy between the excited and ground states. Equally important are oscillator strengths and transition dipole moments, which determine the intensity and probability of absorption or emission events. Theoretical methods must also address radiative and nonradiative decay rates, which dictate excited-state lifetimes, quantum yields, and overall photochemical efficiency. These properties are often sensitive to the interplay between electronic correlation, orbital relaxation, and vibronic coupling, making them stringent tests of any computational approach.

The concepts presented in this section form the foundation for the theoretical frameworks and computational methodologies introduced in the following sections. Together, they establish the language and context needed to explore how density-based approaches can offer fresh insights into the structure, energetics, bonding, and reactivity of excited states.

15.2 Theoretical Frameworks of Excited States

Accurate theoretical description of excited electronic states requires methods that go beyond standard ground-state approaches. This section introduces the primary frameworks developed to compute excited states [3, 7, 9, 11], including wavefunction-based methods, density-functional approaches, state-specific methods such as ΔSCF, and semi-empirical schemes. We also briefly touch upon treatments beyond the Born–Oppenheimer approximation relevant for excited-state phenomena.

15.2.1 Wavefunction-based Methods

Wavefunction methods explicitly construct excited-state wavefunctions as linear combinations of Slater determinants or configuration state functions, enabling systematic improvement and often high accuracy. These methods include configuration interaction singles (CIS), multireference methods such as CASSCF and CASPT2, and coupled-cluster-based methods like equation-of-motion coupled-cluster (EOM-CC).

The simplest excited-state wavefunction beyond Hartree–Fock (HF) is the *CIS* method. Starting from an HF ground-state reference $|\Phi_0\rangle$, excited states $|\Psi_k\rangle$ are expressed as:

$$|\Psi_k\rangle = \sum_i^{\text{occ}} \sum_a^{\text{virt}} c_i^a(k) a_a^\dagger a_i |\Phi_0\rangle, \tag{15.1}$$

where a_i and $a_a^\dagger$ annihilate an electron in occupied orbital i and create one in virtual orbital a, respectively. Solving the Schrödinger equation within this singly-excited space yields excitation energies ω_k from the eigenvalue problem:

$$\mathbf{H}_{\text{CIS}}\ \boldsymbol{c} = \omega_k\ \boldsymbol{c}, \tag{15.2}$$

where $\mathbf{H}_{\text{CIS}}$ is the Hamiltonian projected onto singly excited determinants and $\boldsymbol{c}$ is linear coefficient matrix in Eq. (15.1). This method has low computational cost ($\sim N^5$) and provides reliable qualitative trends for low-lying excitations, but it neglects electron correlation beyond the HF reference, overestimates excitation energies, and is poor in describing charge-transfer and double excitations.

For systems exhibiting strong static correlation (e.g., near-conical intersections or bond dissociation) require multireference approaches such as CASSCF or CASPT2. The CASSCF method partitions orbitals into inactive, active, and virtual subsets, and the CASSCF wavefunction is a full CI in the active space:

$$|\Psi_{\text{CASSCF}}\rangle = \sum_I c_I |\Phi_I\rangle, \tag{15.3}$$

with simultaneous optimization of orbitals and c_I coefficients. To capture dynamic correlation, second-order perturbation theory (CASPT2, NEVPT2) is applied on top of CASSCF. This method is accurate for strongly correlated states and conical intersections, but it is computationally demanding and therefore limited to small active-space sizes ($\sim$ 16 orbitals).

Equation-of-motion coupled-cluster (EOM-CC) methods [20] treat excited states as excitations from a correlated ground state $|\Psi_0\rangle = e^T|\Phi_0\rangle$, where T is the cluster operator (typically truncated at singles and doubles, $T = T_1 + T_2$). Excited states are generated by the excitation operator $\boldsymbol{R}_k$:

$$|\Psi_k\rangle = \boldsymbol{R}_k e^T |\Phi_0\rangle, \quad \boldsymbol{R}_k = r_0 + \sum_{ia} r_i^a a_a^\dagger a_i + \sum_{ijab} r_{ij}^{ab} a_a^\dagger a_b^\dagger a_j a_i + \cdots . \tag{15.4}$$

The energies and excitation amplitudes satisfy:

$$\overline{\boldsymbol{H}}\boldsymbol{R}_k|\boldsymbol{\Phi}_0\rangle = \omega_k \boldsymbol{R}_k|\boldsymbol{\Phi}_0\rangle, \tag{15.5}$$

where $\overline{\boldsymbol{H}} = \boldsymbol{e}^{-\boldsymbol{T}}\boldsymbol{H}\boldsymbol{e}^{\boldsymbol{T}}$ is the similarity-transformed Hamiltonian. EOM-CCSD (singles and doubles) is widely used for valence excited states and has a favorable balance of accuracy and cost. This method is accurate for single excitations and size-extensive and can be systematically improved. However, it requires high computational cost (N^6 scaling) and faces challenges with double excitations and species with multireference character.

15.2.2 Time-dependent DFT

Time-dependent density functional theory (TD-DFT) [21–24] in the linear-response regime has become the workhorse for excited-state calculations, combining affordable scaling with reasonable accuracy. Starting from the time-dependent Kohn–Sham (TDKS) equation:

$$i\frac{\partial}{\partial t}\phi_i(\boldsymbol{r},t) = \left[-\frac{1}{2}\nabla^2 + v_{\text{eff}}(\boldsymbol{r},t)\right]\phi_i(\boldsymbol{r},t), \tag{15.6}$$

where the effective time-dependent Kohn–Sham potential includes the external, Hartree, and exchange–correlation potentials. Its linear-response formalism leads to Casida's equation:

$$\begin{pmatrix} \boldsymbol{A} & \boldsymbol{B} \\ \boldsymbol{B} & \boldsymbol{A} \end{pmatrix}\begin{pmatrix} \boldsymbol{X} \\ \boldsymbol{Y} \end{pmatrix} = \omega\begin{pmatrix} 1 & 0 \\ 0 & -1 \end{pmatrix}\begin{pmatrix} \boldsymbol{X} \\ \boldsymbol{Y} \end{pmatrix}, \tag{15.7}$$

where matrices $\boldsymbol{A}$ and $\boldsymbol{B}$ are constructed from Kohn–Sham orbital energies and the exchange–correlation kernel. This approach has favorable cost scaling $(\sim N^4)$ and offers good accuracy for many valence excitations. Accurate description of Rydberg excitations also requires diffuse basis sets, without which TD-DFT systematically underestimates their energies. Unfortunately, it underestimates charge-transfer excitations, performs poorly for double excitations, and shows strong functional dependence. Range-separated hybrid functionals can improve charge-transfer descriptions by incorporating long-range exact exchange interactions.

15.2.3 Orbital-relaxation-based Approaches: The ΔSCF Method

The ΔSCF method [3, 5, 25, 26] computes excited-state energies as total energy differences between separately optimized SCF states:

$$E_{\text{exc}} = E^{*}_{\text{SCF}} - E^{0}_{\text{SCF}}, \tag{15.8}$$

where E^{*}_{SCF} is the energy of the excited-state SCF solution, obtained by enforcing an excited electron configuration. Unlike linear-response methods, ΔSCF naturally includes orbital relaxation, which is crucial for states with significant rearrangement, such as core excitations and some charge-transfer states.

This approach nevertheless tends to variationally collapse to the ground state and is plagued by spin contamination and convergence difficulties. To mitigate these, the maximum overlap method can be employed, guiding SCF iterations to maintain the target excited configuration by maximally overlapping with previous iteration orbitals. ΔSCF is computationally inexpensive and complementary to TD-DFT, especially valuable for core-level spectroscopy.

While ΔSCF provides state-specific orbital relaxation and good excitation energies, it does not in itself offer conceptual insight into how electron density reorganizes across states. Because density redistribution is the physical origin of bonding changes, stability shifts, and reactivity patterns in excited states, we will turn to density-based frameworks that can complement this energetic picture (see Section 15.4).

15.2.4 Other Excited-states Methods

Semi-empirical methods such as ZINDO/S and CNDO use parametrized Hamiltonians tailored for excitation energies, often focusing on π-conjugated systems. Density-functional tight-binding (DFTB) methods have been extended to excited states via linear-response TD-DFTB, allowing rapid screening of large systems. Though less accurate, these approaches provide access to excited-state properties for very large molecules or materials beyond the reach of ab initio methods.

Also, excited-state phenomena frequently involve coupling between electronic and nuclear degrees of freedom, requiring treatment beyond the Born–Oppenheimer separation with nonadiabatic couplings, which quantify mixing between electronic states due to nuclear motion. Additionally, conical intersections arise where potential energy surfaces cross, allowing ultrafast internal conversion and nonradiative decay. Incorporating these effects is essential for modeling photochemical reactions and excited-state dynamics.

Selecting the appropriate excited-state computational method depends critically on the nature of the target state—whether valence, Rydberg, charge-transfer, or core excitation—as well as the degree of electron correlation involved, ranging from weak to strong. Moreover, practical considerations such as system size and available computational resources further influence method choice, especially given the varying cost and scaling of techniques. Additionally, the desired observables—such as excitation energies, oscillator strengths, excited-state gradients, or nonadiabatic couplings—play a key role in determining suitability. Often, a pragmatic workflow leverages multiple methods: ΔSCF excels for core-level states, TD-DFT serves well for initial broad screening, and more demanding approaches like EOM-CCSD or CASPT2 provide benchmark accuracy for critical cases.

15.3 Case Studies of Excited States

This section presents six selected case studies of excited states where theoretical methods have played a decisive role in understanding and predicting photophysical and photochemical phenomena. Each example highlights the interplay between computational methodology, physical interpretation, and experimental validation.

15.3.1 Organic Chromophores

Organic chromophores [27, 28] such as stilbene, azobenzene, and coumarin derivatives exhibit characteristic $\pi \rightarrow \pi^*$ and $n \rightarrow \pi^*$ transitions in the UV–vis range. TD-DFT is often the first-line approach for such systems, balancing reasonable accuracy with computational affordability. Benchmark calculations with higher-level methods such as EOM-CCSD or CASPT2 are frequently used to assess functional performance.

For example, for stilbene, TD-DFT with long-range corrected functionals reproduces the $\pi \rightarrow \pi^*$ S_1 absorption maximum within 0.1–0.2 eV of experiment, while failing to accurately predict the $n \rightarrow \pi^*$ S_2 ordering without careful functional choice. In studies like this, selection of the exchange–correlation functional and inclusion of solvent effects via PCM models are crucial for matching observed spectral shifts.

15.3.2 Photochemistry of Transition-metal Complexes

Transition-metal complexes such as $[Ru(bpy)_3]^{2+}$ and Fe(II) polypyridyl derivatives display rich photochemistry involving ligand-to-metal charge transfer, metal-to-ligand charge transfer (MLCT), and metal-centered states [29, 30]. These systems often require multireference methods (CASSCF/CASPT2) to treat static correlation, or spin–orbit coupled TD-DFT for large-scale screening. Spin–orbit coupling (SOC) plays a major role in intersystem crossing (ISC) rates and population dynamics among singlet, triplet, and quintet manifolds.

For example, in Fe(II) polypyridyls, ultrafast ISC (<100 fs) from ^{1}MLCT to ^{3}MC states explains the loss of long-lived excited states, limiting photovoltaic performance. SOC matrix elements obtained from CASSCF/NEVPT2 wavefunctions match experimentally derived rates from transient absorption spectroscopy. For these systems, proper treatment of relativistic effects and dynamic correlation is essential for accurate excited-state lifetimes.

15.3.3 Biological Chromophores

Photoactive biological molecules, such as retinal in rhodopsin or the chromophore in photoactive yellow protein (PYP) [31], are embedded in complex protein environments that modulate their photophysics. QM/MM approaches are indispensable, where the chromophore is treated at the quantum level and the surrounding protein environment at a molecular mechanics level. This allows explicit modeling of hydrogen bonding, electrostatic fields, and steric constraints.

In rhodopsin, for example, CASSCF dynamics reveal that photoisomerization of 11-cis retinal to all-trans occurs on a sub-picosecond timescale via a conical intersection, with the protein pocket steering the torsional pathway. Environmental effects can shift absorption maxima by >0.2 eV and alter reaction coordinates, underscoring the necessity of hybrid methods for realistic modeling.

15.3.4 Core-level Spectroscopy in Catalysis

X-ray spectroscopies such as X-ray absorption (XAS) and X-ray emission are powerful probes of catalytic intermediates, especially at metal centers [32]. Core-level excitations are well described by ΔSCF methods, EOM-CCSD for core-ionized states, or transition-potential DFT. For heavier elements, scalar relativistic and spin–orbit corrections are mandatory to reproduce experimental fine structure.

For instance, in cobalt-based water oxidation catalysts, ΔSCF combined with relativistic corrections predicts L-edge XAS spectra in excellent agreement with experiment, enabling the identification of Co(IV)–oxo intermediates. Direct comparison of computed spectra with experiment allows not just energy assignment but also determination of oxidation states and ligand field parameters.

15.3.5 Singlet Fission in Organic Photovoltaics

Singlet fission (SF) [6, 33–36] is a process where one photoexcited singlet splits into two triplets, potentially doubling photocurrent in solar cells. SF requires accurate treatment of multiexcitonic states, often approached with CASSCF/CASPT2, spin-flip TD-DFT, or dimer-based exciton models. Periodic boundary conditions are used for crystalline materials like pentacene and tetracene.

In crystalline pentacene, multireference calculations show that the S_1 state is resonant with the triplet–triplet (TT) state, explaining the near-unity fission yield. Intermolecular coupling constants extracted from electronic structure calculations feed into kinetic models that reproduce experimentally measured rates. Both electronic structure and molecular packing govern SF efficiency, making combined molecular/crystal-level modeling necessary.

Beyond energetics, density-based descriptors could provide new insight into why certain dimers or aggregates exhibit strong coupling conducive to fission, potentially linking electron redistribution patterns to fission propensity. This represents an unexplored but promising direction for integrating density-based methods with excitonic physics.

15.3.6 Excitonic Effects in 2D Materials

Monolayer transition-metal dichalcogenides (TMDCs) [37, 38] such as MoS_2 exhibit exceptionally strong excitonic effects due to reduced dielectric screening and quantum confinement. The GW approximation provides accurate quasiparticle energies, while the Bethe–Salpeter equation (BSE) describes excitonic states and optical spectra. Exciton binding energies in TMDCs can reach 0.5–0.7 eV, far exceeding those in bulk semiconductors.

GW+BSE calculations for MoS_2 predict a prominent A exciton peak at ~1.9 eV, matching photoluminescence experiments. Strain and dielectric engineering simulations show tunability of exciton energies by >0.1 eV. Accurate modeling requires explicit electron–hole interaction; single-particle approaches fail to reproduce key spectral features.

These case studies illustrate the breadth of excited-state theory—spanning organic and inorganic chemistry, biology, and materials science—and the necessity of method selection tailored to the problem at hand. They also demonstrate the power of integrating theory with experiments to gain deep mechanistic insight.

The exceptional exciton binding energies in TMDCs also make them sensitive to subtle changes in electron density and screening. Density-based information measures may offer complementary tools for analyzing delocalization patterns or predicting how strain and dielectric environments reshape excitonic wavefunctions.

15.4 Density-based Insights for Excited States

The theoretical and computational tools surveyed above offer state-resolved energetics and spectroscopic properties with high accuracy, yet they reveal little about how electron density reorganizes upon excitation or how such reorganizations control bonding, stability, and reactivity. This gap is especially important because excited-state chemistry is fundamentally a *density-driven phenomenon*: shifts in electron distribution alter electrostatics, sterics, and exchange–correlation effects, thereby reshaping the chemical landscape. Just as ground-state reactivity arises from density polarization and redistribution, excited-state behavior is likewise governed by how electron density reorganizes under electronic promotion. The density-based reactivity frameworks developed in earlier chapters thus present a natural and powerful complement to traditional excited-state electronic structure methods.

In this section, we extend these frameworks to excited states and show that density-based descriptors—originally formulated for the ground state—capture distinctive patterns and conceptual insights unique to electronic excitation. An important distinction must be made between transition densities obtained from linear-response formalisms and fully state-specific densities generated by ΔSCF, CASSCF, or ESMF-like methods. Density-based reactivity descriptors require the latter, as they reflect the actual electron distribution of the excited state. All analyses in this chapter employ state-specific densities optimized for S_0, S_1, and T_1 spin states.

Extending density-based frameworks to excited states could yield several new conceptual insights. First, electron density redistributions upon excitation produce characteristic and quantifiable patterns in Shannon entropy, Fisher information, and deformation densities. Second, bonding descriptors such as the bonding and noncovalent interaction (BNI) index reveal that the nature of bonding itself can change dramatically between S_0, S_1, and T_1, providing a conceptual handle on excited-state bond orders. Third, energy partitioning analyses show that the stability determinants of S_1 and T_1 differ in fundamental ways, even when their potential energy profiles appear superficially similar. Finally, density-based reactivity descriptors expose systematic—and sometimes inverted—regioselectivity patterns in electrophilic aromatic substitution across S_0, S_1, and T_1. Together, these results demonstrate that density-based theory offers not only numerical prediction but also qualitative understanding of excited-state chemistry.

In a recent study [39], we extended four density-based reactivity frameworks to the realm of excited states. Our objectives were twofold. First, we established that these frameworks can be applied to excited states in a feasible and productive manner. Second, we demonstrated that excited-state results possess distinctive characteristics that may, or may not, differ substantially from those of the ground state. To this end, we investigate the lowest singlet excited state (S_1) and the lowest triplet excited state (T_1) for a set of 80 molecular systems, comparing them with their respective ground states (S_0). The properties examined span three domains—conformational stability, bonding, and reactivity. Our analysis revealed unique patterns of conformational stability for S_1 and T_1, along with characteristic bonding features and regioselectivity trends. Moreover, we uncovered strong correlations between the singlet–triplet energy gap and descriptors derived from OF-DFT, CDFT, and ITA. These correlations open new avenues for predicting the propensity for SF—a photophysical process in which one singlet exciton (S_1) splits into two triplet excitons (T_1). Such predictive capability is crucial for advancing photoelectric conversion efficiency in photovoltaic devices for solar energy applications and beyond.

15.4.1 Densities in Excited States

Figure 15.1 presents an illustrative example comparing the difference maps of total electron density (Figure 15.1a,d), Shannon entropy (Figure 15.1b,e), and Fisher information (Figure 15.1c,f) for carbon monoxide in its first excited singlet (S_1) and triplet (T_1) states, using the ground state (S_0) values as references. In the density difference map for T_1 (Figure 15.1a), electron density accumulation (positive regions) is primarily localized on the terminal oxygen atom (left side). A qualitatively similar pattern appears for Shannon entropy (Figure 15.1b) and Fisher information (Figure 15.1c), although quantitative variations are evident. These quantitative differences are critical when such measures are used to characterize molecular properties. In contrast, for S_1, the density difference map (Figure 15.1d) shows accumulation on both carbon and oxygen atoms: on oxygen, positive regions reside on the terminal side, while on carbon, they are located on both sides of the C—O bond. Shannon entropy (Figure 15.1e) and Fisher information (Figure 15.1f) again produce qualitatively similar trends, but with subtle quantitative discrepancies. Together, these results reveal that (i) the two lowest excited states can differ markedly from the ground state owing to distinct electron density redistributions,

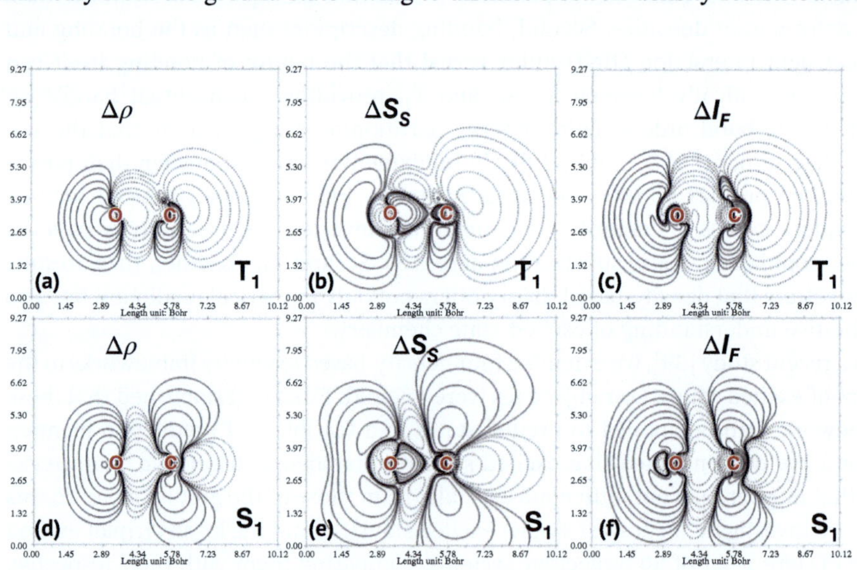

Figure 15.1 The difference map of total electron density $\Delta\rho$ (a and d), Shannon entropy ΔS_S (b and e), and Fisher information ΔI_F (c and f) of the first excited triplet state T_1 (a–c) and the first excited singlet state S_1 with the ground state S_0 as the reference for carbon monoxide. *Source*: Reproduced from Ref. [39] with permission of Springer Nature.

and (ii) information-theoretic functionals of the density can capture fine details of heterogeneous electron distributions, providing essential input for the quantitative evaluation of physicochemical properties.

15.4.2 Bonding in Excited States

Shown in Figure 15.2 are the BNI signature isosurfaces for the two dimers in three states. Also shown in the Figure is the optimized bond length for each species. From the figure, it is unambiguous that N_2 in S_0 is a triple covalent bond, in T_1 it is a double bond, but in S_1 it becomes a single covalent bond. For F_2, the bond/interaction type for S_0, T_1, and S_1 states is a single covalent bond, van der Waals interaction, and single covalent bond, respectively. These results can be confirmed by the bond order analysis for these species. For example, using Laplacian bond order analysis, we obtained that the bond order of N_2 in S_0, T_1, and S_1 states is 3.261, 2.136, and 1.433, respectively, suggesting that they are triple, double, and single covalent bonds. These bond order results are in good agreement with those obtained from the BNI isosurfaces shown in Figure 15.2.

15.4.3 Stability in Excited States

Figure 15.3 presents the total energy profiles of the three states of acetic acid substituted by fluorine and methyl groups as functions of the C—C bond rotation angle, with Figure 15.3a for S_1, Figure 15.3b for T_1, and the blue curve in both panels representing the ground state S_0. The profiles are referenced to the lowest-energy conformation of each state. Two key observations emerge. First, the energy landscapes of

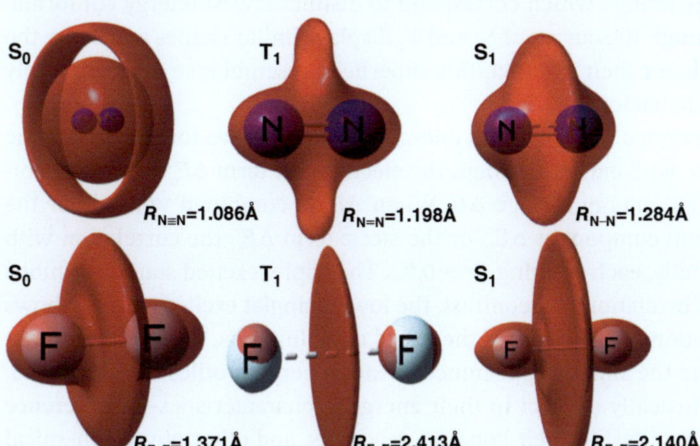

Figure 15.2 The signature isosurface of the BNI (bonding and noncovalent interaction) index for N_2 and F_2 molecules in three electronic states, S_0, T_1, and S_1. Also shown are the bond distances of these species in three different electronic states. The isovalue for N_2 is 0.11 for S_0; 0.65 for T_1; and 0.43 for S_1. For F_2, the isovalue is 0.55 for S_0; 1.10 for T_1; and 0.40 for S_1. Units in a.u. *Source*: Reproduced from Ref. [39] with permission of Springer Nature.

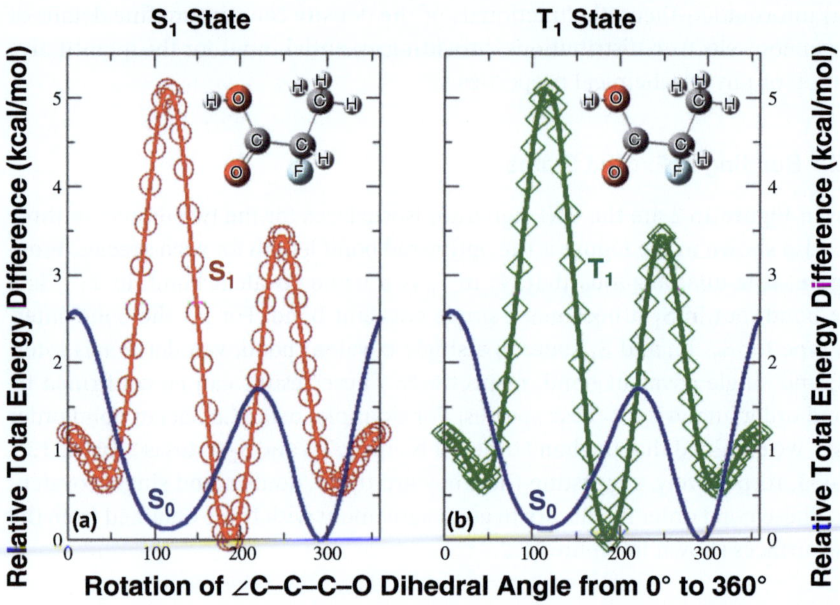

Figure 15.3 Profiles of the relative total energy as a function of the ∠C—C—C—O dihedral angle of the acetic acid derivative, CHFMeCOOH, from 0° to 360° with the step size of 5° for (a) the first excited singlet state S_1 and (b) the first excited triplet state T_1 using the lowest total energy as the reference for each state. The blue curve is the same profile as the ground state S_0. *Source*: Reproduced from Ref. [39] with permission of Springer Nature.

the excited states differ entirely from that of the ground state, including the location of their zero-energy points, which correspond to distinct lowest-energy conformations. Second, although the curves of S_1 and T_1 display similar shapes and share the same dihedral angle for their minima, this superficial resemblance does not imply identical energetic behavior.

Following our previous work on total energy partitioning, we found that for the ground state, in line with earlier findings, the electrostatic term ΔE_e is strongly correlated with the total energy change ΔE. When ΔE_e is combined with either the exchange–correlation component ΔE_{xc} or the steric term ΔE_s, the correlation with ΔE improves markedly, each yielding $R^2 > 0.93$. The triplet excited state T_1 exhibits these same strong correlations. In contrast, the lowest singlet excited state S_1 shows no such robust relationships for any of the tested combinations. These results demonstrate that despite the similar appearance of their energy profiles in Figure 15.3, S1 and T1 are intrinsically distinct in their energetic characteristics—a difference that is expected to manifest in their bonding, reactivity, and other physicochemical properties.

15.4.4 Reactivity in Excited States

We next examine the reactivity of excited states, focusing on regioselectivity in electrophilic aromatic substitution with ortho/para- and meta-directing groups.

Previous studies showed that information gain from ITA and its first-order approximation using Hirshfeld charges can both predict regioselectivity in these reactions. Although Hirshfeld charges are smaller in magnitude than other partial charges, subtle differences between them provide a reliable measure of electrophilicity, nucleophilicity, and regioselectivity. Our study shows that for ortho/para-directing groups, S_0 favors substitution at ortho and para positions, in agreement with experiment. In T_1, the preference generally shifts to the meta position, as this site has the largest negative Hirshfeld charge in most cases. However, exceptions do exist, such as Ar-N(COCl)$_2$, Ar-NH$_2$Ar$^+$, and Ar-OCONH$_2$. These results indicate that ortho/para-directing groups in S_0 often—but not always—become meta-directing in T_1. In S_1, no clear pattern emerges. For meta-directing groups, S_0 favors the meta position, but in T_1 the preference often shifts to ortho/para positions, with some exceptions (Ar-PO$_2$, Ar-PH$_2$, Ar-SiHO). S_1 shows a similar tendency to T_1, again with a few exceptions. Thus, put together, ground-state directing effects can change significantly in excited states, but these shifts are not universal.

Besides Hirshfeld charges, other density-based reactivity descriptors—such as the Fukui function from CDFT, information gain from ITA, and local temperature from OF-DFT—can also identify regioselectivity. Figure 15.4 shows that toluene (left), an ortho/para director in S_0, becomes meta-directing in S_1 and T_1. For nitrobenzene (right), the reverse is seen, which shifts from meta-directing in S_0 to ortho/para-directing in both excited states. Together with the Hirshfeld charge results, these findings demonstrate that excited states can have markedly different regioselectivity from the ground state. Such differences can be exploited in photochemical processes to produce unexpected synthetic outcomes and transformations.

These changes in regioselectivity originate from the distinct density polarization patterns of S_1 and T_1. In S_1, the population of antibonding π orbitals reduces

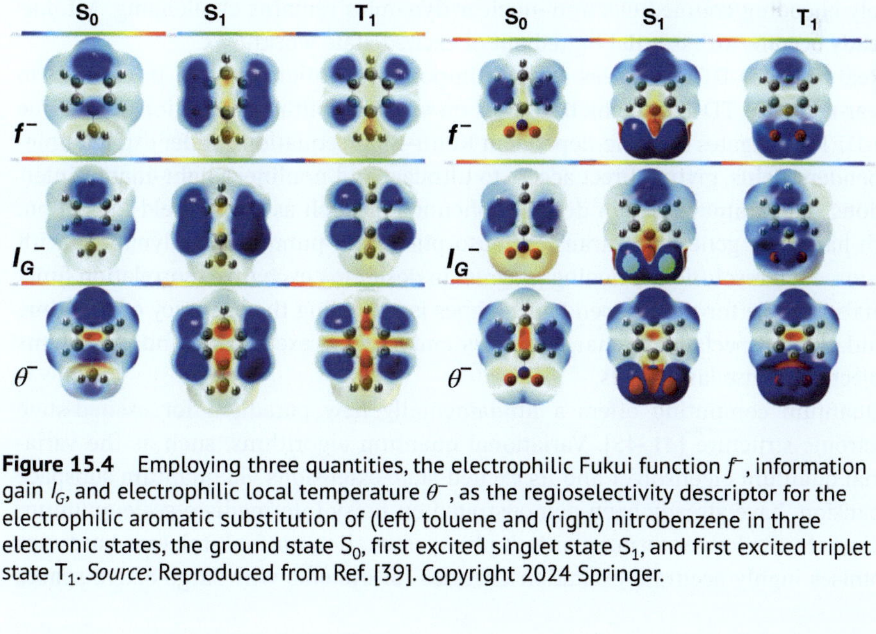

Figure 15.4 Employing three quantities, the electrophilic Fukui function f^-, information gain I_G, and electrophilic local temperature θ^-, as the regioselectivity descriptor for the electrophilic aromatic substitution of (left) toluene and (right) nitrobenzene in three electronic states, the ground state S_0, first excited singlet state S_1, and first excited triplet state T_1. *Source*: Reproduced from Ref. [39]. Copyright 2024 Springer.

aromatic stabilization and weakens classical directing rules. In T_1, the unpaired electron localizes spin density at specific positions, altering electrophilic susceptibility via exchange stabilization. Thus, excited-state regioselectivity can be understood as the direct manifestation of modified electron-density landscapes.

15.5 Frontiers of Excited State Research

Because methods for excited-state modeling are advancing rapidly across physics, chemistry, materials science, and data science, the frontiers of the field are unusually broad. The central challenge is to achieve a simultaneous balance of accuracy, efficiency, and conceptual clarity for large and complex systems. Among the most active directions, several are particularly relevant to the themes of this book: machine learning (ML) models for excited states, real-time time-dependent density functional theory, quantum computing approaches, polariton chemistry, excited states in complex environments, multiexciton processes, nonadiabatic dynamics, and relativistic effects. Together, these developments are reshaping how excited-state phenomena are computed, interpreted, and ultimately controlled.

ML has emerged as a powerful complement to traditional electronic-structure methods for excited states [4, 16, 18]. Neural networks, kernel methods, and graph-based models are now trained on high-level ab initio data or experimental results to reproduce excited-state potential energy surfaces, forces, and spectra at greatly reduced computational cost. Such models can accelerate nonadiabatic molecular dynamics by orders of magnitude and enable high-throughput screening of photophysical and photochemical materials. Active learning strategies further improve robustness by selectively adding reference data in regions of high uncertainty, such as near conical intersections or in strongly multireference regimes. Although accurately encoding coupled electron–nuclear dynamics remains challenging, ML has already become an essential ingredient of excited-state workflows.

Real-time TD-DFT provides another important frontier [23, 40]. In contrast to linear-response TD-DFT, which focuses on small-amplitude excitations, real-time TD-DFT propagates the time-dependent Kohn–Sham equations under explicit time-dependent fields, giving direct access to ultrafast and nonlinear light–matter interactions. This framework can describe phenomena such as strong-field ionization, high-harmonic generation, transient absorption, and pump–probe dynamics with femtosecond resolution. Ongoing progress in designing exchange–correlation functionals suitable for time-dependent regimes is extending the accuracy of real-time simulations, especially for charge-transfer and Rydberg excitations, and for systems subject to intense laser fields.

Quantum computing offers a fundamentally new paradigm for excited-state electronic structure [41–45]. Variational quantum algorithms, such as the variational quantum eigensolver and its excited-state extensions via quantum subspace expansion, have already been demonstrated on noisy intermediate-scale quantum devices. In the longer term, quantum phase estimation on fault-tolerant hardware promises highly accurate excitation energies for systems with large active spaces and strong correlation. Hybrid classical–quantum workflows, which use quantum resources to treat the most correlated part of the problem and classical methods for the remainder, are actively being developed [46]. These approaches are still in their infancy, but they point to a future in which classically intractable excited-state problems may become accessible.

Strong light–matter coupling and polariton chemistry constitute another rapidly developing frontier [47]. When molecules are placed in optical cavities or plasmonic structures, their electronic excitations can hybridize with confined photon modes to form polaritons—mixed light–matter states that reshape potential energy surfaces and modify reaction pathways. Theoretical descriptions that combine electronic-structure methods with quantum electrodynamics (QED) Hamiltonians are being used to analyze how cavity parameters influence barrier heights, branching ratios, and energy-transfer efficiencies. These studies suggest new ways to manipulate photochemical reactivity by engineering electromagnetic environments rather than molecular structures alone.

The study of excited states in complex environments is equally important, since most photophysical and photochemical processes occur in condensed phases, solvated systems, proteins, or solid-state hosts [31, 48, 49]. Embedding strategies such as quantum mechanics/molecular mechanics (QM/MM), polarizable continuum models, and density-based embedding allow accurate quantum treatment of a reactive subsystem while accounting for environmental polarization and fluctuations. In biological systems, for example, QM/MM simulations can capture how protein electrostatics and conformational dynamics modulate chromophore absorption, isomerization pathways, and access to conical intersections. In materials science, similar approaches are applied to describe exciton transport, charge separation, and energy migration in molecular crystals, aggregates, and organic semiconductors.

Multiexciton processes, including SF and triplet–triplet annihilation, remain a major frontier because of their potential impact on solar energy conversion and photon upconversion [33–36]. Accurate modeling of these phenomena requires a consistent description of strongly correlated multiexciton states, their couplings, and their nonadiabatic dynamics. Multireference wavefunction methods, spin-flip approaches, and increasingly ML-accelerated dynamics are being combined to simulate multiexciton formation, separation, and recombination in realistic molecular and crystalline environments. In parallel, advances in nonadiabatic excited-state molecular dynamics—such as surface-hopping and multiple-spawning methods enhanced by machine-learned potential energy surfaces and coupling elements—are extending simulations to larger systems and longer timescales [50]. These tools are crucial for interpreting ultrafast spectroscopies and for understanding how energy flows among electronic states and nuclear degrees of freedom.

Finally, relativistic effects and spin–orbit coupling (SOC) are being incorporated more systematically into excited-state modeling, particularly for heavy-element systems and transition-metal complexes [28, 51]. Two-component and four-component relativistic Hamiltonians, combined with SOC-enabled equation-of-motion coupled-cluster and multireference methods, now permit predictive simulations of phosphorescence, ISC, and core-level spectroscopies. These developments are essential for accurately describing the photophysics of organometallic complexes, lanthanides, and actinides, and for interpreting advanced X-ray spectroscopies.

Across all of these frontiers, there is a clear shift from computing individual excitation energies toward constructing coherent mechanistic pictures of excited-state processes in realistic environments. In this context, density-based reactivity descriptors—extended from the ground state to S_1 and T_1—offer a complementary way to interpret how changes in electron density control stability, bonding, and reactivity. As ML, quantum computing, and multiscale modeling continue to evolve, integrating these technologies with density-based concepts will be central to achieving both predictive power and chemical understanding in excited-state science.

15.6 Summary and Outlook

Excited-state phenomena lie at the heart of countless chemical, biological, and materials processes, from the fundamental photophysics of molecular chromophores to the light-harvesting capabilities of complex nanostructures. In this chapter, we have built a bridge from the essential concepts of electronic excitation to state-of-the-art computational methods, practical case studies, density-based analyses, and emerging frontiers.

We began by examining the fundamentals of excited states, emphasizing the diverse nature of excitations—valence, Rydberg, charge-transfer, and core—and the key physical ingredients that govern them: electron correlation, symmetry and selection rules, nonadiabatic coupling, and spin–orbit effects. Understanding these factors provides the conceptual compass needed to navigate the rich and sometimes counterintuitive landscape of excited-state chemistry.

From there, we surveyed the theoretical frameworks available for excited-state computations, spanning wavefunction-based methods such as CIS, Equation-of-Motion Coupled-Cluster Singles and Doubles (EOM-CCSD), and multireference approaches like CASSCF/CASPT2; density-functional-based approaches, particularly time-dependent density-functional theory (TD-DFT); and orbital-relaxation strategies such as the ΔSCF method. Each approach comes with characteristic strengths, limitations, and computational demands, making method selection a context-dependent decision. In practice, hybrid workflows—such as TD-DFT for screening, followed by high-level wavefunction methods for targeted states—offer a pragmatic balance between accuracy and efficiency.

We then explored case studies illustrating how theory meets experiment across a range of systems: organic chromophores, transition-metal complexes, biological photoactive proteins, core-level spectroscopy in catalysis, SF in photovoltaics, and excitonic effects in 2D materials. These examples highlight both the versatility of excited-state methods and the importance of tailoring the computational strategy to the physics and chemistry of the problem at hand.

Beyond energetics, density-based analyses provide complementary chemical insights into many excited-state properties. Tools such as electron-density difference maps and reactivity descriptors from OF-DFT and ITA capture how bonding, charge distribution, and reactivity patterns change upon electron excitation. These descriptors are essential for interpreting property and reactivity changes associated with photochemical processes and thus rationalizing photochemical pathways.

Finally, we considered frontiers in excited-state research. ML promises to accelerate excited-state property prediction and even uncover hidden structure–property relationships; quantum computing could make tractable the exact treatment of large active spaces; real-time TD-DFT and polariton chemistry open new regimes of light–matter interaction; and nonadiabatic dynamics simulations are approaching chemically relevant timescales for complex systems. Advances in spectroscopic modeling—particularly X-ray and ultrafast techniques—are enabling ever-closer theory–experiment synergy. These developments are expanding both the scope of systems we can study and the depth of mechanistic insight we can achieve.

Looking forward, several challenges and opportunities stand out. (i) *Scalability and accuracy*—developing methods that retain high accuracy while scaling to large, complex systems (e.g., proteins, nanomaterials) remains a central goal. (ii) *Excited-state databases and benchmarking*—community-wide datasets for excited states, analogous to ground-state benchmarks, will improve method validation and ML training. (iii) *Integration with experiment*—closer coupling between computation and advanced spectroscopies—especially ultrafast and high-resolution X-ray methods—will help resolve transient and rare events. (iv) *Multiscale modeling*—incorporating environmental effects (solvent, solid-state, biological) into excited-state dynamics will be critical for predictive modeling of real-world systems. (v) *Automated exploration*—emerging algorithms for automated mapping of excited-state reaction networks could revolutionize our understanding of photochemical pathways.

References

1 Kim J, Oh J, Osuka A, Kim D. Porphyrinoids, a unique platform for exploring excited-state aromaticity. Chem Soc Rev. 2022 Jan 4;51(1):268–92.

2 Hettich CP, Zhang X, Kemper D, Zhao R, Zhou S, Lu Y, et al. Multistate energy decomposition analysis of molecular excited states. JACS Au. 2023 Jul 24;3(7):1800–19.

3 Paetow L, Neugebauer J. Excited state dipole moments from ΔSCF: a benchmark. Phys Chem Chem Phys. 2025 Aug 7;27(31):16354–70.

4 Westermayr J, Marquetand P. Machine learning for electronically excited states of molecules. Chem Rev. 2021 Aug 25;121(16):9873–926.

5 Yang W, Fan Y. Fractional charges, linear conditions and chemical potentials for excited states in ΔSCF theory. arXiv. 2024; arXiv:2408.08443v2.

6 Fallon KJ, Budden P, Salvadori E, Ganose AM, Savory CN, Eyre L, et al. Exploiting excited-state aromaticity to design highly stable singlet fission materials. J Am Chem Soc. 2019 Sep 4;141(35):13867–76.

7 Lischka H, Nachtigallová D, Aquino AJA, Szalay PG, Plasser F, Machado FBC, et al. Multireference approaches for excited states of molecules. Chem Rev. 2018 Aug 8;118(15):7293–361.

8 Nagy Á. Information theoretical and thermodynamic view of the excited-state density functional theory of Coulomb systems. J Chem Phys. 2020 Oct 21; 153(15):154103.

9 González L, Escudero D, Serrano-Andrés L. Progress and challenges in the calculation of electronic excited states. ChemPhysChem. 2012 Jan 16;13(1):28–51.

10 Ottosson H. Exciting excited-state aromaticity. Nat Chem. 2012 Dec;4(12):969–71.

11 Ghosh S, Verma P, Cramer CJ, Gagliardi L, Truhlar DG. Combining wave function methods with density functional theory for excited states. Chem Rev. 2018 Aug 8; 118(15):7249–92.

12 Nagy Á. Relative information in excited-state orbital-free density functional theory. Int J Quantum Chem. 2020 Dec 5;120(23):e26405.

13 Guégan F, Pigeon T, De Proft F, Tognetti V, Joubert L, Chermette H, et al. Understanding chemical selectivity through well selected excited states. J Phys Chem A. 2020 Jan 30;124(4):633–41.

14 Dreuw A, Head-Gordon M. Single-reference ab initio methods for the calculation of excited states of large molecules. Chem Rev. 2005 Nov;105(11):4009–37.

15 Zhao D, He X, Ayers PW, Liu S. Excited-state polarizabilities: a combined density functional theory and information-theoretic approach study. Molecules. 2023;28(6):2576.

16 Cignoni E, Suman D, Nigam J, Cupellini L, Mennucci B, Ceriotti M. Electronic excited states from physically constrained machine learning. ACS Cent Sci. 2024 Feb 29;10(3):637–48.

17 Görling A. Density-functional theory for excited states. Phys Rev A. 1996 Nov;54(5):3912–5.

18 Dral PO, Barbatti M. Molecular excited states through a machine learning lens. Nat Rev Chem. 2021 Jun;5(6):388–405.

19 Lasorne B, Worth GA, Robb MA. Excited-state dynamics. WIREs Comput Mol Sci. 2011 May;1(3):460–75.

20 Krylov AI. Equation-of-motion coupled-cluster methods for open-shell and electronically excited species: the hitchhiker's guide to Fock space. Annu Rev Phys Chem. 2008;59:433–62.

21 Hirata S, Ivanov S, Grabowski I, Bartlett RJ. Time-dependent density functional theory employing optimized effective potentials. J Chem Phys. 2002 Apr 15; 116(15):6468–81.

22 Laurent AD, Jacquemin D. TD-DFT benchmarks: a review. Int J Quantum Chem. 2013 Sep 5;113(17):2019–39.

23 Casida ME, Huix-Rotllant M. Progress in time-dependent density-functional theory. Annu Rev Phys Chem. 2012 Jan 13;63:287–323.

24 Teale AM, Helgaker T, Savin A, Adamo C, Aradi B, Arbuznikov AV, et al. DFT exchange: sharing perspectives on the workhorse of quantum chemistry and materials science. Phys Chem Chem Phys. 2022 Dec 7;24(47):28700–81.

25 Yang W, Ayers PW. Foundation for the ΔSCF approach in density functional theory. arXiv. 2024; arXiv:2403.04604.

26 Vandaele E, Mališ M, Luber S. The ΔSCF method for non-adiabatic dynamics of systems in the liquid phase. J Chem Phys. 2022 Apr 7;156(13):130901.

27 Castellano FN. Altering molecular photophysics by merging organic and inorganic chromophores. Acc Chem Res. 2015 Mar 17;48(3):828–39.

28 Sasikumar D, John AT, Sunny J, Hariharan M. Access to the triplet excited states of organic chromophores. Chem Soc Rev. 2020 Sep 1;49(17):6122–40.

29 Vogler A. Photoreactivity of metal-to-ligand charge transfer excited states. Coord Chem Rev. 1998 Oct;177(1):81–96.

30 May AM, Dempsey JL. A new era of LMCT: leveraging ligand-to-metal charge transfer excited states for photochemical reactions. Chem Sci. 2024 May 8; 15(18):6661–78.

31 Larsen DS, van Grondelle R. Initial photoinduced dynamics of the photoactive yellow protein. ChemPhysChem. 2005 May;6(5):828–37.

32 Norman P, Dreuw A. Simulating X-ray spectroscopies and calculating core-excited states of molecules. Chem Rev. 2018 Aug 8;118(15):7208–48.

33 Smith MB, Michl J. Singlet fission. Chem Rev. 2010 Nov 10;110(11):6891–936.

34 Zimmerman PM, Musgrave CB, Head-Gordon M. A correlated electron view of singlet fission. Acc Chem Res. 2013 Jun 18;46(6):1339–47.

35 Casanova D. Theoretical modeling of singlet fission. Chem Rev. 2018 Aug 8; 118(15):7164–207.

36 He G, Parenti KR, Budden PJ, Niklas J, Macdonald T, Kumarasamy E, et al. Unraveling triplet formation mechanisms in acenothiophene chromophores. J Am Chem Soc. 2023 Oct 11;145(40):22058–68.

37 Manzeli S, Ovchinnikov D, Pasquier D, Yazyev OV, Kis A. 2D transition metal dichalcogenides. Nat Rev Mater. 2017 Jun 13;2(8):17033.

38 Chowdhury T, Sadler EC, Kempa TJ. Progress and prospects in transition-metal dichalcogenide research beyond 2D. Chem Rev. 2020 Nov 25;120(22):12563–91.

39 An X, Zhang W, He X, Li M, Rong C, Liu S. Density-based reactivity theory applied to excited states. AAPPS Bull. 2024 Feb 2;34(1):8.

40 Kolesov G, Grånäs O, Hoyt R, Vinichenko D, Kaxiras E. Real-time TD-DFT with classical ion dynamics: methodology and applications. J Chem Theory Comput. 2016 Feb 9;12(2):466–76.

41 Cao Y, Romero J, Olson JP, Degroote M, Johnson PD, Kieferová M, et al. Quantum chemistry in the age of quantum computing. Chem Rev. 2019 Oct 9; 119(19):10856–915.

42 Alexeev Y, Batista VS, Bauman N, Bertels L, Claudino D, Dutta R, et al. A perspective on quantum computing applications in quantum chemistry using 25–100 logical qubits. J Chem Theory Comput. 2025 Nov 11;21(20):11335–11357.

43 Preskill J. Quantum computing in the NISQ era and beyond. Quantum. 2018 Aug 6;2:79.

44 Motta M, Rice JE. Emerging quantum computing algorithms for quantum chemistry. WIREs Comput Mol Sci. 2022 May;12(3):e1580.

45 Tilly J, Chen H, Cao S, Picozzi D, Setia K, Li Y, et al. The variational quantum eigensolver: a review of methods and best practices. Phys Rep. 2022 Nov;986:1–128.

46 Liu S. Harvesting chemical understanding with machine learning and quantum computers. ACS Phys Chem Au. 2024 Mar 27;4(2):135–42.

47 Hertzog M, Wang M, Mony J, Börjesson K. Strong light-matter interactions: a new direction within chemistry. Chem Soc Rev. 2019 Feb 4;48(3):937–61.

48 Vandewal K. Interfacial charge transfer states in condensed phase systems. Annu Rev Phys Chem. 2016 May 27;67:113–33.

49 Vlček A. The life and times of excited states of organometallic and coordination compounds. Coord Chem Rev. 2000 May; 200–202:933–78.

50 Nelson TR, White AJ, Bjorgaard JA, Sifain AE, Zhang Y, Nebgen B, et al. Non-adiabatic excited-state molecular dynamics: theory and applications for modeling photophysics in extended molecular materials. Chem Rev. 2020 Feb 26; 120(4):2215–87.

51 Marian CM. Spin-orbit coupling and intersystem crossing in molecules. WIREs Comput Mol Sci. 2012 Mar;2(2):187–203.

16

Miscellaneous Applications

TABLE OF CONTENTS

In the preceding chapters, we have explored applying density-based approaches to a variety of well-established chemical problems, from bonding to aromaticity to catalysis. Yet the scope of these methods extends further. In this chapter, we turn to a collection of miscellaneous applications that, while diverse in subject matter, highlight the versatility and adaptability of the density-based perspective. Conformational stability, anomeric effects, and S_N2 reactions reveal how density-based reasoning illuminates subtle energetic and mechanistic balances. Proton-coupled electron transfer (PCET), metal specificity, and oxidation states (OSs) showcase its power in appreciating complex electronic properties. Meanwhile, responses to external electric fields (EEFs), polarizability in macromolecules, and synergies with machine learning (ML) demonstrate how density-based analysis can be carried into new frontiers.

By gathering these seemingly diverse topics under a common lens, our aim is not merely to catalog but to underscore a broader point that density-based thinking offers a unifying language for understanding and predicting chemical phenomena, both classical and emerging.

16.1 Conformational Stability

Conformational flexibility is a defining feature of molecules. Even without breaking bonds, small changes in torsional angles or bond angles can lead to striking differences in stability, reactivity, and function. For decades, the effects driving these preferences have been the subject of continuous debate [1–7]. Steric repulsion and hyperconjugation are the two classic explanations, one emphasizing crowding between functional groups and the other highlighting stabilizing orbital interactions. Yet neither has proved universally satisfactory, and even for a molecule as simple as ethane, the community has long been divided.

To move beyond this stalemate, we turned to density-based approaches in density functional theory (DFT) [8, 9], where the total energy can be partitioned into well-defined components with distinct physicochemical insights. The idea is simple but powerful: instead of choosing between steric or hyperconjugation, one can decompose conformational stability into three contributions—electrostatic, steric, and quantum (the latter capturing exchange and correlation effects). This approach makes it possible to see which effects dominate and which provide subtle refinement, without relying on qualitative orbital interaction arguments.

In the first of our studies [6, 7], we examined bond rotation barriers in six prototypical molecules, ethane, methylamine, methanol, hydrazine, hydroxylamine, and hydrogen peroxide, all with one rotatable single bond. By scanning the full range of dihedral angles of the rotatable bond and then decomposing the corresponding energy changes, we discovered clear trends. Across all systems, electrostatic interactions were found to track the barrier heights most directly: the higher the barrier, the less favorable the electrostatics in the less stable conformer. Steric and exchange–correlation effects also contributed positively, but their magnitudes were smaller. Importantly, including these secondary terms in multivariable models substantially improved quantitative accuracy, showing that while electrostatics dominates, the other effects are indispensable for a complete picture. This resolved the long-standing ethane controversy: both steric hindrance and hyperconjugation are present, but they operate within an electrostatically governed framework.

Another study by us extended this analysis from bond rotations to conformational stability more broadly [10]. Here, we tested nearly 50 small molecules, as well as flexible bond-angle bending in water, carbon dioxide, ammonia, and borane. We compared two explanatory traditions: the orbital-based Walsh diagram approach and the density-based energy partitioning scheme. Walsh diagrams, which track how frontier orbital energies change with geometry, offered at best qualitative predictions. By contrast, the density-based approach revealed strong and quantitative correlations between stability and the electrostatic component of the energy change. Adding steric and quantum contributions refined these correlations even further. In every case examined, electrostatic interaction emerged as the principal determinant of stability, with steric and quantum effects fine-tuning the result, in excellent agreement with the results from the first study above.

The lesson from these studies is clear. Conformational changes are best understood not as battles between steric crowding and hyperconjugation, but as reorganizations of electronic density governed above all by electrostatics. Steric and quantum contributions certainly matter, but they serve to adjust, rather than define, the overall trend. This reframing unifies what had been competing schools of thought, providing a consistent density-based account of conformational energetics.

For readers less familiar with the historical debate, the practical implication is straightforward: when confronted with questions of why one conformation is preferred over another, the first place to look is the electrostatic interaction encoded in the density. Steric effects and quantum exchange–correlation refinements will be present, but they do not overturn the central role of electrostatics.

In summary, a density-based decomposition of energy provides a unified and quantitative picture of conformational stability [6]. Molecular preferences—whether in bond rotations or angle deformations—originate primarily from electrostatic interactions embedded in the electron density, with steric and quantum terms offering essential refinements. This perspective not only resolves long-standing controversies but also establishes a framework that extends naturally to larger systems such as biomolecules, polymers, and condensed-phase materials, where conformational preferences are equally governed by density reorganization.

16.2 Anomeric and Related Effects

Few stereoelectronic effects in chemistry have stirred as much attention as the anomeric effect [11]. First identified in carbohydrate chemistry, it refers to the preference of electronegative substituents at the anomeric carbon to occupy the sterically more hindered axial position, contrary to what simple steric reasoning, which would favor the equatorial position, would predict. For decades, hyperconjugation—particularly the donation from lone pairs into σ^* orbitals—was considered the principal explanation. Others have emphasized dipole–dipole interactions, while still others have pointed to steric crowding. The absence of consensus has kept the anomeric effect at the center of debate, raising a broader question: how can we obtain a unified and quantitative understanding of such stereoelectronic preferences?

Density-based approaches provide a natural framework to address this puzzle [12–14]. By decomposing the total energy into physically interpretable components—electrostatic, steric, and quantum—and by complementing this decomposition with information-theoretic analysis (ITA) quantities, we can move beyond orbital-centric arguments. In our studies related to this topic, density-based approaches were applied to a variety of systems, ranging from simple acyclic molecules to substituted sugars, and extended to the more general question of whether a "generalized anomeric effect" exists [13]. The results point consistently to one conclusion: electrostatics is the principal driving force, with hyperconjugation and steric effects playing subsidiary but indispensable roles.

Our first study focused directly on classical carbohydrate systems [12]. Using energy decomposition analysis, it was shown that the relative stability of axial versus equatorial substituents is overwhelmingly correlated with the electrostatic component. Lone-pair/σ^* hyperconjugation, though present, provided only modest stabilization and could not account for the systematic trends across series. Steric repulsion likewise contributed, but its effect was corrective rather than determinative. This reframed the anomeric effect: not as a curious exception to steric reasoning but as a natural consequence of how electronegative substituents redistribute charge density to favor particular electrostatic alignments.

The second study broadened the scope to generalized anomeric effects [13], asking whether analogous conformational preferences occur in acyclic molecules of the form R_1–X–CH_2–Y–R_2, where X and Y are heteroatoms. Across 45 model systems, a clear pattern emerged: in approximately 80% of cases, the *gauche* conformer was more stable than the *anti*, an inversion of the usual hydrocarbon trend. Energy decomposition analyses revealed that the electrostatic interaction almost single-handedly dictated the stability ordering. Hyperconjugation, as probed through natural bond orbital (NBO) analyses, showed only weak correlations, and dipole arguments fared even worse. Importantly, the same electrostatic origin applied both when the generalized effect held and when it failed, unifying what might otherwise seem like disparate cases.

To further strengthen this conclusion, our third work turned to using both energy and ITA descriptors [14]. Quantities such as Shannon entropy, Fisher information, and Kullback–Leibler divergence were calculated directly from the electron density. These measures provided independent confirmation that conformational preferences track reorganizations of density that minimize electrostatic repulsion and optimize nuclear–electron attractions. By combining energy components with ITA indices, nearly quantitative fits to conformational energy differences were achieved across diverse systems. The implication is striking: stereoelectronic effects like the anomeric effect can be predicted directly from density-based descriptors, without recourse to orbitals at all.

Taken together, these studies reveal a consistent theme: although hyperconjugation and steric effects contribute to anomeric preferences, they do so within an electrostatically optimized density environment. The density-based energy and ITA analyses show that what orbital descriptions label as "donation" or "delocalization" is, at its core, a manifestation of how the electron density reorganizes to minimize electrostatic penalties and maximize stabilizing interactions. This density-based interpretation unifies what previously appeared to be competing explanations.

From a broader perspective, this case study illustrates how long-standing controversies in chemistry can be resolved by recasting the problem in density terms. Just as conformational changes in ethane and related molecules were clarified by decomposing energies into electrostatic, steric, and quantum contributions, so too does the anomeric effect find a natural and quantitative explanation in the density-based framework. This convergence underscores the versatility of density-based thinking, and it hints at broader applications: wherever stereoelectronic effects are invoked—whether in enzymatic catalysis, conformational control in drug molecules, or functional group orientation in polymers—electrostatics embedded in the density will provide the first clue.

16.3 S_N2 Reactions

The bimolecular nucleophilic substitution (S_N2) reaction has long been regarded as a paradigm of chemical reactivity, illustrating in its simplest form the intimate interplay between structure, energetics, and reactivity. At its core, an S_N2 process proceeds through the backside attack of a nucleophile on a saturated carbon center, producing a concerted displacement of the leaving group via a trigonal bipyramidal transition state. Despite its apparent simplicity, the S_N2 mechanism reveals a remarkable complexity: steric hindrance, electronic stabilization, exchange and correlation effects, and substituent influences all cooperate to determine barrier heights and reactivity patterns. Historically, qualitative intuition, such as steric hindrance, has provided useful but limited guidance, while more formal decomposition analyses have struggled to clearly separate the individual contributions shaping the energy landscape.

Density-based approaches in DFT have offered a decisive advance. In early work by us [15], a decomposition scheme was introduced that partitions barrier heights into three physically interpretable components: steric, electrostatic, and quantum effects. Applied to a broad series of substituted methyl halides reacting with fluoride, this analysis revealed that although steric repulsion raises the barrier and quantum stabilization tends to offset it, the decisive factor correlating with activation energies is the electrostatic contribution. The long-held view that steric congestion alone governs S_N2 reactivity thus required revision: electrostatics, not sterics, proved to be the true descriptor of reactivity, modulated by a delicate steric–quantum balance.

Building on this foundational work, we later expanded the use of density-based descriptors with ITA quantities to quantify electrophilicity in S_N2 reactions [16]. We confirmed that a strong correlation exists between the reaction barrier height and the electrostatic interaction, reinforcing our earlier findings. However, we also highlighted that the contributions from steric and exchange–correlation effects, while less significant, are indispensable for a complete understanding of the reaction. We demonstrated that the barrier height could be reliably predicted by various ITA descriptors, such as Shannon entropy, Fisher information, and information gain on the regioselective carbon atom. These findings underscore the effectiveness of a multifaceted, density-based approach in providing a more nuanced conceptual framework for predicting and understanding chemical reactivity.

In a subsequent study on S_N2 reactions, we delved even deeper into the quantification of steric effects by introducing and systematically examining the concept of "steric charge" [17]. This study was built on our earlier work using Weizsäcker kinetic energy to quantify steric energy. The concept of "steric charge"—a derivative quantity rooted in the Weizsäcker kinetic energy—provides a direct density-based measure of steric hindrance. Its strong linear correlation with S_N2 barrier heights gives steric effects a rigorous, quantifiable footing traditionally absent from qualitative steric arguments.

The scope of this density-based perspective was further tested in the long-standing puzzle of the α-effect for S_N2 reactions [18], which refers to the anomalously high nucleophilicity of species with adjacent electron-donating atoms. We employed the density-based decomposition framework to resolve this controversy. Our results demonstrated that the enhanced reactivity of α-effect nucleophiles does not arise from simple steric relief or ground-state destabilization but rather from favorable density redistribution and electrostatic stabilization in the transition state. By isolating this contribution, the study provided a quantitative and unified explanation for a phenomenon that had resisted consistent interpretation for decades.

Together, these studies chart a trajectory from the dissection of steric, electrostatic, and quantum effects in gas-phase S_N2 reactions to the reconceptualization of steric effect, the identification of electrostatic interactions as the dominant controlling factor, and the resolution of the α-effect puzzle. More broadly, we demonstrate how chemical understanding emerges naturally from the density-based partitioning of energy and information. By grounding classical notions of steric hindrance and nucleophilicity in rigorous density-based descriptors, this framework not only sharpens predictive accuracy but also enriches the conceptual foundations of physical organic chemistry. S_N2 reactions, once viewed primarily through the lens of steric hindrance and orbital interactions, now stand as a clear example of how density-based approaches within DFT can unify long-debated concepts into a quantitative understanding.

16.4 Proton-coupled Electron Transfer

PCET represents one of the most fundamental classes of chemical processes, underpinning energy conversion, enzymatic catalysis, and redox chemistry in both biology and materials science. Unlike pure electron or proton transfers, PCET involves concerted or sequential movement of both charge and proton, with mechanisms that vary subtly depending on the balance of driving forces and the electronic structure of the system. Understanding the mechanistic nature of PCET remains a long-standing challenge, as it requires disentangling the relative tendencies for proton transfer and electron transfer, while also accounting for the stabilizing role of coupled charge redistribution.

Conceptual density functional theory (CDFT) offers a natural framework to address this challenge. In particular, the *dual descriptor* [19–21], derived as the difference between electron density response functions for nucleophilic and electrophilic attacks, provides a spatially resolved measure of charge separation propensity [22]. We pioneered this application by showing that the dual descriptor can identify reactive sites where electron density is predisposed to reorganize in concert with proton motion. By quantifying the tendency for charge separation within donor–acceptor pairs, we were able to predict whether a PCET reaction would favor a concerted or stepwise mechanism. This work provided the first density-based criterion for PCET mechanism classification, moving beyond energy-based arguments toward a predictive, conceptual tool rooted directly in the electron density.

The power of this approach was further illustrated in the study of the coenzyme B12–tyrosine complex in adenosylcobalamin-dependent enzymes [23]. These systems are archetypal for PCET, where precise control of charge and proton motion is essential for enzymatic function. Using the dual descriptor, we demonstrated that the charge-separation propensity of the Tyr–B12 complex could be rigorously mapped, providing quantitative insight into how the enzyme environment tunes PCET pathways. This analysis reveals that local density reactivity indicators can rationalize why certain active-site configurations favor synchronous proton–electron movement, while others allow for asynchronous or sequential mechanisms. This density-based characterization aligned closely with experimental observations of PCET in B12 enzymes, offering both validation of the approach and novel mechanistic insight.

These studies show that the dual descriptor functions as an alternative conceptual tool for probing PCET reactivity. Just as descriptors such as electronegativity, hardness, and Fukui functions have clarified the principles of chemical reactivity for simpler systems, the dual descriptor extends this conceptual reach to one of the most intricate and biologically relevant classes of reactions. Our studies demonstrated that even in the most complex coupled reactions, where protons and electrons move hand-in-hand, the key to understanding lies in the electron density and its associated descriptors. As these methods are extended to broader classes of systems, they reinforce the central thesis that chemical reactivity is, at its heart, a property of electron density.

16.5 Metal Specificity

The ability of biological systems to distinguish between metal ions is one of the most remarkable features of nature's design. Proteins and cofactors often bind selectively to a single metal, even in environments where competing ions of similar size and charge are abundant. Heme centers, for example, preferentially incorporate iron over other first-row transition metals, enabling electron transport and oxygen activation. Likewise, metalloproteins discriminate between Mn, Fe, Co, Ni, Cu, and Zn despite their shared d-block chemistry. The origins of this specificity have long been debated, with explanations ranging from hard–soft acid–base (HSAB) [24, 25] arguments to subtle steric or coordination geometry effects. Yet such heuristic rules often fall short of providing predictive, quantitative insight.

CDFT [20, 26–29] provides a natural framework for addressing this problem. Its descriptors—chemical potential, hardness, softness, Fukui function, and spin-polarized reactivity indices—offer direct measures of how electronic structures adapt to metal binding. In one of the earliest systematic applications [30], we studied the metal-binding specificity of porphyrins using CDFT descriptors. Porphyrins serve as versatile ligands across biology, binding iron in hemoglobin, magnesium in chlorophyll, and other metals in synthetic analogs. By evaluating global CDFT descriptors for a range of first-row transition metals bound to porphyrin, we demonstrated that binding specificity could be rationalized through the alignment of chemical potentials and the minimization of hardness mismatch. Metals whose electronegativity and hardness were best matched to the porphyrin framework exhibited the strongest preference, providing a density-based rationale for the experimentally observed selectivity. This analysis went beyond HSAB by supplying a quantitative connection between electronic structure descriptors and ligand preference.

Building upon this foundation, we further examined the more focused question of why iron is the overwhelmingly favored metal in heme systems [31]. Employing a spin-polarized extension of CDFT, the study probed how the spin state of the metal influences binding specificity. Transition metals differ not only in their charge density and hardness but also in their propensity to adopt particular spin configurations. By calculating spin-resolved reactivity indices, we showed that iron uniquely balances spin polarization with electronic compatibility in porphyrin, leading to both stable binding and accessible redox activity. Other metals, especially those from the same group as iron, such as Ru, while forming stable complexes, failed to provide the same combination of spin adaptability and electronic matching. This work highlighted the essential role of spin polarization in metal selection, offering a rigorous explanation for iron's privileged role in biological systems.

These studies illustrate how metal specificity can be reframed in density-based terms. Rather than appealing to empirical chemical intuition, CDFT descriptors provide predictive measures of binding preference. Chemical potential and hardness capture the fundamental electron transfer balance between metal and ligand, while spin-polarized descriptors reveal how magnetic and electronic adaptability govern the fine structure of specificity. These findings not only rationalize why iron dominates in hemoproteins and why porphyrins select certain metals over others but also establish a framework that is extendable to a broad range of metalloproteins and synthetic catalysts.

Just as in other applications presented in this chapter, these results show that seemingly complex selectivity patterns arise naturally from density-encoded properties—here, the interplay of chemical potential matching, hardness compatibility, and spin polarization—providing a generalizable and predictive framework for understanding metal–ligand specificity.

16.6 Oxidation States

The OS, or oxidation number, is a fundamental and widely used concept in chemistry. It serves as a cornerstone of electron bookkeeping, guiding our understanding of redox chemistry, catalysis, inorganic structures, and reaction mechanisms. Defined as the charge an atom would have after an ionic approximation of its bonds, it provides a means to track the degree of oxidation or reduction for an atom within a molecule. Despite its ubiquity in chemical discourse and education, OS has long resisted a rigorous, first-principles definition. While assigning an OS is straightforward for simple molecules using a standard set of rules based on electronegativity rules or Lewis structures, this process can become ambiguous and challenging for more complex systems. For instance, assigning an OS to the nitrogen atom in peroxynitrous acid (HOONO) is not immediately clear, as different assumptions about the oxygen atoms' bonding lead to different possible values for nitrogen. Other examples in transition-metal oxides or organometallic complexes with competing resonance forms and delocalization effects challenge simple OS assignment. This ambiguity of OS assignment is a long-standing issue in chemistry, prompting a need for more robust and universally applicable methods for OS determination.

In our 2019 study [32], we examined this issue using a data set of 49 molecules containing C, N, O, P, Cl, and Br. By applying ITA quantities (e.g., Shannon entropy, Fisher information, information gain, relative Rényi entropies) alongside Hirshfeld charges, we found remarkably strong linear correlations between these density descriptors and conventional OS values. This establishes that OS do leave measurable imprints in the density.

A crucial insight was that the slopes of these correlations differ by element, preventing a single global fit. The missing factor proved to be the crystal radius: when included, per-element differences collapsed into a unified framework, with global fits achieving $R2$ values of ~0.95. Thus, the crystal radius acts as a physically motivated scaling factor, reflecting the spatial extent of the ionic electron cloud central to the oxidation-state concept; once incorporated, element-dependent differences collapse into a unified relationship.

Further refinement showed that multivariate models incorporating multiple ITA descriptors and crystal radius can predict OSs to within about ± 0.5 units across the entire data set. Importantly, the method works not only for simple main-group systems but also in ambiguous cases where multiple OS assignments are possible, providing consistent and chemically meaningful results.

These results reinforce the idea that OS is not merely a bookkeeping device but an emergent property imprinted in the electron density and recoverable through density-based descriptors. By identifying robust density signatures of this ubiquitous concept, we show that OS can be predicted rather than assigned, embedding them firmly in the density-based paradigm. This advance not only resolves long-standing ambiguities but also enables OSs to be tracked quantitatively in redox processes and catalytic cycles—thereby strengthening the bridge between traditional chemical language and rigorous electronic structure theory.

16.7 Impacts of EEFs

As "smart reagents" for directly tuning the electron density, EEFs have recently emerged as powerful, tunable tools capable of reshaping molecular structure, properties, and reactivity [33–35]. Unlike conventional catalysts that operate by introducing fixed steric or electronic perturbations, EEFs provide continuous "knobs" of strength and orientation, offering unprecedented control. In a series of recent studies by us [36, 37], density-based approaches combining CDFT with ITA descriptors have revealed how directionality, additivity, and simultaneous modulation of barrier heights and widths can be captured within the density-based framework. This has transformed the qualitative intuition surrounding EEFs into quantitative, predictive design rules.

The effect of field orientation was first established using benzoic acid as a model system [36], where acidity and aromaticity were mapped against uniform fields applied along three orthogonal axes. The response proved strongly anisotropic: fields aligned with the carboxyl group caused the largest and most systematic changes in acidity, while fields perpendicular to the ring produced quadratic responses, reflecting both structural fragility and resistance of aromatic stabilization until disrupted. These behaviors were traced to frontier orbital alignment, dipole polarization, and field-induced charge redistribution. ITA quantities, such as information gain and relative Rényi entropies, showed excellent correlation with acidity and aromaticity changes, while CDFT indices, such as the electrophilicity index, hardness, and molecular electrostatic potentials, offered complementary mechanistic rationalizations. Together, these results highlighted that acidity is highly sensitive to in-plane field perturbations, while aromaticity is globally diminished by field-induced density reorganization.

The directionality effects were subsequently extended to substituted benzoic acids [37], where a striking additivity principle was uncovered. Regardless of whether the substituents were electron-donating or withdrawing and independent of the field's alignment with the carboxyl axis, the combined effect of multiple substituents could be predicted from single-substituent data using simple sum rules. Local properties such as acidity followed normalized additivity, while global properties like aromaticity required non-normalized forms to avoid double-counting. These findings generalize the classic Hammett additivity rule to field-perturbed systems and establish ITA descriptors such as relative Fisher information and entropy-based indices as robust predictors of substitution patterns under external fields.

Beyond molecular properties, EEFs were also shown to decisively shape chemical reactivity. Using archetypal gas-phase S_N2 reactions, it was demonstrated that fields aligned along the reaction coordinate led to a quadratic lowering of activation barriers, accompanied by a simultaneous narrowing of the intrinsic barrier width. This dual modulation, decreasing barrier height while compressing barrier width, provides a direct means to enhance reaction rates through both classical (barrier lowering) and quantum (tunneling enhancement) channels. The width analysis was enabled by a new partitioning scheme in which extrema in the derivative of Hirshfeld charges along the intrinsic reaction coordinate define natural pre-TS, TS, and post-TS regions. This partitioning reveals that steric and electrostatic contributions dominate in early and late stages, whereas charge transfer governs the transition state, allowing field-induced barrier modulation to be interpreted directly in density terms.

These studies offer a coherent design strategy: align fields with the reaction coordinate to harness quadratic barrier lowering and narrowing; exploit in-plane fields to maximize acidity control; and use substituent additivity to anticipate responses in multiply substituted systems. CDFT and ITA descriptors provide the quantitative language for these predictions, ensuring that the impact of EEFs can be anticipated and tuned in advance. Conceptually, the message is that external fields do not merely stabilize intermediates opportunistically but reshape the electron density in predictable, quantifiable ways. This density-centered perspective positions EEFs as general and tunable levers in synthesis, catalysis, and electrochemistry, opening a new frontier for molecular design.

16.8 Polarizability for Macromolecules

Polarizability, the measure of how a system's electron density redistributes in response to external perturbations [38], lies at the heart of intermolecular interactions, optical response, and electronic transport. While its evaluation is relatively tractable in small molecules, predicting polarizability in macromolecular and condensed-phase systems remains a formidable challenge. The complexity arises from the sheer size of the systems, the interplay of local and nonlocal effects, and the delicate balance between computational cost and accuracy. Traditional post-Hartree–Fock methods, though accurate, scale too steeply with system size to be of practical use in realistic macromolecular environments. This situation calls for approaches that remain grounded in first principles while providing both efficiency and interpretability.

Recent progress in density-based approaches has provided precisely such a pathway. In our 2023 work on condensed-phase systems [39], we showed that polarizabilities at the post-Hartree–Fock level can be reliably reproduced using density-based descriptors derived from CDFT and ITA. These descriptors encode the essential physics of electron redistribution, allowing global response properties to be reconstructed from local density features. When benchmarked against molecular crystals and liquids, this framework consistently delivered accuracy on par with correlated ab initio calculations, but at a fraction of the computational cost. In this way, polarizability—once a bottleneck for large-scale modeling—became accessible for realistic condensed-phase applications.

A complementary study expanded the scope of these ideas to macromolecular systems [40, 41], revealing how ITA descriptors such as Shannon entropy, Fisher information, and information gain correlate with patterns of density redistributions. By decomposing the polarizability into local contributions, it became clear that the property is not simply additive with system size but reflects a hierarchical organization: local functional groups exhibit characteristic responses, structural motifs amplify or dampen these effects, and the macromolecule as a whole expresses an integrated polarizability. This insight is particularly valuable for biomolecular and polymer systems, where modularity and repeating units dominate the structural organization. The finding that fragment additivity can be recovered through appropriately normalized ITA descriptors paves the way for fragment-based modeling strategies that are both efficient and conceptually transparent.

Building on this density-based foundation, our 2024 work [42] introduced a fragment-based deep learning framework for the simultaneous prediction of polarizabilities and NMR shieldings. By integrating ITA descriptors with ML techniques, we were able to overcome the inherent scaling limits of conventional approaches while retaining physical interpretability. Crucially, the model treats macromolecules as assemblies of chemically meaningful fragments, each described by density-based descriptors. The learning algorithm then reconstitutes the full-system response by accounting for fragment couplings. Benchmark tests revealed that this method achieves near–post-Hartree–Fock accuracy for large biomolecules and synthetic polymers, while offering orders-of-magnitude gains in computational efficiency. This approach illustrates how density-based thinking can merge seamlessly with data-driven techniques to create predictive tools for complex systems.

Two other recent contributions further reinforce this framework [43, 44]. One examined why ITA descriptors are so powerful across diverse chemical problems, offering a unifying explanation for their ability to capture nonlocal and collective density responses, of which polarizability is a prime example. The other, published in *Molecules*, presented a broader overview of applying CDFT and ITA descriptors to chemical reactivity and properties in excited states [44]. These studies manifest that the tools used for macromolecular polarizability are part of a consistent and generalizable density-based philosophy.

16.9 Merging with ML

The rise of ML in quantum chemistry offers not just a new computational toolkit, but also a chance to reimagine how density-based concepts can be extended, generalized, and operationalized at scale. Traditional DFT provides both rigorous formalisms and conceptual descriptors but faces practical limits in accuracy and efficiency when applied to large or complex systems. ML, with its ability to extract nonlinear structure–property relationships from data, complements DFT by providing interpolation power where analytic functionals or descriptors struggle. A central theme of our recent work has been to explore how density-based approaches and ML can merge to create new theoretical frameworks, not merely to accelerate calculations, but to deepen understanding of chemical phenomena.

One direction has been the development of fragment-based deep learning architectures that retain the spirit of conceptual DFT [42, 45]. In this approach, molecules are decomposed into chemically meaningful fragments, each described by information-theoretic or other density-based descriptors, and ML is then trained to reconstruct global properties. Unlike black-box learning of raw structures, this method embeds chemical insight directly into the learning process. Applied to the simultaneous prediction of polarizabilities and NMR shieldings, this framework demonstrated that fragment-level density features contain sufficient transferable information to recover global response properties across diverse chemical environments. While in the previous section we highlighted the efficiency and accuracy of this approach for macromolecules, here it should be stressed that the deeper contribution lies in showing how density-based descriptors can serve as a natural bridge between physical theory and modern data-driven inference.

A second direction extends density concepts to multi-property learning [46]. Correlation energy, often viewed as an isolated quantum correction, leaves signatures across multiple observables. By learning these interrelationships, ML can treat correlation energy as a latent variable linking diverse properties—a perspective fully aligned with the density-based philosophy that chemical concepts emerge from underlying fundamental quantities. In a 2025 study, we constructed a deep learning model capable of predicting both correlation energies and a range of physicochemical properties within a unified framework. The key insight is that electron correlation, often treated as an isolated correction in quantum chemistry, in fact links multiple observables that can be simultaneously captured and exploited by ML. By training the model on data sets of correlated wavefunction calculations, the framework was able to generalize across different molecules and properties, thus highlighting how correlation energy can act as a unifying latent variable connecting disparate chemical quantities. This represents a shift from property-by-property prediction toward integrated, system-level modeling, aligned with the density-based philosophy that chemical concepts should emerge from fundamental variables rather than be treated as independent add-ons.

A third application of merging DFT and ML [45] focuses on quantifying Lewis acidity and basicity, classical concepts that have long resisted universal, density-based definitions. In this work, we combined density-based reactivity theory with data-driven regression to identify descriptors that reliably predict Lewis acid and base strengths across diverse systems. By training on both experimental scales and high-level calculations, the model achieved robust transferability, bridging the gap between abstract density descriptors and practical chemical scales. What distinguishes this study is its demonstration that ML can provide a calibration layer that maps density-based quantities onto chemically meaningful scales without sacrificing interpretability. Rather than replacing theory with empirical fitting, ML here acts as a translator, aligning CDFT and ITA outputs with experimental chemical intuition.

These studies illustrate that merging density-based approaches with ML yields more than computational speed-ups. It opens a pathway toward a new hybrid paradigm: density-based descriptors anchor predictions in physical theory, while ML captures correlations and nonlinearities that escape closed-form models. This synergy produces predictive frameworks that are accurate, efficient, and conceptually transparent. It also reinforces the philosophical view that chemical understanding arises from the interplay between fundamental variables (such as density or correlation energy) and emergent properties (such as polarizability, acidity, or reactivity). By integrating ML into this picture, we ensure that density-based theory remains not only a foundation of quantum chemistry, but also a driver of its next generation.

16.10 Summary and Outlook

In this chapter we have explored a wide range of miscellaneous applications of density-based approaches within density functional theory, demonstrating their power to illuminate diverse chemical problems. Despite the varied topics—conformational stability, the anomeric effect, S_N2 reactions, PCET, metal specificity, OSs, EEFs, polarizability of macromolecules, and the merging of density concepts with ML—a unifying theme emerges. In each case, electron density and its associated descriptors serve as the fundamental language for capturing chemical behavior, bridging the gap between abstract quantum theory and intuitive chemical concepts.

The case studies presented here highlight the versatility of density-based thinking. We saw how density measures can rationalize classical structural preferences such as conformational stability and the anomeric effect, and how they reveal hidden factors in complex mechanisms such as S_N2 and PCET. In the realm of inorganic chemistry, density descriptors provided fresh insight into metal specificity and the elusive assignment of OSs, showing that even the most traditional chemical concepts are rooted in quantifiable density signatures. Beyond static molecular systems, the density framework proved equally powerful in predicting responses to perturbations, including the effects of EEFs and the polarizability of macromolecular assemblies. Finally, by merging density-based descriptors with emerging ML methods, we pointed to a new paradigm in which theory and data-driven inference reinforce each other, enabling both accurate prediction and deeper understanding.

Looking ahead, the integration of density-based approaches with ML, multiscale modeling, and experimental data will broaden their reach to complex biological assemblies, functional materials, and systems under extreme conditions. The philosophical appeal endures: density-based methods anchor chemical concepts—traditional or newly emerging—in the most fundamental variable available. By embracing this unifying perspective, density-based reactivity theory remains not only a practical computational framework but also a conceptual engine for deepening chemical understanding.

References

1 Pophristic V, Goodman L. Hyperconjugation not steric repulsion leads to the staggered structure of ethane. Nature. 2001 May 31;411(6837):565–8.
2 Liu S, Govind N. Toward understanding the nature of internal rotation barriers with a new energy partition scheme: ethane and n-butane. J Phys Chem A. 2008 Jul 24;112(29):6690–9.
3 Esquivel RO, Liu S, Angulo JC, Dehesa JS, Antolín J, Molina-Espíritu M. Fisher information and steric effect: study of the internal rotation barrier of ethane. J Phys Chem A. 2011 May 5;115(17):4406–15.

4 Weinhold F. Rebuttal to the bickelhaupt–baerends case for steric repulsion causing the staggered conformation of ethane. Angew Chem Int Ed. 2003 Sep 15;42(35):4188–94.

5 Bickelhaupt FM, Baerends EJ. The case for steric repulsion causing the staggered conformation of ethane. Angew Chem Int Ed. 2003 Sep 15;42(35): 4183–8; discussion 4188.

6 Liu S. Origin and nature of bond rotation barriers: a unified view. J Phys Chem A. 2013 Feb 7;117(5):962–5.

7 Liu S, Govind N, Pedersen LG. Exploring the origin of the internal rotational barrier for molecules with one rotatable dihedral angle. J Chem Phys. 2008 Sep 7;129(9):094104.

8 Liu S. Steric effect: a quantitative description from density functional theory. J Chem Phys. 2007 Jun 28;126(24):244103.

9 Tsirelson VG, Stash AI, Liu S. Quantifying steric effect with experimental electron density. J Chem Phys. 2010 Sep 21;133(11):114110.

10 Liu S, Schauer CK. Origin of molecular conformational stability: perspectives from molecular orbital interactions and density functional reactivity theory. J Chem Phys. 2015 Feb 7;142(5):054107.

11 Mo Y. Computational evidence that hyperconjugative interactions are not responsible for the anomeric effect. Nat Chem. 2010 Aug;2(8):666–71.

12 Huang Y, Zhong A-G, Yang Q, Liu S. Origin of anomeric effect: a density functional steric analysis. J Chem Phys. 2011 Feb 28;134(8):084103.

13 Cao X, Liu S, Rong C, Lu T, Liu S. Is there a generalized anomeric effect? Analyses from energy components and information-theoretic quantities from density functional reactivity theory. Chem Phys Lett. 2017 Nov;687:131–7.

14 Zhou X, Yu D, Rong C, Lu T, Liu S. Anomeric effect revisited: perspective from information-theoretic approach in density functional reactivity theory. Chem Phys Lett. 2017 Sep;684:97–102.

15 Liu S, Hu H, Pedersen LG. Steric, quantum, and electrostatic effects on S(N)2 reaction barriers in gas phase. J Phys Chem A. 2010 May 13;114(18):5913–8.

16 Wu Z, Rong C, Lu T, Ayers PW, Liu S. Density functional reactivity theory study of SN2 reactions from the information-theoretic perspective. Phys Chem Chem Phys. 2015 Oct 28;17(40):27052–61.

17 Liu S, Liu L, Yu D, Rong C, Lu T. Steric charge. Phys Chem Chem Phys. 2018 Jan 17;20(3):1408–20.

18 Poddar A, Zhao D, Ayers PW, Liu S, Chattaraj PK. What dictates the α-effect in gas-phase SN2 reactions? A density functional theory study. J Phys Chem A. 2025 Feb 20;129(7):1847–55.

19 Morell C, Grand A, Toro-Labbé A. New dual descriptor for chemical reactivity. J Phys Chem A. 2005 Jan 13;109(1):205–12.

20 Geerlings P, Chamorro E, Chattaraj PK, De Proft F, Gázquez JL, Liu S, et al. Conceptual density functional theory: status, prospects, issues. Theor Chem Acc. 2020 Feb;139(2):36.

21 Ayers PW, Morell C, De Proft F, Geerlings P. Understanding the Woodward-Hoffmann rules by using changes in electron density. Chem Eur J. 2007;13(29):8240–7.
22 Liu S, Ess DH, Schauer CK. Density functional reactivity theory characterizes charge separation propensity in proton-coupled electron transfer reactions. J Phys Chem A. 2011 May 12;115(18):4738–42.
23 Kumar N, Liu S, Kozlowski PM. Charge separation propensity of the coenzyme B_{12} –tyrosine complex in adenosylcobalamin-dependent methylmalonyl–CoA mutase enzyme. J Phys Chem Lett. 2012 Apr 19;3(8):1035–8.
24 Chattaraj PK, Lee H, Parr RG. HSAB principle. J Am Chem Soc. 1991 Feb;113(5):1855–6.
25 Datta D. On Pearson's HSAB principle. Inorg Chem. 1992 Jun;31(13):2797–800.
26 Parr RG, Yang W. Density-Functional Theory of Atoms and Molecules. Oxford University Press; 1989.
27 Liu S, editor. Conceptual density functional theory: Towards a new chemical reactivity theory. 1st ed. Wiley-VCH GmbH: Wiley; 2022.
28 Liu S. Conceptual density functional theory and some recent developments. Acta Phys-Chim Sin. 2009;25(3):590–600.
29 Geerlings P, De Proft F, Langenaeker W. Conceptual density functional theory. Chem Rev. 2003 May;103(5):1793–873.
30 Feng X-T, Yu J-G, Lei M, Fang W-H, Liu S. Toward understanding metal-binding specificity of porphyrin: a conceptual density functional theory study. J Phys Chem B. 2009 Oct 8;113(40):13381–9.
31 Feng X-T, Yu J-G, Liu R-Z, Lei M, Fang W-H, De Proft F, et al. Why iron? A spin-polarized conceptual density functional theory study on metal-binding specificity of porphyrin. J Phys Chem A. 2010 Jun 3;114(21):6342–9.
32 Wu J, Yu D, Liu S, Rong C, Zhong A, Chattaraj PK, et al. Is it possible to determine oxidation states for atoms in molecules using density-based quantities? An information-theoretic approach and conceptual density functional theory study. J Phys Chem A. 2019 Aug 8;123(31):6751–60.
33 Clarys T, Stuyver T, De Proft F, Geerlings P. Extending conceptual DFT to include additional variables: oriented external electric field. Phys Chem Chem Phys. 2021 Jan 21;23(2):990–1005.
34 Geerlings P, De Proft F. External fields in conceptual density functional theory. J Comput Chem. 2023 Jan 30;44(3):442–55.
35 Chen J, Liu S, Li M, Rong C, Liu S. A density functional theory and information-theoretic approach study of chiral molecules in external electric fields. Chem Phys Lett. 2020 Oct;757:137858.
36 Li M, Wan X, He X, Rong C, Liu S. Impacts of external fields on aromaticity and acidity of benzoic acid: a density functional theory, conceptual density functional theory and information-theoretic approach study. Phys Chem Chem Phys. 2023 Jan 18;25(3):2595–605.
37 Li M, Wan X, Rong C, Zhao D, Liu S. Directionality and additivity effects of molecular acidity and aromaticity for substituted benzoic acids under external electric fields. Phys Chem Chem Phys. 2023 Oct 18;25(40):27805–16.

38 Nochebuena J, Piquemal J-P, Liu S, Cisneros GA. Cooperativity and frustration effects (or lack thereof) in polarizable and non-polarizable force fields. J Chem Theory Comput. 2023 Nov 14;19(21):7715–30.

39 Zhao D, Zhao Y, He X, Li Y, Ayers PW, Liu S. Accurate and efficient prediction of post-Hartree-Fock polarizabilities of condensed-phase systems. J Chem Theory Comput. 2023 Sep 26;19(18):6461–70.

40 Zhao D, Zhao Y, He X, Ayers PW, Liu S. Efficient and accurate density-based prediction of macromolecular polarizabilities. Phys Chem Chem Phys. 2023 Jan 18;25(3):2131–41.

41 Zhao D, Liu S, Chen D. A density functional theory and information-theoretic approach study of interaction energy and polarizability for base pairs and peptides. Pharmaceuticals. 2022;15(8):938.

42 Zhao D, Zhao Y, Xu E, Liu W, Ayers PW, Liu S, et al. Fragment-based deep learning for simultaneous prediction of polarizabilities and NMR shieldings of macromolecules and their aggregates. J Chem Theory Comput. 2024 Mar 26;20(6):2655–65.

43 Zhao Y, Zhao D, Liu S, Rong C, Ayers PW. Why are information-theoretic descriptors powerful predictors of atomic and molecular polarizabilities. J Mol Model. 2024 Oct 3;30(11):361.

44 Zhao D, He X, Ayers PW, Liu S. Excited-state polarizabilities: a combined density functional theory and information-theoretic approach study. Molecules. 2023;28(6):2576.

45 Zhuo L, Zheng Y, Zeng L, Zhao Y, Li M, Rong C, et al. Quantification of Lewis acidity and Lewis basicity: a density-based reactivity theory study. J Comput Chem. 2025 Aug 15;46(22):e70212.

46 Yuan Y, Zhao Y, Lu L, Wang J, Chen J, Liu S, et al. Multiproperty deep learning of the correlation energy of electrons and the physicochemical properties of molecules. J Chem Theory Comput. 2025 Jun 24;21(12):5997–6006.

Part III

Perspectives

17

From Chemical Concepts to Chemical Understanding

TABLE OF CONTENTS

17.1 Introduction

Chemistry is the science of change and transformation, where matter is not merely described but actively reshaped through the interplay of empirical practice and theoretical insight. At its core lie profound philosophical questions: *What* are chemical entities like molecules, bonds, and orbitals? *How* do we come to know their structures and behaviors? And *why* do they exhibit the properties and reactivities we observe? This chapter explores these questions through the lenses of chemical ontology—the study of what exists in the chemical world—and chemical epistemology—the methods by which we acquire and validate chemical knowledge. Together, these dimensions converge to produce chemical understanding, the ability to explain and predict chemical phenomena in a coherent and actionable way [1, 2].

Central to this exploration are chemical concepts—molecule, bond, functional group, aromaticity, and more—which serve as both ontological scaffolds defining the entities of chemistry and epistemic tools shaping how we study and manipulate them [3, 4]. This chapter examines the nature of these concepts and their roles in bridging ontology and epistemology, beginning with an analysis of chemical ontology and its connections to physics and biology. It then delves into chemical epistemology, detailing how experimental techniques, theoretical models, and synthetic practices generate knowledge. We explore how hypotheses are represented in chemistry, from Platonic ideals to pragmatic tools, and compare the ontological perspectives of valence bond theory (VBT) [5], molecular orbital theory (MOT) [6], and density functional theory (DFT) [7]. These theories reveal the pluralistic nature of chemical knowledge, where multiple models coexist to provide complementary insights into molecular reality.

Ultimately, this chapter argues that chemical understanding emerges from the dynamic integration of concepts, ontology, and epistemology, forming an interwoven architecture of chemical knowledge. By examining this structure, we not only clarify the foundations of chemistry but also outline the ontological and epistemological requirements for future theoretical models—models that must balance computational precision with conceptual clarity. These insights facilitate a "*what-how-why*" triad understanding of chemistry: what is believed to exist, how it is known, and why it behaves as it does. This perspective is also set to profoundly influence the application of machine learning and quantum computing to chemistry [1, 2], which will be discussed in the next chapter.

17.2 Ontology, Chemical Ontology, and Relations with Physics and Biology

17.2.1 Ontology

Ontology is a fundamental branch of philosophy concerned with the nature of being, existence, and the classification of entities. It addresses foundational questions such as: What kinds of things exist? What does it mean for something to exist? How should entities be categorized, and what relationships hold between them? Originating in ancient Greek thought—particularly in the works of Parmenides, Plato, and Aristotle—ontology has long shaped how we think about reality. In the scientific context, ontology examines the kinds of entities that scientific theories posit and what it means for those entities to "exist." For instance, are electrons truly real, or are they just theoretical constructions? Do chemical bonds have an independent existence, or are they merely artifacts of our models? Is a molecule a natural kind, or simply a useful fiction in our efforts to describe chemical behavior? These questions illustrate how ontology intersects with scientific practice, shaping how we interpret the reality behind the models we use.

Ontology, as a technical framework for organizing knowledge, can be understood through four fundamental components: *entities*, *properties*, *relations*, and *rules* (or *axioms*). These components provide a structured approach for representing complex domains that demand clarity, consistency, and interoperability.

Entities are the fundamental units an ontology describes—such as atoms, molecules, ions, orbitals, electron densities, or potential energy surfaces. *Properties* capture the characteristics of these entities, both intrinsic (e.g., electronegativity, charge, geometry) and relational (e.g., bond order, acidity, reactivity). *Relations* encode how entities connect to one another—such as "atom is part-of molecule," "bond is-a interaction," or "functional group is-a molecular substructure." *Rules and axioms* define constraints or inferential patterns, such as "electron density integrates to N" or "functional groups confer characteristic patterns of reactivity." These four elements provide a structured foundation for representing chemical knowledge.

Ontologies built on this structure—such as NCIt in oncology, ChEBI in chemistry, HPO in human phenotypes, or QUDT in quantum physics—standardize terminology, clarify complex relationships, and enable automated reasoning. They also accommodate the modeling of emergent or dynamic phenomena (e.g., consciousness, quantum entanglement) by incorporating relations and axioms that evolve from observed patterns, such as "high noise reduces entanglement." In this way, ontologies not only organize existing knowledge but also support the discovery of new insights in complex, data-intensive fields like precision medicine, machine learning, and quantum computing.

17.2.2 Chemical Ontology

Chemical ontology explores the fundamental entities that chemistry posits to exist and examines the nature of their existence. It raises several core questions: (i) What are chemical substances—are they simply collections of atoms, or do they exhibit emergent properties that grant them distinct identities? (ii) Are atoms and molecules real physical entities, or are they idealized constructs used within theoretical models? (iii) What is the ontological status of chemical bonds, orbitals, and other theoretical constructs—do they exist independently, or are they explanatory tools? (iv) How do macroscopic properties such as color, boiling point, or reactivity emerge from or relate to microscopic structures?

These questions are central to debates within chemical ontology concerning realism, reductionism, and emergence. A realist perspective holds that molecules, bonds, and orbitals are real because they consistently explain and predict observable chemical behavior. In contrast, an anti-realist might argue that these entities are conceptual instruments—useful for organizing experience and guiding practice, but not necessarily reflective of an independent ontological reality. In navigating these positions, chemical ontology helps clarify what chemistry is fundamentally about and how its core concepts connect to the real world.

17.2.3 Molecular Ontology

The *ontology of molecules* offers a structured framework for understanding what molecules are, how they exist, and what roles they play in the molecular world. Following the above introduction, we can unpack it into four interrelated components. (i) *Entities*: Molecules are treated as discrete, identifiable units of matter—entities in their own right. While composed of atoms (which are also entities), molecules exhibit emergent characteristics that go beyond their atomic parts. They are not merely theoretical constructs but observable (albeit indirectly) through techniques like spectroscopy, crystallography, and reactivity analysis. Their reproducibility and stability support viewing them as natural kinds, with identity grounded in their specific composition, structure, and reactivity. (ii) *Properties*: Molecules possess both intrinsic and relational properties. Intrinsic properties include mass, geometry, dipole moment, electronic configuration, and more, many of which, like aromaticity or chirality, emerge only at the molecular level. Relational properties, such as solubility, acidity, or binding affinity, depend on how molecules interact with their environment or other molecules, and are central to how they are studied and applied. (iii) *Relations*: Molecules are embedded in networks of ontological relations. Internally, these involve bonding types (covalent, ionic, delocalized, etc.) that define molecular identity and stability. Externally, molecules relate to their surroundings through chemical reactions, physical interactions, and biological functions. These relationships are captured in models such as potential energy surfaces and reaction mechanisms. (iv) *Processes and rules*: Molecules are dynamic, undergoing vibrations, rotations, conformational shifts, and chemical transformations. Their identity persists through many of these changes, provided core structural features remain intact. These processes often demonstrate universal principles or general rules such as the electronegativity equalization principle [8], the maximum hardness principle [9–12], or HSAB [13–15]. Any ontological account of molecules must therefore consider their stable configurations, inherent dynamics, and recurring patterns and regularities.

17.2.4 Ontological Relations Among Physics, Chemistry, and Biology

Each scientific discipline operates with its own implicit or explicit ontology. In physics, ontology includes particles, fields, forces, and space-time. In biology, it involves genes, species, and ecosystems. These ontologies are shaped by their respective epistemologies, including the methods, models, and explanatory goals of each discipline. They define what counts as a legitimate entity, what kinds of properties matter, and what relations are fundamental. Chemistry, as a bridge between the physical and life sciences, presents a particularly rich ontological landscape. Its central entities—atoms, molecules, electrons, bonds, orbitals, functional groups, and reactivity trends—form the conceptual scaffolding of chemical knowledge.

Yet many of these entities are not directly observable, and their status depends on chemical epistemology contexts via experimental measurements or theoretical models.

These natural sciences—physics, chemistry, and biology—form a hierarchical continuum of ontological emergence, where each successive layer builds upon but is not reducible to the one beneath it. Physics, at the foundational level, provides the most universal and fundamental principles governing matter, energy, space, and time. Chemistry emerges from these physical principles but introduces novel concepts such as molecular stability, bonding, and reactivity—properties that are not directly obvious from the physics of particles alone. Biology, in turn, arises from chemical systems but is characterized by organization, functionality, and purposeful (selectivity) behavior that reflect a further level of ontological complexity.

Based on the current understanding of classical and quantum mechanics, this ontological hierarchy relationship among physics, chemistry, and biology can be understood graphically, as shown in Figure 17.1 [1]. At its core, physics is concerned with determining the total energy E of a system—whether through Newton's laws for macroscopic substances or the Schrödinger equation for microscopic particles. Here, E is expressed as a functional of the particle number N and external potential $v(\mathbf{r})$,

$$E = E\,[N, v]. \tag{17.1}$$

In terms of Newton's law, force $\boldsymbol{F}$ acting on a particle can be obtained by differentiating a potential energy function that describes the energetic behavior of the particle. In either case, this total energy defines a multidimensional function. In physics, evaluating this potential energy function is the goal, providing a foundational yet universal description of the system's static and dynamic behavior. For complex systems such as proteins and DNA, though, computing their energies often requires multiscale modeling techniques (left image in Figure 17.1).

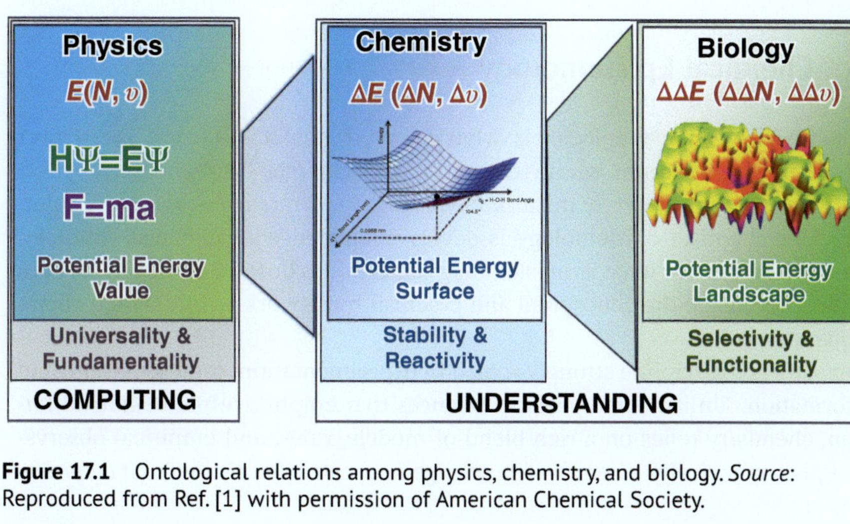

Figure 17.1 Ontological relations among physics, chemistry, and biology. *Source:* Reproduced from Ref. [1] with permission of American Chemical Society.

Chemistry emerges as a discipline where *change* becomes central [1, 2]. We are not just interested in the total energy E but in energy differences, ΔE, that result from variations in N or $v(\mathbf{r})$, ΔN and Δv,

$$\Delta E = E\,[N + \Delta N, v + \Delta v] - E\,[N, v]. \tag{17.2}$$

These differences underpin molecular stability and chemical reactivity, determining whether bonds form, break, or rearrange. The reaction coordinate becomes a path across different potential energy functions, forming a *potential energy surface* with valleys (stable states), peaks (transition states), and slopes (reaction forces) defining the behavior of molecular systems (middle image in Figure 17.1).

Biology, in contrast, introduces yet another ontological leap. Living systems operate not merely by changes in energy, ΔE, but by the change of changes, that is, the second-order variations, $\Delta\Delta E$, due to the change in ΔN and Δv, $\Delta\Delta N$ and $\Delta\Delta v$. These reflect context-sensitive adaptability, such as in enzyme–substrate specificity or signal transduction, where small variations in structure or environment lead to disproportionately large or finely tuned functional outcomes. In effect, biology operates on rugged *potential energy landscapes* [16], where the emergence of biological selectivity and functionalities is encoded in the shape and navigability of these potential energy landscapes (right image in Figure 17.1).

Therefore, we see an ontological progression: physics provides the full potential energy function $E\,[N, v]$; chemistry interprets meaningful changes via stability and reactivity on the potential energy surface (ΔE); and biology exploits the higher-order structure and dynamics of the potential energy landscape ($\Delta\Delta E$) to achieve selectivity and functionality. This is not merely a matter of increasing complexity, but of emergence, where each level introduces new entities, relations, and laws irreducible to those below, thus leading to new insights and understanding in their own right. Each discipline retains autonomy, yet they are deeply interdependent, forming a scaffold of emergence—from particles, to patterns, to purpose—and each revealing a different facet of the same reality [1].

17.3 Chemical Epistemology

How do we know what a molecule is, what it does, or how it will react? The answer is not purely empirical nor wholly theoretical but emerges from chemical epistemology, the study of how chemical knowledge is generated, justified, and interpreted. This chemical epistemology is deeply entangled with chemical ontology. While ontology defines what is, epistemology determines how we can know what is. Together, they shape the conceptual and practical framework within which chemistry operates.

Chemistry is a discipline strongly rooted in experimentation, representation, and transformation. Unlike some physical sciences that emphasize mathematical formalism, chemistry relies on a rich blend of models, rules, and empirical observations. Epistemologically, chemical knowledge is constructed through (i) *empirical observation*: Techniques such as spectroscopy, chromatography, and crystallography provide indirect yet highly informative access to molecular structure and behavior. (ii) *Modeling and visualization*: Molecular knowledge is mediated through representations like Lewis structures, molecular orbitals, resonance forms, and electron density maps. These models are not literal depictions but heuristic devices that support reasoning and prediction. (iii) *Synthesis and manipulation*: Knowing a molecule often involves making it. Synthesis is not just a technical process but an epistemic act that confirms identity, structure, and reactivity. (iv) *Classification and heuristics*: Concepts like functional groups, aromaticity/antiaromaticity, acidity/basicity, and reactivity patterns reflect epistemic generalizations that we heavily rely on. These are often learned through experience and practice rather than formal deduction. Finally, (v) *Computational tools*: Computational models (e.g., VBT, MOT, DFT, molecular dynamics) serve as sources of knowledge about molecules, bridging theoretical predictions with experimental observations. Therefore, chemical epistemology is pluralistic and practice-oriented. It involves a dynamic interplay between theory, experiment, and representation, and emphasizes approximation and utility over formal exactness.

These epistemic practices gain meaning only when anchored to chemical ontology: what we can measure, infer, or compute depends on what kinds of chemical entities we assume to exist. The above features of chemical epistemology have resulted from the uniqueness of chemical ontology, where molecules, as central entities of chemical ontology, are not directly observable. Instead, we come to know molecules through multimodal inference: interpreting spectra, analyzing reaction behavior, modeling electronic structure, and more. The knowledge of a molecule includes (i) *composition* (what atoms and in what ratios), (ii) *structure* (geometry, bonding, stereochemistry), (iii) *properties* (boiling point, reactivity, polarity, etc.), and (iv) *transformations* (how it reacts or decomposes). Each of these facets is supported by different epistemic practices. For example, NMR spectra reveal connectivity; IR spectra show vibrational modes; quantum chemical calculations estimate electronic structures. Thus, molecular knowledge is always partial, contextual, and representation-dependent.

Chemical epistemology and chemical ontology are fundamentally intertwined, each shaping and informing the other. Ontology sets the stage for what we aim to know and how we seek that knowledge. If we accept that molecules, bonds, or electron densities are real entities, this influences the experimental techniques we employ (e.g., spectroscopy, crystallography) and the theoretical tools we use (e.g., quantum chemistry). For instance, treating electron density as real guides the development of computational approaches like DFT, while assuming the existence of discrete molecules justifies efforts to determine their structures and reactivities. At the same time, epistemological practices can shape ontology. The consistent success of tools like MOT lends ontological credibility to entities like frontier orbitals of highest-occupied molecular orbital and lowest-unoccupied molecular orbital, despite their indirect observability.

This dynamic interplay gives rise to both model-dependence and ontological pluralism. Different models of bonding and structure—such as VBT, MOT, and

DFT—offer divergent yet epistemically valid accounts of what a molecule is. This supports a pluralistic ontology in which multiple, model-dependent realities can coexist. Furthermore, the concept of emergence highlights that molecular properties often arise from, but are not reducible to, atomic components. Epistemologically, this suggests that chemistry cannot be fully reduced to physics; it has its own methods and explanatory frameworks. Ontologically, it affirms the status of molecules as genuine entities with emergent properties, not merely as the sum of atomic parts. Together, these insights underscore the reciprocal and evolving relationship between what we know and what we believe to exist.

17.4 Representations of Hypotheses

Relevant to ontology and epistemology is the representation of hypotheses. Scientific and theoretical inquiry often begins with hypotheses: structured assumptions or conceptual models that guide our understanding of complex phenomena. Yet the nature of these hypotheses—what they are, how they function, and whether they reflect real features of the world or merely serve as tools—has long been debated in science. The representation of hypotheses varies across different epistemological and ontological frameworks, from Platonic realism [17] to constructivist pragmatism. A brief introduction to the main representations of hypotheses follows, exploring their underpinnings and implications for chemical modeling and theory-building.

17.4.1 Four Representations

Platonic Representation: Hypotheses as Timeless Truths. In the Platonic tradition, hypotheses are conceived as representations of ideal, abstract truths. Drawing from Plato's theory of Forms, this view holds that mathematical structures, logical relationships, and perfect forms exist in a realm independent of human perception. Hypotheses, under this view, are approximations or glimpses into this ideal world. Our role is to "discover" these truths through reason, abstraction, and mathematical formalism. For example, the Schrödinger equation, the Pythagorean theorem, or the notion of perfect spheres in geometry are often invoked in this Platonic spirit. Hypotheses here are judged not only by empirical adequacy but also by elegance, coherence, and alignment with deep mathematical structure.

Kantian Representation: Hypotheses as Structures of Experience. Immanuel Kant offered a different view: our knowledge of the world is shaped by a priori structures of human cognition. In this Kantian framework, hypotheses do not mirror a noumenal reality (things-in-themselves) but rather structure the phenomena (appearances) that we experience. Hypotheses are mental constructs that reflect the necessary conditions for organizing empirical data. Accordingly, scientific hypotheses are tools for systematizing experience, constrained by the forms of human intuition such as space, time, and causality. Rather than discovering timeless truths, we articulate models that conform to how we are able to perceive and think. This perspective sees scientific progress not as convergence to a mind-independent truth but as a refinement of the conceptual apparatus through which we interact with the empirical world.

Constructivist Representation: *Hypotheses as Human-made Models*. Constructivism takes a more radical turn by asserting that hypotheses are social and cognitive constructions. Knowledge is not discovered but constructed, shaped by cultural practices, linguistic frameworks, historical contingencies, and technological constraints. From this view, hypotheses are contingent models that help us navigate and manipulate the world rather than reveal its underlying essence. Constructivist representations emphasize the instrumental and contextual value of hypotheses. Theories in chemistry, biology, or psychology are seen as tools for problem-solving, prediction, and coordination of practice rather than mirrors of reality. This stance is especially prominent in the social sciences.

Pragmatic Representation: *Hypotheses as Tools for Action*. Closely related to constructivism, but often more moderate, is the pragmatic representation of hypotheses [18]. Influenced by thinkers like William James, John Dewey, and contemporary philosophers like Bas van Fraassen, pragmatism emphasizes the utility of hypotheses. A hypothesis is good not if it is "true" in a metaphysical sense, but if it works—if it leads to successful predictions, interventions, or problem solutions. In the pragmatic view, hypotheses are tested and revised through their consequences. Truth becomes a matter of performance under empirical conditions. This is the dominant stance in many areas of applied science, engineering, and machine learning, where models are chosen based on effectiveness rather than deep ontological commitments.

In chemistry, these representations are not abstract philosophical positions but implicit assumptions embedded in our models. Molecular orbitals often carry a Platonic flavor; VSEPR and heuristic models are deeply pragmatic; and structural formulas reflect a Kantian shaping of experience. Recognizing these representational commitments clarifies why different theories yield different kinds of chemical insights.

17.4.2 Relations with Ontology and Epistemology

The representation of hypotheses is shaped by the dynamic interplay between ontology and epistemology. Ontology determines what kinds of entities and phenomena a hypothesis can meaningfully address (e.g., molecules, bonds, electron density), while epistemology guides how hypotheses are formulated, supported, and tested through methods, evidence, and reasoning. In this way, a hypothesis acts as a bridge between what we believe exists and what we aim to understand, embodying both ontological assumptions and epistemic strategies. Effective hypotheses require alignment between these dimensions, ensuring that scientific inquiry remains both conceptually coherent and empirically grounded.

Moreover, the way hypotheses are represented reflects a range of philosophical commitments—from *Platonic realism*, which treats scientific entities as reflections of ideal forms, to *Kantian structuring*, which emphasizes the role of conceptual frameworks in shaping experience; from *constructivist modeling*, which focuses on how scientific knowledge is contextually built, to *pragmatism*, which values the utility and success of hypotheses in practice. Rather than being in conflict, these perspectives can be seen as complementary, offering a pluralistic lens through which to understand the nature and function of hypotheses.

17.5 Ontological Views of Hypotheses in MOT, VBT, and DFT

Understanding what a molecule *is* involves more than identifying its atoms and bonds; it requires an ontological framework that interprets molecular identity, structure, and behavior. Three central theories in theoretical chemistry, VBT, MOT, and DFT, provide distinct and often complementary approaches to modeling molecules. Each theory embodies a particular ontological commitment, shaping not only how we model molecules but also how we understand their fundamental nature. The roles of these theories in formulating and interpreting molecular ontology are discussed below.

17.5.1 VBT: A Part-based Ontology

VBT [5] conceives of molecules as assemblages of atoms linked by localized electron-pair bonds. In this framework, chemical bonds arise from the overlap of atomic orbitals, hybridized as necessary, and further refined using resonance to account for delocalization. VBT aligns closely with chemical intuition, allowing us to describe molecules in terms of discrete entities—atoms, bonds, and lone pairs—which correspond to familiar concepts from classical structural chemistry. VBT contributes to molecular epistemology by structuring how we *visualize*, *communicate*, and *reason* about molecules, even when it lacks the full predictive power of MOT or DFT.

Ontologically, VBT supports a reductionist, atomistic worldview. Molecules are built from well-defined parts, each playing a specific role in the molecule's identity and function. This lends itself to a kind of chemical realism, in which the features represented in Lewis structures or hybrid orbital models are treated as more than mere approximations: they are real constituents of molecular systems. VBT's interpretive strength lies in its ability to account for chemical reactivity, stereochemistry, and bonding patterns using a language that is both rigorous and conceptually accessible.

17.5.2 MOT: A Holistic Quantum View

In contrast to VBT, MOT [6] contributes to molecular epistemology by offering a quantum-mechanical description in which electrons are delocalized across the entire molecule. Molecular orbitals arise from linear combinations of atomic orbitals and extend over the entire molecule, forming a coherent electron cloud. MOT does not emphasize atomic parts but rather the collective behavior of electrons within molecular symmetry constraints.

MOT's ontology is holistic: a molecule is not merely a sum of its atomic constituents but a new quantum entity defined by its total electronic structure. This theory supports a Platonic realism in which molecular orbitals are treated as ontologically real, even if they are not directly observable. Such a view is particularly powerful in explaining phenomena like aromaticity, molecular spectroscopy, and magnetic properties, which resist description in purely localized terms.

By treating the molecule as a single quantum object, MOT highlights emergent properties—features of molecules that arise only at the whole-system level. These include collective behaviors that cannot be inferred simply by analyzing atomic interactions, suggesting that molecules have identities that transcend their constituent atoms.

17.5.3 DFT: A Field-based Ontology

DFT [7, 19, 20] offers a third perspective, focusing not on wavefunctions or orbitals per se, but on the electron density—a three-dimensional scalar field that represents the probability distribution of electrons in space. According to the Hohenberg–Kohn theorems, all ground-state properties of a system are determined by its electron density. Thus, DFT shifts the ontological focus from particles and orbitals to continuous spatial distributions. DFT revolutionized molecular epistemology by making first-principles electronic structure calculations computationally feasible for large and complex systems.

DFT supports a field-based or topological ontology of molecules. Instead of viewing molecules as assemblies of discrete parts or unified quantum entities, DFT models them as charge distributions governed by energy principles and variational constraints. This perspective is particularly useful in analyzing chemical reactivity through density-based indices (e.g., Fukui functions, hardness/softness), as well as in interpreting molecular structure via the topology of the electron density (e.g., in Bader's QTAIM) [21].

While DFT includes Platonic elements—such as the ideal, exact functional—in practice, it is often interpreted pragmatically. The electron density is a measurable quantity, and approximations of the exchange–correlation functional are constructed based on empirical data or physical reasoning. Thus, DFT occupies a middle ground between ontological realism and methodological pragmatism.

17.5.4 Comparative Ontological Implications

Each theory frames the molecule differently: VBT sees it as a composition of atomic parts connected by real-space bonds; MOT views it as a delocalized quantum system with emergent properties; DFT treats it as a spatial field with topological significance. These differing views carry significant philosophical weight. VBT aligns with classical reductionism and chemical realism. MOT implies a Platonic, quantum-holistic realism where orbitals are more than calculation tools. DFT introduces a novel, field-theoretic conception of molecular identity, bridging quantum theory with measurable observables.

Rather than conflicting, these perspectives are often complementary. VBT provides intuitive and qualitative insight into molecular reactivity; MOT explains delocalization and spectral properties; DFT enables quantitative predictions across a wide range of systems. Each contributes a unique ontological layer to our understanding of molecules, suggesting that molecular ontology is inherently multifaceted and cannot be reduced to a single interpretive framework.

Together, MOT, VBT, and DFT reflect a pluralism in chemical ontology and epistemology: different models contribute in different ways to molecular knowledge. VBT supports conceptual clarity and chemical intuition; MOT provides quantum-mechanical explanations of delocalization; DFT delivers computational power and predictive accuracy. Rather than competing, these frameworks complement one another, each filling epistemological roles the others cannot fully address. This pluralism exemplifies how chemical knowledge is often layered, model-based, and context-dependent, yet coherent and productive in advancing chemical understanding.

17.6 Chemical Understanding

17.6.1 Chemical Understanding as the *Why*

Chemical understanding [1, 2] refers to more than just the acquisition of factual knowledge or technical skills; it is the capacity to explain, predict, and conceptually integrate chemical phenomena. While one may know that a reaction proceeds or that a molecule has a particular structure, understanding chemistry requires knowing *why* the reaction occurs or *why* that structure gives rise to properties. It involves synthesizing knowledge across scales and domains—from the quantum mechanical foundations of bonding to the macroscopic behavior of substances—to form a cohesive explanatory framework. Unlike rote memorization or algorithmic calculation, this form of understanding is both conceptual and relational, emphasizing the connections between observations, principles, and mechanisms.

At its core, chemical understanding is rooted in the interplay between empirical evidence and conceptual reasoning. Patterns observed in experimental data often lead to the development of theoretical constructs—such as electronegativity, hybridization, or resonance—which serve as bridges between observation and explanation. These constructions are not mere heuristics but are embedded in a tradition of rational inquiry that seeks to unify disparate observations under coherent theoretical umbrellas. For example, the periodic trends in reactivity and bonding are not just facts to be memorized; they emerge from deeper quantum mechanical principles and the structure of the atom. Thus, understanding is achieved when one can link experimental results to abstract models in a way that provides not only predictive power but also explanatory depth.

This kind of understanding is fundamentally integrative. It connects domains such as thermodynamics, kinetics, quantum theory, and spectroscopy into a holistic picture that allows us to design new molecules, forecast reactivity, and explain anomalies. Importantly, it emerges when *epistemological tools*—such as abstraction, analogy, idealization, and model-based reasoning—are used effectively to interpret the *ontological structures* of chemical reality, like atoms, molecules, and reaction pathways. In this sense, chemical understanding is both a product of and a guide to scientific inquiry. It enables us not only to work within established paradigms but also to challenge and reshape them, providing the "why" that drives innovation and discovery in the chemical sciences.

17.6.2 Relations with Chemical Ontology and Chemical Epistemology

Chemical understanding is deeply entangled with chemical ontology—the study of what exists in the chemical world. To understand a molecule, for instance, requires an ontological commitment: one must treat it as a coherent entity composed of atoms, governed by structural features such as bond types, electron distributions, or molecular orbitals. Ontology supplies the foundational categories—such as atoms, ions, orbitals, or transition states—that frame what can be meaningfully discussed or explained. However, this relationship is not one-directional. As chemical understanding deepens, particularly in unusual or extreme systems like nonclassical ions or fluxional molecules, it may challenge and revise ontological assumptions. What counts as a "bond" or a "molecule" can evolve in light of new interpretations, making the relationship between understanding and ontology inherently reciprocal and dynamic.

In parallel, chemical epistemology provides the methodological framework for understanding. Epistemology encompasses the tools and practices that we use to build reliable knowledge: experimental measurements, spectroscopic techniques, computational modeling, and mechanistic reasoning. Chemical understanding emerges when these diverse sources of evidence coalesce into a coherent and predictive explanation. For example, to understand a catalytic cycle or a reaction mechanism, we integrate kinetic data, quantum chemical calculations, and chemical intuition to construct plausible pathways and rationalize observed behavior. As new evidence emerges or modeling techniques improve, epistemological practices evolve, and with them, our understanding becomes more refined, provisional, and flexible.

The interplay among ontology, epistemology, and understanding can be illustrated through an example like electrophilic aromatic substitution (EAS) [22, 23]. Ontologically, we begin with the concept of aromaticity, treating the benzene ring as a delocalized π-system with defined reactivity. Epistemologically, we use mechanistic analysis, spectroscopy, and computation to determine where and how substitution occurs. Understanding emerges only when these two domains converge: when we can explain, predict, and manipulate EAS outcomes based on a unified framework of electron-donating/withdrawing groups, reaction conditions, and orbital interactions. As we learn more about how substituents alter electron density and reaction profiles, our ontological notions of functional groups evolve, prompting refinements in epistemic strategies for analyzing structure–reactivity relationships. In this way, chemical understanding drives and is driven by a continual feedback loop between what we consider real and how we come to know it.

Figure 17.2 Schematic representation of the interwoven structure of chemical knowledge. *Source:* Reproduced from Ref. [1] with permission of American Chemical Society.

Ultimately, the relationship between ontology, epistemology, and understanding forms a dynamic, mutually reinforcing triangle. Ontological commitments determine what entities we think are operative in a system. Epistemological practices determine how we investigate and justify claims about those entities. Understanding arises when both dimensions are aligned within a coherent explanatory framework. As this framework expands and evolves, both ontology and epistemology are subject to revision. Chemistry, therefore, is not merely a collection of techniques or facts, but a rich and evolving system of inquiry. It becomes truly transformative when the "what," the "how," and the "why" of the chemical world work in concert to deepen our grasp of nature and enhance our capacity to shape it. The integrated nature of this triad—ontology, epistemology, and understanding—is schematically represented in Figure 17.2, which highlights their mutual interdependence.

17.7 Chemical Concepts as the Foundation

Chemistry is a science, both conceptual and practical—one that does not just describe matter but actively constructs frameworks for understanding and manipulating it. Central to this enterprise are chemical concepts, such as "molecule," "bond," "functional group," "acid/base," and "aromaticity." These are not merely linguistic labels or convenient abstractions. Rather, they play a foundational role in shaping both ontology and epistemology. Chemical concepts are the conceptual glue that binds the ontology of chemical entities with the epistemology of chemical knowledge. They define what molecules and their components are, and they enable us to know, model, and manipulate these entities. Far from passive labels, these concepts are active mediators—tools of thought that shape both what is believed to exist and how that belief is justified.

17.7.1 Chemical Concepts as Ontological Scaffolds

Ontology concerns the fundamental categories of being—what exists and how it is structured. Chemical concepts serve as the bridge between abstract ontological categories and concrete scientific practice by defining which entities are considered real and meaningful. One core concept is the molecule, which functions as an ontological unit: it allows us to treat specific groupings of atoms, such as benzene, as coherent, stable, and persisting entities with identities that transcend their atomic constituents. Similarly, the concept of a chemical bond—covalent, ionic, metallic, or hydrogen—is foundational to molecular ontology. Though not directly observable, bonds provide structural explanations that underpin how we understand molecular stability and reactivity.

Other chemical concepts further extend this ontological framework. Functional groups, such as hydroxyl or carboxyl, allow for the classification of molecules into families that display consistent chemical behavior, effectively functioning as natural kinds within chemical taxonomy. Beyond this, higher-level concepts like aromaticity, chirality, and acidity reflect emergent properties that arise only at the molecular level. These properties are not fully reducible to atomic physics, suggesting that molecules possess their own distinct ontological status. In this way, chemical concepts do more than organize chemical knowledge—they shape what is taken to exist in the chemical world and how that existence is understood.

17.7.2 Chemical Concepts as Epistemic Tools

Epistemology concerns the processes through which we acquire, structure, and justify knowledge. Chemical concepts are central to this process, shaping how phenomena are modeled, interpreted, and communicated. Concepts like hybridization, resonance, and electronegativity are foundational to the construction of models such as Lewis structures and molecular orbital diagrams. These are not literal depictions of molecular reality but rather epistemic tools—frameworks that help us organize observations, predict outcomes, and explain chemical behavior. In this way, concepts act as scaffolding for understanding, allowing abstract theories to be operationalized in practical contexts.

Chemical concepts also function as heuristics—rules of thumb that guide reasoning and simplify complex problem-solving. A concept like nucleophilicity enables us to anticipate reaction regioselectivity and pathways without resorting to full quantum calculations. Furthermore, concepts play a critical role in the interpretation of experimental data. A peak in an IR spectrum, for instance, only gains meaning when interpreted through a conceptual understanding of bond vibrations. Finally, chemical concepts serve as a shared epistemic language that supports education, collaboration, and communication within the chemical community. Whether in textbooks, lab meetings, or industry protocols, these concepts allow us to convey complex ideas efficiently and effectively.

17.7.3 The Interdependence of Concept, Ontology, and Epistemology

Chemical concepts do not operate solely within the bounds of ontology or epistemology—they inherently function in both domains. Their stability, explanatory power, and cross-contextual utility grant them legitimacy as both representations of real entities and tools for acquiring knowledge. This dual role is especially evident in foundational concepts like bonds, orbitals, or functional groups, which persist across various models and practices not just because they are useful, but because they enable us to reliably explain, predict, and manipulate molecular behavior.

The ontological and epistemological roles of chemical concepts are deeply interconnected. On one hand, epistemic success can justify ontological status: when concepts like "bond" or "orbital" consistently yield predictive and explanatory success, they are often treated as reflecting something real about the molecular world. On the other hand, ontological commitment enhances epistemic access. Assuming the existence of entities like delocalized electrons or transition states enables us to build models and design experiments that make these entities more intelligible and operational. Additionally, conceptual pluralism—where different models (e.g., VBT, MOT, and DFT) represent the same phenomenon in distinct ways—enriches both ontology and epistemology by offering multiple, complementary perspectives on what molecules are and how they can be understood.

17.7.4 The Interwoven Architecture of Chemical Knowledge

Figure 17.2 depicts the intrinsic relationships among chemical concepts, chemical ontology, chemical epistemology, and chemical understanding. These elements, while distinct, are intimately interdependent, forming a self-sustaining and evolving framework for how we think, explain, and discover.

At the foundation of this structure lie chemical concepts—the essential intellectual tools used to describe, predict, and rationalize chemical behavior. Concepts such as bond polarity, molecular orbital, acidity, or aromaticity are not merely definitions but heuristic devices that enable us to make sense of complex phenomena. They serve as the interface between the theoretical and the observable, allowing for abstraction, simplification, and explanation across scales.

These concepts gain legitimacy and coherence within a framework of chemical ontology, which is concerned with what kinds of entities exist in the chemical world and how they are structured and interrelated. Ontological commitments in chemistry include beliefs about the existence and nature of atoms, molecules, bonds, orbitals, and even more abstract entities like transition states or electron density distributions. Ontology provides the structural backbone of chemical knowledge: it tells us what kinds of things we are talking about when we use chemical concepts.

Complementing ontology is chemical epistemology, which addresses how we come to know what we claim to know. It examines the sources and validation of knowledge: How do experimental data justify theoretical models? What is the status of a computed property versus a measured one? To what extent are our models approximate, metaphorical, or realist? Epistemology ensures that the conceptual and ontological claims we make are evidentially grounded and methodologically sound.

Chemical understanding emerges at the nexus of these three domains. It is not reducible to the possession of facts or the mechanical application of models. Rather, it reflects a deep integration of concepts, ontological insight, and epistemic justification. Chemical understanding manifests in the ability to explain, predict, and design chemical behavior across novel contexts. It is demonstrated in the capacity to reason about mechanisms, interpret spectroscopic data, construct meaningful models, and recognize when exceptions reveal deeper truths.

The triangle diagram in Figure 17.2, which links chemical ontology, chemical epistemology, and chemical understanding atop a foundation of chemical concepts, illustrates the dynamic structure of chemical knowledge. Concepts serve as the foundational tools of thought, while ontology grounds them in assumptions about what exists, and epistemology governs how they are justified and applied. Understanding arises from the integration of these elements, reflecting our ability to navigate between theory, reality, and evidence. The triangle also implies a recursive process: shifts in one domain (e.g., new epistemic insights or ontological claims) can drive changes in the others, fostering ongoing refinement of chemical knowledge.

Chemical knowledge is not a linear accumulation of facts but a dynamic, interconnected system. The figure illustrates this complexity: chemical concepts form the foundational layer, yet their significance emerges only through their integration within an ontological and epistemological framework. Together, these elements support the evolving process of chemical understanding. This model reflects not only how chemistry is taught and practiced but also reveals the deeper philosophical richness at the heart of a science devoted to transformation and change.

17.8 Ontological and Epistemological Requirements for Future Theories

Solving the Schrödinger equation for complex chemical and biological systems remains a central challenge in theoretical chemistry. While advances in computation and algorithmic precision have steadily improved our ability to approximate solutions, raw numerical accuracy is no longer a sufficient standard. Chemistry is not just about numbers—it is about understanding matter through structured, meaningful, and applicable concepts. To fully serve the discipline, future theoretical models must satisfy stringent criteria rooted in chemical epistemology (how knowledge is gained) and chemical ontology (what exists and how we categorize it). This section outlines key ontological and epistemological requirements that future theoretical models must satisfy.

Conceptual Accessibility and Interpretability. At the core of chemical epistemology is the need for models to support conceptual understanding. We do not merely seek numerical outputs; we require representations that explain phenomena in terms of familiar and useful concepts: bonds, orbitals, functional groups, aromaticity, acidity, and so on. Future theoretical models must produce results that are interpretable within the conceptual framework of chemistry, not just mathematically exact. Additionally, interpretability enables prediction, explanation, and design. A wavefunction that accurately describes a molecule's energy but cannot explain its geometry, reactivity, or behavior in terms of chemical ideas offers little epistemic value. Hence, models must be more than computational engines—they must be cognitive tools that organize and expand chemical knowledge.

Ontological Compatibility and Clarity. Chemical concepts also define chemical ontology. Bonds, lone pairs, hybrid orbitals, delocalized electrons, transition states, and others are part of this ontology. Theoretical models should therefore be ontologically informative: they must either support existing chemical entities or clearly justify new ones. This means future models must map their mathematical structures onto chemically meaningful entities. If a model proposes a novel quantity (e.g., an electron delocalization index or energy decomposition term), it must clarify its ontological status: Is it a calculational artifact, an emergent property, or a fundamental feature of molecules?

Multiscale and Emergent Behavior. One of the most philosophically rich aspects of chemistry is its multiscale nature. Molecules display properties that emerge from the interaction of atoms but are not reducible to them—think of aromaticity, polarity, or catalysis. Future theoretical models must account for this emergent behavior without collapsing into purely atomistic reductionism. Also, good models should respect chemical modularity (e.g., the general behavior of a carboxyl group across different molecules) and support chemical transferability (e.g., patterns of reactivity). These features reflect the ontological and epistemological reality that chemical behavior is often patterned, classifiable, and system-dependent.

Heuristic Power and Predictive Utility. In chemical practice, we use heuristics to make educated guesses—rules such as "electron-withdrawing groups stabilize carbanions" or "resonance increases molecular stability." Future models must support and refine such heuristic reasoning, not obscure it. This requires future models to be transparent, diagnostic, and generalizable. A model that can correctly predict reaction outcomes but gives no clue as to why is epistemically sterile. The ability to reveal causal structure—what factor drives what effect—is more valuable than raw accuracy alone.

Pluralism and Representational Integration. Chemical knowledge is pluralistic: different models and conceptual schemes (e.g., MOT, VBT, and DFT) often coexist. Rather than striving for a singular "correct" model, future frameworks should allow coherence among multiple representations and offer mechanisms to translate across them. Furthermore, models should produce outputs in formats we already use and trust—for example, potential energy surfaces, orbital diagrams, and bonding patterns. These representational interfaces are not superficial; they are deeply epistemological, enabling reasoning, teaching, and communication.

Philosophical Revelations. This criterion emphasizes that the task of theoretical chemistry is not just to solve the Schrödinger equation but to solve it in a way that is chemically meaningful. Theoretical models must speak the language of chemical concepts, respect the chemical ontology of emergent and functional entities, and serve the epistemological function of guiding thought, action, and learning. In this manner, approximate, conceptually rich models (e.g., VSEPR, hybrid orbitals, or qualitative MO diagrams) are not inferior to highly accurate ab initio calculations because they anchor knowledge in usable conceptual form. Epistemology and ontology thus become guiding lights for the next generation of quantum chemical models.

To advance theoretical chemistry, we must go beyond the pursuit of numerical exactitude and embrace a richer set of philosophical and practical criteria. Future theoretical models must be interpretable, ontologically coherent, heuristically powerful, and compatible with the conceptual practices of chemistry. Thus, the future of theoretical chemistry lies not only in computational sophistication but in the continual deepening of chemical ontology, epistemology, and conceptual understanding—the three pillars that sustain chemistry as a coherent and explanatory science. Only then will they fulfill the deeper mission of theoretical and computational chemistry—*not just to compute but also to understand.*

References

1 Liu S. The evolving quest for chemical understanding in the quantum age. J Chem Theory Comput. 2025 Oct 9;21(20):10068–79.

2 Liu S. Harvesting chemical understanding with machine learning and quantum computers. ACS Phys Chem Au. 2024 Mar 27;4(2):135–42.

3 Liu S, Zhang X. Chemical concepts from density functional theory. Acta Phys-Chim Sin. 2018;34(6):563–6.

4 Liu S. Exploring Chemical Concepts Through Theory and Computation. Wiley; 2024.

5 Zhou C, Ying F, Wu W. Chemical concepts from *ab initio* valence bond theory. In Liu S, editor. Exploring Chemical Concepts Through Theory and Computation. Wiley; 2024. pp. 23–42.

6 Long Gu F, Yu J, Yang W. Chemical concepts from molecular orbital theory. In Liu S, editor. Exploring Chemical Concepts Through Theory and Computation. Wiley; 2024. pp. 1–22.

7 Zhao D, He X, Rong C, Liu S. Chemical concepts from density-based approaches in density functional theory. In Liu S, editor. Exploring Chemical Concepts Through Theory and Computation. Wiley; 2024. pp. 71–100.

8 Mortier WJ, Leuven KU. Electronegativity equalization and its applications. In: Sen KD, Jørgensen CK, editors. Structure and Bonding. Berlin, Heidelberg: Springer Berlin Heidelberg; 1987. pp. 125–43.

9 Parr RG, Chattaraj PK. Principle of maximum hardness. J Am Chem Soc. 1991 Feb;113(5):1854–5.

10 Pearson RG. The principle of maximum hardness. Acc Chem Res. 1993 May;26(5):250–5.

11 Chattaraj PK, Ayers PW. The maximum hardness principle implies the hard/soft acid/base rule. J Chem Phys. 2005 Aug 22;123(8):086101.

12 Pan S, Solà M, Chattaraj PK. On the validity of the maximum hardness principle and the minimum electrophilicity principle during chemical reactions. J Phys Chem A. 2013 Feb 28;117(8):1843–52.

13 Chattaraj PK, Lee H, Parr RG. HSAB principle. J Am Chem Soc. 1991 Feb;113(5):1855–6.

14 Datta D. On Pearson's HSAB principle. Inorg Chem. 1992 Jun;31(13):2797–800.

15 Ho T-L. Hard soft acids bases (HSAB) principle and organic chemistry. Chem Rev. 1975 Feb 1;75(1):1–20.

16 Onuchic JN, Luthey-Schulten Z, Wolynes PG. Theory of protein folding: the energy landscape perspective. Annu Rev Phys Chem. 1997;48:545–600.

17 Huh M, Cheung B, Wang T, Isola P. The platonic representation hypothesis. arXiv. 2024; arXiv:2405.07987.

18 Ren W, Han Z. The representation of pragmatic knowledge in recent ELT textbooks. ELT J. 2016 Oct;70(4):424–34.

19 Parr RG, Yang W. Density-Functional Theory of Atoms and Molecules. Oxford University Press; 1989.

20 Geerlings P, Fias S, Boisdenghien Z, De Proft F. Conceptual DFT: chemistry from the linear response function. Chem Soc Rev. 2014 Jul 21;43(14):4989–5008.

21 Bader RFW. Atoms in Molecules. Oxford University Press; 1989.

22 Liu S. Where does the electron go? The nature of ortho/para and meta group directing in electrophilic aromatic substitution. J Chem Phys. 2014 Nov 21;141(19):194109.

23 Liu S. Quantifying reactivity for electrophilic aromatic substitution reactions with Hirshfeld charge. J Phys Chem A. 2015 Mar 26;119(12):3107–11.

18

Chemical Understanding with Machine Learning and Quantum Computers

TABLE OF CONTENTS

18.1 Introduction

In 1929, the late U.K. theoretical physicist and Nobel laureate Paul A. M. Dirac famously remarked [1] that "the underlying physical laws necessary for the mathematical theory of a large part of physics and the whole of chemistry are thus completely known, and the difficulty is only that the exact application of these laws leads to equations much too complicated to be soluble." This statement crystallized a profound tension at the heart of theoretical chemistry: the completeness of fundamental laws versus the intractability of their application. With the advent of artificial intelligence (AI)/machine learning (ML) and quantum computing (QC),

Dirac's dilemma is being revisited not just as a technical obstacle but as a deeper philosophical challenge involving chemical ontology and chemical epistemology.

Chemical ontology concerns the kinds of entities and properties that populate the chemical world, including molecules, bonds, orbitals, reactivity, and aromaticity, and how they are constituted and distinguished. Dirac's assertion implies a form of *reductionism*, suggesting that chemical phenomena are ultimately derivable from quantum mechanical laws. While this view has foundational appeal, it underplays the *emergent* nature of many chemical concepts [2]. Concepts such as acidity, electronegativity, or aromaticity are not explicit in the Schrödinger equation but emerge as stable, useful categories for organizing chemical behavior. From an ontological standpoint, chemistry possesses its own autonomous layer of reality. The failure to derive chemical regularities directly from fundamental equations is not merely a computational inconvenience; it signals that chemical entities and properties have a reality that cannot be reduced away. As philosopher of science Michael Polanyi argued, higher-level scientific concepts often have *irreducible meanings* that are lost when treated solely through lower-level theories.

Chemical epistemology deals with how we come to know and justify claims in chemistry. Dirac's statement privileges *deductive* knowledge from first principles—if only the equations could be solved, chemical knowledge would follow. In practice, however, much of chemistry relies on *heuristics, empirical rules, approximations*, and *models*, from the Woodward–Hoffmann rules [3–5] to orbital-based molecular orbital theory (MOT) [6] and density functional theory (DFT) [7] frameworks. Here lies the heart of the epistemic response to Dirac: the goal is not to solve the equations exactly but to develop approximate, reliable models that capture essential behaviors. The evolution of theoretical and computational chemistry in the twentieth century has been a testament to this pragmatic shift. The question is not whether we can derive chemistry from physics in principle, but whether we can understand and predict chemical phenomena *in practice*, using the right abstractions.

AI/ML has introduced a new epistemic framework: *data-driven knowledge construction*. These tools bypass the need to solve fundamental equations by learning patterns directly from data—experimental or simulated. For instance, ML models can predict reaction outcomes, material properties, or retrosynthetic pathways with surprising accuracy, often without explicit physical models. This raises a pivotal question: *Are we merely accelerating predictions, or are we creating new forms of chemical understanding?* From a philosophical standpoint, ML may be reconfiguring the way epistemic justification works in chemistry—shifting it from deductive or mechanistic explanation to inductive, probabilistic reasoning. Furthermore, AI systems may help us identify *novel concepts* (e.g., latent reactivity motifs, reactivity fingerprints, or nonclassical bonding patterns) that were previously inaccessible, thus enriching chemical ontology itself.

QC, meanwhile, directly tackles the heart of Dirac's dilemma: the insolubility of the Schrödinger equation for many-body systems. Quantum algorithms such as the variational quantum eigensolver (VQE) [8] and quantum phase estimation (QPE) [9, 10] promise to simulate electronic structure problems exponentially more efficiently than classical algorithms. If successful, they may finally bridge the gap between quantum theory and chemical complexity, allowing for accurate predictions of ground and excited states of complex molecules. Yet even if QC succeeds, it will not displace the need for higher-level chemical concepts. Much as Newton's laws did not render "force" or "mass" obsolete, solving the Schrödinger equation will not eliminate the utility of concepts like "resonance" or "oxidation state." Rather, QC will provide a firmer foundation upon which the edifice of chemical understanding stands.

The resolution of Dirac's dilemma, therefore, lies not in a single technological or philosophical breakthrough but in an embrace of *constructive pluralism*. Chemistry thrives not by fully reducing its entities to physics, but by building bridges between levels of understanding—fundamental, emergent, empirical, and computational. AI and ML offer new ways of modeling, explaining, and discovering. QC pushes the boundary of solvable physics. Together, they do not solve Dirac's dilemma in the narrow sense of exact derivation but rather transcend it, transforming the *problem of insolubility* into a *framework for innovation*.

In the twenty-first century, after 100 years of quantum theory invention [11], the central question is no longer "Can chemistry be derived from physics?" but "What combinations of theory, data, and computation lead to the most fruitful understanding of chemical phenomena?" This reorientation—from reduction to integration, from equation to concept—marks the maturation of chemistry in the era of artificial intelligence and QC. In the following sections, we trace the evolution of *constructive pluralism* in theoretical chemistry over the past century, from valence bond theory (VBT) to MOT, and onward to DFT, before proposing how emerging paradigms of AI/ML and QC may foster new forms of chemical understanding in the decades ahead.

18.2 The Ultimate Challenge of In Silico Simulations in Chemistry

Chemical science today confronts a wide array of urgent challenges—from designing novel functional materials and solving health-related problems to harvesting solar energy and achieving sustainable growth. From the perspective of theoretical and computational chemistry, many of these grand challenges can, in principle, be tackled through in silico simulations, particularly within the framework of multi-scale modeling [12, 13]. As illustrated in Figure 18.1a [14], chemical systems span a vast range of space and time scales, from microscopic (quantum events at the atomic level), to mesoscopic (molecular assemblies), to macroscopic (bulk materials and systems biology). Simulating behavior at each of these scales demands different computational methodologies: quantum mechanics at the microscopic scale, classical mechanics at the macroscopic scale, and hybrid or coarse-grained models at the mesoscopic level.

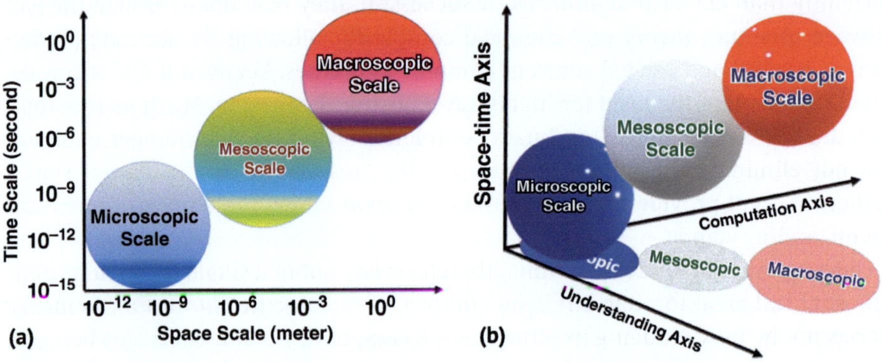

Figure 18.1 (a) Multiscale modeling of chemical systems and (b) the ultimate challenge of simulations. *Source*: Reproduced from Ref. [14] with permission of American Chemical Society.

These methods populate what we refer to as the computation axis in Figure 18.1b. Each computational approach provides different kinds of outputs—structural, thermodynamic, electronic, spectroscopic, and more—across a spectrum of accuracy and interpretability. Together, these results span the property space of the system under study. However, this axis alone does not capture the full depth of what it means to understand chemistry.

Enter the understanding axis, perpendicular to the computation axis in Figure 18.1b [14]. One might ask: why do we need a separate axis for understanding? The answer lies in the fundamental distinction between raw physical observables and chemically meaningful patterns. Computations typically return total values—energy, electron density, force, and so on—based on the laws of physics. Yet in chemistry and biology, our interest lies not in totals but in differences: changes—and even in higher-order changes—in energy, force, or structure that correspond to reactivity, selectivity, or function. These subtle differences are the seeds of chemical concepts—ideas like aromaticity, electrophilicity, regioselectivity, and steric effects—which often cannot be directly computed but must be inferred through comparative and interpretive reasoning.

This distinction points to a deeper epistemological insight: computations are governed by physics, whereas understanding in chemistry is built on abstraction, generalization, and concept formation. Historically, these concepts originated from experimentation, serving as tools for organizing and predicting chemical behavior. They represent an ontological layer that emerges above physical observables and, as such, are not explicitly present in the formalism of quantum mechanics. For example, the notion of a "bond" does not exist in quantum theory—but in wavefunction theory (WFT), bonding can be rationalized through orbital overlap, and in DFT, newer approaches using density-based descriptors can elucidate both covalent and noncovalent interactions [15–19].

These examples illustrate the heart of the challenge: computational chemistry must not only "crank numbers" along the *computation axis* but also translate those numbers into insight along the *understanding axis*. Doing so requires theoretical creativity, interpretive frameworks, and often new conceptual tools tailored to the method being used. Aromaticity [20, 21], acidity [22, 23], steric hindrance [24], and electrophilicity [25, 26]—these are not computational outputs of any single equation, but emergent ideas that require integration across data, theory, and experience.

Figure 18.1b thus represents more than a multiscale modeling scheme; it symbolizes the ultimate challenge of in silico simulations in chemistry: to bridge the gap between physical computation and chemical understanding. Only by embracing both axes—precision in calculation and depth in interpretation—can we hope to realize the full potential of computational chemistry in addressing the challenges ahead.

18.3 Prior Paradigms: Orbital- and Density-based Frameworks

The vastness of chemical space—the set of all conceivable molecular structures—is matched by the richness of property space, which encompasses an equally complex array of physical, chemical, and biological behaviors. The relationship between these two spaces is inherently many-to-many: a single molecular structure may give rise to multiple properties (e.g., acidity, aromaticity, nucleophilicity), while multiple distinct structures may share similar functions (e.g., hydrophobicity, binding affinity, or inhibitory activity). The latter phenomenon—many molecules exhibiting the same property—is often addressed through inverse molecular design, a key objective in AI-driven discovery and rational drug or materials design.

From the standpoint of quantum chemistry, the forward mapping—from chemical space to property space—is governed by the Schrödinger equation, whose exact solution is feasible only for the simplest systems. In practice, computational chemists rely on approximate methods, primarily grouped into two foundational frameworks: orbital-based theory (e.g., VBT and MOT) and density-based theory (i.e., DFT).

18.3.1 Orbital-based Approach: Orbitals and Property Prediction

As illustrated in Figure 18.2a [14], orbital-based WFT begins with a molecular species, represented mathematically by its Hamiltonian operator, $\widehat{H}$, in chemical space. Solving the electronic Schrödinger equation yields the total electronic energy E and the many-electron wavefunction Ψ, often constructed from a set of orbitals $\{\phi_j\}$—molecular orbitals in MOT or localized bond orbitals in VBT. From this wavefunction, a variety of properties P_i can be derived as functionals of the orbitals: $P_i\left[\phi_j\right]$.

Beyond numerical predictions, orbital-based methods offer chemically intuitive insights. The emergence of conceptual constructs—such as bonding/antibonding

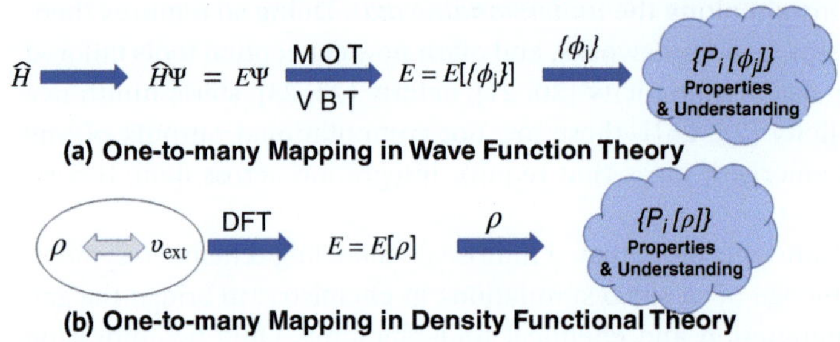

Figure 18.2 The one-to-many mapping from chemical space to property space with (a) orbital-based theory and (b) density-based theory. *Source*: Reproduced from Ref. [14] with permission of American Chemical Society.

interactions, orbital symmetry, and electron delocalization—supports traditional chemical understanding grounded in qualitative reasoning. For example, Fukui's frontier molecular orbital (FMO) theory [27] and the Woodward–Hoffmann rules [3] revolutionized how we predict and rationalize chemical reactivity by linking electronic structure to stereoelectronic effects and reaction mechanisms. These developments exemplify how ontological tools (orbitals) serve epistemological functions (explanation and understanding) in chemistry.

18.3.2 Density-based Approach: From Density to Conceptual Insights

In parallel, DFT [7] provides an alternative, yet equally powerful, framework for the forward mapping (Figure 18.2b) [14]. Rather than solving for the many-electron wavefunction, DFT reformulates the problem in terms of the electron density $\rho(\mathbf{r})$, a physically observable and three-dimensional function. According to the Hohenberg–Kohn theorems [28], the ground-state density uniquely determines the external potential v_{ext}, and thereby all properties of the system. While practical implementations of DFT—such as the Kohn–Sham (KS) approach [29]—introduce auxiliary orbitals to compute the kinetic energy, orbital-free DFT (OF-DFT) [30, 31] seeks to model all relevant quantities directly from the density, and has been gaining renewed attention in recent literature. Like WFT, DFT also supports chemical understanding, though through a distinct set of tools. Conceptual DFT (CDFT) [32–35] was developed to systematically explore reactivity using density-based descriptors such as electronegativity, chemical hardness, Fukui functions, electrophilicity, and dual descriptors. These concepts link electron density to experimentally meaningful phenomena, including molecular acidity, metal-binding specificity, and proton-coupled electron transfer mechanisms.

Recent developments have extended density-based frameworks further [18, 19], utilizing the density gradient, Laplacian, and other real-space descriptors to explore fundamental chemical phenomena. Our own work has introduced several density-derived descriptors capable of distinguishing between covalent bonds and noncovalent interactions [15], quantifying steric effects [24], characterizing electrophilicity and nucleophilicity [26], and predicting regioselectivity and stereoselectivity [36]. These efforts underscore the ontological shift enabled by DFT: from orbitals as abstract constructs to electron density as a directly observable and conceptually rich quantity. Recent reviews [18, 19] have summarized this progress, and a new book [2] has detailed how chemical concepts traditionally derived from experiments—such as bonding, reactivity, and selectivity—can be revisited and reinterpreted through both orbital- and density-based theories.

18.4 Machine Learning as a New Paradigm

18.4.1 ML as a New Framework to Solve the Schrödinger Equation

ML has emerged not merely as a computational tool but as a potentially transformative framework in theoretical chemistry—one that challenges how we approach the Schrödinger equation. Traditionally, we have relied on two main paradigms to solve this equation: WFT (e.g., VBT and MOT) and DFT. Both are rooted in explicit formulations of quantum mechanics. ML, by contrast, offers a striking departure: it can *solve the Schrödinger equation without actually solving it*. That is, ML bypasses the need to compute wavefunctions or densities by learning direct mappings from molecular structure to properties—effectively approximating the results of quantum mechanics through pattern recognition and statistical inference. This phrasing should be interpreted as an epistemological analogy rather than a literal claim: ML does not replace the Schrödinger equation nor derive chemical laws from data. Instead, it approximates the functional mapping from structure to observable properties that quantum mechanics implies. In other words, ML emulates the consequences of the Schrödinger equation without engaging its formal machinery [11, 14].

This approach reshapes the epistemology of theoretical chemistry. Rather than deriving chemical properties from first principles, ML models are trained on large datasets of precomputed quantum mechanical results or high-fidelity experimental data. These models learn to infer relationships between inputs (such as atom types, geometries, or graph-based descriptors) and outputs (such as energies, forces, or spectroscopies). Once trained, an ML model can predict the properties of new molecules or materials, orders of magnitude faster than traditional quantum calculations, often with comparable accuracy. The process circumvents the laborious and sometimes intractable step of solving the Schrödinger equation for each new system. In this sense, ML offers a *shortcut through theory*, solving problems that would traditionally require immense computational resources—without ever engaging the underlying equation directly.

Ontologically, this raises fundamental questions about the nature of explanation in chemistry. In WFT and DFT, chemical behavior is derived from a presumed reality of quantum entities—wavefunctions, orbitals, electron densities. ML, by contrast, operates in an empirical space shaped by observable patterns. It does not assume or require an explicit model of electrons or orbitals to make accurate predictions; it infers structure–property relationships from data alone. Thus, ML does not reach down to the ontological "ground floor" of chemistry but instead builds a bridge from the *surface-level regularities* of chemical behavior back toward the underlying structure—learning the consequences of quantum mechanics without referencing its formalism. Yet, despite this abstraction, the predictions of ML can mirror those obtained by solving the Schrödinger equation, challenging the view that fundamental understanding must always be rooted in direct derivation from physical laws.

Importantly, this epistemological shift does not abandon explanation but reframes it. ML models can be designed to respect known physics and advances in interpretable ML and physics-informed architectures (such as SchNet [37], PhysNet, or equivariant neural networks) that allow them to encode chemical principles implicitly. In doing so, ML is not a black-box detour around theory, but a complementary framework—one capable of both accelerating computation and revealing new patterns that traditional methods might miss. Moreover, by learning from vast and diverse datasets, ML systems can generalize to novel chemical systems, explore uncharted regions of chemical space, and even propose new reactions or materials.

In summary, ML offers a radical reimagining of how we solve the Schrödinger equation. Rather than deriving chemical properties through explicit computation of quantum mechanical objects, ML learns the outcomes of those computations and generalizes them to new cases. It *solves the equation without solving it* [11], providing answers that are consistent with quantum theory without engaging its machinery directly. This shift signals a broader paradigm change in theoretical chemistry: one where knowledge is constructed not just from deduction and simulation, but from data-driven inference, statistical modeling, and adaptive abstraction. ML does not replace quantum mechanics—it reframes our access to it, enabling new forms of understanding, prediction, and discovery at unprecedented speed and scale.

18.4.2 The New Many-to-many Mapping

Figure 18.3 [14] depicts how ML accomplishes a many-to-many mapping between chemical space and property space. A properly trained ML model can take numerous molecular inputs and simultaneously predict a wide variety of

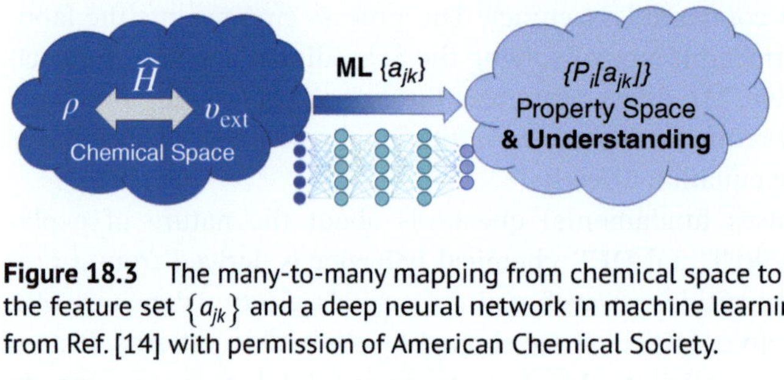

Figure 18.3 The many-to-many mapping from chemical space to property space through the feature set $\{a_{jk}\}$ and a deep neural network in machine learning. *Source*: Reproduced from Ref. [14] with permission of American Chemical Society.

properties—thermodynamic, electronic, structural, or even biological. This mapping, however, hinges critically on the design of the feature set, denoted as $\{a_{jk}\}$, which encodes the molecular system in a machine-readable and physically meaningful form.

To serve both predictive and interpretative goals, an ideal ML feature set should satisfy the following four criteria: (i) size-extensiveness: to ensure scalability and transferability to larger systems; (ii) self-adaptiveness: to account for local environmental variations in atoms or functional groups; (iii) physical interpretability: to retain epistemic transparency and facilitate chemical understanding; and (iv) density-reproducibility: to maintain consistency with DFT's foundational theorems, ensuring that the features are rich enough to encode the ground-state electron density $\rho(\boldsymbol{r})$, from which all observables can, in principle, be derived.

These requirements reflect an epistemological commitment: ML in chemistry is not merely about fitting data—it must also provide explanatory insight and be consistent with the known ontological framework (i.e., quantum mechanics, MOT, and DFT). Descriptors such as atom-condensed shape functions, electronic moments, and information-theoretic measures like Shannon entropy have been proposed to meet some of these requirements, though none has yet fulfilled all four criteria simultaneously [14].

Importantly, the many-to-many architecture of ML is especially well-suited for *inverse design*—the process of identifying molecular structures that yield desired properties. This many-to-one problem is typically intractable for conventional WFT and DFT methods, which are inherently forward-looking. In contrast, ML can invert the mapping between structure and function, facilitating breakthroughs in catalysis, materials discovery, and drug development.

18.4.3 Ontological and Epistemological Challenges of ML

The integration of ML into theoretical chemistry marks more than a methodological shift—it represents a profound epistemological and ontological turning point. While ML accelerates discovery, it also brings challenges of epistemic opacity, model bias, and limited generalizability, prompting us to reconsider what it truly means to "understand" a chemical system.

ML reshapes chemical ontology by redefining what counts as a chemical entity. Traditional constructs—atoms, bonds, and orbitals—are being supplemented or even replaced by data-driven features and latent variables in high-dimensional space. These learned representations challenge the status and role of conventional concepts, raising questions about whether ML-extracted patterns are chemically real or merely predictive tools. As such, chemical ontology is expanding, though often at the expense of interpretability.

ML also transforms chemical epistemology, shifting the basis of knowledge from mechanistic explanation to empirical inference. Unlike theory-driven approaches rooted in physical laws, ML often produces results that are accurate but inscrutable, making it difficult to distinguish prediction from explanation. Still, emerging tools—like model-agnostic interpretation, information theory, and counterfactual analysis—offer new epistemic pathways, provided we maintain standards of generalizability and conceptual clarity.

Finally, ML compels a reevaluation of chemical understanding itself. Understanding in chemistry has historically meant uncovering mechanisms and transferable principles. ML tends to reveal correlations rather than causes, but, when used judiciously, it can also help formulate new chemical concepts—latent reactivity classes, learned descriptors, or synthetic heuristics. The path forward lies in blending data-driven models with theoretical insight, forming a new synthesis that expands, rather than replaces, the scope of chemical theory.

It is essential to emphasize that the current role of ML in chemistry is predominantly predictive rather than explanatory. While ML models can learn remarkably accurate structure–property mappings, and while latent features may reveal patterns inaccessible to traditional theories, these learned representations are not yet equivalent to mechanistic or concept-driven understanding. ML does not currently generate self-standing chemical concepts in the same sense as orbitals, electron densities, or functional groups. Rather, ML offers *proto-concepts*—statistical regularities and coarse-grained abstractions—that require human interpretation to become chemically meaningful. This distinction preserves conceptual continuity with the last chapter: prediction is not yet understanding, and understanding requires epistemic justification beyond statistical correlation.

18.5 The Coming Era of Quantum Computers

QC represents a transformative shift in how we may come to simulate, understand, and ultimately know the chemical world [10, 38, 39]. Rooted in the foundational insights of Feynman, Manin, and Benioff, QC was originally conceived as a solution to a profound epistemological dilemma: classical computers are fundamentally ill-suited to efficiently simulate quantum systems, including those that underpin all of chemistry. This mismatch, as Dirac presciently noted, lies not in our ignorance of the fundamental physical laws but in the computational intractability of solving them for realistic systems.

Whereas classical computers operate using bits—units of information that are either 0 or 1—quantum computers leverage qubits, which can exist in superpositions of states and become entangled with one another. These inherently quantum-mechanical features enable QC to explore a vast configuration space exponentially faster than classical machines, opening new doors to tractable quantum simulations of molecules and materials.

At the ontological level, QC is more than a technological upgrade—it embodies an alignment between the substrate of computation and the substrate of nature itself. It is a shift in the *being* of chemical modeling: from simulating quantum systems with fundamentally classical logic (a mismatch of ontologies) to simulating quantum systems with a quantum-mechanical framework that mirrors the systems being modeled. In other words, QC allows us to build epistemological tools that are ontologically consistent with the quantum nature of reality [11].

Despite this promise, current quantum computers remain in the noisy intermediate-scale quantum era [39]—devices limited by small qubit counts, short coherence times, and error-prone operations. Yet progress is swift. Algorithms like the VQE [8] have already been implemented for small molecules (e.g., H_2, LiH, H_{12}), coupling classical optimization routines with quantum subroutines to estimate ground-state energies via trial wavefunctions. Though rudimentary, these early demonstrations mark a significant epistemic shift: the solution of the Schrödinger equation is no longer entirely symbolic or numerical—it becomes hybrid and variational, carried out by an interplay between quantum logic and classical control.

Looking ahead, the paradigm shift brought by QC will likely unfold along multiple axes: (i) Hardware scaling: Future quantum devices will host millions of high-fidelity qubits with fault-tolerant architectures, enabling the simulation of complex, correlated many-body systems previously unreachable with any method. (ii) Algorithmic evolution: Today's VQE may be replaced or substantially improved by more expressive ansätze, better optimization routines, or entirely new quantum-native algorithms for solving electronic structure problems. (iii) Quantum machine learning (QML) [40]: Perhaps the most revolutionary avenue is the confluence of QC and machine learning. QML seeks to combine the representational richness of ML with the computational power of quantum mechanics, enabling not only faster training and better generalization, but potentially more physically grounded feature extraction and model interpretability. This synergy holds profound implications for inverse design, reaction prediction, and systems-level chemical reasoning.

From a chemical epistemology standpoint [11, 14], QC challenges long-standing assumptions about what it means to "solve" a problem in quantum chemistry. It introduces new forms of knowledge production that are neither purely deductive (as in WFT), nor purely empirical (as in machine learning), but something more integrated, constructive, probabilistic, and co-evolved with our tools. QC does not simply accelerate existing methods; it reframes the epistemic landscape of chemical theory. It may enable us to access chemical truths through quantum-native computation, thereby approaching the long-sought ideal of solving the Schrödinger equation on its own terms—that is, with accuracies and system sizes unattainable by classical resources alone.

Despite its promise, QC does not itself yield chemical understanding unless supplemented by conceptual frameworks. Solving the Schrödinger equation on a quantum computer—whether via VQE, QPE, or future quantum-native algorithms—produces accurate energies and wavefunctions, but these outputs are not intrinsically explanatory. As with high-level ab initio methods, understanding emerges only when quantum results are interpreted through chemical constructs such as bonding, electron delocalization, aromaticity, or reactivity patterns. Thus, QC strengthens the ontological foundations of chemical theory while still requiring epistemic mediation to transform computation into genuine chemical insight.

18.6 How to Harvest Chemical Understanding from Theories

To compute is necessary—but to understand is essential [14]. This distinction underscores the ultimate aspiration of in silico simulations, as highlighted in Figure 18.1b. Over the past decades, computational chemistry has made tremendous strides along the axis of predictive capability, driven by multiscale modeling techniques and an explosion of data. However, the *epistemic axis*—the dimension that concerns how we *understand* chemical phenomena—has not received comparable attention. If chemistry is to progress not only in accuracy but in insight, we must ask: How can we extract genuine chemical understanding from computational results?

In Figure 18.4 [14], we introduce a conceptual framework for doing precisely this. Each square represents a projected subspace of chemical reality under a particular theoretical framework—each defined by its own *basic variable*. From the standpoint of chemical ontology (what exists in chemical theory) and chemical epistemology (how we know and explain chemical phenomena), these projections capture partial but meaningful representations of the full chemical space. Understanding arises from interpreting these projections using human-interpretable concepts such as bonds, reactivity, and selectivity.

18.6.1 Orbital-based Frameworks: Understanding via Orbitals

In WFT, such as Hartree–Fock (HF) and post-HF methods, the basic variables are the molecular orbitals, $\{\phi_i\}$. This orbital-based picture gives rise to constructs of chemical ontology such as bonding, antibonding, hybridization, and frontier orbitals. The epistemic tools of this framework include theories like the FMO theory and Woodward–Hoffmann rules, which link molecular reactivity to the structure and symmetry of orbitals. In Figure 18.4, the square shaded in green reflects this ontological commitment to orbitals and their explanatory utility in understanding covalency, electronic transitions, and chemical transformations.

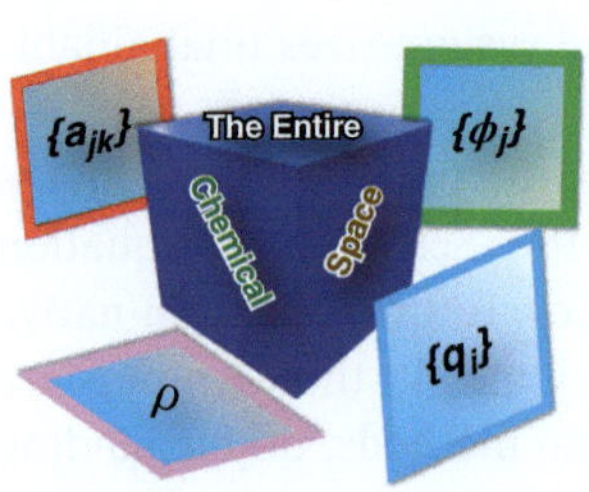

Figure 18.4 Schematic representation of how chemical understanding can be harnessed from wavefunction theory, density functional theory, machine learning, and quantum computer using orbitals $\{\phi_i\}$, electron density ρ, features $\{a_{jk}\}$, and qubits $\{q_i\}$, respectively. *Source*: Reproduced from Ref [14] with permission of American Chemical Society.

18.6.2 Density-based Frameworks: Understanding via Electron Density

DFT shifts the ontological basis from orbitals to the electron density $\rho(\mathbf{r})$, treating it as the fundamental variable that encodes all observable ground-state properties. The density-based framework enables a different kind of epistemology: one focused on global and local electron distribution. Using electron localization functions [41, 42], the Laplacians of the density, energy decomposition schemes, and information-theoretic measures (e.g., Shannon entropy, Fisher information), DFT facilitates chemically meaningful interpretation of bonding, reactivity, aromaticity, steric effects, electrophilicity, and other emergent properties. The square shaded in purple in Figure 18.4 symbolizes this projection.

These two paradigms—WFT and DFT—offer complementary and orthogonal insights into chemical space. They are not mutually exclusive; rather, they exemplify the constructive pluralism at the heart of chemical theory, where multiple, coexisting models contribute to a more complete understanding.

18.6.3 Machine Learning: Understanding via Features

In the emerging framework of machine learning, the basic variables are the feature sets $\{a_{jk}\}$—abstract representations of chemical systems fed into ML models. The ontological status of features is more ambiguous: they can represent geometric, electronic, or topological properties, but often lack direct interpretability. To transform ML from a black box into a source of chemical understanding, we must rigorously evaluate feature sets against four essential criteria: (i) descriptive power—features must capture chemically relevant information; (ii) interpretability—features should be relatable to known chemical concepts; (iii) transferability—features must generalize across chemical systems; (iv) causality or mechanistic relevance—features should link to underlying physical or chemical causes.

We envision two pathways to chemical understanding from ML: (i) an *FMO-like* path that identifies dominant features governing behavior and (ii) a *DFT-like* path that interprets global patterns using information-theoretic metrics applied to feature space. Once feature sets satisfy the previous four criteria, ML can serve not just to predict but to explain and conceptualize—thus creating new chemical understanding, possibly even new ontological entities grounded in learned structures. This is symbolized in Figure 18.4 by the red square.

18.6.4 QC: Understanding via Qubits

QC introduces yet another foundational shift. Its basic variables are qubits $\{q_i\}$, which, due to their superposition and entanglement, represent a radically different projection of chemical space. In principle, QC allows us to explore exponentially large Hilbert spaces and solve the Schrödinger equation for larger, more complex systems. However, the epistemological question remains: How do we understand results derived from quantum circuits and entangled states?

As with ML, we must develop ways to interpret outcomes from hybrid quantum-classical algorithms like the VQE and emerging paradigms in QML. Here, qubits are not merely computational devices but potential ontological units through which we may reconceptualize chemical states. Analogous to feature space in ML, the qubit space must also be organized and mapped to chemically meaningful quantities. This projection—represented by the blue square in Figure 18.4 [14]—remains in its infancy but holds the potential for a transformative expansion of our chemical imagination.

18.6.5 Epistemic Integration and Future Outlook

Taken together, the green, purple, red, and blue planes in Figure 18.4 embody the pluralistic and multidimensional nature of chemical understanding in the computational era. Each framework offers a partial yet valuable lens onto chemical space, determined by its basic variable, its epistemic strategies, and its ontological commitments. Importantly, these modes of understanding are complementary, not competitive.

As we transition into an era dominated by ML and QC, we must not abandon the lessons of orbital and density-based theories. Instead, we should strive to synthesize insights across paradigms—constructing a layered, interoperable ontology of chemistry supported by interpretable, epistemologically robust models.

In this sense, the future of in silico chemistry lies not only in solving equations faster or more accurately but in developing frameworks that make the results understandable, generalizable, and conceptually meaningful. Only then can we claim to have fulfilled the deeper promise of chemical simulation: not just computing but also understanding.

18.7 From Multiscale Modeling to Hierarchical Modeling

As we look toward the future of theoretical and computational chemistry, a new paradigm is emerging—*hierarchical modeling*—one that promises to transform how we simulate, interpret, and ultimately understand chemical systems [11, 14]. Unlike traditional multiscale modeling, which adopts a bottom-up framework by aggregating fine-grained information to coarser levels, hierarchical modeling offers a top-down, relational structure that prioritizes interactions among entities within and across nested levels of complexity.

In this framework, ML and QC play central roles. Their capacity to perform many-to-many mappings between chemical structures, descriptors, properties, and phenomena enables us to identify patterns and dependencies not just within a single level of theory (e.g., atomic or molecular) but across different levels of abstraction—from electrons and orbitals to functional groups, to macromolecular assemblies, and even to chemical ecosystems. In doing so, hierarchical modeling begins to address the ultimate challenge Dirac highlighted nearly a century ago: though the fundamental physical laws are known, the equations they imply remain far too complex to be directly solvable. ML and QC may not solve these equations analytically, but they offer new representational strategies that let us sidestep the intractability—without sacrificing understanding.

From the perspective of chemical ontology, hierarchical modeling embraces the layered structure of chemical reality, where entities at each level (atoms, bonds, molecules, supramolecular systems) are both autonomous and emergent from lower levels. Crucially, this modeling recognizes that higher-level chemical properties—like chirality, selectivity, cooperativity, or reactivity—are not always reducible to individual electrons or nuclei. Instead, they emerge from complex interdependencies and symmetries, often governed by weak interactions and noncovalent forces. Such emergence cannot be fully captured by reductionist multiscale models; it requires a holistic appreciation of nested structure and function, which hierarchical modeling enables.

From the standpoint of chemical epistemology, hierarchical modeling also forces a rethinking of how we know and understand chemical phenomena. Traditional understanding has long relied on explanatory constructs (orbitals, densities, mechanistic steps) derived from established physical theory. In the era of ML and QC, we now face the challenge of extracting interpretable, generalizable chemical insight from black-box predictions and high-dimensional feature spaces. This calls for a new epistemic framework, one grounded in feature-based reasoning, informational metrics (e.g., Shannon entropy, Fisher information), and data-driven inference that respects chemical intuition while going beyond it.

Hierarchical modeling thus provides a flexible scaffold to bring these strands together. It can accommodate pattern recognition and functional classification, discover latent variables at intermediate levels, and uncover hierarchies of causality or control (e.g., identifying which features in a catalyst most influence selectivity across reaction classes). Importantly, the ability to identify invariant descriptors across levels opens the door to scale-free representations of chemical systems, analogous to those seen in protein interaction networks, gene regulation, and complex adaptive systems.

AlphaFold [43, 44] exemplifies hierarchical modeling by processing protein structure prediction across multiple levels of abstraction—from amino acid sequences (1D), to residue-residue relationships (2D), to full 3D atomic structures—while integrating domain knowledge such as biochemical constraints and physical plausibility. Its architecture uses iterative refinement and internal feedback loops, enabling emergent structural concepts like folding motifs and functional domains to arise from data-driven representations rather than being predefined. This hierarchical, multiscale approach not only achieves high predictive accuracy but also fosters mechanistic interpretability, making AlphaFold a powerful model of how ML can support conceptual emergence and deepen chemical understanding in complex biological systems.

Additional examples of hierarchical structures have already appeared and been discussed within earlier chapters of this book. The *principle of cooperativity and frustration* (Chapter 8) reveals how local interactions propagate into higher-order patterns, where nonadditive many-body interactions give rise to mesoscopic stabilization or destabilization. Likewise, the *principle of chirality hierarchy* (Chapter 9) demonstrates a structured progression—0D point chirality to 1D axial and helical chirality, to 2D planar chirality, and ultimately to 3D supramolecular chirality—in which each level constrains, conditions, and enriches the next. Catalysis (Chapter 14) provides yet another example: the catalytic cycle reorganizes a single prohibitive activation barrier into a sequence of manageable energetic steps, and the waterwheel model illustrates how this multistep partitioning of uphill work constitutes a genuinely hierarchical organization of energy flow and mechanism, in which each intermediate functions as a lower-level energetic subproblem and their ordered coordination gives rise to emergent behavior—reduced effective activation and the possibility of globally uphill transformations—that no individual step alone can produce. Together, these phenomena exemplify the essence of hierarchical modeling: chemically meaningful organization that emerges across multiple representational layers. Alongside large-scale breakthroughs such as *AlphaFold*, they show that hierarchical structure is not merely a computational strategy but a fundamental mode of chemical structure and behavior.

In this light, hierarchy is no longer synonymous with reductionism. Rather, it becomes a tool for navigating complexity—allowing both decomposition and recomposition, enabling us to ask not just *what is happening* in a system, but *why*, *where*, and *how* different layers of chemical phenomena interact. Concepts such as self-organization, cooperation, frustration, synergy, and emergent order—long studied in biological and condensed matter systems—now become central to chemical modeling, especially for soft matter, biocatalysis, molecular assembly, and adaptive chemical networks.

Hierarchical modeling will unify simulation and understanding across frameworks—from orbitals in VBT and MOT, densities in DFT, features in ML, to qubits in QC. These foundational variables define different planes of chemical projection, and hierarchical modeling links them, not by forcing a common resolution, but by establishing mappings, analogies, and complementary views—each shedding light on the same chemical space from a unique epistemic angle. Ultimately, this modeling framework may also serve as the long-awaited bridge between computation and understanding. As we have emphasized throughout, calculation is not understanding. The mere availability of accurate data on energies or properties does not automatically yield conceptual insight. Instead, chemical understanding arises from seeing structure in complexity, recognizing regularities across systems, and organizing knowledge into interpretable frameworks. Hierarchical modeling provides the scaffolding for such synthesis—with ML and QC acting not just as computational engines but as ontological and epistemological amplifiers.

To conclude, returning to Dirac's 1929 insight cited at the start of this chapter, we suggest that the goal may no longer be to solve these equations exactly, but to understand the systems they describe. Hierarchical modeling powered by ML and QC offers a new paradigm for exploring chemical complexity through layered approximation, abstraction, and pattern recognition. This shift calls for a synthesis of physical theory, data-driven inference, and conceptual innovation, recasting theoretical chemistry as a search not just for solutions but for emergent structure, organizing principles, and deeper understanding.

18.8 Concluding Synthesis: Toward a Unified View of Chemical Understanding

Chemistry, more than any other science, thrives at the intersection of representation, emergence, and explanation. Across this book, we have traced how chemical understanding arises from the dynamic interplay among ontology (what exists in chemistry), epistemology (how we know it), and the conceptual frameworks that connect computation to meaning. The four paradigms examined here—orbital-based theory, density-based theory, machine learning, and QC—each supply a distinct projection of chemical space, grounded in different basic variables (orbitals, electron density, features, and qubits). None of these perspectives is complete on its own; each captures a different aspect of molecular reality, and each offers unique epistemic tools for explanation and prediction.

What unifies these paradigms is not a single representation or formalism, but their capacity to generate *concepts*—the scaffolding through which we think, reason, and design. Orbitals provided mechanistic intuition; electron density offered real-space conceptual clarity; features in ML hint at new modes of abstraction; and qubits promise quantum-native descriptions of many-body systems. In the emerging framework of hierarchical modeling, these diverse methods are no longer competing approaches but complementary layers of a multilevel architecture of understanding.

Ultimately, the resolution of Dirac's dilemma lies not in exact solutions or brute-force computation but in cultivating a conceptual ecosystem in which theories, data, models, and abstractions enrich one another. The future of theoretical chemistry will be shaped not only by the speed or accuracy of ML and QC but by our ability to extract, refine, and integrate chemical concepts across these paradigms. The quest of chemistry, now as ever, remains the pursuit of understanding: to see structure in complexity, meaning in data, and principles in the deep organization of matter.

References

1 Simões A. Dirac's claim and the chemists. Phys Perspect. 2002 Aug;4:253–66.
2 Liu S. Exploring Chemical Concepts Through Theory and Computation. Wiley; 2024.
3 Woodward RB, Hoffmann R. The conservation of orbital symmetry. Angew Chem Int Ed. 1969 Nov;8(11):781–853.

4 Ayers PW, Morell C, De Proft F, Geerlings P. Understanding the Woodward-Hoffmann rules by using changes in electron density. Chem Eur J. 2007;13(29):8240–7.

5 Geerlings P, Ayers PW, Toro-Labbé A, Chattaraj PK, De Proft F. The Woodward-Hoffmann rules reinterpreted by conceptual density functional theory. Acc Chem Res. 2012 May 15;45(5):683–95.

6 Gu FL, Yu J, Yang W. Chemical concepts from molecular orbital theory. In Liu S, editor. Exploring Chemical Concepts Through Theory and Computation. Wiley; 2024. pp. 1–22.

7 Parr RG, Yang W. Density-Functional Theory of Atoms and Molecules. Oxford University Press; 1989.

8 Tilly J, Chen H, Cao S, Picozzi D, Setia K, Li Y, et al. The Variational Quantum Eigensolver: a review of methods and best practices. Phys Rep. 2022 Nov;986:1–128.

9 Dorner U, Demkowicz-Dobrzanski R, Smith BJ, Lundeen JS, Wasilewski W, Banaszek K, et al. Optimal quantum phase estimation. Phys Rev Lett. 2009 Jan 30;102(4):040403.

10 Alexeev Y, Batista VS, Bauman N, Bertels L, Claudino D, Dutta R, et al. A perspective on quantum computing applications in quantum chemistry using 25-100 logical qubits. J Chem Theory Comput. 2025 Nov 11;21(22):11345–57.

11 Liu S. The evolving quest for chemical understanding in the quantum age. J Chem Theory Comput. 2025 Oct 9;21(20):10068–79.

12 Tozzini V. Multiscale modeling of proteins. Acc Chem Res. 2010 Feb 16;43(2):220–30.

13 Horstemeyer MF. Multiscale modeling: a review. In: Leszczynski J, Shukla MK, editors. Practical aspects of computational chemistry. Dordrecht: Springer Netherlands; 2010. pp. 87–135.

14 Liu S. Harvesting chemical understanding with machine learning and quantum computers. ACS Phys Chem Au. 2024 Mar 27;4(2):135–42.

15 Zhong S, He X, Liu S, Wang B, Lu T, Rong C, et al. Toward density-based and simultaneous description of chemical bonding and noncovalent interactions with Pauli energy. J Phys Chem A. 2022 Apr 21;126(15):2437–44.

16 Liu S, Rong C, Lu T, Hu H. Identifying strong covalent interactions with Pauli energy. J Phys Chem A. 2018 Mar 22;122(11):3087–95.

17 Huang Y, Liu L, Rong C, Lu T, Ayers PW, Liu S. SCI: a robust and reliable density-based descriptor to determine multiple covalent bond orders. J Mol Model. 2018 Jul 21;24(8):213.

18 Rong C, Zhao D, He X, Liu S. Development and applications of the density-based theory of chemical reactivity. J Phys Chem Lett. 2022 Dec 8;13(48):11191–200.

19 He X, Li M, Rong C, Zhao D, Liu W, Ayers PW, et al. Some recent advances in density-based reactivity theory. J Phys Chem A. 2024 Feb 22;128(7):1183–96.

20 Feixas F, Matito E, Poater J, Solà M. Quantifying aromaticity with electron delocalisation measures. Chem Soc Rev. 2015 Sep 21;44(18):6434–51.

21 Solà M. Aromaticity rules. Nat Chem. 2022 Jun;14(6):585–90.

22 Liu S, Schauer CK, Pedersen LG. Molecular acidity: a quantitative conceptual density functional theory description. J Chem Phys. 2009 Oct 28;131(16):164107.

23 Cao X, Rong C, Zhong A, Lu T, Liu S. Molecular acidity: an accurate description with information-theoretic approach in density functional reactivity theory. J Comput Chem. 2018 Jan 15;39(2):117–29.

24 Liu S. Steric effect: a quantitative description from density functional theory. J Chem Phys. 2007 Jun 28;126(24):244103.

25 Wang B, Rong C, Chattaraj PK, Liu S. A comparative study to predict regioselectivity, electrophilicity and nucleophilicity with Fukui function and Hirshfeld charge. Theor Chem Acc. 2019 Dec;138(12):124.

26 Liu S, Rong C, Lu T. Information conservation principle determines electrophilicity, nucleophilicity, and regioselectivity. J Phys Chem A. 2014 May 22;118(20):3698–704.

27 Fukui K. Role of frontier orbitals in chemical reactions. Science. 1982 Nov 19;218(4574):747–54.

28 Hohenberg P, Kohn W. Inhomogeneous electron gas. Phys Rev. 1964 Nov 9; 136(3B):B864–71.

29 Kohn W, Sham LJ. Self-consistent equations including exchange and correlation effects. Phys Rev. 1965 Nov 15; 140(4A):A1133–8.

30 Mi W, Luo K, Trickey SB, Pavanello M. Orbital-free density functional theory: an attractive electronic structure method for large-scale first-principles simulations. Chem Rev. 2023 Nov 8;123(21):12039–104.

31 Xu Q, Ma C, Mi W, Wang Y, Ma Y. Recent advancements and challenges in orbital-free density functional theory. WIREs Comput Mol Sci. 2024 May;14(3):e1724.

32 Geerlings P, De Proft F, Langenaeker W. Conceptual density functional theory. Chem Rev. 2003 May;103(5):1793–873.

33 Geerlings P, Chamorro E, Chattaraj PK, De Proft F, Gázquez JL, Liu S, et al. Conceptual density functional theory: status, prospects, issues. Theor Chem Acc. 2020 Feb;139(2):36.

34 Liu S. Conceptual density functional theory and some recent developments. Acta Phys-Chim Sin. 2009;25(3):590–600.

35 Liu S, editor. Conceptual Density Functional Theory: Towards a New Chemical Reactivity Theory. 1st ed. Wiley-VCH GmbH: Wiley; 2022.

36 Liu S, Rong C, Lu T. Electronic forces as descriptors of nucleophilic and electrophilic regioselectivity and stereoselectivity. Phys Chem Chem Phys. 2017 Jan 4;19(2):1496–503.

37 Schütt KT, Kessel P, Gastegger M, Nicoli KA, Tkatchenko A, Müller KR. Schnetpack: a deep learning toolbox for atomistic systems. J Chem Theory Comput. 2019 Jan 8;15(1):448–55.

38 Cao Y, Romero J, Olson JP, Degroote M, Johnson PD, Kieferová M, et al. Quantum chemistry in the age of quantum computing. Chem Rev. 2019 Oct 9;119(19): 10856–915.

39 Preskill J. Quantum computing in the NISQ era and beyond. Quantum. 2018 Aug 6;2:79.

40 Zeguendry A, Jarir Z, Quafafou M. Quantum machine learning: a review and case studies. Entropy (Basel). 2023 Feb 3;25(2):287.

41 Becke AD, Edgecombe KE. A simple measure of electron localization in atomic and molecular systems. J Chem Phys. 1990 May;92(9):5397–403.

42 Savin A, Nesper R, Wengert S, Fässler TF. ELF: the electron localization function. Angew Chem Int Ed. 1997 Sep 17;36(17):1808–32.

43 Jumper J, Evans R, Pritzel A, Green T, Figurnov M, Ronneberger O, et al. Highly accurate protein structure prediction with AlphaFold. Nature. 2021 Aug;596(7873):583–9.

44 Abramson J, Adler J, Dunger J, Evans R, Green T, Pritzel A, et al. Accurate structure prediction of biomolecular interactions with AlphaFold 3. Nature. 2024 Jun;630(8016):493–500.

Index

Note: Page numbers in *italics* and **bold** refers to figures and tables respectively.

f

g